P9-DVN-004

# OPERATION OF MUNICIPAL WASTEWATER TREATMENT PLANTS

## Volumes in Manual of Practice No. 11

# OPERATION OF MUNICIPAL WASTEWATER TREATMENT PLANTS

## Manual of Practice No. 11
## Sixth Edition

### Volume I  Management and Support Systems

*Prepared by*
*Operation of Municipal Wastewater Treatment Plants Task Force of the Water Environment Federation*

**WEF Press**

New York  Chicago  San Francisco  Lisbon  London  Madrid
Mexico City  Milan  New Delhi  San Juan  Seoul
Singapore  Sydney  Toronto

*The* **McGraw·Hill** *Companies*

Cataloging-in-Publication Data is on file with the Library of Congress.

McGraw-Hill books are available at special quantity discounts to use as premiums and sales promotions, or for use in corporate training programs. For more information, please write to the Director of Special Sales, Professional Publishing, McGraw-Hill, Two Penn Plaza, New York, NY 10121-2298. Or contact your local bookstore.

*Operation of Municipal Wastewater Treatment Plants, Volume I*



1 2 3 4 5 6 7 8 9 0 DOC/DOC 0 1 2 1 0 9 8 7

ISBN: P/N 978-0-07-154368-2 of set
978-0-07-154367-5

MHID: P/N 0-07-154368-6 of set
0-07-154367-8

This book is printed on acid-free paper.

**IMPORTANT NOTICE**

The material presented in this publication has been prepared in accordance with generally recognized engineering principles and practices and is for general information only. This information should not be used without first securing competent advice with respect to its suitability for any general or specific application.

The contents of this publication are not intended to be a standard of the Water Environment Federation (WEF) and are not intended for use as a reference in purchase specifications, contracts, regulations, statutes, or any other legal document.

No reference made in this publication to any specific method, product, process, or service constitutes or implies an endorsement, recommendation, or warranty thereof by WEF.

WEF makes no representation or warranty of any kind, whether expressed or implied, concerning the accuracy, product, or process discussed in this publication and assumes no liability.

Anyone using this information assumes all liability arising from such use, including but not limited to infringement of any patent or patents.

## Task Force

Prepared by the Operation of Municipal Wastewater Treatment Plants Task Force of the Water Environment Federation

Michael D. Nelson, *Chair*

Douglas R. Abbott
George Abbott
Mohammad Abu-Orf
Howard Analla
Thomas E. Arn
Richard G. Atoulikian, PMP, P.E.
John F. Austin, P.E.
Elena Bailey, M.S., P.E.
Frank D. Barosky
Zafar I. Bhatti, Ph.D., P. Eng.
John Boyle
William C. Boyle
John Bratby, Ph.D., P.E.
Lawrence H. Breimhurst, P.E.
C. Michael Bullard, P.E.
Roger J. Byrne
Joseph P. Cacciatore
William L. Cairns
Alan J. Callier
Lynne E. Chicoine
James H. Clifton
Paul W. Clinebell
G. Michael Coley, P.E.
Kathleen M. Cook
James L. Daugherty
Viraj de Silva, P.E., DEE, Ph.D.
Lewis Debevec
Richard A. DiMenna
John Donnellon
Gene Emanuel
Zeynep K. Erdal, Ph.D., P.E.
Charles A. Fagan, II, P.E.
Joanne Fagan
Dean D. Falkner
Charles G. Farley
Richard E. Finger
Alvin C. Firmin
Paul E. Fitzgibbons, Ph.D.
David A. Flowers
John J. Fortin, P.E.
Donald M. Gabb
Mark Gehring
Louis R. Germanotta, P.E.
Alicia D. Gilley, P.E.
Charlene K. Givens
Fred G. Haffty, Jr.
Dorian Harrison
John R. Harrison
Carl R. Hendrickson
Webster Hoener
Brian Hystad
Norman Jadczak
Jain S. Jain, Ph.D., P.E.
Samuel S. Jeyanayagam, Ph.D., P.E., BCEE
Bruce M. Johnston
John C. Kabouris
Sandeep Karkal
Gregory M. Kemp, P.E.

Justyna Kempa-Teper
Salil M. Kharkar, P.E.
Farzin Kiani, P.E., DEE
Thomas P. Krueger, P.E.
Peter L. LaMontagne, P.E.
Wayne Laraway
Jong Soull Lee
Kurt V. Leininger, P.E.
Anmin Liu
Chung-Lyu Liu
Jorj A. Long
Thomas Mangione
James J. Marx
Volker Masemann, P. Eng.
David M. Mason
Russell E. Mau, Ph.D., P.E.
Debra McCarty
William R. McKeon, P.E.
John L. Meader, P.E.
Amanda Meitz
Roger A. Migchelbrink
Darrell Milligan
Robert Moser, P.E.
Alie Muneer
B. Narayanan
Vincent L. Nazareth, P. Eng.
Keavin L. Nelson, P.E.
Gary Neun
Daniel A. Nolasco, M. Eng., M. Sc., P. Eng.
Charles Norkis
Robert L. Oerther
Jesse Pagliaro
Philip Pantaleo
Barbara Paxton
William J. Perley
Beth Petrillo
Jim Poff
John R. Porter, P.E.
Keith A. Radick
John C. Rafter, Jr., P.E., DEE
Greg Ramon
Ed Ratledge
Melanie Rettie
Kim R. Riddell
Joel C. Rife
Jim Rowan
Hari Santha
Fernando Sarmiento, P.E.
Patricia Scanlan
George R. Schillinger, P.E., DEE
Kenneth Schnaars
Ralph B. (Rusty) Schroedel, Jr., P.E., BCEE
Pam Schweitzer
Reza Shamskhorzani, Ph.D.
Carole A. Shanahan
Andrew Shaw
Timothy H. Sullivan, P.E.
Michael W. Sweeney, Ph.D., P.E.
Chi-Chung Tang, P.E., DEE, Ph.D.
Prakasam Tata, Ph.D., QEP
Gordon Thompson, P. Eng.
Holly Tryon
Steve Walker
Cindy L. Wallis-Lage
Martin Weiss
Gregory B. White, P.E.
George Wilson
Willis J. Wilson
Usama E. Zaher, Ph.D.
Peter D. Zanoni

Under the direction of the MOP 11 Subcommittee of the Technical Practice Committee

## *Water Environment Federation*
## *Improving Water Quality for 75 Years*

Founded in 1928, the Water Environment Federation (WEF) is a not-for-profit technical and educational organization with members from varied disciplines who work toward the WEF vision of preservation and enhancement of the global water environment. The WEF network includes water quality professionals from 79 Member Associations in more than 30 countries.

For information on membership, publications, and conferences, contact

Water Environment Federation
601 Wythe Street
Alexandria, VA 22314-1994 USA
(703) 684-2400
http://www.wef.org

# Contents

## Volume I: Management and Support Systems



Chapters 11, 23, and 32 were not updated for this edition of MOP 11. Chapters 11, 23, and 32 included herein are taken from the 5th edition of the manual, which was published in 1996.

## Volume II: Liquid Processes

## Volume III: Solids Processes

## *Manuals of Practice of the Water Environment Federation*

The WEF Technical Practice Committee (formerly the Committee on Sewage and Industrial Wastes Practice of the Federation of Sewage and Industrial Wastes Associations) was created by the Federation Board of Control on October 11, 1941. The primary function of the Committee is to originate and produce, through appropriate subcommittees, special publications dealing with technical aspects of the broad interests of the Federation. These publications are intended to provide background information through a review of technical practices and detailed procedures that research and experience have shown to be functional and practical.

# Acknowledgments

This manual was produced under the direction of Michael D. Nelson, *Chair*.

The principal authors and the chapters for which they were responsible are

George Abbott (2)
Thomas P. Krueger, P.E. (2)
Thomas E. Arn (3)
Greg Ramon (3)
Gregory M. Kemp (4)
Roger A. Migchelbrink (5)
Melanie Rettie (6)
Michael W. Sweeney, Ph.D., P.E. (6)
David Olson (7)
Russell E. Mau, Ph.D., P.E. (8)
Peter Zanoni (9)
Louis R. Germanotta, P.E. (10)
Norman Jadczak (10)
James Kelly (11)
Richard Sutton, P.E. (12)
Alicia D. Gilley, P.E. (13)
Jorj Long (14)
Keavin L. Nelson, P.E. (15)
Ed Ratledge (16)
Kathleen M. Cook (17)
Frank D. Barosky (17)
Jerry C. Bish, P.E., DEE (18)
Ken Schnaars (19)
Donald J. Thiel (20, Section One)
William C. Boyle (20, Section Two)
Donald M. Gabb (20, Section Three)
Samuel S. Jeyanayagam (20, Section Four)
Kenneth Schnaars (20, Section Five)
Alan J. Callier (20, Section Six)
Jorj Long (20, Appendices)
Howard Analla (21)
John R. Harrison (22)
Anthony Bouchard (23)
Sandeep Karkal, P.E. (24)
Daniel A. Nolasco, M. Eng., M. Sc., P. Eng. (25)
Donald F. Cuthbert, P.E., MSEE (26)
Carole A. Shanahan (27)
Kim R. Riddell (28)
Peter L. LaMontagne, P.E. (29)
Patricia Scanlan (30)
Webster Hoener (30)
Gary Neun (30)
Jim Rowan (30)
Hari Santha (30)
Elena Bailey, M.S., P.E. (31)
Rhonda E. Harris (32)
Peter L. LaMontagne, P.E. (33)
Wayne Laraway (33)

Contributing authors and the chapters to which they contributed are

David M. Mason (3)
David W. Garrett (9)
Earnest F. Gloyna (23)
Roy A. Lembcke (23)
Jerry Miles (23)
Michael Richard (23)
John Bratby, Ph.D., P.E. (29)
Curtis L. Smalley (32)

Michael D. Nelson served as the Technical Editor of Chapters 10 and 25.

Tony Ho; Dr. Mano Manoharan, Standards Development Branch, Ontario Ministry of the Environment, Canada; and Dr. Glen Daigger, CH2M HILL collaborated in the successful development of the methodologies presented in Chapter 25.

Efforts of the contributors to this manual were supported by the following organizations:

Abbott & Associates, Chester, Maryland
Alberta Capital Region Wastewater Commission, Fort Saskatchewan, Alberta, Canada
Allegheny County Sanitary Authority, Pittsburgh, Pennsylvania
Alliant Techsystems, Lake City Army Ammunition Plant, Independence, Missouri
American Bottoms Regional Wastewater Treatment Facility, Sauget, Illinois
Archer Engineers, Lee's Summit, Missouri
Baxter Water Treatment Plant, Philadelphia, Pennsylvania
Bio-Microbics, Inc., Shawnee, Kansas
Biosolutions, Chagrin Falls, Ohio
Black & Veatch, Kansas City, Missouri
Brown and Caldwell, Seattle, Washington; Walnut Creek, California
Burns and McDonnell, Kansas City, Missouri
Capitol Environmental Engineering, North Reading, Massachusetts
Carollo Engineers, Walnut Creek, California
CDM, Albuquerque, New Mexico; Manchester, New Hampshire
CH2M Hill, Cincinnati, Ohio; Santa Ana, California
City of Delphos, Delphos, Ohio
City of Edmonton, Alberta, Canada
City of Fairborn Wastewater Reclamation District, Fairborn, Ohio
City of Frankfort, Frankfort Sewer Department, Frankfort, Kentucky
City of Phoenix Water Service Department, Phoenix, Arizona
City Of Tulsa Water Pollution Control Section, Tulsa, Oklahoma
City of Tulsa, Tulsa, Oklahoma
Clayton County Water Authority, Jonesboro, Georgia
Consoer Townsend Envirodyne Engineers, Nashville, Tennessee
Consolidated Consulting Services, Delta, Colorado
CTE Engineers, Spokane, Washington
David A. Flowers, Cedarburg, Wisconsin
DCWASA, Washington, D.C.
Department of Biological Systems Engineering, Washington State University, Pullman, Washington
Earth Tech Canada, Inc., Markham, Ontario, Canada
Earth Tech, Sheboygan, Wisconsin
East Norriton-Plymouth-Whitpain Joint Sewer Authority, Plymouth Meeting, Pennsylvania

EDA, Inc., Marshfield, Massachusetts
Eimco Water Technologies Division of GLV, Austin, Texas
El Dorado Irrigation District, Placerville, California
Electrical Engineer, Inc., Brookfield, Wisconsin
EMA, Inc., Louisville, Kentucky; St. Paul, Minnesota
Environmental Assessment and Approval Branch, Ministry of the Environment, Toronto, Ontario, Canada
Eutek Systems, Inc., Hillsboro, Oregon
Fishbeck, Thompson, Carr & Huber, Inc., Farmington Hills, Michigan; Grand Rapids, Michigan
Floyd Browne Group, Delaware, Ohio
Gannett Fleming, Inc., Newton, Massachusetts
Givens & Associates Wastewater Co., Inc., Cumberland, Indiana
Grafton Wastewater Treatment Plant, South Grafton, Massachusetts
Greater Vancouver Regional District, Burnaby, British, Columbia, Canada
Greeley and Hansen, L.L.C., Chicago, Illinois; Philadelphia, Pennsylvania; Phoenix, Arizona
Hazen and Sawyer, PC, New York, New York; Raleigh, North Carolina
HDR Engineering, Charlotte, North Carolina
Hillsborough County Water Department, Tampa, Florida
Hubbell, Roth, & Clark, Inc., Detroit, Michigan
Infrastructure Management Group, Inc., Bethesda, Maryland
ITT/Sanitaire WPCC, Brown Deer, Wisconsin
Kennedy/Jenks Consultants, San Francisco, California
Los Angeles County Sanitation Districts, Whittier, California
MAGK Environmental Consultants, Manchester, New Hampshire
Malcolm Pirnie, Inc., Columbus, Ohio
Massachusetts Institute of Technology, Cambridge, Massachusetts
Metcalf & Eddy, Inc., Philadelphia, Pennsylvania
Metro Wastewater Reclamation District, Denver, Colorado
Mike Nelson Consulting Services LLC, Churchville, Pennsylvania
Ministry of Environment, Toronto, Ontario, Canada
MKEC Engineering Consultants, Wichita, Kansas
MWH, Cleveland, Ohio
New York State Department of Environmental Conservation, Albany, New York
NOLASCO & Assoc. Inc., Ontario, Canada
Nolte Associates, San Diego, California
Northeast Ohio Regional Sewer District, Cleveland, Ohio

Novato Sanitary District, Novato, California
NYSDEC, Albany, New York
Orange County Sanitation District, Fountain Valley, California
Parsons Engineering Science, Inc., Tampa, Florida
Pennoni Associates, Inc., Philadelphia, Pennsylvania
Peter LaMontagne, P.E., New Britain, Pennsylvania
Philadelphia City Government, Philadelphia, Pennsylvania
Philadelphia Water Department, Philadelphia, Pennsylvania
PSG, Inc., Twin Falls, Idaho
R.V. Anderson Associates, Limited, Toronto, Ontario, Canada
Red Oak Consulting—A Division of Malcolm Pirnie, Inc., Phoenix, Arizona
Research and Development Department, Metropolitan Water Reclamation District of Greater Chicago, Cicero, Illinois
Rock River Water Reclamation District, Rockford, Illinois
Rohm and Haas Co., Croydon, Pennsylvania
Sanitary District of Hammond, Hammond, Indiana
Seacoast Environmental, L.L.C., Hampton, New Hampshire
Siemens Water Technologies, Waukesha, Wisconsin
Simsbury Water Pollution Control, Simsbury, Connecticut
Tata Associates International, Naperville, Illinois
Thorn Creek Basin Sanitary District, Chicago Heights, Illinois
Trojan Technologies Inc., London, Ontario, Canada
U. S. Environmental Protection Agency, Philadelphia, Pennsylvania
U.S. Filter Operating Services, Indian Harbour Beach, Florida
United Water, Milwaukee, Wisconsin
University of Wisconsin, Madison, Wisconsin
USFilter, Envirex Products, Waukesha, Wisconsin
USFilter, Inc., Norwell, Massachusetts
Veolia Water North America, L.L.C., Houston, Texas; Norwell, Massachusetts
West Point Treatment Plant, Seattle, Washington
West Yost & Associates, West Linn, Oregon
Weston & Sampson Engineers, Inc., Peabody, Massachusetts

# Chapter 1

# Introduction

This sixth edition of *Operation of Municipal Wastewater Treatment Plants* (MOP 11), an update of the edition published in 1996, aims to be the principal reference for the superintendent or chief operator of municipal wastewater treatment plants (WWTPs) and thereby helps maintain compliance. Based on the practices, experiences, and innovations of many who have operated the thousands of plants built or upgraded since 1996, this updated manual reflects the state of the art in plant management and operation. It emphasizes principles of treatment plant management, troubleshooting, and preventive maintenance and recognizes that the plant's success over the long run depends largely on sound management.

The revised version of Chapter 2 discusses the evolution of the Clean Water Act from a point source approach to a holistic watershed strategy, changes in wastewater treatment plant financing, and the positive trend of beneficial use of biosolids and treated effluent. This chapter also discusses improvements in toxicity testing and advances in wastewater treatment technology, which include membrane filtration (e.g., microfiltration and ultrafiltration). The Environmental Management System approach is also discussed; in particular, the way it can be used to improve management practices and outcomes of utility operations.

Chapter 3 was reorganized with extensive editing throughout to reflect new headings and topics. It contains extensive new text, figures, and tables in all subject areas. The new version's perspective includes a shift from programmatic content (e.g., a compendium of "how to") to more of an overview (broader discussion on general management considerations and things to think about for covered topics).

Chapter 4 has been updated to focus on the U.S. Pretreatment Program requirements defined in 40 *CFR* 403. The chapter offers a brief history of the National Pretreatment Program, an overview of the program requirements, and a brief discussion of the program responsibilities for publicly owned treatment works (POTWs) and industrial users. New information briefly discusses local limits, including maximum allowable headworks loading and removal credits. References are provided that offer more-in-depth discussion of the program and its requirements for POTWs and industrial users.

The revised version of Chapter 5 contains basic safety for the operations, maintenance, and administration of wastewater treatment plants and collection systems. It includes new material on parasites and updated information on regulations.

Because wastewater utilities rely on vast amounts of historical, transactional, and real-time information to efficiently operate and maintain various systems, to ensure compliance with regulations, and to make informed decisions, Chapter 6 comprehensively explores the various areas comprising information management that are unique and vital to utilities and their customers. They include operations, administration, laboratory, process control, and recordkeeping and other functional area perspectives. While the entire chapter has been updated from the previous edition, the areas of risk management, regulation updates (e.g., the capacity, management, operations, and maintenance rule), and information technology integration and security have been expanded to reflect current requirements.

The revised version of Chapter 7 provides information on sensors and incorporates a new section on facility personnel protection and security, including fire, card access, intrusion detection, closed-circuit television, and page party systems.

Chapter 8 provides a brief review of pump and pumping fundamentals, then presents operation and maintenance (O&M) guidance for liquid and solids pumping, and finally discusses pumping station considerations. The revised version of this chapter has been updated to reduce the repetitiveness of the O & M considerations listed in previous version (i.e., general O&M considerations that are applicable to all pumps are presented first in a dedicated section, then more specific O&M practices are outlined under each type of pump or pumping system). Also, the revised version of this chapter provides updated O&M information, recognizing improvements in pump driver and control technologies, and includes more information about computerized O&M systems.

The revised version of Chapter 9 is a complete overview and guide to the handling, storage, and feeding of various chemicals used in wastewater treatment. The chapter can serve as a refresher to anyone experienced in chemical handling, as well as an introduction to those who have little or no knowledge in this area. This revised version has updated references, tables, and figures. One new table includes chemical feeding information (e.g., feed concentrations, feeding methods, and equipment type). Methods for choosing the appropriate storage system are also discussed. New text additions include a discussion of the U.S. Environmental Protection Agency's Risk Management Regulations and Spill Prevention Considerations. Technical additions, which are applicable to both small and large treatment plants, feature discussions on sodium hypochlorite storage and feed systems, including onsite generation. Lastly, a "Helpful Hints" section has been added that provides a checklist of items for use by operations personnel and design engineers.

Chapter 10 provides information about the electrical distribution system and its components. A section on basic terminology has been added to increase reader understanding of electrical topics. Sections on "Staffing and Training", "Maintenance and Troubleshooting", and "Cogeneration" have been expanded. The chapter has been restyled to increase readability and selected references have been updated.

Chapter 11, which was not updated for this edition, outlines the vital support role performed by in-plant utilities in the operation of wastewater treatment facilities. Some utilities assist with equipment and process operations, and others support the safety and welfare of plant personnel. Some utilities (e.g., water supplies; compressed air; communications systems; heating, ventilating, and air conditioning systems; fuel supply systems; and roadways) continuously serve as an integral part of the plant's daily functions. Others (e.g., fire protection, storm drainage, and flood protection) serve on an occasional, seasonal, or emergency basis. Loss of any of these services can impair operations, cause facility failure, or impose risks and discomfort on plant staff.

The revised version of Chapter 12 abandons calendar-based maintenance strategies in favor of condition-based practices. This paradigm shift occurred long ago in world-class maintenance groups in industries of all kinds. Additionally, emphasis is placed on basic maintenance fundamentals over computer tools.

Chapter 13 is intended to be a reference of practice for operators involved in managing air emissions from wastewater conveyance and treatment facilities. This chapter provides information on odor measurement and characterization, as well as the mechanisms of odor generation in both the collection system and treatment processes. Significant changes in the revised version of this chapter include the discussion of appropriate odor-control methods and technologies, because many innovative treatment technologies have been introduced in the past several years. Operation and maintenance requirements for control equipment are also included. Finally, odor-control strategies for operators are discussed, providing a sequence of steps to consider when approaching an odor-control problem.

In the revised edition of Chapter 14, the process-control section now includes two examples of standard operating procedures (SOPs). The list-style SOP provides information for certified operators. The narrative-style SOP provides a detailed description of the procedures to be followed. This approach is good for full documentation and training, and utilities that use cross-training. In addition, the process-control section now contains a detailed explanation of how to read and apply a process and instrumentation diagram.

This update of Chapter 15 has been crafted to review the current state of the outsourcing industry and present the options that have evolved over the past decade. It is not intended to be complete in every detail because every outsourcing opportunity presents unique requirements and goals. Rather, it is intended as a guide, providing a broad overview of the various options available, critical issues that should be considered, and strategies to make an outsourcing effort successful. The outsourcing market is a dynamic one, affected by many considerations, and it is continually evolving. Municipal or private entities considering outsourcing their treatment facility or utility operations are encouraged to openly explore existing options with various service providers and consultants in the industry to define which solution is best for each individual case.

The revised version of Chapter 16 offers a different approach to training development, a simple yet comprehensive look at training and the role of the trainer in an organization, and answers such questions as: What makes training work, and what common mistakes guarantee that it will fail? Why is skill stratification an issue now, and what can be done about it? What can a trainer really do for the organization, and what should trainers never be expected to do? What are policies, procedures, and operating manuals, and how should these very different documents be organized and written? How can a trainer know what kind of training will work? Are there substitutes for training that can achieve the same results?

As detailed in Chapter 17, it is important for facility operators to understand the physical and chemical characteristics of wastewater and the way to properly sample and test it to provide the treatment necessary to discharge clean water to the environment. Chapter 17 gives an overview of the composition of wastewater; explains the different types of sampling needed, how to collect and preserve samples, what equipment is needed, and why the chain of custody is important; and briefly discusses airspace monitoring relative to worker safety and odor control. In particular, sampling and testing are important to provide the data that operators need to treat the wastestream properly to meet legal requirements. The revised version of this chapter provides more detail and adds tests that were not included in the previous edition of MOP 11. The added material will help operators better understand wastewater and how to provide effective treatment.

Chapter 18 has been revised and expanded to include more descriptions of commonly used equipment and systems and important O&M aspects involved in keeping those systems in good working order. It also includes equipment and systems, such as vortex grit chambers and screening presses, that are currently used to treat raw wastewater that were not included in the last version. The chapter serves as a practical guide to understanding important O&M issues and concerns related to good operations, maintenance, and management practices related to preliminary treatment systems.

This edition of Chapter 19 includes several full-scale and current operational modifications and improvements that have been made to the primary treatment process since 1996. The chapter includes expanded discussions on primary sludge fermentation, enhanced primary treatment, dissolved air flotation primary systems, and combined grit removal/primary treatment processes. The chapter also provides a brief description about the ways that computer systems can assist with the operation of the primary treatment process. The primary treatment process appears to be a simple process to operate compared with other processes in a wastewater treatment plant; however, new advances to the primary treatment process are being made to improve both removal efficiencies in primary tanks and process reliability in downstream biological processes. For example, modifying the primary treatment process to a primary sludge fermentation system can increase the production of volatile fatty acids, which are known to play an important role in the success of the biological phosphorus removal process.

As outlined in Chapter 20, the activated sludge process is the most widely used biological process for reducing the concentration of organic pollutants in wastewater. Although well-established design standards based on empirical data have evolved over the years, poor process performance can still present problems for many municipal WWTPs. The objective of this chapter is to help operators and other wastewater treatment professionals better understand the process, solve performance problems, and improve operations. It describes activated sludge process variations and focuses on operation, addressing pertinent process theory, alternative process control strategies, energy conservation, and troubleshooting. This chapter doubles as the stand-alone manual, *Activated Sludge* (Manual of Practice No. OM-9). This chapter/manual now serves as the Water Environment Federation's primary reference for operating the activated sludge process.

As detailed in Chapter 21, tricking filters, biotowers, and rotating biological contactor (RBC) processes are generally known as fixed-film treatment processes. Of these three processes, the trickling filter predates biotowers, RBCs, and combined fixed-film and suspended growth (FF/SG) processes. New types of filter media are now used; therefore, rock media systems are labeled *trickling filters*, and plastic media systems are labeled *biotowers*. The trickling filter process is being incorporated in wastewater utilities using new methods or process modes, and many rock filters are being refurbished for continued use. This chapter aids both operators and engineers in grasping the O&M requirements of trickling filters in both existing and new systems. This chapter discusses many of the changes in the design and operation of RBCs. It includes discussions of predicting operating problems or plant overload and emphasizes methods of upgrading or improving the operation of RBCs. Combined processes [i.e., coupling trickling filters, biotowers, or RBCs (fixed-film) with suspended-growth (activated sludge) processes] are attempts to take advantage of the strengths and minimize the weaknesses of each process. In many cases, the practice is used to reduce construction costs by avoiding the building of additional tankage. This chapter addresses O&M concerns associated with combining of biological processes.

The revised version of Chapter 22 on biological nutrient removal (BNR) incorporates the best training information available from recent workshops or texts on nitrogen and phosphorus removal. It is intended to provide operators, managers, and engineers with both the fundamentals and practical examples of design and operation of BNR facilities. Chapter 22 is a complete rewrite on nutrient removal and contains figures and examples never before published.

Chapter 23, which was not updated for this edition, emphasizes that stabilization lagoons and land-treatment processes are natural systems commonly used to treat the municipal wastewater of small communities (populations less than 20,000). Stabilization lagoons typically provide secondary treatment; land-treatment systems often provide higher levels of treatment. This chapter provides information that will enable operators of lagoons and land treatment systems to prevent problems, solve them, and improve the performance of these natural systems.

Chapter 24 provides information on physical and chemical treatment processes. Examples of chemical processes are heavy metal precipitation, phosphorus precipitation, acid or alkali addition for pH control, and chlorine or hypochlorite disinfection. Physical processes include flow measurement, grit removal, primary and secondary clarification/sedimentation, filtration, and centrifugation. At the time of this writing, most WWTPs in the United States provide preliminary, primary, and secondary levels of wastewater treatment, where the focus is on reducing carbonaceous biochemical oxygen demand (CBOD) and total suspended solids (TSS) to meet discharge permit parameters. However, to further enhance the quality of receiving waters, regulators are increasingly targeting reductions in nitrogen, phosphorus, metals, and non-biodegradable soluble organics in addition to the conventional BOD, TSS, and fecal coliform parameters. Another area of regulatory and environmentalist focus is in providing additional (tertiary) treatment to secondary effluent so it can be recycled for applications traditionally served by potable water (e.g., irrigation and industrial uses). Both objectives can be addressed by physical and chemical processes. Chemical addition can be used to enhance the reduction of conventional parameters, such as suspended solids, or new processes can be added to reduce the levels of such parameters as phosphorus and soluble, non-biodegradable chemical oxygen demand (COD). Accordingly, this chapter addresses the use of physical and chemical treatment as it relates to (a) enhanced suspended solids reduction in primary treatment, (b) further reduction of solids in advanced treatment processes following conventional treatment, and (c) reductions in such parameters as phosphorus and soluble, non-biodegradable COD.

Chapter 25 outlines the steps of process performance improvement. An important relevant topic, flow meter accuracy, is discussed. Next, tools used to define the process performance of the existing facility are presented. They include hydraulic capacity analysis, tracer testing, and aeration system analysis. Embedded in the tools are approaches to improve process performance. A section then follows on chemical addition. Chemical selection steps are detailed, in addition to examples of process performance improvements. Next, a section on bioaugmentation is presented. Appendices are included to fully explain flow meters and aeration system analysis, and present a tracer case history.

While the understanding of fundamental wastewater disinfection mechanisms has changed little over the years, applications of the technologies have evolved considerably. Chapter 26 has been revised to include current industry trends associated with the post-September 11, 2001, emphasis on security and safety when selecting technologies; the recent focus on receiving water effects caused by disinfection byproducts; the challenges associated with wet weather treatment applications; and new and evolving technologies.

The revised version of Chapter 27 has been expanded to include a discussion of the final disposition of biosolids and associated costs. Emphasis is placed on cooperating with regulators, networking with organizations, and developing information about biosolids constituents, quality control, and public education.

Chapter 28 describes the types of residuals present in wastewater. It details where the residuals are found and their typical percentages in each stage of the wastewater treatment process. The chapter also presents information on the biological and chemical makeup of the different types of residuals. The chapter further describes how to obtain samples of these residuals and what typical tests must be performed from regulatory and process-control perspectives. Testing procedures are described and referenced for further information and reasoning is given as to which test should be performed and what information it can provide. The tables have been updated to reflect the latest regulatory requirements. This chapter is a good general reference regarding the types of residuals typically found in wastewater and the types of testing procedures typically used to analyze them.

For the 6th edition of Chapter 29, the sections on centrifuges and dissolved air flotation have been completely rewritten to be more specific and user friendly.

Chapter 30 serves as a resource for plant staff to increase their knowledge of anaerobic digestion for biosolids treatment. While this chapter focuses on operational and troubleshooting guidance, it also includes a general discussion of digestion theory, such as information on available digestion process equipment and the common benefits and drawbacks that can affect the digestion process or its O&M. Significant changes and additions to the text include information on digestion pretreatment technologies, advanced digestion processes, regulatory issues associated with anaerobic digestion, and an expanded section on gas-cleaning technologies. Discussions of all technologies and equipment have been updated to reflect most recent developments.

As addressed in Chapter 31, aerobic digestion was initially used in plants that typically treated waste activated sludge from treatment systems that did not have a primary settling process—only waste activated or trickling-filter sludge, or mixtures of waste activated or trickling-filter sludge. Typically, if a primary settling process was included, anaerobic digestion was the process of choice because reliable techniques to thicken and aerobically digest flows containing more than 4% solids were not established at the time. Because of tighter effluent nitrogen and phosphorus standards in the United States in the late 1990s, primary clarifiers have slowly been eliminated from the process train to preserve the good carbon-to-nitrogen ratio typically required to achieve successful biological nitrogen removal. Both the new effluent limits and new techniques for controlling aerobic digestion processes and accurately predicting system performance have made aerobic digestion attractive once again. A number of anaerobic digesters have been converted to aerobic digesters because of their relatively easy operation, lower equipment cost, and ability to produce a better quality supernatant with lower nitrates and phosphorus (that protects the liquid side upstream). Another benefit of aerobic digestion is fact that these systems can achieve comparable volatile solids reduction with shorter retention periods, they have less hazardous cleaning and repairing tasks, and they do not produce an explosive digester gas. The revised chapter was prepared in response to the new performance requirements of the rules and regulations for beneficial reuse, resulting in the identification of techniques that improved the process performance of aerobic digestion. These techniques are grouped into the following categories: (1) pre-thickening, (2) staged operation, (3) aerobic–anoxic operation, and (4) temperature control. These techniques are explored in depth in this chapter.

As described in Chapter 32 (which was not updated for this edition), because WWTPs have become more efficient at producing high-quality effluents, more solids are being generated. Solids treatment processes—thickening, dewatering, stabilization, and disposal—represent a significant portion of the cost of wastewater treatment. Stabilization processes further treat solids to reduce odors or nuisances, reduce the level of pathogens, and facilitate efficient disposal or reuse of the product. This chapter focuses on the following stabilization methods:

- Composting,
- Lime stabilization,
- Thermal treatment,
- Heat drying, and
- Incineration.

All of these stabilization methods (other than incineration) are widely used for treating sludge to Class A or Class B levels for beneficial reuse and disposal, as referenced in the U.S. Environmental Protection Agency's 40 *CFR* 503 regulations.

The 6th edition of Chapter 33 reflects the changes in dewatering technology over the last decade or more. Vacuum filters are uncommon now, supplanted by belt filters and centrifuges. Drying beds are less popular but reed beds are gaining popularity as an example of "green" technology. The industry is more dependent on polymers, and this section is greatly expanded. This chapter explains how the various dewatering processes work so readers can understand them and perhaps better operate the process in their own plants.

# Chapter 2

# Permit Compliance and Wastewater Treatment Systems

# INTRODUCTION

This chapter reviews the various laws and regulations applicable to the safe treatment and disposal of materials from wastewater treatment facilities, describes the typical processes used to provide safe and acceptable treatment of wastewater, and discusses treatment and disposal alternatives for the solids removed from wastewater. For consistency with federal and state terminology, this manual uses the term *sludge* to refer to the solids separated from wastewater during treatment. For information on permit-required sampling, process control, and compliance sampling, see Chapter 17.

# PERMIT COMPLIANCE

Municipal wastewater treatment plants buffer the natural environment from the wastewater generated in urban areas. Because uncontrolled releases of wastewater would degrade the water, land, and air on which life depends, the government has developed a comprehensive set of laws and regulations on safely treating and disposing of wastewater and sludge. They are summarized here. (Government agencies continually modify laws and regulations, so water and wastewater treatment professionals should review the current discharge rules when preparing and submitting the permit application for a specific plant.)

**CLEAN WATER ACT.** Congress enacted the Water Pollution Control Act Amendments of 1972 in response to growing public concern about water pollution. When amended again in 1977, this law became known as the Clean Water Act (CWA). It established the basic structure for regulating discharges of pollutants into U.S. waters. The Act's four guiding principles are:

- No one has the right to pollute U.S. waters, so a permit is required to discharge any pollutant;
- Permits limit the types and concentrations of pollutants allowed to be discharged, and permit violations can be punished by fines and imprisonment;
- Some industrial permits require companies to use the best treatment technology available regardless of the receiving water's assimilative capacity; and
- Pollutant limits involving more waste treatment than technology-based levels, secondary treatment (for municipalities), or best practicable technology (for industries) are based on waterbody-specific water quality standards.

The Act's primary objective is to restore and maintain the chemical, physical, and biological integrity of U.S. waters. So, the CWA set two national goals: achieve fishable and swimmable waters (wherever possible) by 1983 and eliminate all pollutant discharges to navigable waters by 1985. The Act's 1977 and 1987 amendments reaffirm the importance of achieving these goals and give states the primary responsibility of implementing CWA through the permitting system established in Section 402.

Anyone discharging pollutants to U.S. waters must have a National Pollutant Discharge Elimination System (NPDES) permit. Dischargers without permits or those who exceed their permit limits are violating the law and are subject to civil, administrative, or criminal penalties. Maximum federal criminal penalties range from $25,000 per day and 1-year imprisonment for negligent violations to $50,000 per day and 3-year imprisonment for knowing violations to $250,000 per day and 15-year imprisonment for knowing endangerment. Federal penalties for violations of permit conditions are pursuant to Title 18, United States Code, Section 1001.

The NPDES permits typically specify the discharge location, the allowable discharge flows, the allowable concentrations (mass loads) of pollutants in the discharge, the limits of the mixing zone (if any), and monitoring and reporting requirements. Before it can be discharged to U.S. waters, municipal wastewater must have received secondary treatment—or more stringent treatment if necessary to meet water quality standards. Effluent from a secondary treatment system typically must contain no more than 30 mg/L of biochemical oxygen demand ($BOD_5$) and 30 mg/L of total suspended solids (TSS), and be between 6.0 and 9.0 pH (on a 30-day average basis), according to 40 *CFR* 133. This regulation also requires that the secondary treatment plant remove at least 85% of $BOD_5$ and TSS from municipal wastewater, except combined sewer overflows (CSOs). In addition, treatment plants designed to handle 19 000 $m^3/d$ (5 mgd) or more (and smaller ones with interference and pass-through problems) must establish pretreatment programs to regulate industrial and other nondomestic wastes discharged into sewers.

Permit applications and required reports must be signed by the appropriate authority. According to 40 *CFR* 122.22, all permits and reports submitted by a corporation must be signed by a responsible corporate officer; those submitted by a partnership or sole proprietorship must be signed by one of the general partners; and those submitted by a municipality, state, federal, or other public agency must be signed by a principal executive officer or ranking elected official.

Between 1970 and 1988, the U.S. Environmental Protection Agency (U.S. EPA) provided $61.1 billion in Federal Construction Grants Program funds to help build new or upgrade existing publicly owned treatment works (POTWs). Meanwhile, states, municipalities, and the private sector have invested more than $200 billion on POTWs, and a comparable amount on operations and maintenance (O&M) of these plants.

Subsequent amendments have modified various CWA provisions. In 1981, for example, Congress amended the CWA to streamline the municipal construction grants program, improving the capabilities of treatment plants built via this funding. In 1987, U.S. EPA began replacing the construction grants program with the State Water Pollution Control Revolving Fund (commonly known as the Clean Water State Revolving Fund):

> With the passage of the Amendments to the [CWA] in 1987, Congress provided for the replacement of the federal Construction Grants program with the Clean Water State Revolving Fund program (CWSRF). The program provides capitalization grants to the states to be used as the basis (along with a required twenty percent state match), to create revolving loan funds which provide low-interest loans to municipalities to finance wastewater infrastructure projects, and to fund water quality projects such as nonpoint source and estuary management.
>
> The states set the loan terms, which may be interest-free to market rates, with repayment periods up to twenty years. Terms may be customized to meet the needs of small and disadvantaged communities. Loan repayments are recycled to perpetuate the funding of additional water protection projects.
>
> Public involvement is an important element of the SRF Programs. Before applying for a capitalization grant, a state is required to provide information about the respective programs and the projects to be funded in an Intended Use Plan which is available for public review and comment. The Intended Use Plan is a requirement in both the Clean Water and Drinking Water State Revolving Fund programs (http://www.epa.gov/region7/water/srf.htm#cwsrf).

Other laws also have changed parts of the CWA. For example, the U.S.–Canadian Great Lakes Critical Programs Act of 1978, which became the Great Lakes Water Quality Agreement of 1980, required U.S. EPA to establish water quality criteria for the Great Lakes. The law specifically requires that the agency determine the maximum levels at which 29 toxic pollutants are safe for humans, wildlife, and aquatic life.

The Act has helped improve water quality tremendously. Many lakes and rivers that were grossly polluted in the 1970s, when the CWA was enacted, now support aquatic life. In the 1960s, worst-case dissolved oxygen levels ranged from 1 to 4 mg/L in heavily polluted waterways; testing in 1985 and 1986 showed that dissolved oxygen levels had risen to between 5 and 8 mg/L.

As improvements have occurred, regulators' focus has changed. For example, regulators initially focused on the chemical aspects of the "integrity" goal. In the last decade, however, they began paying more attention to physical and biological integrity. Regulators originally concentrated on discharges from "point-source" facilities, such as municipal treatment plants and industrial facilities. In the late 1980s, they also began to manage "nonpoint sources," such as stormwater runoff from streets, farms, and construction sites. Some of the runoff programs are voluntary; others follow the traditional regulatory approach.

In the last 10 years, CWA programs have been evolving from a program-, source-, or pollutant-specific approach to a more holistic, watershed-based approach in which protecting healthy waters is as important as restoring impaired ones. Stakeholder involvement is an integral part of this new approach for achieving and maintaining state water quality and other environmental goals—not just those specified in the CWA.

***Solids Management.*** Between 1972 and 1998, annual sludge production at U.S. wastewater treatment plants increased from 4.6 million to 6.9 million U.S. dry tons. In other words, sludge production has increased 50% since CWA was enacted, while the U.S. population only grew 29%, according to the Council of Economic Advisers (Washington, D.C.). The U.S. Environmental Protection Agency expected annual sludge production to rise to 8.2 million U.S. dry tons by 2010.

Much of this sludge is stabilized to produce biosolids—a primarily organic material—and beneficially used as a soil amendment (land-applied). The U.S. Environmental Protection Agency estimates that by 2010 about 70% of sludge will be beneficially used (Table 2.1). The agency regulates all biosolids that are landapplied, incinerated, or surface-disposed under 40 *CFR* 503 (58 *FR* 9248-9415). Sludge that is landfilled with other waste is regulated under 40 *CFR* 258.

A number of states have more restrictive solids management programs. For example, Oklahoma does not permit surface disposal (disposing of sludge in sludge-only landfills, surface impoundments or lagoons, or waste piles of sludge).

In July 2002, the National Research Council (NRC) published the results of its 18-month study on *Biosolids Applied to Land: Advancing Standards and Practices.*

TABLE 2.1 U.S. Environmental Protection Agency projections for beneficial use and disposal.

| | Beneficial use | | Disposal | | | |
|---|---|---|---|---|---|---|
| Year | Land-applied | Advanced treatment | Other use | Landfill | Incineration | Other |
| 1998 | 41% | 12% | 7% | 17% | 22% | 1% |
| 2005 | 45% | 13% | 8% | 13% | 20% | 1% |
| 2010 | 48% | 13.50% | 8.50% | 10% | 19% | 1% |

Although there is no documented scientific evidence that the regulations have failed to protect public health, the NRC noted, uncertainty persists about the health effects of land-applied biosolids. So, the NRC made approximately 60 recommendations for addressing the uncertainties. In response, U.S. EPA began developing a strategy to address these issues and published *Standards for the Use or Disposal of Sewage Sludge* (Final Agency Response to the National Research Council Report on Biosolids Applied to Land and the Results of [U.S.] EPA's Review of Existing Sewage Sludge Regulations [*Federal Register* Doc. 03-32217, Filed 12-30-03]). In it, U.S. EPA identified 15 pollutants for possible regulation. These pollutants will undergo a more refined risk assessment and risk characterization, which may lead to a notice of proposed rule making under the CWA (http://www.epa.gov/EPA-WATER/2003/April/Day-09/w8654.htm).

***Toxics.*** The Clean Water Act authorizes U.S. EPA to regulate the discharge of toxic chemicals to the environment. It also declares that U.S. policy is to prohibit toxic amounts of these pollutants from being discharged. Soluble toxics can threaten human health if they are in water used for drinking or swimming. Insoluble toxics can adsorb to sediment, be consumed by aquatic life, and thus enter the human food chain (via bioaccumulation).

The Act currently lists 126 priority pollutants (40 *CFR* 423, Appendix A) that must be controlled. These pollutants are divided into four classes: heavy metals and cyanide, volatile organic compounds, semivolatile organic compounds, and pesticides and polychlorinated biphenyls (PCBs). Also, certain nontoxic organic compounds in wastewater can become toxic chlorinated organic compounds, such as trihalomethanes, when the water is disinfected via chlorine.

Analysts use whole effluent toxicity (WET) testing to determine whether treatment plant effluent is toxic to humans or the aquatic environment. The test is a bioassay in which sensitive aquatic organisms are put in a container of effluent and monitored for several days to see how well they survive. If the effluent is toxic, regulators require that treatment plant staff systematically reduce the toxic components via a toxics reduction evaluation and appropriate treatment-plant or pretreatment-program modifications.

***Stormwater.*** Under 40 *CFR* 122, point sources are prohibited from discharging industrial-related stormwater into a U.S. waterbody without an NPDES stormwater permit. These permits protect receiving waterbodies by requiring the permittee to implement appropriate stormwater controls and best management practices (BMPs). Treatment plants that discharge stormwater into U.S. waters or a municipal separate storm sewer system typically must have a "Sector T—Treatment Works" NPDES multisector general permit, as do 1-mgd or larger treatment plants and those with an approved industrial pretreatment program under 40 *CFR* 403. More information on stormwater discharges and permits may be found on U.S. EPA Web site at cfpub1.epa.gov/npdes/stormwater/indust.cfm.

**OCCUPATIONAL SAFETY AND HEALTH ACT.** The Occupational Safety and Health Act (OSHA) addresses the safety and health of industrial workers. It provides guidelines for developing regulations on handling and storing hazardous materials, such as chlorine, acids, caustics, and other chemicals used to treat wastewater. (Although OSHA was not originally written for wastewater treatment plants, several states have applied its provisions to them.)

The Act also protects communities via a material safety data sheet (MSDS) provision that requires wastewater treatment plants to report inventories of hazardous chemicals to state and local emergency response agencies (40 *CFR* 370). Under the Hazard Communication Act, which is part of the Superfund Amendments and Reauthorization Act (SARA) Title III, treatment plants must train employees on proper use and handling of chemicals, provide them with the appropriate MSDSs, and notify state and local officials if they have extremely hazardous materials onsite in quantities exceeding certain thresholds; for example, the threshold quantity for chlorine is 1135 kg (2500 lb). Also, Section 313 of SARA Title III (40 *CFR* 372) requires treatment plants to report significant releases of certain toxic substances not covered by NPDES permits. As an example, the release quantity for chlorine is 4.5 kg (10 lb). (For more information on OSHA, see Chapter 5.)

**STATE AND LOCAL PROGRAMS.** Under Section 402(b) of the CWA, the U.S. EPA administrator delegated administration of the NPDES program to qualifying states. The state-run programs typically are called SPDES [(State Name) Pollution Discharge Elimination System] programs. These programs address effluent limits and monitoring requirements, sanitary sewer overflow (SSO) reporting, and sludge management requirements.

The effluent limits include conventional and unconventional pollutants, as well as year-round, seasonal, WET, and site-specific requirements. Examples of conventional pollutants include BOD and TSS, whereas unconventional pollutants may include metals, pesticides, and radioactive materials. Typical permit parameters include average and maximum flowrates, $BOD_5$, TSS, fecal coliform, pH, ammonia, phosphorus, and WET testing. The NPDES/SPDES permit establishes discharge limits (both concentration and mass loading) and monitoring requirements for each parameter.

The permit also requires permittees to develop and comply with approved sludge management plans. The plans establish "acceptable management practices" or numerical limits for pollutants in sludge, as promulgated in CWA Section 405(d)(2).

The permit may also include pollution prevention and pretreatment requirements. Pollution prevention requirements involve assessing treatment plant facilities, equipment, and staff duties to determine whether and how they could be modified to minimize waste and avoid pollution. Pretreatment requirements involve establishing and operating an industrial pretreatment program that complies with CWA Section 402(b)(8), General Pretreatment Regulations (40 *CFR* 403). (More stringent state and local requirements may also apply.) The pretreatment program's goals are to avoid introducing pollutants that will interfere with treatment plant operations or pass through the plant untreated, and to promote opportunities for recycling treated wastewater and sludge.

In addition to NPDES/SPDES permits, local and state ordinances may address environmental protection, pretreatment of waste streams discharged to treatment plants, residuals management, and occupational safety and health. These ordinances are sometimes more stringent than the federal rules for the same activity. For example, some local ordinances only allow Class A biosolids to be land-applied, while the federal regulations permit land application of both Class A and Class B biosolids. Other localities have developed wellhead-protection programs because of concerns about increasing aquifer-contamination rates. And some local agencies regulate the use of treated effluent to irrigate golf courses and other landscapes.

State and local agencies also develop the requirements for certifying POTW staff, including collection system workers, treatment plant operators, and laboratory technicians. The certification requirements typically are based on plant size, complexity, and required reliability. The agencies often establish similar requirements for privately owned or operated systems that discharge to a POTW or to state waters. For more information, check your state's employment requirements.

**ENVIRONMENTAL MANAGEMENT SYSTEMS.** An environmental management system (EMS) is designed to integrate environmental issues into an organization's management processes. It enables any organization to control the effect of its activities, products, or services on the natural environment, allowing the organization to not only achieve and maintain compliance with current environmental requirements, but also proactively manage future environmental issues that might affect daily activities. Wastewater utilities will find that implementing an EMS improves its environmental, financial, and other results.

Implementing an EMS does not require a utility to create an entirely new system. It simply provides a structure to help organizations incorporate the responsibilities, practices, procedures, processes, and resources needed to implement and maintain an effective environmental management system into their routines. Most of an EMS' required elements are already part of existing management programs.

For more information on EMSs, access U.S. EPA's Web site (www.epa.gov/ems/index.htm) or type "environmental management system" into your favorite Web browser.

## LIQUID TREATMENT PROCESSES

Wastewater treatment is a multi-stage process designed to clean water and protect natural waterbodies. Municipal wastewater contains various wastes; it typically consists of about 99.94% liquid and 0.06% solids. A typical U.S. city, including its private dwellings, commercial establishments, and industrial contributors, produces between 379 and 455 L/d/person (100 and 120 gpd/person) of wastewater—not including the water that infiltrates and exfiltrates the collection system. If untreated or improperly collected and treated, this wastewater and its related solids could hurt human health and the environment.

Stormwater runoff and CSOs also produce large volumes of wastewater that can affect public health and the environment. For more information on stormwater and CSO controls, see Water Environment Federation's *Prevention and Control of Sewer System Overflows*.

**OBJECTIVES.** A treatment plant's primary objectives are to clean the wastewater and meet the plant's permit requirements. Treatment plant personnel do this by reducing the concentrations of solids, organic matter, nutrients, pathogens, and other pollutants in wastewater. The plant must also help protect the receiving waterbody, which can only absorb so many pollutants before it begins to degrade, as well as the human health and environment of its employees and neighbors.

When establishing permit requirements for a treatment plant, regulators may consider the following issues in addition to the minimum statutory requirements:

- Preventing disease,
- Preventing nuisances,
- Protecting drinking water supplies,
- Conserving water,
- Maintaining navigable waters,
- Protecting waters for swimming and recreational use,
- Maintaining healthy habitats for fish and other aquatic life, and
- Preserving pristine waters to protect ecosystems.

One of the challenges of wastewater treatment is that the volume and physical, chemical, and biological characteristics of wastewater continually change. Some changes are the temporary results of seasonal, monthly, weekly, or daily fluctuations in the wastewater volume and composition. Other changes are long-term, the results of alterations in local populations, social characteristics, economies, and industrial production or technology.

The quality of the receiving water and the public's health and well-being may depend on a treatment plant operator's ability to recognize and respond to potential problems. These responsibilities demand a thorough knowledge of existing treatment facilities and wastewater treatment technology.

**COLLECTION SYSTEM.** A collection system is a network of pipes, conduits, tunnels, equipment, and appurtenances used to collect, transport, and pump wastewater. Typically, the wastewater flows through the network via conventional gravity sewers, which are designed so each pipe's size and slope will maintain flow toward the discharge point without pumping. Lift stations are used to move wastewater from lower to higher elevations.

There are three principal types of municipal sewers: sanitary sewers, storm sewers, and combined sewers. *Sanitary sewers* convey wastewater from residential, commercial, institutional, or industrial sources, as well as small amounts of groundwater infiltration and stormwater inflow. *Storm sewers* convey stormwater runoff and other drainage. *Combined sewers* convey both sanitary wastes and stormwater.

The type of collection system may profoundly affect treatment plant operations. For example, the high flows and heavy sediment loads discharged by a combined system during and after a storm make effective treatment more challenging. Excessive infiltration and inflow in a poorly maintained sanitary system can have similar results. Extensive collection systems may discharge foul-smelling, septic wastewater to the plant unless the odors are controlled en route.

Today, municipalities rarely construct combined sewers, and most have made efforts to separate stormwater from sanitary wastewater. Their efforts may be inexpensive, such as requiring homeowners to disconnect roof drains from municipal sewer systems, or major construction projects, such as rerouting stormwater from sanitary sewers to nearby creeks and rivers.

**PRELIMINARY TREATMENT.** Preliminary treatment typically begins with removing materials, such as hydrogen sulfide, wood, cardboard, rags, plastic, grit, grease, and scum, that might damage the plant headworks or impair downstream operations. These materials may be removed via chemical addition, pre-aeration, bar racks, screens, shredding devices, or grit chambers. Preliminary treatment may also include coagulation, flocculation, and flotation to remove particles and biological solids from the wastewater. (For more information on preliminary treatment, see Chapter 18.)

**PRIMARY TREATMENT.** Primary treatment involves removing suspended and floating material from wastewater. Well-designed and -operated primary treatment facilities may remove as much as 60 to 75% of suspended solids and between 20 and 35% of total $BOD_5$. They do not, however, remove colloidal solids, dissolved solids, and soluble $BOD_5$. (For more information on primary treatment, see Chapter 19.)

**SECONDARY TREATMENT.** Secondary treatment reduces the concentrations of dissolved and colloidal organic substances and suspended matter in wastewater. Generally, secondary treatment reduces 85% of TSS and $BOD_5$, resulting in concentrations between 10 and 30 mg/L.

Most secondary treatment processes involve biological treatment—typically, attached- or suspended-growth systems. Both rely on a mixed population of microorganisms, oxygen, and trace amounts of nutrients to treat wastewater. The microorganisms consume organic material in the waste to sustain themselves and reproduce. They also convert nonsettleable solids to settleable ones. In *attached-growth systems*, such as trickling filters, packed towers, and rotating biological contactors, the microorganisms are attached to supporting media. In *suspended-growth systems*, such as lagoons and activated sludge processes, the microorganisms are drifting throughout the wastewater.

Secondary treatment process effluent contains high levels of suspended biological solids that must be removed before the effluent is further treated or discharged to a receiving waterbody. Most treatment plants use settling tanks to separate the solids from the liquid, although flotation and other methods may be used.

**PHYSICAL–CHEMICAL TREATMENT.** Physical and chemical treatment processes typically are used to remove oil, grease, heavy metals, solids, and nutrients from wastewater. For example, screening, sedimentation, and filtration are used to physically separate solids from wastewater. Chemical coagulation and precipitation is used to promote sedimentation. Activated carbon adsorption is used to remove organic pollutants. Breakpoint chlorination and lime addition are used to reduce nitrogen and phosphorus concentrations, respectively. (For more information on physical–chemical treatment, see Chapter 24.)

**ADVANCED WASTEWATER TREATMENT.** Advanced wastewater treatment (AWT) processes typically are used to further reduce the concentrations of nutrients (nitrogen or phosphorus) and soluble organic chemicals in secondary treatment effluent. These processes may be physical, chemical, biological, or a combination.

For example, membrane filtration—microfiltration, ultrafiltration, nanofiltration, and reverse osmosis—is used to remove organics, nutrients, and pathogens from wastewater. (It traditionally was used for industrial wastewater treatment but has been gaining popularity at municipal treatment facilities.)

A treatment plant's permit requirements typically influence which, if any, AWT processes are used.

**DISINFECTION.** Disinfection inactivates or destroys pathogenic bacteria, viruses, and protozoan cysts typically found in wastewater. These pathogens cause such waterborne diseases as bacillary dysentery, cholera, infectious hepatitis, paratyphoid, poliomyelitis, and typhoid.

The growth of water reuse for irrigation and other purposes, and changes in requirements for toxics, including chlorine, to protect aquatic life are altering disinfection policies and, subsequently, disinfection practices. Chemical disinfection processes involving halogens, particularly chlorine, historically dominated the wastewater treatment field. However, concerns about chlorine safety and the requirement to dechlorinate some discharges have made other disinfection methods, such as ozonation and ultraviolet irradiation, more popular. *Ozonation* involves using ozone radicals to destroy pathogen cell walls. *Ultraviolet irradiation* involves using electromagnetic energy to destroy an organism's genetic material (DNA and RNA). Both are more costly than chlorine disinfection, but effective treatment alternatives. (For more information on disinfection, see Chapter 26.)

**EFFLUENT DISCHARGE.** Effluent matters because where it will be discharged or how it will be reused influences the permit requirements set and the treatment processes needed to meet those requirements. Treatment plant effluent can be discharged to a surface waterbody or wetlands, used to recharge groundwater aquifers via percolation through the ground or deep-well injection, or land-applied. It may be considered an alternative water source (because of its nitrogen and phosphorus content) and used to irrigate golf courses, parks, plant nurseries, and farms. It also could be the source water for a constructed wetland, adding nutrients that support the aquatic environment, thereby enhancing wildlife habitat and public recreation. In addition, effluent can be used by industries as cooling or makeup water for chemical processes.

## RESIDUALS MANAGEMENT

Sludge treatment processes are often the most difficult and costly part of wastewater treatment. Untreated sludge is odorous and contains pathogens. Sludge stabilization processes reduce odors, pathogens, and biodegradable toxins, as well as bind heavy metals to inert solids, such as lime, that will not leach into the groundwater. The resulting biosolids can be used or disposed of safely.

**TYPES OF RESIDUALS.** Wastewater residuals include primary, secondary, mixed, and chemical sludge, as well as screenings, grit, scum, and ash. Between 40 and 60% of influent TSS is primary sludge. It typically has a concentration of 2 to 6% solids when removed from the primary clarifiers. Secondary (biological) sludge is composed largely of microorganisms. Between 0.5 and 2.0% of TSS is biological sludge. Mixed residuals—commingled primary and secondary sludge—typically makes up 1.0 to 3.5% of influent TSS. The concentration and characteristics of chemical sludge depend on the treatment chemicals (alum, ferric salts, or lime) used. It typically is found at treatment plants that have tertiary treatment, such as phosphorus removal.

The use or disposal method for residuals depends on how much treatment they have received. *Biosolids* are residuals that have been stabilized so they can be beneficially used as a soil amendment. Combustible residuals, such as screenings, may be incinerated or landfilled. Noncombustible residuals, such as grit, may be landfilled.

**TREATMENT PROCESSES.** Typical solids treatment processes include thickening, stabilization, digestion, chemical, composting, dewatering, incineration, and heat drying. These processes are briefly described below.

***Thickening.*** Thickening processes remove water to reduce the volume of liquid sludge, but the material retains the characteristics of a liquid (e.g., it flows). A thickened sludge typically contains between 1.5 and 8% solids. Thickening is intended to reduce the volume of sludge so sludge treatment, storage, and hauling processes, equipment, and costs can also be reduced.

There are three types of thickening:

- Pre-thickening (thickening before stabilization and dewatering),
- Post-thickening (thickening after stabilization but before beneficial use), and
- Recuperative thickening (thickening biosolids and returning them to the stabilization process).

Pre-thickening processes include gravity thickeners, dissolved air floatation (DAF), centrifuges, gravity belt thickeners, and rotary belt thickeners. Gravity thickeners work best on primary and chemical sludges; they do not work well with combined sludges. DAF is typically used to thicken WAS. Mechanical thickeners, such as centrifuges, gravity belt thickeners, and rotary belt thickeners, are used on all types of sludges. They remove more water from sludges than a gravity thickener or DAF does.

***Stabilization.*** Stabilization via digestion or chemical stabilization reduces the sludge's pathogen content, making the material a biosolids suitable for beneficial use.

*Digestion.* Aerobic and anaerobic digestion reduce the volatile solids and pathogen content of sludge, thereby reducing odors and producing an environmentally acceptable soil amendment. Aerobic digestion uses microbes in open or closed vessels or lagoons to oxidize the sludge's organic matter into carbon dioxide, water, and ammonia. It is becoming popular to operate these digesters at temperatures higher than 55 °C (131 °F) because doing so reduces the pathogen level in biosolids.

Anaerobic digestion uses microbes in a closed tank containing little or no oxygen (Figure 2.1). The tank is typically operated at temperatures between 35 and 38 °C (95 and 100 °F), but may be operated at more than 55 °C (131 °F) to further reduce pathogens

**FIGURE 2.1** An example of anaerobic digestion in a closed tank.

and solids. Anaerobic digestion destroys more than 40% of volatile solids and produces an offgas containing about 65% methane, which may be collected and used as a fuel.

Both types of digestion can produce a Class A or Class B biosolids suitable for beneficial use, depending on time, temperature, and flow configurations involved (40 *CFR* 503).

*Chemical Stabilization.* Chemical stabilization typically involves using lime to raise the sludge's pH to 12.0 for 2 hours. This reduces pathogens and odors. Rules regarding chemical stabilization can be found in 40 *CFR* Part 503 rules. This stabilization method can also produce a Class A or Class B biosolids suitable for beneficial use (40 *CFR* 503).

*Composting.* Composting—windrow, aerated windrow, static pile, aerated static pile, and in-vessel—involves using microorganisms, a bulking agent (such as wood chips, leaves, or sawdust), and a controlled environment [typically 55 to 60 °C (131 to 140 °F)] to decompose organic matter in sludge, as well as reduce its volume and odors. The moisture and oxygen levels are also controlled to minimize odors during the process. Biosolids are typically used in windrow systems, while sludge can be composted in aerated static piles or an in-vessel system. Composting is gaining popularity because the finished material is an excellent soil conditioner.

***Dewatering.*** Dewatering further reduces the volume and weight of biosolids via air-drying (on sand); vacuum-assisted drying beds; or mechanical dewatering equipment, such as belt filter presses, centrifuges, plate-and-frame filter presses, and vacuum filters. When used with polymers, belt filter presses and centrifuges can produce a biosolids "cake" that contains 15 to 25% solids. Sand drying beds typically produce a cake containing between 10 and 50% solids. Vacuum-assisted beds produce a cake containing 10 to 15% solids (when polymers are used). Plate-and-frame presses can produce a cake containing 30 to 60% solids (when lime, ferric chloride, or fly ash is used). Vacuum filters can produce a cake containing 12 to 30% solids (when the biosolids have first been conditioned with lime, ferric chloride, or polymers).

***Heat Drying.*** Heat drying processes, such as low-temperature heat drying, flash drying, rotary kiln drying, indirect drying, vertical indirect drying, direct–indirect drying, and infrared drying, reduce the volume of secondary sludge and destroy pathogens. They typically produce a commercially marketable biosolids. Digestion is typically not a prerequisite.

***Incineration.*** Incineration typically involves firing biosolids at high temperatures in a multiple-hearth or fluidized-bed combustor, turning them into ash, and destroying volatile solids and pathogens in the process. It degrades many organic chemicals but can form others (e.g., dioxin), so products of incomplete combustion must be controlled. Air emissions also must be controlled. Metals are not degraded; they concentrate in the ash. Most incinerated municipal sludge will produce nonhazardous ash, which can be landfilled (40 *CFR* 258). It also can be used as an aggregate in concrete or a fluxing agent in ore processing. If inorganic or organic constituents exceed Appendix II to Part 258, disposal in a hazardous waste landfill will be required (http://www.epa.gov/tribalmsw/pdftxt/40cfr258.pdf).

***Beneficial Use.*** Unstabilized sludge that is landfilled with other waste must comply with 40 *CFR* 258. Once stabilized into biosolids, however, 40 *CFR* 503 requires that the material be used or disposed of via one of the following environmentally acceptable alternatives: land application, surface disposal, or incineration. Treatment plant staff should assess local land availability, public acceptance, and transportation constraints before selecting a use or disposal alternative.

Land application involves spreading biosolids on the soil surface or injecting it into soil (Figure 2.2). The material, which adds organic matter and nutrients to soil, must

**FIGURE 2.2** An example of land-applying biosolids.

meet Class A or Class B standards to be land-applied (40 *CFR* 503). Class A biosolids require more stabilization to reduce pathogens below detectable limits, but can be commercially marketed or land-applied without any pathogen-related restrictions.

## SUGGESTED READING


Criteria for Municipal Solid Waste Landfills (1996) *Code of Federal Regulations*, Part 258, Title 40. http://www.epa.gov/tribalmsw/pdftxt/40cfr258.pdf (accessed April 2007).

National Research Council (2002) *Biosolids Applied to Land: Advancing Standards and Practices*; National Academy Press: Washington, D.C.

*Standards for the Use or Disposal of Sewage Sludge*; Agency Response to the National Research Council Report on Biosolids Applied to Land and the Results of EPA's Review of Existing Sewage Sludge Regulations (2003) *Fed. Regist.*, **68** (68), 17379–17395. http://www.epa.gov/EPA-WATER/2003/April/Day-09/w8654.htm (accessed April 2007).

State Revolving Fund. http://www.epa.gov/region7/water/srf.htm#cwsrf (accessed 2007).

U.S. Environmental Protection Agency (1999) *Biosolids Generation, Use, and Disposal in the United States*; EPA530-R-99-009; U.S. Environmental Protection Agency: Washington, D.C.

U.S. Environmental Protection Agency (2003) Watershed Rule; 40 *CFR* 122, 124, and 130; U.S. Environmental Protection Agency: Washington, D.C.

U.S. Environmental Protection Agency, *Municipal Technologies* Technology Fact Sheet. http://www.epa.gov/OW-OWM.html/mtb/mtbfact.htm (accessed April 2006).

U.S. Environmental Protection Agency Law & Regulations, Clean Water Act. http://www.epa.gov/r5water/cwa.htm (accessed April 2006).

U.S. Environmental Protection Agency (2000) Progress in Water Quality: An Evaluation of the National Investment in Municipal Wastewater Treatment; EPA/832-R/00/008; June. http://www.epa.gov/OW-OWM.html/wquality/wquality.pdf (accessed April 2006).

U.S. Environmental Protection Agency, Environmental Management Systems. http://www.epa.gov/ems/index.html (accessed April 2006).

Water Environment Federation (1999) *Prevention and Control of Sewer System Overflows*, 2nd ed.; Manual of Practice No. FD-17; Water Environment Federation: Alexandria, Virginia.

# Chapter 3

# Fundamentals of Management

## INTRODUCTION

The art of good management is an evolutionary process, and volumes of books have been written on the topic as witnessed by visiting the management section within your local bookstore or, more specifically, publications offered by the Water Environment Federation (WEF), the American Water Works Association, and other relevant trade organizations such as the International City and County Management Association and the American Management Association. This chapter provides an overview of general facility management considerations and does not attempt to convey the full level of detail associated with the topic of management.

Perhaps your organization has just appointed you "facility manager" or perhaps you have selected a career path toward achieving such a critically important position. Maybe you just want to learn what it means to be the person responsible for an expensive and complex treatment plant, or facility, or you might be a member of a large facility's management team. Regardless of your point-of-view, the facility manager is the singular focus for successful facility performance. By "successful performance," we mean the effective orchestration of staff, work processes, and technologies that result in treating wastewater liquid, solids, and airstreams that comply with regulatory limits and meet neighbor and stakeholder expectations.

To manage a facility that continually meets organizational, environmental, and stakeholder expectations is indeed a challenge. As the focal point and voice of the treatment facility, the plant manager must deal effectively with everyone and everything including municipal officials, regulators, media representatives, consultants, design engineers, equipment manufacturers, suppliers, public interest groups, the plant's neighbors and other concerned citizens, and the plant's employees (Figure 3.1). A successful treatment plant manager must also build effective relationships with employees and maintain stakeholder and public interests; possessing technical competency of the plant's unit operations is simply not enough. The definitive task of the manager is to marshal

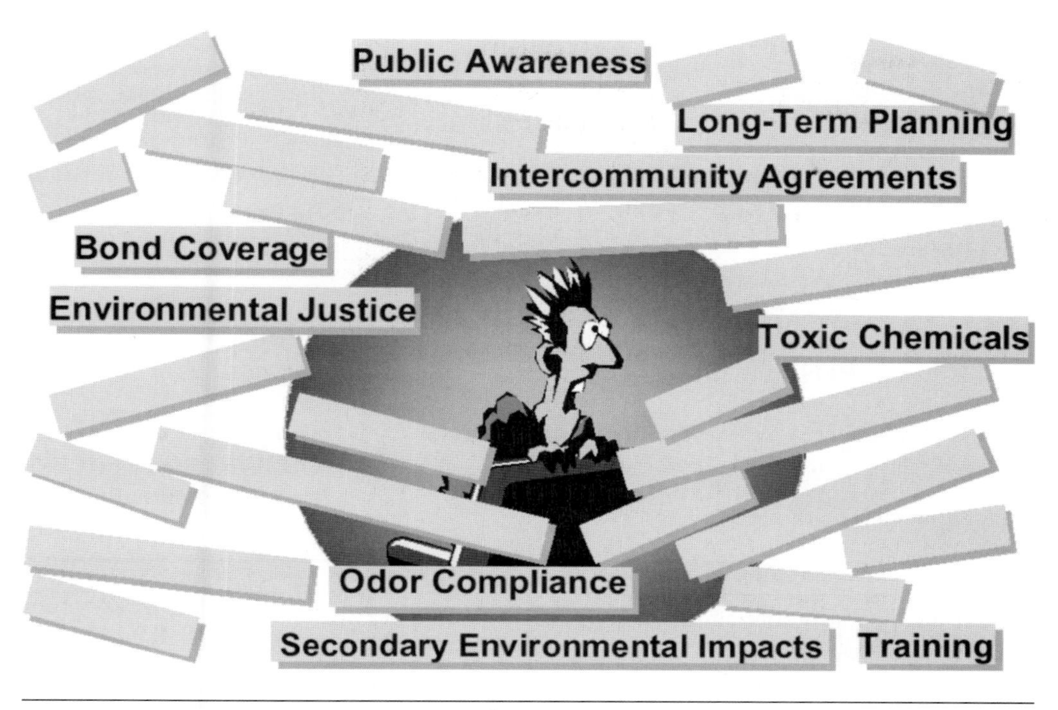


FIGURE 3.1 Wastewater management challenges.

and effectively apply all available resources and to make the most of them in achieving organization.

## AN OVERVIEW OF TREATMENT PLANT MANAGEMENT RESPONSIBILITIES

Facility management is a difficult balancing act; however, when executed properly, it results in a facility and staff performing at their peak potential as well as responding well to changes and challenges that arise. As depicted in Figure 3.2, there are many elements associated with facility management. The following six sections describe general areas of responsibility that the plant manager should consider.

**ENVIRONMENTAL AND PUBLIC HEALTH CONSIDERATIONS.** The fundamental purpose, or mission, of a treatment facility is *to protect the health of the public and the environment.* The plant manager must never forget about this mission. Although the mission is simple, it is much more difficult to achieve. The major respon-

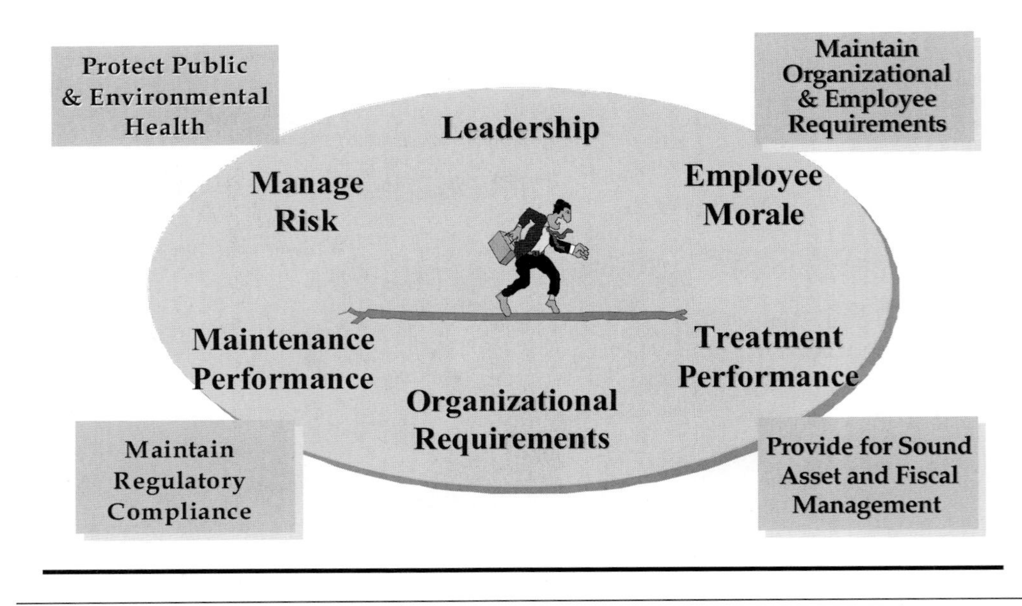


FIGURE 3.2 Facility management is a difficult balancing act.

sibility of ensuring that the facility functions in accordance with specified regulatory and other local requirements lies squarely on the manager's shoulders. The manager's functions include planning, organizing, directing, and controlling to accomplish a plant's mission. A typical mission for the manager includes operating a safe facility in continuous compliance with permit and other applicable legal requirements, yet doing so in an environment with budget limitations as well as organizational policies and procedures. This can be a difficult balancing act as these functions are often carried out in a political atmosphere under the scrutiny of a public keenly aware of environmental and financial concerns.

**PHYSICAL ASSET MANAGEMENT RESPONSIBILITIES.** The phrase, "physical asset management," refers to the strategy of managing the life-cycle cost (both capital and operations and maintenance expenses), use, and reliability of the systems, equipment, and other physical assets associated with the treatment facility to optimize their value in support of facility operations (Figure 3.3). The bottom line is that the manager must maintain a long-term perspective on the total life-cycle costs of the facility's assets, and must use effective and proactive procedures that support informed system design, equipment selection, proper installation, sound operation strategies, and effective maintenance programs and processes.

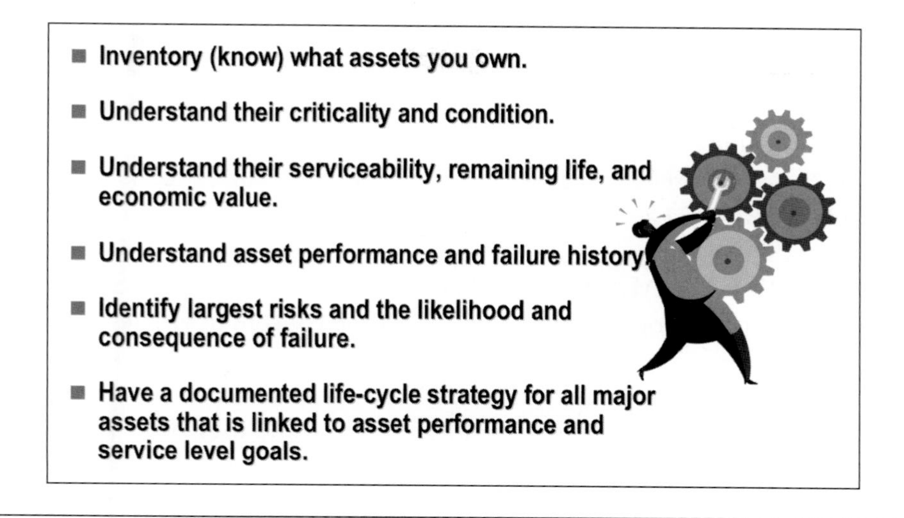


FIGURE 3.3 Asset management strategies.

**ORGANIZATIONAL AND LEADERSHIP RESPONSIBILITIES.** Organizations have a responsibility to their employees to clearly convey their mission statements. Conversely, the mission statement must communicate the organization's purpose. Managers must take a leadership role and provide the direction necessary for success. Good leadership skills allow managers to get things done through people.

Effective leaders use a number of resources to achieve their goals. They must set the course, energize the group, provide the means, and develop the plan in order to be successful. The role of the wastewater facility manager is to ensure that the organization's mission is met while providing employees with an opportunity for job growth and satisfaction. The organization's goals must be clear, obtainable, and have a general direction for all to follow. Effective managers must communicate the organization's mission to their audience. This includes using language that can be understood; that is, language that is neither too simple nor too complex to bore or threaten those receiving the information.

The manager must also align the goals of the facility with those of the greater organization and governance. As a leader, the manager must balance near-term goals with long-term interests and plans. Good leadership requires the manager to look beyond the daily technical issues of normal plant operation and maintain a perspective of long-term goals and directions. This includes considering organizational and staff requirements necessary to meet these goals.

The work environment is an important element within the manager's control to provide a safe workplace that contributes to employee satisfaction, effective performance (Figure 3.4), and good morale. Employees work harder for managers who are concerned and willing to make changes to put people first. Managers must instill the notion that "we are all in this together" to be successful. Finally, success is the result of every employee working as a team for one common cause.

**OPERATIONS AND MAINTENANCE RESPONSIBILITIES.** At the foundation of facility operations and maintenance is the concept of efficiency. Although it is often used in association with both positive and negative cost and budgetary effects, the term "efficiency" embodies much more. Efficiency improvement processes are a basic part of the management continuum and, while they often focus on near-term results, they also affect total life-cycle costs. The successful manager leads the organization in recognizing that virtually all processes, activities, or functions can be improved, whether they be treatment process control, energy use, employee performance, or maintenance management. At the heart of this process is the development of, and adherence to, the operations plan described later in this chapter. Achieving improvements requires constant attention to discrete practices, functions, subgroups, personalities, and technologies within the facility. Not being satisfied with the status quo and using effective monitoring strategies associated with recordkeeping and performance measurement are core elements used to manage operations and maintenance activities.

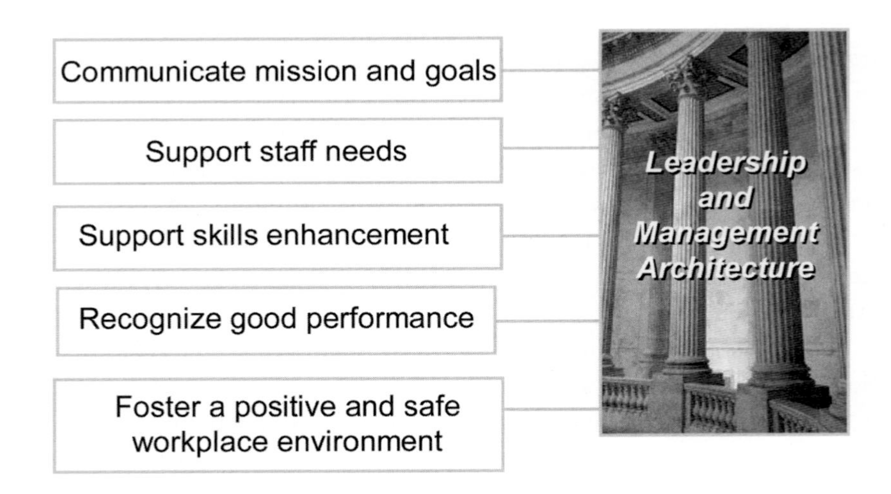


FIGURE 3.4 Maximizing staff performance.

**FISCAL MANAGEMENT RESPONSIBILITIES.** Wastewater utilities must provide high quality, environmentally responsible services at reasonable costs. Under today's cost control environment, the treatment manager faces a significant economic challenge to fulfill the organization's mission at the lowest practical cost (the term "practical" is used deliberately here). Practical cost management considers the long-term total life-cycle cost component and embodies sound operating strategies coupled with informed maintenance management activities. Cost management (Figure 3.5) is, indeed, a difficult balancing act that pits the constraints of near-term operating budgets against long-term asset management objectives and protection of public and environmental health issues.

All facility managers must advocate for and directly address the challenge of obtaining necessary funds for operating, maintaining, and improving the plant as necessary. The organization's budget process is central to meeting this responsibility, and includes clear definition of operation and maintenance expenses as well as long-term capital improvements necessary to fulfill the facility mission. The core of a sound budget is informed planning, justification and management of cost elements, and the involvement of community stakeholders in setting wastewater fees or rates to support the needs of the organization. In essence, the annual budget may be regarded as the mechanism for administering the operations plan for the facility. Tracking and managing the budget to the organization's goals can be the key to effective facility performance and help balance near and long-term demands. Moreover, tracking progress against this plan as the year progresses is an essential management responsibility.

- **An Iterative Process:**
  - **Define service standards and performance levels.**
  - **Make informed decisions on asset investments.**
  - **Understand the financial implications.**
  - **Assess customer impacts, i.e., rate increases.**
  - **Ensure that funding requirements are identified and impacts are understood.**

FIGURE 3.5 Fiscal management is a critical responsibility.

**REGULATORY, NEIGHBOR, AND OTHER STAKEHOLDER RELATIONSHIP RESPONSIBILITIES.** The facility manager has a critical responsibility in developing and maintaining effective relationships with external stakeholders (e.g., those that fall outside of the facility "fence line"). The acceptance of a facility as a good neighbor by the public cannot be understated. The term "effective" implies a relationship management strategy that recognizes and provides constructive means to manage differences in points-of-view and opinion as well as maintain positive relationships as much as possible. External stakeholders include neighbors, regulatory agencies, other groups and departments within the organization (e.g., police, fire, and public works), the media, bargaining units, and special interest groups. Given this diversity, as well as the general nature and perception of wastewater, the facility manager must seek to maintain effective relationships and interactions with external stakeholders. To achieve this requires a sound communications strategy and program that is known and practiced by all staff. In many cases, there is a distinct advantage for the manager to act as the primary "voice" of the plant to facilitate consistent and effective communications with external stakeholders.

## MANAGEMENT CONSIDERATIONS

The environment that a facility operates in is constantly changing. Consider the many internal and external pressures and influences that a facility faces (Figure 3.6). These include political interests, regulatory mandates, security, economic development, changing workforce demographics, workforce retention (e.g., succession management), aging infrastructure, labor/management relations, and the organization's regulating policies.

How the facility reacts to these influences is driven by the plant manager's ability to maintain an informed and well-thought-out operations strategy that is supported by sound documentation and performance measurement practices. Responding to any of the aforementioned influences can be made easier and effective by having information available to make sound decisions as well as develop response strategies that are defensible and logical.

Do you find yourself in a reactionary mode dealing with daily issues, with little time to pause and consider things from a long-term perspective? If so, you are not alone. Consider, however, that you may be engaging in a tactical rather than a strategic operating mode and, if that is all you are doing, you may be missing the opportunity to build a strong organization that is better suited to meet the challenges of tomorrow. The phrase, "integrated strategic management approach," describes pathways to strategic thinking and how they can be used to maximize the potential of the organization in the near-term as well as into the future (Figure 3.7). The major pathways to the

FIGURE 3.6 Internal and external influences.

integrated strategic management approach are to develop an outstanding leadership team, engage in thoughtful planning, develop a long-term operating strategy, manage assets from a business-like perspective, and engage mechanisms that develop stakeholder support and alignment.

Of primary importance to the facility is that its strategy be aligned with the objectives of the organization. Linking the facility plan with the overall organizational strategic plan, mission, and goals will ensure alignment and consistency in the organization.

FIGURE 3.7 Pathways to an integrated strategic management approach.

## OPERATIONS AND MAINTENANCE PLAN

The operations and maintenance plan is the foundation of an effective operations strategy and serves as the basis for managing the entire operation of the facility. The operations and maintenance plan is considered the core management tool for the plant manager to organize, administer, and respond to daily operational elements at the facility. The operations and maintenance plan not only defines the roles and responsibilities of staff, but also establishes procedures for handling everyday situations and emergencies. By formalizing standard operating procedures, the operations and maintenance plan also provides a framework that facilitates periodic reviews and updates in response to changes.

The operations and maintenance plan should be considered a compilation of individual plans covering specific planning and management elements. As such, it is a living document. Each of the plan's elements should be reviewed and updated annually or when significant events or changes occur that require adjustment. While components of an operations and maintenance plan can vary according to the specific requirements and issues at a given facility, all operations and maintenance plans have the following elements in common: staffing, personnel management, external relations and communication, reporting and recordkeeping, and emergency operations. Each of these elements is described in the following sections.

## STAFFING

The staffing section of a facility's operations and maintenance plan should describe the structure of the organization, what reporting relationships exist, and who is responsible for what task. An organization chart should be used to convey this information. The staffing section should include a narrative describing the elements of the organization's structure, with an appendix listing the position descriptions of staff. Position descriptions should be updated as plant equipment, processes, or other conditions change. Each position description typically includes a brief description of the character of the work, a list of the tasks required, the supervisory reporting position (supervision received), the positions to be supervised (supervision exercised), education and experience required for hiring or placement, and typical tasks the incumbent must accomplish if the plant's mission is to be achieved.

Although an organization chart can provide a snapshot of a facility's organizational structure, in reality the organization works in a much more fluid and dynamic way. Consider that the organization that you have laid out is perfect for normal situations as well as ensuring accountability and clarity of roles. However, what happens during periods when normal operations are disrupted? For example, how much independent responsibility should a shift operator have in establishing process directives during plant upsets? Who defines the hierarchy designated to respond to employee absences or emergencies? Because the way a facility responds to atypical situations is often significantly different from the way it handles normal circumstances (including input from or subordination to external agencies such as firefighters, police, or the mayor), it is important for an organization plan to cover various states of operation including normal, stressed, and emergency operations.

The roles and responsibilities section of the operations and maintenance plan can define the details of the three operating states previously identified. For this section of the plan, think of your facility as a multiple-response team or one that must function under routine circumstances as well as abnormal conditions (e.g., equipment malfunctions, power outages, and service interruptions) and emergencies (e.g., fire, adverse weather conditions, process upsets, and terrorist threats). Staff roles and responsibilities under each of these scenarios should be considered and described. This section of the plan should also designate who assumes responsibility in the event of staff absences, either planned or unplanned. The position descriptions prepared by plant management are not negotiable with employees because they represent the tasks required for each position in the plant's overall organization (Figure 3.8). Position descriptions

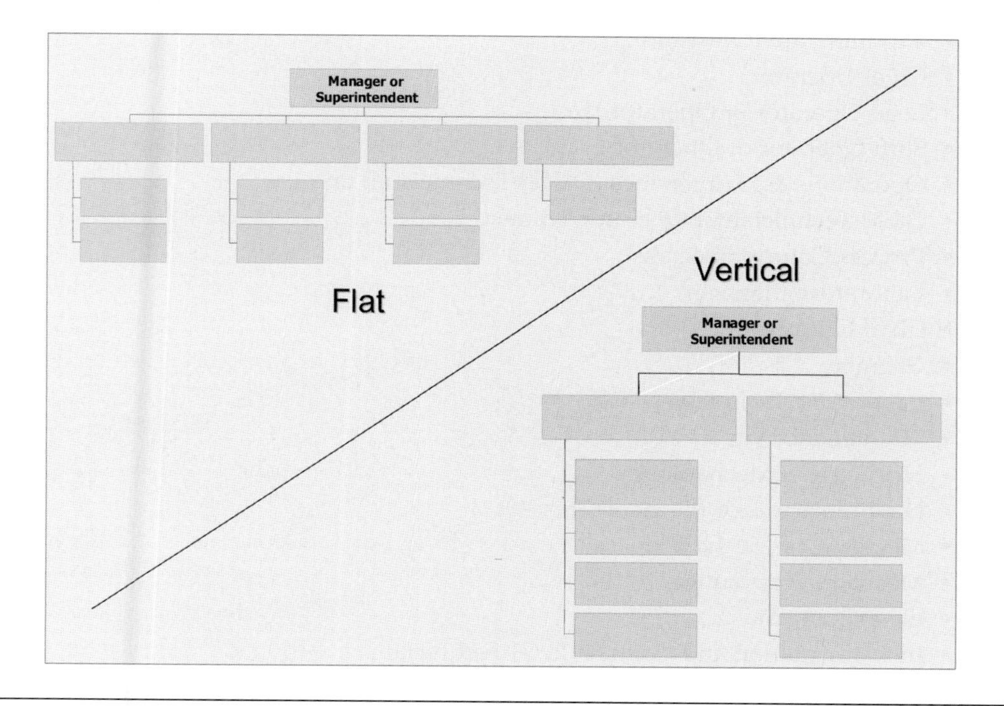


**FIGURE 3.8** Types of organizational structures.

need to be updated, however, as plant equipment, processes, technologies, or other conditions change, and an incumbent employee may request such changes in the position description. Each position description includes a brief description of the character of the work, a list of the tasks required, the supervisory reporting position (supervision received), the positions to be supervised (supervision exercised), education and experience required for hiring or placement, and typical tasks the incumbent must accomplish if the plant's mission is to be achieved.

Consider the maintenance function and roles and responsibilities of the maintenance team. Recognize that maintenance and operations are both "part of the team" and must work together and not separately. Thus, strive to foster a workplace atmosphere that supports the accomplishments of both groups and does not put one group subservient to the other. Seek means to maintain an effective communication and cooperative interaction between operations and maintenance personnel.

Consider the following titles that are commonly used in treatment facility position descriptions:

- Superintendent,
- Assistant Superintendent,
- Administrative Assistant,
- Chief Operator,
- Lead Operator or Operator II,
- Shift Operator or Operator I,
- Operations & Maintenance (O&M) Technician II and I,
- O&M Technician or Operator Trainee,
- Process Control Specialist,
- Laboratory Manager,
- Chief Chemist,
- Chemist,
- Laboratory Technician,
- Maintenance Supervisor,
- Heavy Duty Mechanic,
- Lead Mechanic or Mechanic II,
- Mechanic or Mechanic I,
- Mechanic Apprentice,
- Electrician, and
- Instrumentation and Control (I&C) Technician.

While these titles may be common to many organizations, facility managers should recognize that their own organization may use other designations and define varying responsibilities for each role. In addition to these conventional operations and maintenance titles, many facilities may have professional engineers and engineering technicians on staff. Regardless of the titles used by the facility, the facility manager needs to ensure that the organization complies with regulatory-driven roles. One example role is the "operator in responsible charge" or "direct responsible charge." These roles are required by the facility's regulatory rating and discharge monitoring report effluent permit process, and are based on the regulatory mandated minimum levels of certification required for individuals at the facility on a shift as well as an overall basis.

## PERSONNEL MANAGEMENT

As the introduction to this chapter points out, effective personnel management is the key to successful operation of a wastewater facility and the manager's primary role centers around people, whether through delegation, motivation, direction, evaluation, or training. For all except small plants, the manager must depend on his or her staff to accomplish the activities necessary to meet the plant's mission. Therefore, to be effective, a manager must master and be comfortable with employee relations and their many facets. These begin with recruiting people who can be trained and motivated, and include development, motivation, evaluation, and discipline (WPCF, 1986). The following are common elements to consider that can support effective personnel management: personnel management system, recruitment, training and development, retention and succession planning, meaningful performance appraisal, and discipline.

**PERSONNEL MANAGEMENT SYSTEM.** A personnel management system (PMS) includes a records system and other elements required to provide structure, direction, and control for the plant's employees. The PMS encompasses personnel policies and procedures such as compliance with the Occupational Safety and Health Administration's Right-to-Know Act and the Americans with Disabilities Act, as well as control procedures such as disciplinary actions and grievance procedures. A good PMS will help decrease misunderstandings and improve the trust and working relationships between management staff and employees. Although the plant's parent agency may administer the PMS, the plant manager will need to maintain personnel records in a secure and locked location at the facility. In addition, the plant manager should develop effective working relationships and coordination with the organization's human resources or personnel department.

The PMS specifies written policies and procedures covering personnel administration. Each employee should receive a written copy of the following policies and procedures:

- Listing of employee benefits such as vacation, sick leave, leave of absence, jury duty, and related benefits.
- Policies and procedures on work hours, time accounting, attendance monitoring, overtime, holidays, safety and health, and others.
- Employee "Right-to-Know" information.
- Employment practices of the organization such as equal employment opportunities, outside hiring, termination procedures, retirement, benefit programs, absenteeism and tardiness controls, employee complaints, and problems and appeals procedures.
- Policies and procedures for professional development, training, and continuing education.
- Policies and procedures governing disciplinary actions.
- An employee record file is one of the most important elements of a PMS. This file should contain a complete employment record from the time of entry of an employee until termination of service. The file may contain the following:
    - Employment application;
    - Interview record;
    - Hiring letter;
    - Wage record;
    - Sick leave, vacation, and other attendance-related records;
    - Disciplinary actions (if any);
    - Training and certification records;
    - Honors and awards;
    - Emergency notification information;
    - Promotion and demotion records;
    - Retirement program records;
    - Benefits selection records;
    - Annual performance review records;
    - Termination record; and
    - A copy of the exit interview.

**RECRUITMENT.** Recruiting and selecting good employees is the beginning of a successful operation. Effective recruitment fosters the interest of qualified applicants and encourages development of the existing staff. The objective of recruitment is to ensure that the number and quality of applicants needed to meet staffing requirements are available. Because recruiting and hiring are controlled by many laws and regulations and often are administered by the human resources or personnel department, obtaining the advice of experienced people on "do's" and "don'ts" should precede recruitment. However, selection criteria must be based on actual job needs without consideration of race, creed, sex, marital status, national origin, religion, or age, unless specific affirmative action programs are in place. If the information sought is related to specific job requirements, determining fitness for a job can include written tests, interviews, police records, former employers, and medical records.

**TRAINING AND DEVELOPMENT.** The operations and maintenance plan should also include a section on staff training and development. Training is a key component to unlocking the potential of staff. Training exemplifies the philosophy of continuous improvement and is essential to motivating employees. The type and intensity of training vary widely from one facility to the next, ranging from the basics in standard problem-solving tools to the skills required for cross-functional teamwork (Figure 3.9).

**FIGURE 3.9** Types of training.

An individual's unique skills are developed over time and are based on a blend of the following two educational elements:

- *Job-specific experience*—This element is "earned" over time and, without the benefit of a program designed to develop job-specific skills, often occurs through "doing". There is generally little formal documentation of this type of knowledge, and documentation of this blend of skills is best captured through a job-skills matrix developed from a profile of individuals who are considered successful within their present position.
- *Formalized instruction*—This element provides an individual with the basis for understanding the "why" and "how" something occurs. Formal instruction can reduce the time an individual requires to become "skilled" in a profession or task. There is generally available documentation for supporting this type of knowledge. Of formalized instruction, there are three basic types: *vocational,* which encompasses secondary education including college, vocational school, or other trades-related formal instruction that provides an individual with his or her core competency; *regulatory,* which encompasses specialized training that focuses on compliance to regulatory requirements including operator certification and health and safety training (this training is provided by trade organizations, the state, and other entities); and *employer-specific,* which covers training required and delivered by the employer to support organization performance and meet the specific requirements of a position (this training includes job-specific and other training to support organizational and individual performance).

The aforementioned elements are required to produce an individual possessing all of the qualities that the plant manager is looking for. Some of the elements for a developmental program that the plant manager should consider implementing include

- Hands-on training by experienced staff;
- Cross-training to learn other jobs;
- Formal schooling (night, weekend, or day classes);
- Special project assignments;
- Rotational job assignments;
- Participation in professional associations;
- Regular training sessions such as weekly "tailgate" safety classes or vendor demonstrations;
- Visits to other plants, industries, and city operations;
- Active participation in meetings;

- Increased delegation of authority; and
- State operator certification.

Refer to the sample leadership development plan shown in Figure 3.10.

An experienced operator or supervisor often trains others in a less-formal process called "hands-on instruction." Some hands-on training considerations are discussed here.

- Define and document standard operating procedures (SOPs) for process control, equipment operation and maintenance, laboratory procedures, and data recording. The SOPs become the standard by which consistent practices can be developed and adhered to.

**Employee Name**______________________________

**Supervisor**____________________

____________________ ____________________
Employee Name Date

____________________ ____________________
Supervisor Date

The Professional Development Plan is your guide to professional development. In it you state what you want to learn, how you will go about learning it, and how you will demonstrate what you have learned.

**I. Self-Assessment**

*What are your leadership strengths?*

*What are your areas for growth in leadership?*

*How do you learn best?*

**II. Statement of Plan**

**A. Select the competencies that you wish to focus your learning for this year.**
**(This field should be populated with skills training identified as needed during the organizational gap assessment. The courses listed below are examples)**
*Check those applicable:*

__ *Conflict Resolution/Negotiation*
__ *Critical Thinking/Decision Making*
__ *Customer Service*
__ *Communication*
__ *Employee Development and Coaching*
__ *Team Leadership*
__ *Strategic Management*
__ *Interpersonal Relations*
__ *Performance Management*
__ *Policy and Procedure Development and Implementation*
__ *Financial and Resource Management*

**III. Learning Strategies**
*(Indicate the Courses or Trainings that you would like to take)*

**IV. How will you demonstrate that the competencies have been achieved?**
*(Examples of sources for this information are certificates, performance appraisals reflecting improvement, agenda from staff meeting that shows team skills achieved, etc.)*

FIGURE 3.10 Sample leadership development plan.

- Set up a checklist of items to be learned and have the trainer and trainee set a reasonable schedule to learn them. Initial each item after confirmation that the trainee has demonstrated his or her competency.
- Organize an orientation, particularly for new employees, to provide an overview of the whole operation, not just the employee's role.
- Emphasize safety elements.
- Provide technical reading and access to trades magazines. Establishment of a plant "library," where employees can go to access reference documents, is suggested. A library could also include all drawings, reports, equipment manuals, and other documentation.
- Document all training delivered and received into the PMS

**RETENTION AND SUCCESSION PLANNING.** The manager should recognize an individual's needs for satisfaction and use this knowledge to motivate the employee. In addition, the manager may generally stimulate motivation by creating a work environment that includes the following elements:

- Make responsibilities known and reasonable. Match authority to responsibilities. Eliminate the frustrations of slow approvals, long meetings, and lack of follow-up.
- Find rewards for good performance. Often, prompt recognition by the boss is the easiest, most effective reward.
- The reasons for work must be clear to those who do the work. With shift work, this requires supervisors to visit all shifts and leave instructions that include reasons for the work.
- Working relationships must be comfortable and encouraging for teamwork. Social events, time out for some fun event at work, and fair and equal treatment can help build an environment that encourages good performance.
- Recognition that continued education and training are essential in the modern work environment, which is characterized by exponential growth in technical knowledge required to improve productivity with decreased resources.

Methods used to develop the teamwork that is essential for a successful plant include:

- Arranging for group effort to solve a problem, to develop a budget for the next year, or to undertake another project. As a requirement for such a team effort, everyone must be able to contribute and must have a stake in the outcome.

- Having employees share their knowledge and goals with one another.
- Providing group and individual awards based on group success and individual contributions in the group.
- Rotating jobs or sharing jobs so employees gain a better understanding of other situations.
- Documenting experiences and lessons learned for future reference and knowledge retention.
- Instilling an atmosphere that is rewarding and enjoyable to work in.

Table 3.1 describes how to develop a succession management program.

**MEANINGFUL PERFORMANCE APPRAISAL.** Organizations use employee performance appraisals as a means to clarify job duties and responsibilities, evaluate employees, provide input into specific development areas, and determine if the employee has achieved the objectives set forth. Generally, the performance appraisal process is directed and administered by the organization's personnel department.

The manager must fully understand and follow the requirements of the performance appraisal process. For example, it is important for the performance appraisal process to clearly define individual duties and responsibilities. The employee must understand what their role is as well as the expectation for their performance. This is done by developing specific, measurable, and reasonable goals that meet the overall objectives of the organization. Managers should encourage employee participation and input when establishing performance criteria. Employees can offer many insights that present opportunities for their success and "buy-in" of the process.

The manager should ensure that a fair and consistent system of employee evaluation is used at all subordinate levels. The system should include prompt recognition and rewards for successful performance, identification of poor performance, and corrective actions needed to improve performance, when necessary. The plant manager uses the performance appraisal to improve day-to-day work procedures and determine long-range needs for further employee development and motivation improvement. The most important part of the performance appraisal is the day-to-day informal communication between the manager and employees. If employees are given timely feedback on their performance throughout the rating period, they should not be surprised with their performance evaluation during the review process.

**DISCIPLINE.** Tardiness and absenteeism are among the most common and persistent problems; as such, special attention to these issues is essential. Concurrent with the performance appraisal process, the manager should also establish policies or guide-

TABLE 3.1 Developing a succession management program.

| | |
|---|---|
| **1. Institutionalize the process and gain stakeholder support** | • Gain commitment from top management and board members<br>• Gather resources<br>• Identify the strategic vision and goals<br>• Incorporate as part of values of the organization<br>• Clearly define objectives for the program |
| **2. Conduct assessments of organizational need** | • Conduct "as is" assessment of the organization<br>  • Workforce<br>  • Processes<br>  • Systems and resources<br>• Conduct "to be" assessment of the organization<br>• Conduct gap analysis |
| **3. Develop the succession planning model** | • Determine employee involvement in program<br>• Build a leadership pipeline<br>• Develop successors for critical functions<br>• Identify training and development strategies<br>• Develop retention strategies<br>• Create knowledge management and transfer strategies |
| **4. Implement succession planning strategies** | • Determine resource needs for implementation<br>• Identify barriers and develop strategies to overcome<br>• Design needed templates, forms, and systems<br>• Develop or update job descriptions<br>• Prepare organization for change<br>• If needed, implement some strategies on a pilot basis<br>• Link succession strategies with human resource processes<br>• Train staff as necessary |
| **5. Continuously measure, evaluate, and adapt** | • Define measures of program success<br>• Design reporting process<br>• Track and communicate progress, celebrate program successes<br>• Capture stakeholder feedback on strategy effectiveness<br>• Adjust program based on evaluative results<br>• Keep top management engaged<br>• Make a 3- to 5-year succession plan part of the organization's strategic planning process |

lines that govern employee behavior and attendance. Poor attendance often serves as a warning for a situation that could get worse. Immediate attention by a supervisor may prompt an employee to address the problem without further supervisory action. Managers should avoid diagnosing personal problems for employees that have external issues that affect their work performance. However, they should point out to the employee that there are sources available to them for assistance, such as an employee assistance program. A program that includes professional counseling is effective in helping employees overcome difficult personal problems and become motivated to work.

Managers must use discipline as a means to correct employee behavior that does not comply with polices and procedures. Discipline should be viewed as a tool to improve performance and/or change behavior. Progressive discipline systems, which are common, may include successive steps of oral and written warning, suspension, and discharge. Returning to work following suspension may be conditional on an employee's written statement of performance commitments. In a union environment, employee rights to representation, appeals, and arbitration may apply.

The plant manager should follow a disciplinary process that includes the following:

- Train all supervisors in disciplinary procedures;
- Ensure that discipline is administered evenly and respectfully;
- Require that investigations, notice, and hearings occur in a timely fashion;
- Ensure that employees understand performance requirements and disciplinary steps; and
- Document performance and discipline and file the records.

## EXTERNAL RELATIONS AND COMMUNICATION

The issue of external relations, that is, how your facility interacts with and responds to neighbors and other organizations under normal and emergency conditions, is a key functional role that is often neglected. Many facilities have addressed the issue of responding to emergencies that involve firefighters and police. For example, they have invited both departments to the facility to review scenarios and to ensure that these agencies are familiar with the facility and that there are clear lines of communication open. Likewise, many facilities have established lines of communication required for mutual assistance among themselves and adjoining utilities in locations that support this type of arrangement. The manager should leverage and build relationships with external stakeholders, including the following:

- Plant neighbors;
- Vendors;
- Police department;
- Fire department (especially for emergency response planning and preparation);
- Regulatory agencies;
- Other city or utility departments (e.g., streets and roads, parks, and legal);

- Media (an external media communications policy and strategy should be developed prior to engaging with the media); and
- Professional organizations (e.g., community groups, and local and state WEF affiliates).

Communication is the foundation to developing and maintaining relationships with external entities as well as those within a facility. Consider the need to define protocol for communicating with agencies such as offices of emergency preparedness or disaster coordinators. Furthermore, consider the need to define a methodology for dealing with inquiries from the public and the media. In an era when misinformation can have dire consequences, these issues need to be addressed. Protocol should be designed to ensure that messages to the public are managed with coordination, clear authority, and accountability.

The manager should develop a sound external communications plan for managing general external communications. Indeed, public consciousness and vigilance about environmental issues, tax abatement movements, and reaction to governmental regulation have created an atmosphere in which public support cannot be taken for granted, no matter how beneficial a service may be. Cultivation and retention of support for specific projects, in particular, may benefit from the services of public relations professionals. Communications with the public help the manager understand the public's concerns and viewpoints. Some plant managers have found the following elements to be useful ways to communicate with the public:

- *Elected officials*—Whether directly or through another manager, elected officials need regular reports on the status of major plant financial and performance issues. Newly elected officials should be invited to see the plant. Presentations at meetings of superior policy-making bodies are important, and the presenter should be well prepared.
- *News media*—When a negative event happens, plant personnel should be as honest and open as possible, be sure of their facts, and inform public officials prior to discussions with the media. It is useful to provide a fact sheet to the media if statistics are important.
- *Plant tours*—School and community groups appreciate tours. A tour requires a safe route, a knowledgeable spokesperson, meaningful handouts, and a show of respect for the visitors and their interests.
- *Talks*—Schools and community groups seek informed speakers from the wastewater treatment plant (WWTP) during environmental weeks or when the press covers an issue related to operations. A talk on plant operations requires preparation, practice, and visual aids. A videotape of the plant, photographs, props

(e.g., water samples showing improvement in water quality from raw through final effluent), and handouts are helpful for such talks.

- *Neighborhood communication*—It is worthwhile to pay special attention to plant neighbors and their interests. Once a problem occurs, the plant should promptly provide an honest explanation of the problem (or at least information about the status of any investigation) and make visible efforts to reduce the problem.
- *Special events*—The manager may choose to invite the media and public officials to celebrate installation of new equipment, permit compliance, public works week, or another event. At least one event should be celebrated each year to keep positive aspects visible.
- *Advisory groups*—Both a community environmental problem and a long-term planning effort provide an opportunity to use public opinion to develop recommendations acceptable to the community. The manager must include representatives of concerned groups and provide the technical support and administrative help necessary for their involvement.
- *Community involvement*—Membership in local service clubs can help a manager understand community issues, meet community leaders, and gain public respect.

During emergency situations, nothing is as critical as clear and effective communication. Part of effective management of emergency communications is the establishment of an emergency call list that includes names and responsible parties to call during an emergency. This list should include contact numbers for local police, fire, and other emergency response entities. A schedule for on-call employees should also be defined and posted for immediate access and should be known by all staff. Finally, a media response plan should be established that includes clear procedures for contacting and responding to media inquiries and interviews. Media contact should be cleared by senior management and coordinated with the organization's external relations/public communication staff.

## REPORTING AND RECORDKEEPING

Although recordkeeping and reporting responsibilities are well-defined in most utilities, they are often not included in an operations and maintenance plan. A facility's operations and maintenance plan should outline the quality assurance and quality control procedures used in reviewing records and reports before their submittal (i.e., who reviews what and what sign-off procedures are required) and should designate who has primary responsibility for responding to inquiries from the regulatory community.

Reports, operations logs, e-mail, and other documents produced during the course of operations and maintenance are part of the public record and must be filed and kept accessible. Archiving of files and data should be conducted in accordance with the organization's policies and procedures for records retention and storage.

A manager should keep in mind the following phrase: "The job is not done until the paperwork is finished." Accordingly, the manager needs to ensure that records are completed, reviewed for accuracy, and maintained in a library or storage center for the required time as defined by the policies of the organization and by law. The following are some records management strategies and concepts to consider:

- *Knowledge management*—Knowledge management is the strategy used to capture, retain, and make available collective information that an organization and its employees possess. There are many elements to knowledge management including administrative, standard and emergency procedures, records and library management, incident recordkeeping and other types of documentation (e.g., photos, videos, reports, and so on), training, daily logs, mentoring, performance assessment, and informal knowledge transfer (e.g., "on-the-job"). Whatever the means used, knowledge management is about capturing and retaining the experience and skills of individuals that may or may not be documented. Some utilities are investing in information technologies that support "capture and access" to information and various media. These systems generally are Web-based and work with or include electronic document management platforms.
- *Records and file management*—Records need to be filed and maintained using a logical file management system. Records include all reports, files, and printed information produced or used by the facility. Administrative staff are generally responsible for this activity. Some organizations have been including electronic document management tools to maintain records and files.
- *Library management*—The plant library is the knowledge center of the facility and, while it may be closely linked to the file system, is an important element to house and locate all printed (and electronic) information (Figure 3.11). This information includes plans, contract documents, manufacturer information and manuals, books, periodicals, and training materials. One or more individuals should bear the responsibility for maintaining the library. A library should have a location that is secure and environmentally sound for long-term paper storage. One effective way to ensure that a library is current is to allocate a senior staff operator or maintenance person to "own" the library and ensure that all documentation is inventoried, indexed, and available.

FIGURE 3.11 A centralized facility library is a sound knowledge management strategy (MSDS = material safety data sheet).

## EMERGENCY OPERATIONS

Emergency planning for plants is defined as the continued development and documentation of actions and procedures aimed at dealing with all hazards—both natural ones and those caused by humans—that could adversely affect the environment or the efficient operation of the facilities. The emergency operating plan (EOP) component of the facility operations and maintenance plan covers the entire facility and involves all employees. Everyone concerned, however, must realize that emergencies do not follow a standard pattern, and personnel must be prepared to adapt to various emergencies (WPCF, 1989). Providing pre-assigned damage assessment teams, each with the responsibility to react to particular types of emergencies, is recommended.

In some ways, the phrase "emergency planning" is misleading because it implies that planning is a one-time effort done before a disaster (FEMA, 1985). Instead, the plan itself may be less important than the process that produces it. The planning process identifies hazards and needs, sets goals, determines objectives, sets priorities, designs action programs, evaluates results, and repeats the steps.

Emergency activities occur in the following four phases (Hulme, 1986):

- *Preparedness (planning)*
  - Develop EOPs and test them.
  - Inventory local resources.

   - Initiate emergency management contacts (individuals, state and federal programs, and private and public organizations).
- *Mitigation*
   - Train personnel in emergency preparedness procedures.
   - Correct improper operations and maintenance practices such as deferred preventive maintenance.
- *Response*
   - Alert the public when necessary.
   - Mobilize emergency personnel and equipment.
   - Evacuate plant personnel and nearby residents when necessary.
- *Recovery*
   - Reconstruct or rehabilitate structures and equipment.
   - Conduct public information and education programs.
   - Develop hazard-reduction programs.

The first step in the planning process is to identify the hazards and dangers faced by the plant. Typical natural and caused hazards and the resultant dangers are shown in Table 3.2. Goals, objectives, and priorities for a particular plant for each process can be established based on the identified dangers. A vulnerability analysis (Figure 3.12) provides a useful tool for formulating an EOP for each potential situation. An outline for a typical EOP is given here.

- *Emergency flow chart*—This chart should be the first page of the binder so that anyone responding to an emergency can proceed to resolve the emergency problem.

**Table 3.2** Hazards and dangers leading to emergencies.

| Hazards | Dangers |
|---|---|
| Natural | |
| Earthquake | Sewer collapse, building collapse, hazardous material release, possible flooding, power failure. |
| Food | Electrocution, fire from electrical shorts, hazardous material release, power failure. |
| Tornado | Building collapse, hazardous material release, power failure. |
| Winter storm | Power failure, plant inaccessible to employees. |
| Human | |
| Chemical release | Damage to the environment, skin and mucous membrane burns; death by inhalation, explosions, fire, or a combination of these. |
| Supply shortage | Shutdown of operations. |
| Fire | Death or injury to employees, shutdown of plant processes. |
| Strike | Shutdown of plant processes. |

Vulnerability Analysis Worksheet

Treatment System Capital City Wastewater Treatment Plant

Assumed Emergency Flood (100 years)

Description of Emergency The flood will cause considerable damage to low lying areas. Bridges will be closed, utility poles downed and electrical power interrupted.

| System component | Effects of emergency | Prevention recommendations |
|---|---|---|
| | Type and extent | |
| Collection lines | 60" interceptor could be washed out at Back Creek crossing | 1. Encase line in concrete<br>2. Maintain pipe and fitting to repair damaged section.<br>3. Provide portable pumps to bypass break.<br>4. Contract for major emergency repair services |

FIGURE 3.12 Example of a vulnerability analysis.

- *Contact lists*—All contact lists should contain names, organizational positions, location telephone numbers (including home, cellular phones, and pagers, if appropriate), and radio call numbers/names, if assigned.
- *Chain-of-command*—This item identifies the line of authority in an emergency.
- *Organization chart of duties*—This chart identifies each group and its emergency activities.
- *Damage assessment forms*
- *List of facilities*—This list includes names, addresses, and telephone numbers of all WWTPs, administrative offices, field offices, pumping plants, and other installations.
- *Emergency equipment list*—This list identifies all heavy equipment and vehicles by their locations.
- *Contractors*
- *Mutual aid agreements*—This information should include the name of the organization that will assist, the means of contact by telephone or radio, and the type of mutual aid to be provided.

- *Public information procedures*—These procedures cover public communications about the emergency and response activities.
- *Emergency operations center*—The center should have the following supplies: telephones (including cellular telephones); two-way radios; log books for operations and communications records; maps, drawings, photographs, and video recordings of the plant and surrounding area; operations manuals; first-aid supplies; cameras and video recorders for recording damage and operations; and living facilities including water, food, and bedding.

## OPERATING EFFICIENTLY

The concept of operating efficiently is a relative term. As stated earlier, although the term efficiency is often used in association with both positive and negative cost and budgetary effects, the term embodies much more. Seeking improvements in efficiency, whether they be in process, treatment performance, staff effectiveness, or cost control, can be argued to be a fundamental and core focus of management in which acceptance of the status quo is never good enough. Achieving efficiency improvements requires constant attention to discrete practices, functions, subgroups, and technologies within the organization.

Facility operations are mandated by a number of factors including (1) compliance with governing and regulatory requirements; (2) compliance with personnel policies and labor rules; (3) conformance with organizational business practices; and (4) the need to balance all activities with a commitment to minimizing costs. How much the manager departs from the status quo (e.g., to improve efficiency) depends largely on the amount of risk he or she is willing to assume. Most public utilities are inherently risk-averse. However, risk can be mitigated by careful and well-informed thought and planning prior to implementing an improvement. Thus, the key lies in defining perceived versus actual risks. Accordingly, recognizing and managing risk elements are important components of efficiency improvement activities, including decisions about how far the organization wants to go in pursuing optimization.

At the heart of all efficiency management activities is the focus on optimizing resource allocation and consumables consumption. The following general elements govern efficiency at a facility:

- *Individual performance*—An employee's ability or treatment process to perform as efficiently as possible affects overall efficiency at the utility (Figure 3.13). In the case of an individual employee who does not have the appropriate tools or

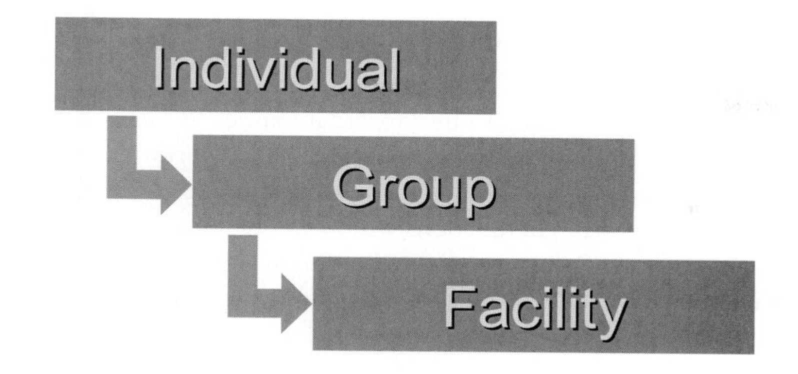


FIGURE 3.13 Efficiency elements.

is not properly trained, the time required to complete a task satisfactorily will most likely be extended. As a result, labor effort to conduct work (e.g., labor costs) will increase. Accordingly, the performance of a single individual can have a direct impact on the cost efficiency of the work unit, the facility, and the organization as a whole. Similarly, a treatment process that is not operating efficiently can substantially affect the performance and efficiency of downstream processes and incur other costs such as increased energy or chemical dosages.

- *Group performance*—Taking the next step from individual employees or treatment processes to groupings, the impacts of inefficient performance can incur even greater cost and risk elements. The performance of teams or work groups affects overall efficiency, and group performance is affected by the efficiency of individual group members. High-performing, efficient work teams excel through their ability to work together while using their individual abilities to the fullest extent practical. Treatment trains (e.g., wet stream or solid streams) can be considered as a group and measured as such.
- *Facility performance*—Facility performance is based on the overall aggregate of staff and/or processes at the facility to operate efficiently in an integrated and coordinated fashion. Although management assumes chief responsibility for organizing and coordinating the interactions and work processes of teams and groups within the facility or unit, a team's composition is usually the most significant factor in determining performance. For example, an energy management team can play a critical role in overall efficiency because team members are focused on a facility-wide initiative, thereby reducing energy costs.

## EXTERNAL BENCHMARKING AND PEER REVIEW

In general terms, benchmarking is a systematic approach for comparing an organization's internal processes to those of other external entities. By using external benchmarking to compare and define how other organizations perform similar processes, internal procedures can then be modified to improve efficiencies and performance.

In benchmarking, key target functions within an organization that are candidates for improvement are identified and compared to similar functions within other organizations and competitors. While a relatively simple concept, benchmarking can be a complex process involving many different approaches and methods of comparison. As with other quality improvement endeavors, benchmarking can be costly or not, depending on the type and extent of benchmark comparisons being sought. The many varied approaches and types of benchmarking activities have resulted in a significant difference of opinion among experts and managers alike in what benchmarking really should consist of as well as how it should be applied within the public sector. Despite the many approaches, there are essentially two types of benchmarking categories, metric benchmarking and process benchmarking. These are outlined here.

- *Metric benchmarking*—In metric benchmarking, one organization's costs or other numerical values, such as staff numbers per million gallons and so on, are compared to the same values of others. Metric benchmarks do not, by themselves, identify how to improve a process or increase efficiency. Rather, they are metric parameters that measure the performance of a specific entity. Confusion may arise because of pressure to use performance benchmarks to compare efficiencies between different entities. However, because each entity has its own unique set of controlling variables, direct comparisons may be inaccurate. Considerable effort is required to account for these variables so that equivalent and fair comparisons can be made.
- *Process benchmarking*—Process benchmarking is a more effective way to apply benchmarking efforts in the public arena, as it focuses on improving internal programs and processes by learning how "the best" organizations conduct such activities. In process benchmarking, a utility's practices and other "non-numerical" items such as procedures and systems are compared to others. With this focus, benchmarking does not get bogged down in accounting for control variables that affect performance benchmarking outcomes. Comparing practices and procedures, such as one plant's odor complaint response mechanism to another, helps identify better methods and procedures of conducting business. Process benchmarking is not difficult. However, before looking externally to see how

others achieve best practice, the organization should understand its own process as a baseline for comparison.

## FINANCIAL PLANNING AND MANAGEMENT

Using sound financial planning and fiscal management strategies are core responsibilities of the effective plant manager (Figure 3.14). All organizations and management have a fiscal responsibility to manage their budgets within the limits of reasonable rate-setting, grant monies, and other sources of operating and capital funds. The extent of funding sources and the capacity of rate payers will govern the strategies and creativity of the manager to maximize the facility's budget to meet near- and long-term cost objectives. Consider the following elements of sound fiscal management:

- ***Define service standards and performance levels.*** The basis for any budget expenditure decision lies in determining the service standards to which the organization and facility must operate. Producing effluent quality within National Pollutant Discharge Elimination System limits as well as biosolids and air quality that fall within standards are basic expectations and the primary reason the facility exists. However, performance levels come at a cost, that is, should the facility be operated in a manner just to get by (e.g., limited staffing and scrimping on maintenance activities) or should it be operated in a manner without excessive regard to cost of service (e.g., operated, maintained, and staffed conservatively). Although the former may result in a lower annual operating and

- **An Iterative Process:**
  - **Define service standards and performance levels.**
  - **Make informed decisions on asset management.**
  - **Understand the financial implications of budgeting decisions (e.g., assess political and customer impacts from needed rate increases).**
  - **Ensure that the budget is monitored and managed.**

FIGURE 3.14 Fiscal management is a critical responsibility.

maintenance cost, it also may forego long-term maintenance expenditures and other investments that may result in additional or premature future costs to limit present expenditures. The latter has its own cost of operating conservatively and may result in unnecessary expenditures and waste. Thus, clearly defining performance standards that are embraced by the organization (e.g., city management and council) and managing to meet these will provide the best defense against management, political, and ratepayer concern over how much a facility requires in its budget.

- ***Make informed decisions on asset management.*** A solid asset investment and management program includes the development of strategies to maintain assets over time and to ensure that assets perform according to established service levels and design criteria throughout their useful lives. This requires a well-managed life-cycle plan for the plant's assets. Development of this plan includes an upfront risk and criticality assessment, condition assessment, and analysis of historic failures and events. This is perhaps the most involved phase of asset management and includes the integration of many considerations such as spare parts and equipment inventories, preventive and corrective maintenance strategies, renewal and replacement criteria, and rehabilitation requirements. Criticality, capacity, risk, and condition analyses identify focus areas toward which preventive maintenance budgets should be targeted. In an environment where budgets are limited, available dollars must be directed toward assets that have a high probability and/or consequence of failure. Part of this strategy may be accepting that some noncritical components can be run to failure, as it is cost-prohibitive and inefficient to implement a zero-failure preventive maintenance program.
- ***Understand the financial implications of budgeting decisions.*** The budget is a roadmap that ideally is based on informed cost-projecting and sound information as to what capital and operating costs the facility will incur during the fiscal year. The goal of financial forecasting is to provide clear vision regarding the potential financial outcomes of current management decisions. The process of developing a budget, while consuming and difficult, should result in a budget that the manager can accept and use that is based on informed financial forecasting to effectively manage the facility. Both capital and operating budgets impact the bottom line, that is, the rates customers pay to receive the services provided by the organization and the facility.
- ***Ensure that the budget is monitored and managed.*** Most organizations track expenditures against their budget and issue standard reports on a periodic basis

(e.g., monthly and quarterly). If the frequency of reporting is insufficient to effectively track and manage expenditures against the facility budget, it is the manager's responsibility to seek this information. This might require requests for more timely reports or specific information or manual tracking of expenditures. A budget needs to be flexible to accommodate changes or unanticipated costs during the year. Most organizations have provisions to make adjustments to a budget through transfers within a budget or by amending a budget if significant increases (or decreases) are required.

## INFORMATION AND AUTOMATION TECHNOLOGIES

Information and automation technologies support efficient and effective operations. They can be powerful tools within the manager's arsenal to manage the facility. Moreover, they are tools to assist both routine and nonroutine activities and can greatly strengthen the performance of the facility and staff. A modern facility cannot be run without plant automation and information management technologies. The following are some of the more common types of automation and information management technologies:

- *Automation technologies*—Process instrumentation and control system technologies provide access to real time data. Supervisory control and data acquisition (SCADA) systems and field instrumentation are tremendous tools to monitor, control, and provide early warning to impending process and equipment problems. When coupled with a data management and reporting system (e.g., data warehouse), SCADA systems can centralize monitoring and control, reduce labor costs for operations, improve the accuracy and timeliness of data used for performance reporting, and furnish critical data to support equipment maintenance.
- *Data management technologies*—At the facility level, data management technologies can include spreadsheets, laboratory information management systems (LIMSs), and computerized maintenance management systems (CMMSs). Of these technologies, organizations seem to struggle most with fully implementing and using the capabilities of a CMMS. Some facilities are investing in more refined commercially available data management technologies such as decision–support systems and sophisticated data reporting technologies. The key idea is that you will need a data warehouse to gather summary-level information together in one place as well as a reporting product to produce the charts and graphs. Consider a general

purpose product that can serve a variety of needs and report data from SCADA, LIMS, CMMS, and any other well-behaved system.

- *Content management systems*—Electronic content management technologies provide the means to store, access, and manage all types of printed information such as manuals, photographs, SOPs, reports, policies, and training materials. They generally are Web-based and can be linked with an electronic document management system, a common technology that many organizations have invested in. Many facilities use their own content management platforms as electronic operations and maintenance manuals.

Technology is constantly evolving and it is beyond the scope of this book to fully review the topic. The facility manager, however, should consider the following in managing and using automation and information technologies:

- A manager needs a structure and process for making the most important decisions about information technologies (or what can be termed "information technology governance"). Do not let every group or the "information technology guru" make isolated decisions. However, the manager should seek input and support from all stakeholders within the organization, including the organization's own information technology group or department. The facility's information technology investments should, as much as possible, "fit" or align with the organization's overall information technology standards and requirements.
- If you are selecting new systems, use a formal process to evaluate software in which the qualities of the vendor are as important as the features of the software. Spend most of your time figuring out how you will do your work and what staff roles/responsibilities are critical. Do not focus on the software until you get this part right.
- Develop an information technology master plan for the facility. Define how the parts are going to fit together, especially how separate systems are going to be integrated so data flows freely. The manager should get help doing this. Do not expect that your (central) information technology group has the expertise or desire to create this sort of plan.
- More and more, control systems are looking like regular computer systems, that is, the same network types, the same computer types, and conventional Windows® and Web-browser software. A manager should stay away from older proprietary systems.
- Key consideration is security. The manager should keep the organization's control network separate from the general purpose network and put a firewall be-

tween them. However, they should not be separated so much that data cannot "flow" because operations data need to flow up in the organization to support regulatory reporting and management decisions.

## PUBLIC–PRIVATE PARTNERSHIPS

The water and wastewater industry has witnessed an increased focus on engaging private services as a tool to bolster efficiency and lower costs of service. The term "privatization" has been applied to the broad range of options that involve private sector provision of traditional government-provided goods and services. The privatization movement is based on the premise that there are goods and services that can be effectively and efficiently provided by private entities, either under contract to or independent of the public sector. From a publicly owned utility management perspective, privatization is better defined as *selective outsourcing* and encompasses the host of private-sector services that may be engaged by a public utility.

## REGULATORY COMPLIANCE CONSIDERATIONS

Overall, wastewater facilities are subject to regulation by many local and national agencies that are concerned with some aspect of their operation. Regulatory agencies and the public expect compliance with permits that may include wastewater discharges, air emissions, building modifications, underground tanks, and other items. The manager needs to learn which laws and regulations affect the plant and set up systems and controls to ensure that applicable regulatory requirements are met. Violations of some regulations can result in civil and criminal penalties for which the manager can be held liable. A manager must ensure that accurate reports are filed on a timely basis.

A manager should personally meet with representatives of the regulatory agencies. Representatives can help recognize problems and suggest actions to prevent or reduce problems. Important regulatory agencies may be expected to inspect the operation at least once a year. When regulatory problems occur, the manager must inform all regulatory agencies with jurisdiction as well as the appropriate municipal officials. The manager also needs to document the facts, corrective action, and individuals contacted. If the problem becomes severe, records can help protect the community and the manager. Above all, the manager should maintain an attitude of courtesy and cooperation with regulatory agencies.

Facility managers may have opportunities to influence the establishment of regulations under which they must operate. While such opportunities may arise by invitation from the regulatory agency, those who are regulated should not hesitate to make recommendations, especially through group action of professional organizations. Finally, regulation should be considered to be in a state of evolution.

## MAINTENANCE MANAGEMENT

Maintenance processes and activities offer significant opportunities for maximizing the total life-cycle cost and performance of equipment processes and other assets improving efficiency. Historically, utilities have often overlooked the tremendous value of tracking and monitoring maintenance activities, both in terms of controlling costs and efficiently allocating and managing facility resources. Better understanding of maintenance staff perspectives and issues permits facility managers to choreograph maintenance activities in close coordination with operations requirements to better assess how these activities are accomplished and how resources are applied. At the core of this management process is the use of performance measures to drive decision-making. As Figure 3.15 shows, an efficient maintenance process continuously measures management, scheduling, and work output and provides feedback on alternatives for improvement.

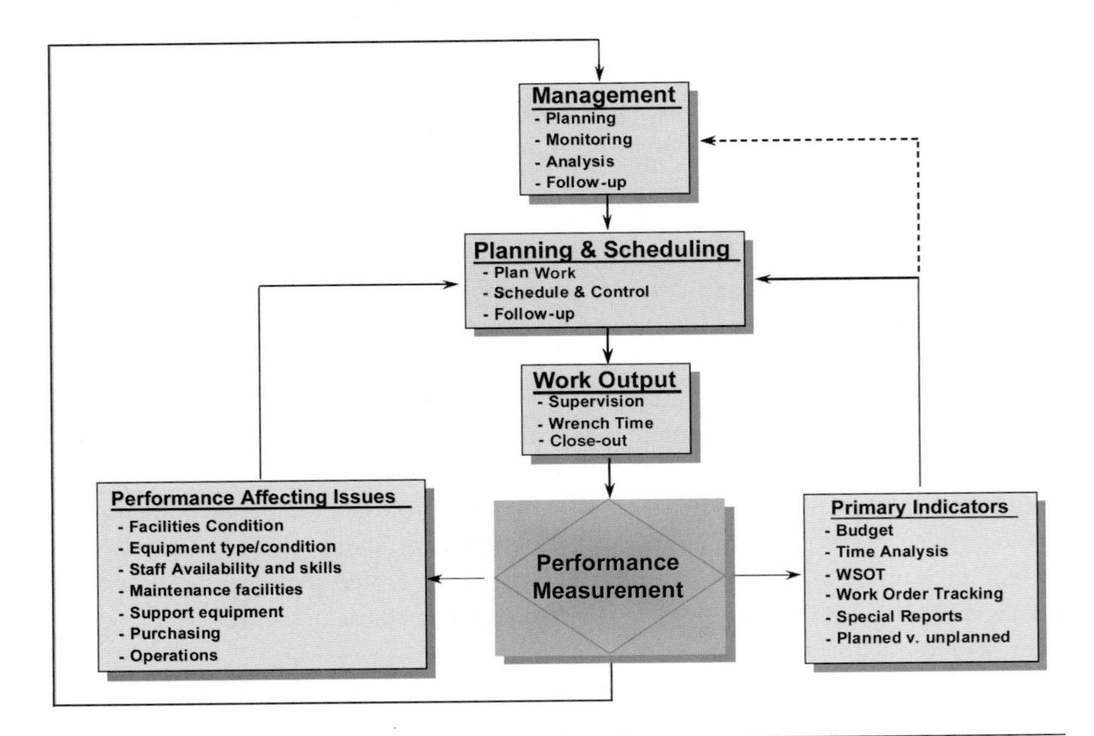


**FIGURE 3.15** The maintenance process (WSOT = work spread over time).

Simply put, a wastewater treatment facility depends on good maintenance as much as effective operations. The two functions must be tightly coordinated, which can be difficult when the groups in charge of these functions must sometimes work toward competing goals. Suppose, for example, a sedimentation basin needs to be removed from service for routine maintenance. The maintenance staff, which have planned and allocated tools, supplies, and labor to accomplish the task, can only perform their work once the operations staff have removed the basin from service and made the unit available. The goal of the operations staff is to minimize the time the basin is out-of-service; the goal of the maintenance staff is to complete all the maintenance tasks as planned. If the operations staff fails to make the filter available, or if the time allotted for the maintenance staff to complete its work is too short, the maintenance activity and the effectiveness of the maintenance staff will be compromised. From a long-term perspective, not completing all maintenance activities as planned will likely result in the inefficient use of resources and increase the frequency with which a piece of equipment must be removed from service for maintenance.

**SELECTING A MAINTENANCE MANAGEMENT STRATEGY.** Managing maintenance activities involves choosing from a number of possible approaches (Figure 3.16). By following equipment manufacturers' recommendations for routine and preventive maintenance, a plant could greatly exceed its annual maintenance budget. Thus,

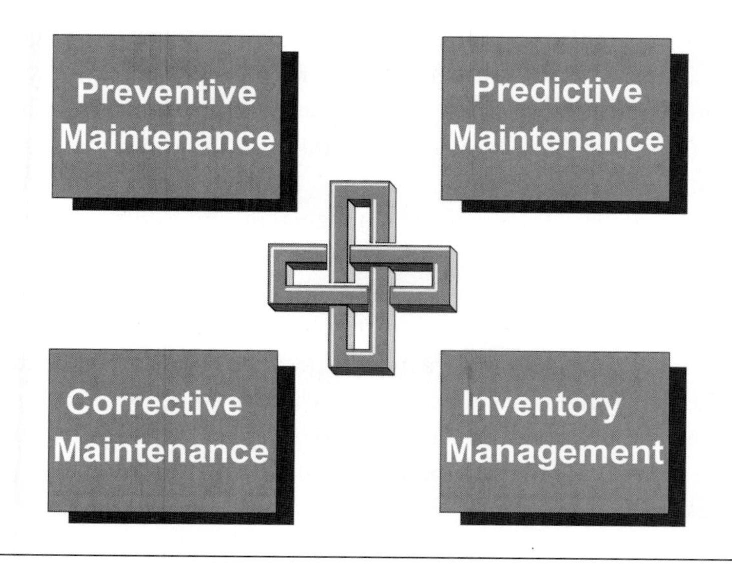


FIGURE 3.16 Components of effective maintenance programs.

the manager must decide what maintenance strategies work best to balance the risks and benefits of deviating from manufacturer recommendations. A classic challenge is balancing preventive maintenance activities, corrective and predictive maintenance practices, and inventory management strategies, which are defined as follows:

- *Preventive maintenance*—Preventive maintenance activities require the allocation of labor and associated costs. Preventive maintenance can be performed efficiently (e.g., by an individual who can do lubrication rounds in record-setting time); however, the challenge lies in deciding which preventive maintenance activities should be performed on what equipment to maximize equipment reliability and life expectancy and to manage risk. Along with risk assessment, cost is an important consideration in selecting preventive maintenance strategies.
- *Corrective maintenance*—Maintenance of equipment that has failed or is nearing failure constitutes corrective maintenance. If corrective maintenance dominates a utility's maintenance activities, this is a hallmark of ineffectiveness and inefficiency. Conversely, letting selected equipment operate to failure can be the least costly strategy if easy service conditions and an evaluation of life-cycle costs suggest that risks and associated corrective maintenance costs are lower than the life-cycle cost of preventive maintenance for that equipment.
- *Predictive maintenance*—Predictive maintenance activities involve monitoring the condition of equipment and selecting the most appropriate time to service it. Although predictive maintenance generally focuses on major equipment that is expensive to service, this strategy represents a highly efficient approach to maintenance in general. Predictive maintenance processes include the application of sensing technologies such as infrared light, vibration, sound, and oil analysis.
- *Inventory management*—Inventory management represents a major cost item in a utility's budget because significant assets can be tied up in inventory. Inventory management can be either a source of or solution to efficiency problems. Consider a maintenance crew that has disassembled a piece of equipment only to find that a key part is not where it is supposed to be in the warehouse. The effect is obvious—a productive allocation of resources (i.e., the crew) is transformed into an inefficient, nonproductive task (i.e., searching for parts within the storage area). Inventory management represents a cost to the utility, and the manager must weigh the costs of creating and maintaining an effective warehousing operation against the costs and risks associated with not formalizing the warehousing process.

An emerging best practice associated with asset management is reliability centered maintenance (RCM). Reliability centered maintenance is a strategy that manages the maintenance function by assessing the replacement cost, criticality, and life expectancy of an asset to determine the desired maintenance strategy. The foundation of RCM is to achieve the lowest life-cycle cost of the asset. For example, the criticality of a small sump pump may not be the same as that for a large aeration blower. The choice to use a rigorous preventive maintenance program on the sump pump (as would logically be used for the blower) may not be the best expenditure of labor and materials for this small piece of equipment with a low criticality. Accordingly, the manager may choose to simply monitor the pump routinely, allow it to operate to failure, and then repair or replace the unit.

**MANAGING AND TRACKING THE MAINTENANCE FUNCTION.** Performance monitoring and measurement lie within the facility's CMMS. A valuable tool in the manager's arsenal, the CMMS is a powerful means of tracking maintenance activities and costs. Most often, the CMMS is a computerized database with high-level tracking and management capabilities. Table 3.3 highlights some of the primary functions and capabilities of a CMMS. In short, a CMMS can serve as a core information resource for the maintenance function; additionally, it can meet asset management and tracking requirements.

**TABLE 3.3** Primary functions and capabilities of a CMMS.

| Major Functions |
|---|
| **General:** allows records of inventory items and utility resources to be entered and maintained, monitors on-hand balances at single or multiple warehouse sites, and processes work orders; can be integrated with other utility management systems as needed. |
| **Time tracking:** allows employees to record activities by electronic entry; needs to be validated against project numbers and accounts. System should be able to track time by activity, location, or unit; provides summary (roll-up) information down to employee-specific details. |
| **Fleet maintenance:** schedules, prioritizes, records, and tracks all preventive and corrective maintenance activities related to vehicle maintenance; provides cost analysis and support for decisions about vehicle repair and replacement. |
| **Work orders:** enters and tracks to closure all work orders, including internal and external resources (time and materials) with a universal work order Web-deployed interface; issues, routes, prioritizes, schedules, and tracks preventive and corrective maintenance activities; reports scheduled and completed activities to managers. |
| **Purchasing:** makes and tracks purchases (e.g., receipt of goods, returns) of equipment, tools, and "original equipment manufactured" parts; codes purchases to work orders and job numbers and by employee; allows purchases to be made by credit cards, purchase cards, and purchase orders. |

## REFERENCES

Federal Emergency Management Agency (1985) *Fire Service Emergency Management Handbook*; Federal Emergency Management Agency: Washington, D.C.

Hulme, H. S., Jr. (1986) Integration of Emergency Management: Key to Success. *Emerg. Manage. Q.*, 4.

Water Pollution Control Federation (1986) *Plant Managers Handbook;* Manual of Practice No. SM-4; Water Pollution Control Federation: Alexandria, Virginia.

Water Pollution Control Federation (1989) *Emergency Planning for Municipal Wastewater Facilities;* Manual of Practice No. SM-8; Water Pollution Control Federation: Alexandria, Virginia.

# Chapter 4

# Pretreatment Program Requirements for Industrial Wastewater

# INTRODUCTION

Many industries discharge their wastes to municipal wastewater treatment works rather than directly to the nation's waterways. These indirect discharges may contain significant quantities of toxic pollutants or other substances that adversely affect the municipality's wastewater treatment system operations or interfere with its performance. Also, some of these pollutants may pass unchanged through the wastewater treatment works and into the receiving waterway, causing the municipality to exceed its National Pollutant Discharge Elimination System (NPDES) permit requirements. The pollutants may also enter the wastewater treatment plant (WWTP) sludge and complicate ultimate disposal of the removed solids. These undesirable consequences can be prevented with effective pretreatment techniques and management practices that reduce or eliminate the offending pollutants from these indirect wastewater discharges. Consequently, an effective program requiring pretreatment of certain nondomestic wastewater discharges can be an important component of operations of a publicly owned treatment works (POTW).

In the United States, the federal government has established the National Pretreatment Program (U.S. EPA, 1999), which provides, among other things, pretreatment standards for nondomestic indirect dischargers and the regulatory basis to require these dischargers to comply with the pretreatment standards. These regulations are designed to preserve and restore the quality of the nation's waterways, to ensure that this valuable resource is available for all generations. While the United States' National Pretreatment Program is not universally mimicked by other countries around the world, it provides a basis for discussion for those nations that do use a similar regulatory approach and enforcement practices. Consequently, the discussion in this chapter is based, in large part, on the United States' National Pretreatment Program.

This chapter offers a brief history of the National Pretreatment Program, an overview of the program requirements, and a brief discussion of the program responsibilities for POTWs and industrial users (IUs). For a more detailed discussion of industrial wastewater pretreatment programs, the reader is referred to the references provided at the end of this chapter and to state and local requirements specific to their locality. Because regulations and requirements are subject to change, the reader should consult the proper authorities before taking any actions that may be affected by laws or regulations.

# NATIONAL PRETREATMENT PROGRAM

The National Pretreatment Program is a joint regulatory effort by local, state, and federal authorities that requires the control of pollutants that may pass through or interfere with POTW treatment processes or contaminate wastewater sludge. Control of pollutants before the discharge of wastewater to the sewer minimizes the possibility of adverse effects to POTWs from controlled pollutants.

**REGULATORY HISTORY.** In 1972, the U.S. Congress passed the Federal Water Pollution Control Act Amendments of 1972, which became known as the Clean Water Act, to restore and maintain the integrity of the nation's waters. This was a national milestone in the management of wastewater and in the United States' efforts to clean up water pollution in the nation's waterways. Although previous legislation had been enacted to address water pollution, those efforts were developed with other goals in mind. For example, the 1899 Rivers and Harbors Act protected navigational interests, while the 1948 Water Pollution Control Act and the 1956 Federal Water Pollution Control Act merely provided limited funding for state and local governments to address water pollution concerns on their own.

The Clean Water Act established a water quality regulatory approach and empowered the U.S. Environmental Protection Agency (U.S. EPA) (Washington, D.C.) to establish industry-specific, technology-based effluent limitations. The Clean Water Act also required U.S. EPA to develop and implement the NPDES permit program to control the discharge of pollutants from point sources and served as a vehicle to implement the industrial technology-based standards.

U.S. EPA's first attempt to implement pretreatment requirements was promulgated in late 1973 in the *Code of Federal Regulations* (*CFR*) as 40 *CFR* Part 128, which established general prohibitions against treatment plant interference and pass-through and pretreatment standards for the discharge of incompatible pollutants from specific industrial categories. However, as a result of legal complications, in June 1978, U.S. EPA replaced 40 *CFR* Part 128 with the General Pretreatment Regulations for Existing and New Sources of Pollution, 40 *CFR* Part 403 (U.S. EPA, 2007c). These regulations now provide the framework for the National Pretreatment Program, which regulates the introduction of industrial wastes to POTWs. As changes to the NPDES regulations are needed, U.S. EPA issues proposed and final rules related to the NPDES permit program.

**REGULATORY STRUCTURE.** The General Pretreatment Regulations establish responsibilities of federal, state, and local government; industry; and the public to implement Pretreatment Standards to control pollutants that pass through or interfere with POTW treatment processes or that may contaminate wastewater sludge. U.S. EPA, in coordination with individual states, the regulated community, and the public, develops, implements, and conducts oversight of the NPDES permit program based on statutory requirements contained in the Clean Water Act and regulatory requirements contained in the NPDES regulations. Enforcement of the regulations generally becomes the responsibility of local control authorities. The General Pretreatment Regulations define the term "control authority" as a POTW that administers an approved pretreatment program because it is the entity authorized to control discharges to its system. However, the regulations provide states with the authority to implement POTW pretreatment programs in lieu of POTWs or for U.S. EPA to administer the pretreatment program in states not authorized to do so.

Among other things, the General Pretreatment Regulations require all larger POTWs (i.e., those designed to treat flows of more than 19 000 $m^3/d$ [5 mgd]) receiving discharges subject to the pretreatment standards and smaller POTWs with significant industrial discharges to establish local pretreatment programs. These local programs must enforce all national pretreatment standards and requirements in addition to any more stringent local requirements necessary to protect site-specific conditions at the POTW.

**GENERAL PRETREATMENT REQUIREMENTS (40 *CFR* PART 403).** The General Pretreatment Regulations apply to all nondomestic sources that introduce pollutants to a POTW. These sources of "indirect discharge" are more commonly referred to as IUs. However, because not all IUs create problems for POTWs, U.S. EPA developed four criteria to define a significant industrial user (SIU). Many of the General Pretreatment Regulations apply only to SIUs, based on the fact that control of SIUs should provide adequate protection of the POTW.

The four criteria that generally define an SIU are as follows:

- An IU that discharges an average of 95 $m^3/d$ (25 000 gpd) or more of process wastewater to the POTW,
- An IU that contributes a process waste stream making up 5% or more of the average dry-weather hydraulic or organic capacity of the POTW treatment plant,
- An IU designated by the control authority as such because of its reasonable potential to adversely affect the POTW's operation or violate any pretreatment standard or requirement, or
- An IU subject to federal categorical pretreatment standards.

However, the control authority may determine that an IU, which may otherwise be classified as an SIU, is not an SIU because it satisfies the criteria for exemption.

The National Pretreatment Program identifies specific requirements that apply to all IUs, additional requirements that apply to all SIUs, and certain requirements that only apply to categorical industrial users (CIUs). The objectives of the National Pretreatment Program are achieved by applying and enforcing three types of discharge standards:

1. Prohibited discharge standards, which are designed to protect against pass-through and interference generally;
2. Categorical pretreatment standards, which are designed to ensure that IUs implement technology-based controls to limit the discharge of pollutants; and
3. Local limits, which address the specific needs and concerns of a POTW and its receiving waters.

U.S. EPA is constantly developing new guidelines and revising or updating existing guidelines. Section 304(m) of the 1987 Water Quality Act requires U.S. EPA to publish a biennial plan for developing new effluent guidelines and a schedule for the annual review and revision of existing promulgated guidelines.

***Prohibited Discharges (40 CFR Part 403.5).*** Prohibited discharge standards are general, national standards applicable to all IUs of a POTW, regardless of whether or not the POTW has an approved pretreatment program or the IU has been issued a permit. These standards are designed to protect against pass-through and interference, protect the POTW collection system, and promote worker safety and beneficial biosolids use. Refer to *Guidance Manual for Preventing Interference at POTWs* (U.S. EPA, 1987) for additional information on managing prohibited discharges.

Specifically, the discharge standards include the following prohibitions:

- Pollutants that will create a fire or explosion hazard in the POTW;
- Pollutants that will cause corrosive structural damage to the POTW, but never discharges with pH lower than 5.0, unless the plant is specifically designed for such discharges;
- Solid or viscous pollutants in amounts that will cause obstruction of the flow in the POTW;
- Any pollutant, including oxygen-demanding pollutants (biochemical oxygen demand [BOD]) and suspended solids, discharged at a flowrate or pollutant concentration that will cause interference in the POTW;
- Heat in amounts that will inhibit biological activity in the plant, resulting in interference, but never heat in such quantities that the temperature at the POTW is higher than 40 °C (104 °F), unless specifically authorized;
- Petroleum oil, nonbiodegradable cutting oil, or products of mineral oil origin in amounts that will cause interference or pass-through;
- Pollutants that result in the presence of toxic gases, vapors, or fumes within the POTW, in a quantity that may cause acute worker health and safety problems; and
- Any trucked or hauled pollutants, except at discharge points designated by the POTW.

These prohibited discharge standards are intended to provide general protection for POTWs. However, their lack of specific pollutant limitations creates the need for additional controls, namely categorical pretreatment standards and local limits.

***Categorical Pretreatment Standards (40 CFR Parts 405 to 471).*** Categorical Pretreatment Standards are limitations on pollutant discharges to POTWs, promulgated by U.S. EPA, that apply to specific process wastewaters of particular industrial categories.

These are national, technology-based standards that apply regardless of whether the POTW has an approved pretreatment program or the IU has been issued a permit. Industries with such discharges are called CIUs.

Categorical standards apply to regulated wastewaters, which are wastewaters from an industrial process that is regulated for a particular pollutant by a categorical pretreatment standard. Compliance with categorical pretreatment standards is intended to be based on measurements of waste streams containing only the regulated process wastewater. Therefore, compliance is measured at the process discharge point and not at the end of the pipe discharge point for the industrial facility. To monitor the regulated process in situations where the discharge from the regulated process cannot readily be isolated from the nonregulated waste streams, U.S. EPA developed the combined waste stream formula (CWF) and the flow-weighted average (FWA) approach for determining compliance with combined waste streams. The CWF is applicable where a regulated waste stream combines with one or more unregulated or dilute waste streams before treatment. Where nonregulated waste streams combine with regulated process streams after pretreatment, the more stringent approach (whether CWF or FWA) is used to adjust the allowable discharge limits.

As of January 2007, there are 56 regulated industrial categories listed in 40 *CFR* Subchapter N, Effluent Guidelines and Standards. U.S. EPA has published a separate regulation for each industrial category that specifies the effluent guidelines (for direct dischargers) and/or categorical pretreatment standards (for indirect dischargers) for that category (refer to 40 *CFR* Parts 405 to 471). Direct dischargers are regulated by NPDES permits, and indirect dischargers are regulated by the National Pretreatment Program. Table 4.1 lists the 30 currently identified industrial categories, with specific contaminant pretreatment standards, the general regulatory citation, and primary contaminants of concern. Note that each industrial category may have several subcategories with differing pretreatment requirements. Also, individual standards may be expressed in concentration units (milligrams per liter or micrograms per liter), production-based units (i.e., milligrams per kilogram of metal produced), or mass units (i.e., kilograms per day).

Table 4.2 lists toxic pollutants, known as priority pollutants, which are subject to regulation.

***Local Limits [40* CFR *Part 403.5(c)].*** Local limits are specific requirements developed and enforced by the control authority (often a POTW) to protect the wastewater treatment system infrastructure and the POTW's receiving waters. The National Pretreatment Program requires the use of local limits, where necessary, to manage pollutants that would otherwise pass through or interfere with the performance of the POTW.

**Table 4.1** Selected categorical industrial users and potential effects on POTWs.

| Industry category | 40 *CFR* Part | Pollutants of concern |
|---|---|---|
| Aluminum forming | 467 | Chromium, cyanide, zinc, oil and grease, total toxic organics (TTO) |
| Battery manufacturing | 461 | Cadmium, chromium, cobalt, copper, cyanide, lead, manganese, mercury, nickel, silver, zinc |
| Carbon black manufacturing | 458 | Oil and grease |
| Centralized waste treatment | 437 | Antimony; arsenic; cadmium; chromium; cobalt; copper; lead; mercury; nickel; silver; tin; titanium; vanadium; zinc; bis(2-ethylhexyl)phthalate; carbazole; o-cresol; p-cresol; n-decane; fluoranthene; n-octadecane; 2,4,6-trichlorophenol; oil and grease; pH; total suspended solids (TSS) |
| Coil coating | 465 | Chromium, copper, cyanide, fluoride, phosphorus, zinc, oil and grease, TTO |
| Copper forming | 468 | Chromium, copper, lead, nickel, zinc, oil and grease, TTO |
| Electrical and electronic components | 469 | Antimony, arsenic, cadmium, chromium, fluoride, lead, zinc, TTO |
| Electroplating | 413 | Cadmium, chromium, copper, cyanide, lead, nickel, silver, zinc, total metals, TSS, pH, TTO |
| Fertilizer manufacturing | 418 | Ammonia, organic nitrogen, nitrate, phosphorus |
| Glass manufacturing | 426 | Fluoride, mineral oil, oil, pH, TSS |
| Inorganic chemicals manufacturing | 415 | Chromium, copper, cyanide, fluoride, iron, lead, mercury, nickel, silver, zinc, pH, chemical oxygen demand (COD) |
| Iron and steel manufacturing | 420 | Ammonia, chromium, cyanide, lead, nickel, zinc, benzo(a)pyrene, dioxins, naphthalene, phenols, tetrachloroethylene |
| Leather tanning and finishing | 425 | Chromium, sulfides, pH |
| Metal finishing | 433 | Cadmium, chromium, copper, cyanide, lead, nickel, silver, zinc, TTO |
| Metal molding and casting | 464 | Copper, lead, zinc, oil and grease, phenols, TTO |

**TABLE 4.1** Selected categorical industrial users and potential effects on POTWs (*continued*).

| Industry category | 40 *CFR* Part | Pollutants of concern |
|---|---|---|
| Nonferrous metals forming and metal powders | 471 | Antimony, cadmium, chromium, copper, cyanide, fluoride, molybdenum, lead, nickel, silver, zinc, ammonia |
| Nonferrous metals manufacturing | 421 | Antimony, arsenic, beryllium, cadmium, chromium, cobalt, copper, cyanide, fluoride, gold, indium, lead, mercury, nickel, selenium, silver, tin, titanium, zinc, TSS, pH, benzo(a)pyrene, phenol |
| Organic chemicals, plastics, and synthetic fibers | 414 | Cyanide, lead, zinc, numerous organics |
| Paving and roofing materials (tars and asphalt) | 443 | Oil and grease |
| Pesticide chemicals | 455 | Priority pollutants |
| Petroleum refining | 419 | Ammonia, oil and grease |
| Pharmaceutical manufacturing | 439 | Acetone, ammonia, n-amyl acetate, benzene, cyanide, ethyl acetate, isopropyl acetate, methylene chloride, toluene, xylenes, and many other organics |
| Porcelain enameling | 466 | Chromium, lead, nickel, zinc |
| Pulp, paper, and paperboard | 430 | Zinc, adsorbable organic halides, chloroform, dioxins, pentachlorophenol, trichlorophenol |
| Rubber manufacturing | 428 | Chromium, lead, zinc, COD, oil and grease |
| Soap and detergent manufacturing | 417 | COD, 7-day BOD |
| Steam electric power generating | 423 | Copper, polychlorinated biphenyls, priority pollutants except chromium and zinc in cooling tower blowdown |
| Timber products processing | 429 | Arsenic, chromium, copper, oil and grease |
| Transportation equipment cleaning | 442 | Cadmium, chromium, copper, lead, mercury, nickel, zinc, fluoranthene, phenanthrene, silica gel treated n-hexane extractable material (SGT-HEM; nonpolar) |
| Waste combustors | 444 | Arsenic, cadmium, chromium, copper, lead, mercury, silver, titanium, zinc |

**TABLE 4.2** Priority pollutants (U.S. EPA, 2007a).

| | | |
|---|---|---|
| 1,1,1-Trichloroethane | Anthracene | Endrin |
| 1,1,2,2-Tetrachloroethane | Antimony | Endrin Aldehyde |
| 1,1,2-Trichloroethane | Arsenic | Ethylbenzene |
| 1,1-Dichloroethane | Asbestos | Fluoranthene |
| 1,1-Dichloroethylene | Benzene | Fluorene |
| 1,2,4-Trichlorobenzene | Benzidine | gamma-BHC (Lindane) |
| 1,2-Dichlorobenzene | Benzo(a)Anthracene | Heptachlor |
| 1,2-Dichloroethane | Benzo(a)Pyrene | Heptachlor Epoxide |
| 1,2-Dichloropropane | Benzo(b)Fluoranthene | Hexachlorobenzene |
| 1,2-Diphenylhydrazine | Benzo(ghi)Perylene | Hexachlorobutadiene |
| 1,2-Trans-Dichloroethylene | Benzo(k)Fluoranthene | Hexachlorocyclopentadiene |
| 1,3-Dichlorobenzene | Beryllium | Hexachloroethane |
| 1,3-Dichloropropene | beta-BHC | Indeno(1,2,3-cd)Pyrene |
| 1,4-Dichlorobenzene | beta-Endosulfan | Isophorone |
| 2,3,7,8-TCDD Dioxin | Bis2-ChloroethoxyMethane | Lead |
| 2,4,6-Trichlorophenol | Bis2-ChloroethylEther | Mercury |
| 2,4-Dichlorophenol | Bis2-ChloroisopropylEther | Methyl Bromide |
| 2,4-Dimethylphenol | Bis2-EthylhexylPhthalate | Methyl Chloride |
| 2,4-Dinitrophenol | Bromoform | Methylene Chloride |
| 2,4-Dinitrotoluene | Butylbenzyl Phthalate | Naphthalene |
| 2,6-Dinitrotoluene | Cadmium | Nickel |
| 2-Chloroethylvinyl Ether | Carbon Tetrachloride | Nitrobenzene |
| 2-Chloronaphthalene | Chlordane | N-Nitrosodimethylamine |
| 2-Chlorophenol | Chlorobenzene | N-Nitrosodi-n-Propylamine |
| 2-Methyl-4,6-Dinitrophenol | Chlorodibromomethane | N-Nitrosodiphenylamine |
| 2-Nitrophenol | Chloroethane | Pentachlorophenol |
| 3,3′-Dichlorobenzidine | Chloroform | Phenanthrene |
| 3-Methyl-4-Chlorophenol | Chromium III | Phenol |
| 4,4′-DDD | Chromium VI | Polychlorinated Biphenyls |
| 4,4′-DDE | Chrysene | (PCBs) |
| 4,4′-DDT | Copper | Pyrene |
| 4-Bromophenyl Phenyl Ether | Cyanide | Selenium |
| 4-Chlorophenyl Phenyl Ether | delta-BHC | Silver |
| 4-Nitrophenol | Dibenzo(a,h)Anthracene | Tetrachloroethylene |
| Acenaphthene | Dichlorobromomethane | Thallium |
| Acenaphthylene | Dieldrin | Toluene |
| Acrolein | Diethyl Phthalate | Toxaphene |
| Acrylonitrile | Dimethyl Phthalate | Trichloroethylene |
| Aldrin | Di-n-Butyl Phthalate | Vinyl Chloride |
| alpha-BHC | Di-n-Octyl Phthalate | Zinc |
| alpha-Endosulfan | Endosulfan Sulfate | |

Among the factors a POTW should consider in developing local limits are the POTW's efficiency in treating wastes; its compliance with its NPDES permit limits; the condition of the water body that receives its treated effluent; any water quality standards that are applicable to the water body receiving its effluent; the POTW's retention, use, and disposal of wastewater sludge; and worker health and safety concerns.

Local limits may be used to translate general prohibitions in the General Pretreatment Regulations into site-specific needs and to control industrial discharges of pollutants not regulated by the categorical standards. A local limit supersedes the categorical pretreatment standard if it is more stringent than the categorical standard for a given pollutant. However, a local limit typically applies at the point of discharge to the sewer system, while a categorical pretreatment standard generally applies at the end of the regulated process.

In evaluating the need for local limit development, it is recommended that control authorities

- Conduct an industrial waste survey to identify all IUs that might be subject to the pretreatment program;
- Determine the character and volume of pollutants contributed to the POTW by these industries;
- Determine which pollutants have a reasonable potential for pass-through, interference, or sludge contamination;
- Conduct a quantitative evaluation to determine the maximum allowable POTW treatment plant headworks (influent) loading for at least arsenic, cadmium, chromium, copper, cyanide, lead, mercury, molybdenum, nickel, silver, and zinc;
- Identify additional pollutants of concern;
- Determine contributions from unpermitted sources to determine the maximum allowable treatment plant headworks loading from "controllable" industrial sources; and
- Implement a system to ensure that these loadings will not be exceeded.

The POTWs should routinely reevaluate their local limits to ensure that they continue to provide adequate protection for POTW operation. An ongoing monitoring program is recommended to support review and revision of local limits and verification/identification of pollutants of concern. An effective ongoing monitoring program should include periodic sampling of influent, effluent, sludge, domestic/commercial wastewaters, and industrial wastewater discharges. U.S. EPA recommends annual reviews of local limits. U.S. EPA has developed a number of guidance documents to assist control authorities in the development of local limits. A guidance manual entitled

*Local Limits Development Guidance* and a separate document entitled *Local Limits Development Guidance Appendices* are available on the U.S. EPA Web site (http://www.epa.gov) (U.S. EPA, 2004a; U.S. EPA, 2004b). In addition, many U.S. EPA regions and states have developed local limits guidance to address regional and state issues.

One of the methods commonly used to derive local limits is the maximum allowable headworks loading (MAHL) method. A summary of the method is presented below.

***Maximum Allowable Headworks Loading Method Summary.*** First, pollutant by pollutant, treatment plant data are used to calculate actual removal efficiencies through the POTW. This allows the back-calculation of the MAHL for each pollutant based on applying the most stringent criterion (i.e., water quality, sludge quality, NPDES permit, or pollutant inhibition levels) for the POTW allowable effluent concentration for each pollutant. After subtracting contributions from domestic sources (which are uncontrollable), the allowable industrial loading for each pollutant is then calculated. This is then either evenly distributed among the industrial users or allocated on an as-needed basis to those industrial users discharging the particular pollutant. Industrial users exceeding their allocation must pretreat their wastewater before discharge to comply with their local limit allowance. However, the regulations also provide a means for POTWs to potentially reduce the pretreatment requirements of their IUs based on the POTW's effectiveness in removing certain pollutants from its waste stream. The "removal credits" potentially available to IUs from the POTW, at its discretion, are summarized below.

***Removal Credits (40 CFR Part 403.7).*** If a POTW can treat certain pollutants to high levels of removal, it can obtain permission from U.S. EPA (or a state that has been granted authority by U.S. EPA) to issue credit to IUs, providing the IU with adjusted pretreatment requirements for those pollutants (40 *CFR* Part 403.7) that factor in the removal achieved by the POTW. A POTW can only issue removal credits if it has an approved local pretreatment program, consistently removes the pollutant for which the removal credit will be issued, and the granting of removal credits will not result in the violation of any local, state, or federal sludge requirements [as defined in 40 *CFR* Part 403.7(a)(1)(ii)]. Removal credits for several pollutants are available through 40 *CFR* Part 403 (Appendix G, Section I) when disposal practices used by a POTW meet the requirements of 40 *CFR* Part 503 for the intended practice (land application, surface disposal, or incineration). Additionally, a removal credit can be obtained for pollutants listed in 40 *CFR* Part 403 (Appendix G, Section II) if their respective concentration in the POTW's wastewater sludge is less than the threshold concentration listed in Appendix G. If a POTW sends all of its wastewater residuals to a municipal solid waste landfill that meets the requirements of 40 *CFR* Part 258, the POTW can issue removal credits for any pollutant in wastewater sludge.

## PRETREATMENT PROGRAM RESPONSIBILITIES FOR PUBLICLY OWNED TREATMENT WORKS

Any POTW or municipal authority that has a total design wastewater flow greater than 19 000 $m^3/d$ (5 mgd) and receives pollutants from industrial users that pass through or interfere with the operation of the POTW or are otherwise subject to pretreatment standards will be required to establish a POTW Pretreatment Program, unless the NPDES state exercises its option to assume local responsibilities as provided for in the General Pretreatment Regulations. Those POTWs with a design flow of 19 000 $m^3/d$ (5 mgd) or less may also be required to develop a POTW Pretreatment Program, if the circumstances warrant. Such circumstances include the nature or volume of the industrial influent, history of treatment process upsets, violations of POTW effluent limitations, and contamination of municipal sludge. Furthermore, the POTW's NPDES permit will be reissued or modified by the NPDES state or U.S. EPA to incorporate the approved pretreatment program as enforceable conditions of the NPDES permit.

**ESTABLISHING LEGAL AUTHORITY.** The POTW Pretreatment Program must be administered to ensure that industrial users comply with applicable pretreatment standards and requirements. Generally, the POTWs act as the pretreatment control authorities, with respect to the IUs that discharge to their systems, while either the state or U.S. EPA generally acts as the pretreatment approval authority. Approval authorities review and approve the pretreatment program and oversee the implementation of the program by the control authorities. In the absence of an approved POTW pretreatment program, the state or U.S. EPA approval authority serves as the control authority.

***Control Authority (Local).*** Before a control authority can implement a pretreatment program, it must develop policies and procedures to administer the program and be given the legal authority to implement and enforce the program requirements. A control authority's legal authority is based on state law; therefore, state laws that confer the minimum federally mandated legal authority requirements on a control authority must be in effect. The policies and procedures for administering the program and the reference to the legal authority to implement and enforce the program are generally embodied in local regulations established by the control authority. Where the control authority is a municipality, legal authority is generally detailed in a Sewer Use Ordinance, which is typically part of city or county code. Regional control authorities frequently adopt similar provisions in the form of "rules and regulations." Likewise, state agencies implementing a statewide program set out pretreatment requirements as state regulations, rather than as Sewer Use Ordinances.

Specific control authority responsibilities include the following:

- Develop, implement, and maintain the approved pretreatment program;
- Evaluate the compliance of regulated IUs;
- Initiate enforcement action against industries, as appropriate;
- Submit reports to approval authorities;
- Develop local limits (or demonstrate why they are not needed); and
- Develop and implement an Enforcement Response Plan (ERP).

U.S. EPA's 2007 guidance document, *EPA Model Pretreatment Ordinance* (U.S. EPA, 2007b), provides a model for POTWs that are required to develop pretreatment programs.

***Approval Authority (State).*** Local pretreatment programs must be approved by the approval authority—either U.S. EPA or the authorized state—that is also responsible for overseeing implementation and enforcement of these programs. The approval authority is responsible for ensuring that local program implementation is consistent with all applicable federal requirements and is effective in achieving the National Pretreatment Program's goals. Specific responsibilities include the following:

- Notify POTWs of their responsibilities,
- Review and approve requests for POTW pretreatment program approval or modification,
- Review requests for site-specific modifications to categorical pretreatment standards,
- Oversee POTW program implementation,
- Provide technical guidance to POTWs, and
- Initiate enforcement actions against noncompliant POTWs or industries.

To carry out these responsibilities, the approval authority monitors local program compliance and effectiveness and dictates or takes corrective actions, where needed, to meet these goals. The approval authority generally uses the following three oversight mechanisms to make these determinations: (1) the program audit, (2) the Pretreatment Compliance Inspection, and (3) the control authority's annual pretreatment program performance report.

As noted in Table 4.3, a total of 46 states/territories are authorized to implement state NPDES permit programs as of January 2007, but only 35 are authorized to be the

**TABLE 4.3** State program status (U.S. EPA, 2007e).

| State | Approved state NPDES permit program | Approved to regulate federal facilities | Approved state pretreatment program | Approved general permits program |
|---|---|---|---|---|
| Alabama | 10/19/1979 | 10/19/1979 | 10/19/1979 | 06/26/1991 |
| Alaska | | | | |
| American Samoa | | | | |
| Arizona | 12/05/2002 | 12/05/2002 | 12/05/2002 | 12/05/2002 |
| Arkansas | 11/01/1986 | 11/01/1986 | 11/01/1986 | 11/01/1986 |
| California | 05/14/1973 | 05/05/1978 | 09/22/1989 | 09/22/1989 |
| Colorado | 03/27/1975 | | | 03/04/1982 |
| Connecticut | 09/26/1973 | 01/09/1989 | 06/03/1981 | 03/10/1992 |
| Delaware | 04/01/1974 | | | 10/23/1992 |
| District of Columbia | | | | |
| Florida | 05/01/1995 | 05/01/2000 | 05/01/1995 | 05/01/1995 |
| Georgia | 06/28/1974 | 12/08/1980 | 03/12/1981 | 01/28/1991 |
| Guam | | | | |
| Hawaii | 11/28/1974 | 06/01/1979 | 08/12/1983 | 09/30/1991 |
| Idaho | | | | |
| Illinois | 10/23/1977 | 09/20/1979 | | 01/04/1984 |
| Indiana | 01/01/1975 | 12/09/1978 | | 04/02/1991 |
| Iowa | 08/10/1978 | 08/10/1978 | 06/03/1981 | 08/12/1992 |
| Kentucky | 09/30/1983 | 09/30/1983 | 09/30/1983 | 09/30/1983 |
| Louisiana | 08/27/1996 | 08/27/1996 | 08/27/1996 | 08/27/1996 |
| Maine | 01/12/2001 | 01/12/2001 | 01/12/2001 | 01/12/2001 |
| Maryland | 09/05/1974 | 11/10/1987 | 09/30/1985 | 09/30/1991 |
| Massachusetts | | | | |
| Michigan | 10/17/1973 | 12/09/1978 | 04/16/1985 | 11/29/1993 |
| Midway Island | | | | |
| Minnesota | 06/30/1974 | 12/09/1978 | 07/16/1979 | 12/15/1987 |
| Mississippi | 05/01/1974 | 01/28/1983 | 05/13/1982 | 09/27/1991 |
| Missouri | 10/30/1974 | 06/26/1979 | 06/03/1981 | 12/12/1985 |
| Montana | 06/10/1974 | 06/23/1981 | | 04/29/1983 |
| Nebraska | 06/12/1974 | 11/02/1979 | 09/07/1984 | 07/20/1989 |
| Nevada | 09/19/1975 | 08/31/1978 | | 07/27/1992 |
| New Hampshire | | | | |
| New Jersey | 04/13/1982 | 04/13/1982 | 04/13/1982 | 04/13/1982 |
| New Mexico | | | | |
| New York | 10/28/1975 | 06/13/1980 | | 10/15/1992 |
| North Carolina | 10/19/1975 | 09/28/1984 | 06/14/1982 | 09/06/1991 |
| North Dakota | 06/13/1975 | 01/22/1990 | 09/16/2005 | 01/22/1990 |
| Northern Mariana Islands | | | | |
| Ohio | 03/11/1974 | 01/28/1983 | 07/27/1983 | 08/17/1992 |
| Oklahoma | 11/19/1996 | 11/19/1996 | 11/19/1996 | 09/11/1997 |
| Oregon | 09/26/1973 | 03/02/1979 | 03/12/1981 | 02/23/1982 |

**TABLE 4.3** State program status (U.S. EPA, 2007e) (*continued*).

| State | Approved state NPDES permit program | Approved to regulate federal facilities | Approved state pretreatment program | Approved general permits program |
|---|---|---|---|---|
| Pennsylvania | 06/30/1978 | 06/30/1978 | | 08/02/1991 |
| Puerto Rico | | | | |
| Rhode Island | 09/17/1984 | 09/17/1984 | 09/17/1984 | 09/17/1984 |
| South Carolina | 06/10/1975 | 09/26/1980 | 04/09/1982 | 09/03/1992 |
| South Dakota | 12/30/1993 | 12/30/1993 | 12/30/1993 | 12/30/1993 |
| Tennessee | 12/28/1977 | 09/30/1986 | 08/10/1983 | 04/18/1991 |
| Texas | 09/14/1998 | 09/14/1998 | 09/14/1998 | 09/14/1998 |
| Trust Territories | | | | |
| Utah | 07/07/1987 | 07/07/1987 | 07/07/1987 | 07/07/1987 |
| Vermont | 03/11/1974 | | 03/16/1982 | 08/26/1993 |
| Virgin Islands | 06/30/1976 | | | |
| Virginia | 03/31/1975 | 02/09/1982 | 04/14/1989 | 04/20/1991 |
| Washington | 11/14/1973 | | 09/30/1986 | 09/26/1989 |
| West Virginia | 05/10/1982 | 05/10/1982 | 05/10/1982 | 05/10/1982 |
| Wisconsin | 02/04/1974 | 11/26/1979 | 12/24/1980 | 12/19/1986 |
| Wyoming | 01/30/1975 | 05/18/1981 | | 09/24/1991 |

pretreatment program approval authority. In all other states and territories, U.S. EPA is the approval authority.

***Federal Authority (U.S. Environmental Protection Agency).*** U.S. EPA, both at the national and the regional level, has responsibilities for the National Pretreatment Program. At the regional level, U.S. EPA must

- Fulfill approval authority responsibilities for states without a state pretreatment program;
- Oversee state program implementation; and
- Initiate enforcement actions, as appropriate.

At the national level, U.S. EPA

- Oversees program implementation, at all levels;
- Develops and modifies regulations for the program;
- Develops policies to clarify and further define the program;
- Develops technical guidance for program implementation; and
- Initiates enforcement actions, as appropriate.

**IDENTIFYING AND MONITORING INDUSTRIAL USERS.** The cornerstone for achieving the objectives of an industrial waste pretreatment program is the identification and classification of the IUs in a community. From the output of this process, the municipality assigns monitoring and enforcement priorities based on each industry's potential for affecting the POTW. The quality of the pretreatment program depends on the personnel involved and the information on which enforcement decisions are based. Therefore, the municipality must frequently update its industrial information base to achieve local and national pollution abatement goals.

***Industrial Survey.*** An industrial survey begins with compiling an inventory of industries and their waste streams. Ideally, the municipal wastewater treatment authority should survey all industries and waste streams in its service area. The survey starts with a preliminary inventory that may be obtained from a review of sources, such as the following:

- Industrial directories;
- Property tax records;
- Municipal agency files (i.e., sewer and water billings);
- Local telephone directories;
- Toxic Release Inventory reports;
- State and federal wastewater discharge monitoring reports; and
- Permit application forms.

To complete the survey, the preliminary inventory is verified and updated through a questionnaire, followed, where appropriate, by a physical inspection of the facility. The questionnaire usually seeks specific information on the following:

- Raw materials used;
- Basic steps involved in plant operations;
- Industrial processes in the plant;
- North American Industry Classification System codes;
- Water usage information;
- Number of connections to the sanitary sewer system;
- Types and quantities of wastes discharged;
- Type of pretreatment technologies used;
- Sludges or process residues disposed of (type and volume);
- Types and quantities of hazardous materials used or stored; and
- Description of chemical storage areas, including information regarding connection to sanitary or storm sewers.

U.S. EPA's 1983 document *Guidance Manual for POTW Pretreatment Program Development* is still a good resource and presents an example of a survey questionnaire that may be suitable for use by control authorities.

The database developed from the survey can be used to identify those industries subject to categorical pretreatment standards and to assign industrial monitoring priorities for efficient use of available sampling equipment and personnel. This information is also necessary for compliance with the POTW's reporting requirements for NPDES permits and approved pretreatment programs.

Whether the POTW has only a few industries or thousands, it needs the database and a system to collect and process the information for future reference or routine access. Selection of the system type, ranging from manual filing to complex data processing equipment, may depend on the amount of data to be stored and available funding.

***Inspection of Industrial Facilities.*** Sampling and analyses of industrial discharges are important elements of an effective enforcement program. In addition to providing a database, sampling serves as the basis for determining compliance with local, state, and national discharge requirements. Before doing a sampling study, inspections of industries are needed to select the type and duration of sampling. Inspections can be conducted as scheduled, unscheduled, or random, depending on the situation.

Scheduled inspections are generally used to obtain information relating to production schedules and cleanup operations and to sketch the facilities' process layout. The sketch should show the location of buildings, process areas, water meters, wastewater treatment facilities, and sampling points. It can also provide other information about the sampling program. Pre-inspection arrangements may be advisable to obtain water-usage data and production records, if the industry is subject to production- or mass-based discharge standards. Once the background data have been obtained, unscheduled inspections or monitoring may follow.

Unscheduled monitoring of an industry may range from a complete industrial plant and recordkeeping inspection to inspection and sampling of only the industrial wastewater treatment facilities. Observations during an unscheduled inspection will likely show the normal operating conditions at the industrial facility and may lead to the municipality's reassessment of its enforcement priorities. Thus, spot checks are useful for determining the reliability of data collected through scheduled monitoring or self-reporting, detecting unreported discharges to the sewer system, or assessing routine operation of wastewater treatment equipment.

U.S. EPA has published a guidance manual entitled *Control Authority Pretreatment Audit Checklist and Instructions* (U.S. EPA, 1992), which presents U.S. EPA-recommended procedures to be used by POTW personnel when conducting an inspection or sampling visit at an IU.

**SAMPLING.** The type of sampling selected for a facility should be based on the type of discharge and the nature of the operations. This section presents a brief discussion of sampling of industrial wastewater discharges. Chapter 17 of this manual provides a more detailed discussion of characterization and sampling of wastewater at POTWs.

Some basic sampling decisions to characterize wastewater include the following:

- Grab or composite sampling,
- Time-based or flow-proportioned sampling, and
- Automated or manual sampling.

A grab sample is a single sample taken at any time that is representative of the flow only at the time it was taken. A composite sample, which is composed of two or more samples taken from the same source and combined into one container to form a single sample, is more representative of the average quality of a discharge over a period of time. The composite sample may be made up of equal portions taken at fixed intervals (time-based composite), or the size and numbers of the portions or subsamples in the composite sample may be based on the discharge rate at the time each subsample is taken (flow-proportioned composite). This type of composite sample best represents the total mass of various pollutants in the waste discharge, particularly when the pollutant concentrations or flowrates change with respect to time.

In general, flow-proportional sampling is recommended wherever possible, except for those parameters that cannot be held and must be collected as grab samples. However, physical constraints at facilities may preclude the collection of flow-proportional sampling, in which case, time-proportional sampling, if executed properly, could be an acceptable substitute. Please refer to Chapter 17 of this manual for additional discussion of sampling methodology and sample-handling protocol.

U.S. EPA has published a guidance manual entitled *Industrial User Inspection and Sampling Manual for POTWs* (U.S. EPA, 1994), which presents U.S. EPA-recommended procedures to be used by POTW personnel when conducting an inspection or sampling visit at an IU.

Also, *CFR* Title 40, Part 136 (U.S. EPA, 2007d), presents guidelines establishing test procedures for the analysis of pollutants, including those pertinent to wastewater sampling.

**ENFORCEMENT IMPLEMENTATION.** A successful industrial pretreatment program stems from a consistent, effective enforcement response policy. The policy must be based on reasonable discharge standards contained in a control mechanism, such as an ordinance, permit, or contract. These standards should protect the municipality's sewer system, employees, wastewater treatment facilities, and the receiving waterway. Where the municipality has an approved pretreatment program, it must enforce the federal categorical pretreatment standards, the general discharge prohibitions, and its own local limits. The policies and terms used by the municipality to deal with noncompliance are contained in the municipality's ERP. To assist control authorities in their enforcement obligations, U.S. EPA has published the guidance document entitled *Pretreatment Compliance Monitoring and Enforcement Guidance* (U.S. EPA, 1986).

***Enforcement Response Plan.*** The General Pretreatment Regulations require control authorities to adopt and implement an ERP. The ERP regulations establish a framework for POTWs to formalize procedures for investigating and responding to instances of IU noncompliance. With an approved ERP in place, POTWs can enforce against IUs on a more objective basis.

To ensure that enforcement response is appropriate and that the control authority actions are not arbitrary or capricious, U.S. EPA strongly recommends that an Enforcement Response Guide (ERG) be included as part of the approved ERP. The ERG identifies responsible control authority officials, general time frame for actions, expected industrial user responses, and potential escalated actions based on the following:

- Nature of the violation (pretreatment standards, reporting [late or deficient], and compliance schedules);
- Magnitude of the violation;
- Duration of the violation;
- Frequency of the violation (isolated or recurring);
- Potential effect of the violation (i.e., interference, pass-through, or POTW worker safety);
- Economic benefit gained by the violator; and
- Attitude of the violator.

U.S. EPA has published a guidance manual entitled *Guidance for Developing Control Authority Enforcement Response Plans* (U.S. EPA, 1989a), which presents U.S. EPA-recommended procedures to be used by POTW personnel when conducting an inspection or sampling visit at an IU.

***Compliance Review of Industrial User Data/Reports.*** Review of monitoring data and supporting documentation provides information regarding the compliance status of an IU's discharge, the quality and quantity of its contribution to the municipal WWTP, and the costs of treating the waste contribution. Compliance with pretreatment program regulations requires IUs to submit reports of self-monitoring activities, including the Baseline Monitoring Report (BMR), semiannual reports, and compliance schedule progress reports. The municipality must acknowledge and act on the industry's self-reported departures from applicable standards. Failure of individual industries to submit required reports should result in prompt enforcement action. In addition, any industry contending that it has no discharge from a regulated process should be required to affirm this claim semiannually and to notify the municipality immediately of any change in its discharge status. Follow-up inspections are needed for those companies subject to regulations and those that claim to be exempted.

***Enforcement Mechanisms.*** The control authority will have at its disposal a number of mechanisms, embodied in the Sewer Use Ordinance, to enforce the requirements of the pretreatment program. These mechanisms, which generally represent an escalating sequence of actions that may be applied, if necessary, to ensure compliance, may range from a telephone call to a criminal lawsuit. Common enforcement mechanisms include the following:

- Notice of violation—may be issued for noncompliance and requires the user to identify the cause of the violation and provide a plan and schedule for remediating the violation.
- Consent agreement—a voluntary agreement between the user and the control authority that identifies a specific action required of the user and a specified date to achieve the action.
- Show cause order.
- Compliance order.
- Cease and desist order.
- Emergency suspension.
- Termination of discharge.

In addition, most control authorities can levy administrative fines for violations.

***Enforcement Tracking.*** It is essential for control authorities to effectively manage information to demonstrate proper implementation of the pretreatment program. The Clean Water Act allows citizens to file suit against a control authority that has failed to implement its approved pretreatment program as required by its NPDES permit. The control authority may be fined and required to enforce against noncompliance with pretreatment standards and requirements in a court order. Furthermore, unresolved noncompliance issues may result in the approval authority enforcing directly against the IU and/or the control authority, and U.S. EPA may also take enforcement action where it deems action by the state or the control authority is inappropriate.

An approval authority will routinely review the overall performance of a control authority in monitoring IUs, identifying noncompliance, and enforcing regulations. Performance will be evaluated based on POTW self-monitoring data, written ERPs, audits, inspections, and pretreatment program reports. Therefore, it is incumbent on the control authority, regardless of the response action taken, to document and track all contact, notices, and meetings with IUs and the responses received.

***Publication of a List of Noncompliant Industries.*** An approved federal pretreatment program requires the municipality to publish, at least annually, a list of industries that were in significant noncompliance with the applicable pretreatment standards during the previous 12 months. The list is to be published in the daily newspaper of largest circulation within the municipal authority's jurisdiction. The definition of *significant noncompliance* can be found in U.S. EPA's General Pretreatment Regulations [40 *CFR* Part 403.8(f)(2)(viii)].

**RECORDKEEPING AND REPORTING.** ***Data Management System.*** The type and complexity of the system required to manage the information generated in operating an industrial pretreatment program is governed by the size of the industrial community to be monitored and the amount of information and data to be stored for future retrieval. For a municipality with a small industrial community, manual files may suffice for the necessary records, including those for inspections, sampling, and storage of industrial self-monitoring reports. Also, the manual approach may be appropriate for fulfilling NPDES and federal pretreatment reporting requirements. A municipal wastewater treatment authority monitoring a large industrial community would certainly need a computerized data management system. In either case, the information to be stored should include the following:

- Industry survey data;
- Self-monitoring report data;
- Company enforcement history;
- Reports of POTW inspections;

- Analytical data from POTW sampling studies; and
- POTW influent, effluent, and residuals.

A computerized data management system offers a calendar as a valuable tool for monitoring the required periodic performance of all necessary functions, such as self-reporting by industrial dischargers, inspection and sampling by the municipality, automated detection of delinquent reports, review and comparison of municipal and industry analytical data, and generation of municipal reports for NPDES permit and federal pretreatment program reporting requirements. A computerized data system can also flag violations of permit limits, an important capability for municipalities with many industries.

***Publicly Owned Treatment Works Reporting.*** A POTW with an approved pretreatment program is required to submit an annual report to U.S. EPA briefly describing the pretreatment program activities for the preceding year. Information to be submitted includes a summary of program changes; a listing of IUs and their compliance status; and the compliance activities initiated, including enforcement actions.

## INDUSTRIAL USER PRETREATMENT PROGRAM RESPONSIBILITIES

The responsibilities of IUs under the National Pretreatment Program are briefly highlighted in this section. Additional details regarding IU responsibilities can be found in *Industrial Wastewater Management, Treatment, and Disposal* (WEF, in press) and in 40 *CFR* Parts 403.12, 403.16, and 403.17. The IUs need to constantly monitor the regulations, as they are subject to periodic revisions. Control authorities need to know which responsibilities IUs have so they can properly monitor their IU performance.

The IUs are required to comply with all applicable pretreatment standards and requirements. Demonstration of compliance requires certain IUs to submit reports, self-monitor, and maintain records. Because control authorities are responsible for communicating applicable standards and requirements to IUs and for receiving and analyzing reports, it is essential for control authority personnel to understand IU reporting and notification requirements contained in the General Pretreatment Regulations. These requirements are summarized below.

### CATEGORICAL INDUSTRIAL USER REPORTING REQUIREMENTS.

An IU that is subject to a categorical pretreatment standard is identified as a CIU. Principal reporting requirements for CIUs are highlighted below.

***Baseline Monitoring Report [40* CFR *Part 403.12(b)].*** The CIUs are required to submit a BMR within 180 days after the effective date of the categorical standard or at least 90 days before commencement of discharge for new sources. The BMR is a one-time report, based on sampling of the CIU's discharge or estimating the discharge characteristics for new sources.

***Compliance Schedule Progress Report [40* CFR *Part 403.12(c)(3)].*** A CIU that is not in compliance with the applicable categorical standards by the time the standards are effective often will have to modify process operations and/or install end-of-pipe treatment to comply. Federal regulations require that the control authority develop and impose a compliance schedule for the CIU to install technology to meet applicable standards. In no case can the final or completion date in the schedule be later than the final compliance date specified in the categorical standards. The CIU must submit progress reports to the control authority no later than 14 days following each date in the compliance schedule.

***90-Day Compliance Reports [40* CFR *Part 403.12(d)].*** Section 403.12(d) of the General Pretreatment Regulations requires a CIU to submit a final compliance report to the control authority. An existing source must file a final compliance report within 90 days following the final compliance date specified in a categorical regulation or within 90 days of the compliance date specified by the control authority, whichever is earlier. A new source must file a compliance report within 90 days from commencement of discharge to the POTW.

***Upset Reports [40* CFR *Part 403.16].*** The CIUs are allowed an affirmative defense for noncompliance with categorical standards if they can demonstrate that the noncompliance was the result of an upset. Conditions necessary to demonstrate that an upset has occurred are detailed in 40 *CFR* Part 403.16 and require the CIU to submit at least an oral report to the control authority within 24 hours of becoming aware of the upset.

***Signatory and Certification Requirements [40* CFR *Part 403.12(l)].*** Pursuant to 40 *CFR* Part 403.12(l), BMRs, 90-day compliance reports, and periodic compliance reports from CIUs must be signed by an authorized representative of the facility and contain a certification statement attesting to the integrity of the information reported.

**CATEGORICAL AND SIGNIFICANT INDUSTRIAL USER REPORTING REQUIREMENTS—PERIODIC COMPLIANCE REPORTS [40 *CFR* PARTS 403.12(E) AND (H)].** After the final compliance date, CIUs (except nonsignificant CIUs) are required to report, during the months of June and December, the self-monitoring results of their wastewater discharge(s), although control authorities may require more frequent monitoring. Also, the control authority must also require semiannual reporting from SIUs not subject to categorical standards. All the results for self-monitoring performed must be reported to the control authority, even if the IU is monitoring more frequently than required.

A control authority may choose to monitor IUs in lieu of the IU performing the self-monitoring.

**REPORTING REQUIREMENTS FOR ALL INDUSTRIAL USERS.** ***Notification of Potential Problems [40 CFR Part 403.12(f)].*** All IUs are required to notify the control authority immediately of any discharges that may cause potential problems. These discharges include spills, slug loads, or any other discharge that may cause a potential problem to the POTW.

***Bypass [40 CFR Part 403.17].*** The General Pretreatment Regulations define *bypass* as the intentional diversion of waste streams from any portion of an IU's pretreatment facility. If a bypass results in noncompliance—even if it was the result of essential maintenance—the IU must provide a report to the control authority detailing a description of the bypass; cause and duration of the bypass; and steps being taken and/or planned to reduce, eliminate, and prevent reoccurrence of the bypass. Oral notice must be provided to the control authority within 24 hours of the detection of an unanticipated bypass, with a written follow-up due within 5 days. For an anticipated bypass, the IU must submit notice to the control authority, preferably 10 days before the intent to bypass.

***Noncompliance Notification [40 CFR Part 403.12(g)(2)].*** If monitoring performed by an IU indicates noncompliance, the IU is required to notify the control authority within 24 hours of becoming aware of the violation. In addition, the IU must repeat sampling and analysis and report results of the resampling within 30 days. The repeat sampling is not required if the control authority samples the IU at least once per month or if the control authority samples the IU between the time of the original sample and the time the results of the sampling are received.

***Notification of Changed Discharge [40 CFR Part 403.12(j)].*** All IUs are required to promptly notify the control authority in advance of any substantial changes in the volume or character of pollutants in their discharge.

***Notification of Discharge of Hazardous Wastes [40 CFR Part 403.12(p)].*** All IUs discharging more than 15 kg per month of a waste that, if otherwise disposed of, would be a hazardous waste pursuant to the Resource Conservation and Recovery Act (RCRA) requirements under 40 *CFR* Part 261, are required to provide a one-time written notification of such discharge to the control authority, state, and U.S. EPA. All IUs discharging any amount of waste, that, if disposed of otherwise, would be an acutely hazardous waste pursuant to the RCRA, must also provide this notification.

**SELF-MONITORING REQUIREMENTS.** All SIUs, including CIUs, must conduct self-monitoring as part of several different reporting requirements. For CIUs, this includes the BMR, 90-day compliance report, and periodic compliance reports. Noncategorical SIUs are required to self-monitor as part of the periodic reporting requirements. Sample collection and analysis for all required pretreatment program reports must be conducted using 40 *CFR* Part 136 procedures and amendments thereto.

Self-monitoring for periodic compliance reports must be conducted in accordance with the IU's discharge permit requirements. The control authority must ensure that these permits specify sampling location(s), required sampling frequencies, sample types to be collected, sampling and analytical procedures, and associated reporting requirements. U.S. EPA's *Industrial User Permitting Guidance Manual* (U.S. EPA 1989b) provides guidance on managing the permitting process.

At a minimum, CIUs must monitor for all categorically regulated pollutants at least once every 6 months, although permits issued by the local control authority may require more frequent monitoring.

**INDUSTRIAL USER RECORDKEEPING REQUIREMENTS.** All IUs are required to maintain records of their monitoring activities. Information, at a minimum, should include the following:

- Sampling methods, dates, and times;
- Identity of the person(s) collecting the samples and of the sampling location(s);
- Dates the analyses were performed and the methods used; and
- Identity of the person(s) performing the analyses and the results of the analyses.

These records should be retained for at least 3 years or longer in cases where there is pending litigation involving the control authority or IU or when requested by the approval authority. These records must be available to the control authority and approval authority for review and copying. Historically, most control authorities do not dispose of any records; rather, older records are archived at an off-site location.

## OVERVIEW OF INDUSTRIAL WASTEWATER PRETREATMENT

This section briefly reviews basic considerations in the pretreatment of industrial wastewater before its discharge to the municipal wastewater treatment system. Additional details regarding industrial wastewater pretreatment can be found in *Industrial Wastewater Management, Treatment, and Disposal* (WEF, in press) and the other references cited in this section.

**INDUSTRIAL WASTEWATER CHARACTERISTICS.** Industrial wastewaters may contain organic and/or inorganic constituents that may be detrimental to the operation of a POTW. The composition of industrial wastewater may change as raw materials and processes are changed. Fluctuations may occur in the flow and composition of wastewater. Wastewater characteristics vary widely from one type of industry to another and even among plants within the same industrial category. Because the list of regulated pollutants is broad, almost any industry may be a source of one or more of these pollutants. Refer to Table 4.1 for a partial list of industries with categorical discharges and potential contaminants of concern for these industries.

**INDUSTRIAL WASTEWATER PRETREATMENT PROCESSES.** The pretreatment requirements for industrial wastewater depend on the wastewater characteristics, particular pollutants requiring removal, flowrates, and degree of pretreatment needed. The various wastewater treatment processes are generally classified as physical, chemical, and biological, and each is described briefly below. Detailed discussions of industrial wastewater characteristics, specific pretreatment applications, and design of industrial wastewater treatment systems can be found in the literature (Eckenfelder, 1999; Nemerow, 1978; WEF, in press). Such information is beyond the scope of this chapter.

Reductions in the volume or strength of raw wastewater may often be achieved through industrial recycling or reuse systems, improved housekeeping practices, and production process modifications to reduce constituents of concern. Application of these industrial plant controls should be considered and can make a significant difference in the pretreatment requirements for an industry.

Physical processes are designed to remove targeted constituents from waste streams by removing the constituents without producing any change in the constituent. Commonly used physical processes include filtration, adsorption, stripping, and solids separation using screening, sedimentation, or flotation. The result of physical pretreatment is a waste stream with lower concentrations of the targeted constituents, while the physically removed constituents must be managed separately for disposal.

Chemical processes produce a change in the targeted constituents in the waste stream by chemically altering the constituents or chemically binding the constituents to other chemical compounds. Neutralization, oxidation, reduction, and coagulation are examples of chemical processes. The result of physical pretreatment is a waste stream with lower concentrations of the targeted constituents, while the chemically removed constituents must be managed separately for disposal. Examples of commonly used coagulants include aluminum and iron salts. Oxidants typically used include chlorine, peroxide, ozone, or oxygen.

Biological processes are primarily designed to reduce dissolved organic materials. Aerobic biological processes include trickling filters and other fixed-film systems and activated sludge systems, including sequencing batch reactors. Anaerobic processes, such as anaerobic filters and anaerobic reactors, have been used for high-strength industrial wastewater. Many industrial wastewaters may require a pretreatment step to reduce toxic pollutants that would otherwise interfere with the operation of biological units.

Treatment technologies for heavy metals removal generally include a precipitation process. Sometimes, effective treatment may require a chemical oxidation step before precipitation. The pH of the wastewater and the oxidative state of the metal influence treatment effectiveness. Sulfide addition precipitates arsenic, and lime or caustic addition can precipitate many metals as hydroxides. Caustic converts hexavalent chromium to the trivalent state and then precipitates the trivalent chromium. Other incompatible pollutants, such as cyanide, are oxidized under alkaline conditions using chlorine.

One of the following four types of treatment processes can remove organic compounds:

- Stripping by air or steam;
- Advanced oxidation processes using various combinations of ozone, hydrogen peroxide, and UV light;
- Adsorption on activated carbon or a synthetic resin; and
- Biological degradation.

The performance of any of those processes for a specific application varies greatly and depends on the nature of the organic compound and the presence of other pollutants. Therefore, a small-scale model study on the specific waste to be treated is recommended before developing any of these processes.

For reference, Figure 4.1 presents a general list of common contaminants and the general treatment systems that may be used to pretreat wastewaters containing these contaminants before discharge to the POTW.

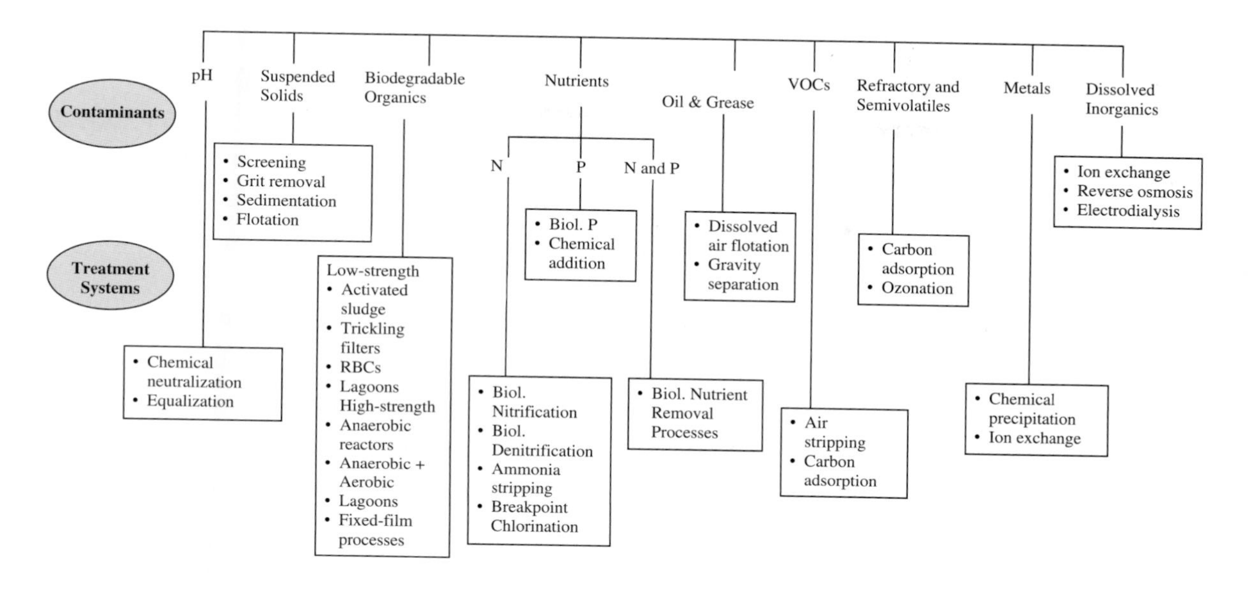


FIGURE 4.1 General treatment systems commonly used to remove major contaminants.*

# REFERENCES

Eckenfelder, W. W. Jr. (1999) *Industrial Water Pollution Control*, 3rd ed.; McGraw-Hill: New York.

Nemerow, N. L. (1978) *Industrial Water Pollution*; Addison-Wesley: Reading, Massachusetts.

U.S. Environmental Protection Agency (1992) *Control Authority Pretreatment Audit Checklist and Instructions*. U.S. Environmental Protection Agency: Washington, D.C.

U.S. Environmental Protection Agency (1983) *Guidance Manual for POTW Pretreatment Program Development*, EPA-833/B-83-100; U.S. Environmental Protection Agency: Washington, D.C.

U.S. Environmental Protection Agency (1986) *Pretreatment Compliance Monitoring and Enforcement Guidance*. U.S. Environmental Protection Agency: Washington, D.C.

U.S. Environmental Protection Agency (1987) *Guidance Manual for Preventing Interference at POTWs*, EPA-833/B-87-201; U.S. Environmental Protection Agency: Washington, D.C.

U.S. Environmental Protection Agency (1989a) *Guidance for Developing Control Authority Enforcement Response Plans*. U.S. Environmental Protection Agency: Washington, D.C.

U.S. Environmental Protection Agency (1989b) *Industrial User Permitting Guidance Manual*, EPA-833/B-89-001; U.S. Environmental Protection Agency: Washington, D.C.

U.S. Environmental Protection Agency (1994) *Industrial User Inspection and Sampling Manual for POTWs*, EPA-831/B-94-001; U.S. Environmental Protection Agency: Washington, D.C.

U.S. Environmental Protection Agency (1999) *Introduction to the National Pretreatment Program*, EPA-833/B-98-002; U.S. Environmental Protection Agency: Washington, D.C.

U.S. Environmental Protection Agency (2004a) *Local Limits Development Guidance*, EPA-833/R-04-002A; U.S. Environmental Protection Agency: Washington, D.C.

U.S. Environmental Protection Agency (2004b) *Local Limits Development Guidance Appendices*, EPA-833/R-04-002B; U.S. Environmental Protection Agency: Washington, D.C.

U.S. Environmental Protection Agency (2007a) Current National Recommended Water Quality Criteria. U.S. Environmental Protection Agency: Washington, D.C., http://www.epa.gov/waterscience/criteria/wqcriteria.html (accessed March 2007).

U.S. Environmental Protection Agency (2007b) *EPA Model Pretreatment Ordinance*, EPA-833/B-06-002; U.S. Environmental Protection Agency: Washington, D.C.

U.S. Environmental Protection Agency (2007c) General Pretreatment Regulations. *Code of Federal Regulations*, Part 403, Title 40.

U.S. Environmental Protection Agency (2007d) Guidelines Establishing Test Procedures for the Analysis of Pollutants. *Code of Federal Regulations*, Part 136, Title 40.

U.S. Environmental Protection Agency (2007e) Specific State Program Status. U.S. Environmental Protection Agency: Washington, D.C., http://cfpub.epa.gov/npdes/statestats.cfm?view=specific (accessed March 2007).

Water Environment Federation (in press) *Industrial Wastewater Management, Treatment, and Disposal*, 3rd ed.; Manual of Practice No. FD-3; Water Environment Federation: Alexandria, Virginia.

# Chapter 5

# Safety

# INTRODUCTION

Safety, by definition, is (1) freedom from danger, risk, or injury; or (2) a device designed to prevent accidents (*American Heritage Dictionary*, 1997). This is a book definition; however, in the field, safety can mean many different things. It can mean working in a manner that will not endanger the lives and limbs of workers and passersby. It can also mean working to avoid losing work days. Unfortunately, safety can also be spoken about, but not practiced.

How does this happen? Everyone knows that there should be work practices used where no one gets hurt or sick. However, often, there are pressures to produce more work on time and more cheaply, and this can result in careless mistakes by workers and even uncorrected bad work habits. Pressures to produce more work on time and more cheaply actually create an atmosphere or culture in which administrators may say the right things but really do not mean them. The term *buy-in* has become popular in management training, and it is used to signify an emotional investment in a topic. Management tries to get workers to "buy-in" to a program. The programs can range from profit sharing, to disciplinary actions, to esprit de corps, to safety. It is the responsibility of management to ensure that personnel understand the idea of working safely. Management can accomplish this in many ways, including through training, disciplinary action for unsafe work, on-site inspections, bonuses for employees, and fostering willing participation in safety programs by paying for them. It is easy for management to talk about safety without paying for it, especially in current times, when there seems to be no extra income to establish or maintain such programs. Some examples of this would be failing to provide the following: compensation for safety shoes for people who require steel-toed shoes, on-site training before workers use new equipment, or annual hearing tests for workers who regularly work where there is noise greater than 85 dB. Management must make sure that any initiative that is started is well-funded and continued. One of management's many responsibilities is to establish a work culture that embraces and supports safe work practices.

Workers bear much responsibility in this attitude also. After all, when the worker is out in the field, he or she is the one doing the work, and he or she bears the results of his or her own actions. If he or she does not, those working closely with him or her will. The wastewater treatment profession is fraught with danger. Wastewater treatment professionals work around many different kinds of hazards, including electrical, bacteriological, viral, confined space, engulfment, mechanical hazards, and simple hazards, such as oncoming traffic and traffic control. The job requires the worker to maintain awareness at all times.

It would be nice to believe that everyone in the world wished protection, advancement, fulfillment, and the best for their fellow humans. It is sad that this is not always the case. Workers are sometimes careless because "it is not their problem", or a worker may think "nobody will know if I do . . . ", or workers may be thinking about personal problems while working. Management sometimes will have financial constraints or be apathetic toward their laborers. Other times, poor judgment is involved that leads to injuries. A biosolids hauler once washed out his trailer with a power washer and entered the tank without thinking. He died within minutes from lack of oxygen. He may have been insufficiently trained or simply made a mistake; either way, he paid for this mistake with his life. It is essential that workers make sure that their place of work actually has a safety culture and does not just discuss one.

To protect workers who sometimes can be taken advantage of by administrative policies that may not have the worker's health as the primary concern, the federal government created the Occupational Safety and Health Administration (OSHA) by passing the Occupational Safety and Health Act of 1970 (*CFR* 1910; Occupational Safety and Health Hazards, 2005) (there is a construction section of law, *CFR* 1926, that is pertinent to the wastewater treatment industry because it deals with trenching, ladders, and other things that collection system maintenance and infrastructure repair crews deal with). It is OSHA's responsibility to ensure the safety of the worker by setting up rules and regulations that protect the worker.

Oddly enough, there are many workers who are not be covered by the OSHA safety rules. Some federal and state workers do not fall under the jurisdiction of the regulation. Also, the employer (town, special district, or municipality) must have a minimum of nine employees to be held to these regulations. As an administrator or supervisor, it is important to realize that *all subcontractors who enter your facility to perform work must meet your safety and training standards.* They are your responsibility. Still, the OSHA regulations are the industry safety standards for much litigation. They will be considered as the baseline for safety guidelines in this text. Most often, employers must realize that they fall under the general duty clause of the OSHA regulations.

The drafters of the Occupational Safety and Health Act also realized that they could not cover every set of circumstances that might arise in a safety and health hazard situation. So, they included a General Duty Clause, which states, in Section 5(a)1, "Duties", the following:

(a) Each employer
   (1) Shall furnish to each of his employees employment and a place of employment which are free from recognized hazards that are causing, or likely to cause, death or serious physical harm to his employees.
   (2) Shall comply with occupational safety and health standards promulgated under this Act.

(b) Each employee shall comply with occupational safety and health standards and all rules, regulations, and orders issued pursuant to this Act which are applicable to his own actions and conduct.

The OSHA General Duty Clause is the catch-all for employer compliance. If OSHA cannot issue a citation based on a specific regulation, it will cite under the General Duty Clause. Remember, as an employer, you must provide a safe workplace for your employees. Mere compliance with all OSHA regulations may not be enough. You must endeavor to ensure that your employees are safety conscious in all activities at your facility.

## COMMUNITY RIGHT-TO-KNOW

There is another set of rules enacted by the federal government for the protection, not of workers, but for other stakeholders that may be affected by chemicals used at wastewater treatment plants. These stakeholders are the surrounding communities that may be affected by any chemical releases. This set of rules is called the Superfund Amendments and Reauthorization Act of 1986, also known as SARA or the Emergency Planning and Community Right-to-Know Act (EPCRA). The latter is actually a free-standing law included in SARA, commonly known as SARA Title III. The purposes of the regulations are to encourage and support emergency planning efforts at the state and more localized levels, so that information concerning potential chemical hazards is known and emergency response plans can be formulated.

Why is this necessary? One would think that somewhere in state records are the design criteria and process records for all wastewater plants and the chemicals they use; however, this is not the case. In one instance, a wastewater treatment professional attempted to obtain authorization that would have changed the treatment process of a small mountain wastewater treatment plant. The county had lost all records that such a facility existed. Later, when going to the state, the professional found that the original site and process applications had been lost and discharge permit renewal applications contained none of this information because they were incomplete as filed. Before computers were used, many databases were manually kept on hand-designed graphs, charts, and checklists, and many regulatory agencies do not have the budget to transfer hand-kept files into state-of-the-art relational databases; these conditions can create problems when attempting to access information. In addition, problems can arise in plants where the operations staff may have changed the processes internally and failed to mention changes on permit update paperwork.

The EPCRA rules are an amendment of the Comprehensive Environmental Response, Compensation, and Liability Act (commonly known as Superfund) regulations of 1980 (CERCLA, 1980). The U.S. Environmental Protection Agency (U.S. EPA) (Washington, D.C.) administers the EPCRA. States create emergency response commissions who, in turn, set up emergency planning committees and districts. These committees typically consist of a representative of the emergency response personnel, state and county officials, environmental personnel, fire departments, and law enforcement. It is their responsibility to ensure that all potentially hazardous materials used in the district are accounted for and that there is a solution in the event of a hazardous release. For example, very large municipal plants may have the trained personnel to deal with chlorine gas leaks safely, but many small communities cannot afford the training of personnel, let alone the materials, tools, and staff to handle emergencies, so they coordinate with districts to make sure that trained personnel exist nearby and that they know who to contact. Title III exists so that communities know what hazardous materials are produced, transported, or used by industries and services so that emergency planning and response activities can be coordinated. The law covers industries and businesses that produce, store, buy, ship, or use hazardous materials.

The first installment of these regulations, or Tier 1, was enacted in the 1980s and required that producers and users of over 4500 kg (10 000 lb) of chemicals report the location to local regulatory agencies. Later, the second, more inclusive series of regulations, Tier 2, were enacted. The level of chemicals that a plant used were much lower, requiring reporting to appropriate local special emergency response districts and the state.

Wastewater treatment facilities are typically involved in the use, storing, and shipping of hazardous substances. U.S. EPA publishes a list of Extremely Hazardous Substances (EHS) (U.S. EPA, 2001), including the quantitative limits, which are typically 230 kg (500 lb) or the threshold planning quantity, whichever is lower. For example, chlorine gas has a limit of 45 kg (100 lb), which is far less than the typical 230-kg (500-lb) limit. Quantitative limits for other hazardous chemicals (non-EHS) are typically 4500 kg (10 000 lb).

Some examples of hazardous chemicals used at wastewater treatment plants are chlorine gas, alum, ammonia, sulfuric acid, sulfur dioxide, methanol, lime, and sodium hydroxide. Complete lists can be found on the Web or by contacting the local authority.

It is important to become familiar with your local district regulatory interface personnel. They will help you set up emergency plans that are economical and reflect a concern for public safety. They will also typically be the first contact in case of emergency. Under SARA and EPCRA, local officials have the decision-making power to lower quantitative limits for hazardous chemicals if they feel that, for some reason, they pose a specific risk because of geographic or localized conditions. In many areas, this district, fire district, and law enforcement departments all have specific laws and jurisdictions that may overlap and may not be the same. For example, a fire district may have more stringent regulations for chlorine than does the county emergency response district or state regulations. In one case, a water treatment facility had a much lower chlorine storage limit than another because a grade school was built next to the plant and the amount of chlorine gas that could be stored there was drastically reduced when the school was ready to open. It is important to know your local regulatory agencies, find the most restrictive regulation, and be sure that you are in compliance with it.

Notification of hazardous chemicals was mandated in 1986. Before that, specific sheets were designed; a Material Safety Data Sheet (MSDS) was allowed to be submitted. Currently, there are forms that have been specifically designed for SARA Title III Tier 2 regulations (Figure 5.1). These forms are typically submitted annually to the ap-

Page 1 of 1

**Tier Two Emergency and Hazardous Chemical Inventory**

*Specific Information by Chemical*

**Facility Identification**

Name ____
Street ____
City ____ County ____ State ____ Zip ____

SIC Code __ __ __ __ Dun & Brad Numbers __ __ - __ __ __ - __ __ __ __

**For Official Use Only** ID # ____ Date Received ____

**Owner/Operator Name**

Name ____ Phone (___)____
Mail Address ____

**Emergency Contact**

Name ____ Title ____
Phone (___)____ 24 Hr. Phone (___)____

Name ____ Title ____
Phone (___)____ 24 Hr. Phone (___)____

*Important: Read all instructions before completing form*

Reporting Period From January 1 to December 31, 1999 ☐ Check if information below is identical to the information submitted last year

| Chemical Description | Physical and Health Hazards (Check all that apply) | Inventory | Container Type | Temperature | Pressure | Storage Codes and Locations (Non-Confidential) *Storage Locations* | Optional |
|---|---|---|---|---|---|---|---|
| CAS __ __ __ __ __ __ __ __ __ Trade Secret ☐<br>Chem. Name ____hypochlorite____<br>Check all that apply ☐ Pure ☐ Mix ☐ Solid X Liquid ☐ Gas ☐ EHS<br>EHS Name ____ | ☐ Fire<br>☐ Sudden Release of Pressure<br>☐ Reactivity<br>☐ Immediate (acute)<br>☐ Delayed (chronic) | __ Max. Daily Amount (code)<br>__ Avg. Daily Amount (code)<br>365 No. of Days On-site (days) | | | | | ☐ |
| CAS __ __ __ __ __ __ __ __ __ Trade Secret ☐<br>Chem. Name ____<br>Check all that apply ☐ Pure ☐ Mix ☐ Solid ☐ Liquid ☐ Gas ☐ EHS<br>EHS Name ____ | ☐ Fire<br>☐ Sudden Release of Pressure<br>☐ Reactivity<br>☐ Immediate (acute)<br>☐ Delayed (chronic) | __ Max. Daily Amount (code)<br>__ Avg. Daily Amount (code)<br>___ No. of Days On-site (days) | | | | | ☐ |
| CAS __ __ __ __ __ __ __ __ __ Trade Secret ☐<br>Chem. Name ____<br>Check all that apply ☐ Pure ☐ Mix ☐ Solid ☐ Liquid ☐ Gas ☐ EHS<br>EHS Name ____ | ☐ Fire<br>☐ Sudden Release of Pressure<br>☐ Reactivity<br>☐ Immediate (acute)<br>☐ Delayed (chronic) | __ Max. Daily Amount (code)<br>__ Avg. Daily Amount (code)<br>___ No. of Days On-site (days) | | | | | ☐ |

Certification (Read and sign after completing all sections)
I certify under penalty of laws that I have personally examined and am familiar with the information submitted in pages one through ___, and that based on my inquiry of those individuals responsible for obtaining information, I believe that the submitted information is true, accurate, and complete.

Name and official title of owner/operator OR owner's/operator's authorized representative ____ Signature ____ Date Signed ____

Optional Information
☐ I have attached a site plan
☐ I have attached a list of site coordinate abbreviations
☐ I have attached a description of dikes and other safeguard measures

**FIGURE 5.1** Tier 2 document (http://www.co.ha.md.us/lepc/tier2form.html).

propriate agencies. If the regulatory agency does not provide a copy of the form, it can be found on the Web at sites such as local county Web sites (i.e., http://www.co.ha.md.us/lepc/tier2form.html).

There are responsibilities that go beyond basic reporting that typically must be fulfilled. An emergency communication protocol needs to be established. This includes who should be called, in which order, in case of emergency. This contact information and an emergency action plan should be laminated and displayed in a prominent place that will be easy to get to, near a telephone. Emergency response must be practiced so that, in case of emergency, there will be few loose ends that could lead to problematic situations. It is up to management to coordinate these practices with the agencies designated in the emergency response plan. Reviews and practices will lead to the elimination of potential problems before they can happen. It is important to note that, while the topic of emergency response plans is being discussed here under chemical emergencies, there are many other emergencies, such as flood, power outages, avalanche, and effects of vandalism, that need to have established action plans. The plan should include specific actions to alleviate the emergency, identification of who should perform the actions, a communication tree for handling the emergency and notifying the public, and a method for resolving the situation. A basic list of contents for an emergency response plan follows:

- Identification of the facilities and the transportation routes for extremely hazardous chemicals. This would also include transport routes for biosolids and sludge, when being transported near bodies of water.
- Emergency response procedures.
- Designation of community and facility coordinators to carry out the emergency response plan.
- Emergency notification procedures.
- Methods for determining if a chemical release has occurred and the probable affected area and population.
- Description of the community's emergency response equipment and identity of responsible programs.
- Evacuation plans.
- Description and schedules of a training program for emergency response personnel.
- Methods and schedules for carrying out emergency response plans.

It is important to remember that any paperwork of this nature becomes public record. Often, diagrams and maps of where hazardous chemicals are kept are submitted with emergency response plans and can be quite detailed. As reported on National Public Radio KCFR 90.1 FM in Denver in April 2003, many of these maps were found on confiscated terrorist computers. Some of the sites detailed railcars of chlorine gas at wastewater plants near sensitive population centers. Whenever details of hazardous chemical storage become part of the public record, plans for security of these chemicals need to be made. In 2004, the U.S. Congress passed a law that requires drinking water facilities to perform vulnerability assessments of key chemical holding areas, water reservoirs, treatment facilities, and pumping stations and submit them to regulatory agencies. Because of primacy issues in many states, these laws have become state code in many areas. Please be aware that the states may have primacy in vulnerability assessment issues and that, nationally, the Homeland Security Department has taken the lead, instead of U.S. EPA. These are fairly detailed and are designed to thwart terrorists. With this in mind, there will be a good amount of security around these vulnerability assessments. The draft before Congress in 2005 of the Wastewater Treatment Works Security Act (Jim Jeffords, I-VT) included the same type of legislation for wastewater facilities.

In designing emergency response plans, most of the material mentioned is related to the EPCRA. However, when making your emergency response plan, it is important to remember that there are other types of emergencies. These can vary, from sanitary sewer overflows and potential watershed contamination, to floods, tropical storms, earthquakes, and extended power outages. All of these are very real possibilities and, while some are unlikely in some geographic areas, there is a need to be creative when considering all the things that could go wrong so that you will not be unprepared as a public health professional in an emergency situation.

## OCCUPATIONAL SAFETY AND HEALTH ADMINISTRATION HAZARDOUS COMMUNICATION STANDARD

Having the "right-to-know" has another meaning that is important to wastewater workers. This is the OSHA Hazardous Communication Standard, or HAZCOM. Basically, workers need to know the chemicals with which they will be dealing. They need to know the proper ways to handle chemicals and proper personal protective equipment (PPE) to use when handling them, any hazards the chemical may pose, limits of exposure, and other significant topics. These can be found in a number of places. Primarily, these are included in the MSDS. The MSDSs of all chemicals used in a facility are to be kept in a labeled notebook or binder in a specific location near the entrance to the facility. In fact, in some large facilities, by code, all of the outbuildings need their own MSDS volumes. Many emergency protection or response districts specify a color for this notebook (sometimes yellow or, often, safety orange).

An MSDS will include a list of the specific information pertaining to the chemical. This information will include the following:

- Identification of composition, Chemical Abstract Services number, formula, chemical weight, and synonyms.
- 24-hour emergency number.
- Physical data on boiling, freezing, and melting points; specific gravity; solubility; and vapor pressure.

- Reactivity, such as incompatibility, decomposition products, and polymerization potential.
- Health-hazard data on effects of exposure (acute and chronic), permissible exposure limits (PELs), and warning signals.
- Environmental effects, such as toxicity effects on the environment.
- Instructions for shipping and compliance with other pertinent federal regulations.
- Exposure control methods, such as PPE and engineering and administrative controls.
- Work practices, such as handling and storage procedures, cleanup, and waste disposal methods.
- Emergency procedures for handling spills, fires, and explosions.
- First-aid procedures.

The MSDS forms the basis of an employee training program on chemical handling. It should *never* be considered a substitute for training, particularly because there is no readability standard for the MSDS. Some are written in easily understood language, and others are written in technical language, as used by chemists and industrial hygienists. As a reference source, the MSDS must be acceptable to supervisors and all employees. Training to read and understand the information presented in the MSDS must be provided by the administration. The MSDS book must have routine checks and updates as newer MSDSs are received or changes in process require new chemicals. *No chemical should ever be received, labeled, stored, or used on-site without the safety and health information being recorded and filed.* There are other reference books that are not required but that list important information, such as threshold limit values and PELs, which are developed by the National Institute for Occupational Safety and Health (NIOSH) (Washington, D.C.) and the American Society of Safety Engineers (Des Plaines, Illinois). These are found in *The NIOSH Pocket Guide to Chemical Hazards* (NIOSH, 2005).

## HAZARDOUS CHEMICALS

We have mentioned HAZCOM, but why is this so important? One of the most serious and problematic hazards that wastewater workers have to deal with is working with hazardous chemicals. There are many that are used regularly in the wastewater treatment industry, including chlorine gas, hydrogen peroxide, sulfuric acid, hydrochloric acid, acetic acid, and caustic soda. There are many others with varying degrees of immediate hazard. Without discussing all of the hazards, the following sections discuss ways of controlling hazards for workers who use these chemicals daily.

**CONTROL MEASURES.** When dealing with chemicals, the economic feasibility of eliminating the use of that chemical must be a primary consideration. Too often, the use of hazardous chemicals continues because "it was always done that way". Other methods of controls are engineering controls that lower the liability in using chemicals or ensuring safe work practices.

***Elimination.*** The use of specific chemicals may be eliminated by changing the process (i.e., the use of anoxic zones to release some of the alkalinity during denitrification to lower the amount of chemical needed); changing equipment; or substituting chemicals that may be less hazardous to use. There are many resources that offer help, and often these are free. Chemical suppliers can suggest alternatives that may be safer to use that may not be familiar to managers. Often, state inspectors may be a helpful resource; they may suggest alternative controls that would lower liability. For example, the City of Anchorage, Alaska, reviewed its chlorine use and found that, because of chemical delivery schedules, it had a high risk factor. Chlorine was delivered infrequently because of the city's location, and large amounts of chlorine were stored on-site. Rather than maintain such a huge possible environmental liability, the city decided to switch to an alternative method of disinfection that uses salt for on-site chlorine generation. While the new method was more expensive, the amount of training and continuous monitoring that was reduced offset any economic advantages of delivering cheaper chlorine gas.

***Engineering Controls.*** Primary measures include ventilation (both general and local exhaust), isolation, enclosure, and workplace redesign. Ventilation controls primarily airborne hazards to separate the hazard from the worker. Again, it is important for managers to work with local stakeholders. The local fire department or hazardous materials division may be able to suggest inexpensive solutions to safety issues. If you work easily with them, they are willing to share their experience with you.

***Work Practices.*** The following are common work practices used to help safeguard employees:

- Initial training for proper safe handling and work practices.
- Using proper PPE.
- Proper housekeeping and storage procedures.
- Developing and using standard operational procedures for routine jobs that have some hazard associated with them.
- Using vacuums or sweeping compounds to keep chemical storage areas clean.
- Prohibiting smoking in areas of possible exposure.
- Using separate eating and washing facilities where ingestion hazard is identified.

- Labeling containers properly, including appropriate National Fire Protection Agency (NFPA) (Quincy, Massachusetts) signs or other detailed information. This is particularly important when the original chemical is transported in another container for use. It is possible to obtain NFPA signs and stickers to put on the container. At the very least, the container should be labeled with the correct contents.
- Posting warning signs to alert employees to hazardous conditions and special precautions.
- Posting emergency instructions at critical operations.
- Having emergency procedures in place for fires, chemical emergencies, spills, and first-aid requirements and practicing them.
- Training and maintaining training records for safe use and handling of all hazardous chemicals.
- Using job safety analyses to help perform operation and maintenance tasks.

The training of personnel to safely handle these chemicals and the provisions of PPE cannot be stressed enough once it is determined that it is not possible to lower exposure to these chemicals by elimination or engineering controls.

**COMMON CHEMICALS.** Table 5.1 is a list of common chemicals found in wastewater treatment, their characteristics, and the most common immediate dangers they can pose to a wastewater employee.

As indicated in the table, many of the gases are colorless, odorless, and tasteless, necessitating the use of gas sensors. Gas sensor use is covered in the section on confined spaces.

**STORAGE GUIDELINES.** ***Standard for Fire Protection Practice for Wastewater Treatment Plants and Collection Facilities.*** *Standard for Fire Protection Practice for Wastewater Treatment Plants and Collection Facilities* (NFPA, 1995) lists information regarding the storage and handling of chemicals found in wastewater treatment facilities. Sections of the Uniform Fire Code (NFPA, 2006; managers should use the current code that has been adopted and has regulatory jurisdiction in their areas) also discuss the storage, dispensing, and use of most hazardous chemicals used in a wastewater treatment plant. Once the manager becomes familiar with these standards, it is important to contact the local fire and building inspectors and other appropriate officials to determine who has jurisdiction and who has the most stringent standard to meet. Sometimes, officials may not agree; however, by initiating dialogue, the manager may facilitate a process whereby all inspectors and regulatory agencies can agree on practices for a particular facility.

**TABLE 5.1** List of common chemicals found in wastewater treatment, their characteristics, and the most common immediate dangers they can pose to a wastewater employee.*

| Common name | Physical characteristics | Danger | Safety precaution |
|---|---|---|---|
| Chlorine gas | Greenish-yellow gas, amber liquid under pressure, highly irritating odor, corrosive in moist atmosphere | Respiratory irritant: 30 mg/L coughing; 40 to 60 mg/L dangerous in 30 minutes; 1000 mg/L lethal in a few breaths | SCBA |
| Caustic soda | Viscous liquid, high pH | Chemical burn | Gloves, face shield or goggles, chemical wash station |
| Sulfuric acid | Liquid, low pH, distinctive odor | Chemical burn | Gloves, face shield or goggles, chemical wash station |
| Hydrochloric acid | Liquid, low pH, distinctive odor | Chemical burn | Gloves, face shield or goggles, chemical wash station |
| Carbon dioxide | Odorless gas | Asphyxiation | Ventilation, SCBA |
| Carbon monoxide | Colorless, odorless, tasteless gas | Asphyxiation, flammable, explosive | Ventilation, SCBA |
| Methane | Colorless, tasteless, odorless, nonpoisonous | Flammable, explosive | Ventilation |
| Hydrogen sulfide | Rotten egg smell, deadens sense of smell | Death in few minutes at 0.2% | Ventilation, SCBA |
| Hydrogen | Colorless, odorless, tasteless gas | Asphyxiation, flammable, explosive | Ventilation |

*Note that this list is not comprehensive of the chemicals found in wastewater plants or all of the dangers that they may pose but is only an example of the immediate dangers. For all of the dangers presented by chemicals on-site, please consult the MSDS sheets and the current issue of the *NIOSH Guide to Chemical Hazard* (NIOSH, 2005).

***Chlorine Storage and Chlorine Room Design.*** Chlorine rooms are worthy of special mention because chlorine is probably the single most dangerous chemical commonly used in wastewater treatment plants, and the storage rooms for chlorine have specific design standards. Many of these design standards can be found in design or standard practice manuals. It is a good idea to be familiar with the following basic principles, but always remember to check local building and fire codes, as they may be different and more stringent:

- Chlorine rooms should be adequately heated (typically higher than 10 °C [50 °F] to keep the chlorine from forming chlorine ice crystals).

- There should be adequate light so that the operator can easily see what he or she is doing.
- There should be an adequate controlling and measuring method so that the chlorine can be dispensed safely. Taking too much chlorine from the cylinder in one day can also form ice crystals. Maximum flow from a 45- or 68-kg (100- or 150-lb) cylinder is 18 kg/d (40 lb/d) and from a 900-kg (1-ton) cylinder is 180 kg (400 lb).
- The chlorine cylinders should be safely fastened in place. This is typically done by using chains or clamps. It should be easy to handle and store cylinders; 45- and 68-kg (100- and 150-lb) cylinders are always stored in a vertical position.
- There should be appropriate carts, clamps, chain falls, or hoists available so that the cylinders can be moved without undue stress on the worker's body.
- Chlorine room doors should have safety glass placed in them so that an outside observer can watch any work being performed or simple periodic visual checks can be made. When chlorine room doors are opened, they should trigger a ventilation fan so that the room is ventilated with fresh air whenever anyone is in the room.
- Because chlorine gas is heavier than air, the ventilation inlet is near the floor to dilute any chlorine gas that may have leaked. The outlet of the ventilation system should be away from the inlet at the top of the room and also away from any traffic patterns that may occur outside the room. This would reduce the threat of accidentally gassing pedestrian traffic. Chlorine rooms in larger plants should have chlorine detection alarms that will typically have battery backup in case of power failure.
- Outside the chlorine room, there should be a self-contained breathing apparatus (SCBA) to be worn in case of emergency entrance to the chlorine room while there is a leak. These need to be tested regularly and fit-tested on the employees that wear them.

During an emergency is not the time to find out that routine maintenance and worker practice have been forgotten. This could easily create a life-threatening situation.

**CHEMICAL DELIVERY AND DISTRIBUTION.** While many wastewater treatment professionals work in larger plants, where chemicals are delivered, there is a small percentage of operators who work in small systems, where chemicals must be brought in or moved from place to place by workers, because delivery is not immediately possible. For example, there are small mountain communities where chlorine gas delivery trucks could not even make it up the road to get there. This means chemicals must be moved in the plant's service truck. All chemical delivery of this type falls under the jurisdiction of the Department of Transportation (DOT) (Washington, D.C.).

When hauling chlorine gas or sulfur dioxide cylinders, the plant's service truck must have the appropriate DOT placards (Figure 5.2). Placarding examples can be found in DOT manuals and purchased from many safety and industrial suppliers.

There will be a need for a chain-of-custody form that shows the quantity, chemical name, emergency response telephone number, class of chemical, UN number, and quantity and class of container. It also requires the signature of the driver for the load. This paperwork needs to be kept on a clipboard on the dash, in the front seat, or in the door panel for immediate accessibility. The MSDS sheet for the chemical also should be attached to the clipboard. Drivers need to have hazardous chemical transportation safety classes.

When transporting liquid chemicals, such as 10% sodium hypochlorite or caustic soda, there cannot be more than 450 kg (1000 lb) on the truck. That basically means one barrel, or possibly two, if the weight is known. The same protocol as above must be followed, with the exception of placarding. In either case, NFPA signage must be on the containers. When carrying containers containing less than 18 L (5 gal), no chain-of-custody paperwork or extensive protocols need to be followed. By using common sense and good common safety procedures, no chemicals that will react with each other should be carried together. All small containers that are carried still need the appropriate marking, class of container, and NFPA decals on the container.

**NATIONAL FIRE PROTECTION ASSOCIATION.** Below is a brief description of the symbols system used by NFPA for marking chemicals (Figure 5.3; NFPA, 2005). These diamond symbols should be on all chemical storage drums and smaller containers and can be purchased from many safety suppliers. For fire protection, these symbols should be displayed outside of buildings that house chemicals.

Other symbols, abbreviations, and words that some organizations use in the Special Hazards section of Figure 5.3 are shown in Figure 5.4 (NFPA, 2005). These uses are **not** compliant with NFPA 704; they are presented to help clarify their meaning when they appear on an MSDS or container label.

## MICROBIOLOGICAL HAZARDS

One thing that unites wastewater workers, schoolteachers, and nursery workers are the microbial milieu found in the workplace. While medical personnel in an emergency room will see severe cases of infectious diseases, they work in a sterile environment with many ready safety precautions. Much of their hazard is based on

**FIGURE 5.2** Typical examples of DOT placards. (Courtesy of Environmental Chemistry.com).

bloodborne pathogens. In a wastewater plant, there are commonly either stagnant anaerobic conditions or an aerated mass of heated microbial material. Wastewater workers are often exposed to low-level aerosolized versions of microbes, some of which may be infectious. The immune systems of many wastewater treatment professionals build up antibodies to a variety of bacterial and viral infectious agents. They become what are nicknamed "universal carriers" because they are often in contact with low levels of infectious agents that will not make them ill, but that they can build immunity to, much like vaccination theory. However, if wastewater workers' bodies do get run down or they come into contact with a significant infectious agent, they can easily become ill.

In an early 1970s text, it was noted that, in testing, approximately 14% of wastewater employees had some type of parasitic infection (Geldreich, 1972). In almost all cases, these were noninvasive, asymptomatic, or dormant, and the body had built defenses against them. As many current wastewater treatment professionals have become more aware of the microbial hazards of the work environment, this number has probably decreased. In any case, it is important to become familiar with some of the more common infectious agents.

**AMOEBIC PARASITES.** ***Entamoeba hystolitica.*** *Entamoeba hystolitica* is an amoebic intestinal parasite that enters the body through oral–fecal contamination as a cyst. The cyst splits into eight amoebic cells in the small intestine that often pass into the large intestine to live without damage to cell walls. At other times, the amoebas may attack the walls of the large intestine and cause severe cramps. In severe cases, the amoebas may pass into the liver or other organs. There are prescription drugs available to treat this parasite. The cyst can be long-lived and is somewhat chlorine-resistant. The symptoms are severe cramping and can escalate to dysentery.

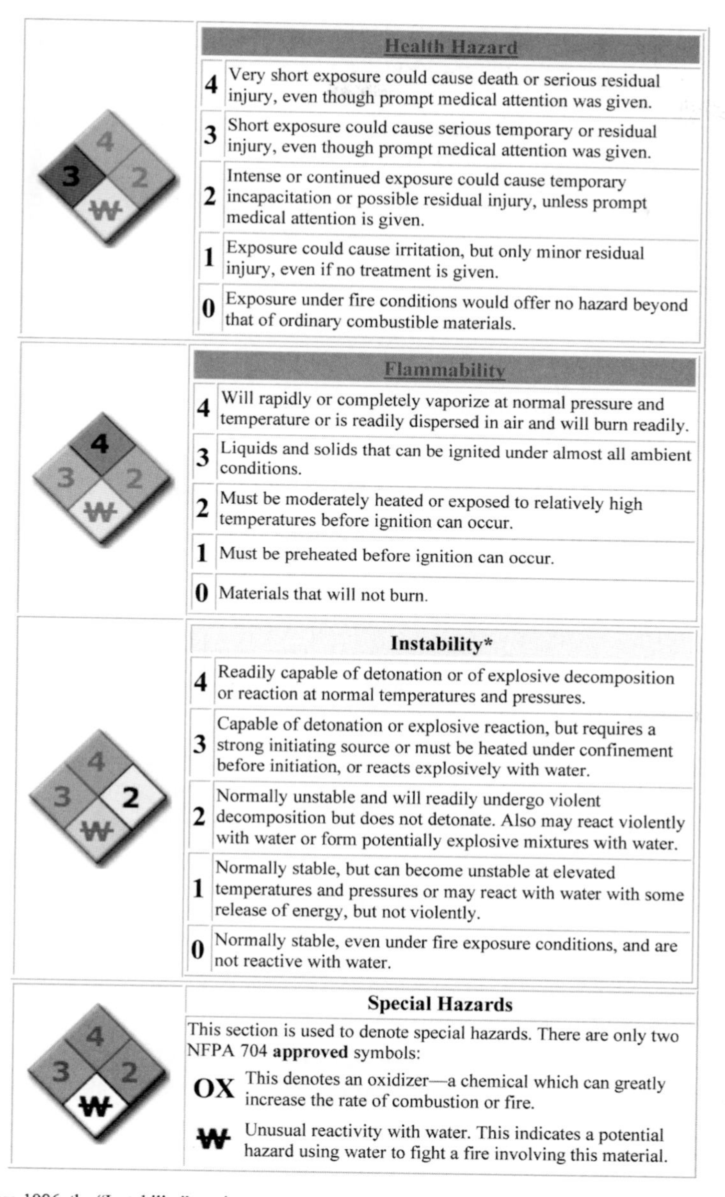

| Health Hazard | |
|---|---|
| 4 | Very short exposure could cause death or serious residual injury, even though prompt medical attention was given. |
| 3 | Short exposure could cause serious temporary or residual injury, even though prompt medical attention was given. |
| 2 | Intense or continued exposure could cause temporary incapacitation or possible residual injury, unless prompt medical attention is given. |
| 1 | Exposure could cause irritation, but only minor residual injury, even if no treatment is given. |
| 0 | Exposure under fire conditions would offer no hazard beyond that of ordinary combustible materials. |

| Flammability | |
|---|---|
| 4 | Will rapidly or completely vaporize at normal pressure and temperature or is readily dispersed in air and will burn readily. |
| 3 | Liquids and solids that can be ignited under almost all ambient conditions. |
| 2 | Must be moderately heated or exposed to relatively high temperatures before ignition can occur. |
| 1 | Must be preheated before ignition can occur. |
| 0 | Materials that will not burn. |

| Instability* | |
|---|---|
| 4 | Readily capable of detonation or of explosive decomposition or reaction at normal temperatures and pressures. |
| 3 | Capable of detonation or explosive reaction, but requires a strong initiating source or must be heated under confinement before initiation, or reacts explosively with water. |
| 2 | Normally unstable and will readily undergo violent decomposition but does not detonate. Also may react violently with water or form potentially explosive mixtures with water. |
| 1 | Normally stable, but can become unstable at elevated temperatures and pressures or may react with water with some release of energy, but not violently. |
| 0 | Normally stable, even under fire exposure conditions, and are not reactive with water. |

| Special Hazards | |
|---|---|
| This section is used to denote special hazards. There are only two NFPA 704 **approved** symbols: | |
| OX | This denotes an oxidizer—a chemical which can greatly increase the rate of combustion or fire. |
| W̶ | Unusual reactivity with water. This indicates a potential hazard using water to fight a fire involving this material. |

*Before 1996, the "Instability" section was entitled "Reactivity". The name was changed because many people did not understand the distinction between a "reactive hazard" and the "chemical reactivity" of the material. The numeric ratings and their meanings remain unchanged.

FIGURE 5.3 Brief description of NFPA symbol system (NFPA, 2005).

|  |  |
|---|---|
| ACID | This indicates that the material is an acid—a corrosive material that has a pH lower than 7.0 |
| ALK | This denotes an alkaline material, also called a base. These caustic materials have a pH greater than 7.0 |
| COR | This denotes a material that is corrosive (it could be either an acid or a base). |
|  | This is another symbol used to indicate that a material is corrosive. |
|  | The skull and crossbones symbol is used to denote a poison or highly toxic material. |
|  | The international symbol for radioactivity is used to denote radioactive hazards; radioactive materials are extremely hazardous when inhaled. |
|  | Indicates an explosive material. This symbol is somewhat redundant because explosives are easily recognized by their instability rating. |

FIGURE 5.4 Other symbols, abbreviations, and words that some organizations use in the Special Hazards section of Figure 5.3 (note that only those appearing in Figure 5.3 are NFPA 704 approved symbols). The exact guidelines by which you can place a *chemical* in one of these categories are available in the NFPA standard (NFPA, 2005).

***Giardia lamblia.*** Often seen in natural water sources, this parasite is also cystic in nature and enters the body via oral–fecal transmission. The effects of this amoebic infection are severe cramps and diarrhea. The effects have been nicknamed beaveritis or beaver fever from its suspected historical host (many mammals can be a source of this disease, including humans). This can be treated with a course of antibiotics. Recent research has shown that cysts that have passed through UV disinfection in wastewater effluent have become noninfectious (Thompson et al., 2003). Prescription drugs have been used to eliminate infections.

***Cryptosporidium.*** *Cryptosporidium* is a protozoan—a single-celled parasite that lives in the intestines of animals and people. This microscopic pathogen causes a disease called *cryptosporidiosis*. The dormant (inactive) form of *Cryptosporidium*, called an oocyst, is excreted in the feces (stool) of infected humans and animals. The tough-walled oocysts survive under a wide range of environmental conditions. Infection occurs through oral–fecal contamination. The most common symptom of cryptosporidiosis is watery diarrhea. There may also be abdominal cramps, nausea, low-grade fever, dehydration, and weight loss. Symptoms typically develop 4 to 6 days after infection, but may appear anytime from 2 to 10 days after infection. People with healthy immune systems are typically ill with cryptosporidiosis for several days, but rarely more than 2 weeks. Some infected individuals may not even get sick. Some people with cryptosporidiosis seem to recover, then get worse again. Those who are infected may shed oocysts in their stool for months, even after they no longer appear ill. This infection can be life-threatening to people with compromised immune systems. Cryptosporidia became prominent in the vernacular of water and wastewater workers in the late 20th century, when more than 400 000 people became infected in Milwaukee, Wisconsin, when the rivers flooded out the filters of the water plant and entered the distribution system. There is no drug to cure the disease. The body's immune system alone must fight the infection.

***Other Parasites.*** Tapeworm and roundworm have been found in wastewater, but their prevalence is low compared to other parasites. The former grows generally in the large intestine and is transmitted by eggs. The route of contamination is oral–fecal. Remember that people infected with all of the above-mentioned parasites can pass over 1000 000 eggs or cysts daily in their stool. Roundworm, in various forms, lives in the large intestine and can cause severe cramps and diarrhea. These worms can be controlled and eliminated though prescription drugs.

**HEPATITIS.** ***Hepatitis A and E.*** These two types of hepatitis are being grouped together in this discussion because of the common methods of infection. Both of these are spread fecal to oral, by touching contaminated material and then touching oral pathways of entry or by drinking contaminated water or eating contaminated food. Hepatitis A is more prevalent in the United States and Canada than Hepatitis E. The latter is more prevalent in Mexico, India, and Africa. Both cause infectious hepatitis (an acute inflammation of the liver). Symptoms can include fatigue, nausea, vomiting, fever and chills, loss of appetite, yellowing of the eyes and skin, and tender liver. Many times, the carrier can be asymptomatic. Infectious hepatitis is an extremely environmentally resistant disease, where the carrier releases 1000 000 000 virions/g feces. It can survive months or years outside of a host. There is a vaccine for prevention of Hepatitis A, and ozone and chlorine can inactivate it. Like Hepatitis A, Hepatitis E never develops into a chronic or long-term illness, but shares the same symptoms as Hepatitis A. Currently, there is no preventative vaccine developed for Hepatitis E. The Center for Disease Control (Atlanta, Georgia) states that wastewater workers are no more at risk than any other occupation.

***Hepatitis B, C, D, and G.*** The symptoms of these viral infections are much the same as other forms of hepatitis. However, the means of transmission of these infections is bloodborne. This means that the virus must come into contact with blood to cause infection. This can happen through puncture wounds, open sores on the inside of the mouth, and open cuts on hands and other parts of the body. This means that the probability of contacting these diseases as a wastewater worker is less than the previously discussed forms of hepatitis. Some of the infections are long-term or chronic, however, with cirrhosis of the liver and liver cancer being the results of infection. There is a vaccination for Hepatitis B that, many times, is part of a preventive health care program provided by employers and is often provided for free or at low cost in rural counties.

**HUMAN IMMUNODEFICIENCY VIRUS.** Human Immunodeficiency Virus (HIV) was a shocking discovery in human health over the past 30 to 40 years. People were not recovering from what seemed to be less-than-fatal infections. In fact, their bodies were getting numerous infections simultaneously. It took time to isolate the cause of the disease, commonly called Acquired Immunodeficiency Syndrome. For wastewater workers, it is good to keep in mind that, at times, there could be infectious HIV particles in wastewater. The levels of HIV in wastewater are generally less than levels of polio. It is a bloodborne infectious agent and must enter an open sore and contact blood in sufficient quantity to establish itself (Casson and Hoffman, 1999). Standard safety and hygiene practices should prevent any chance of infection. At this time, there is no cure or vaccine. There are drugs that seem to stabilize the condition in case of infection.

**SEVERE ACUTE RESPIRATORY SYNDROME.** Severe Acute Respiratory Syndrome (SARS) is a very new infectious agent. It is a viral infection that can survive up to four days in excrement, according to the World Health Organization (2003). There is no known vector of transmission. It is thought to be transmitted by **direct** person-to-person contact through droplets spread when an infected person coughs or sneezes and the droplets are spread to a nearby contact. It probably is not transmitted through a fecal–oral route. In one case, leaking wastewater pipes within a building helped spread the disease. In Taipei (Taiwan), sections of the collection system were chlorinated as a preventative measure for collection system workers (SARS Prevention, 2003).

**LEPTOSPIROSIS.** Leptospirosis is a bacterial disease that affects humans and animals. It is caused by bacteria of the genus *Leptospira*. In humans, it causes a wide range of symptoms, although some infected persons may have no symptoms at all. Symptoms of leptospirosis include high fever, severe headache, chills, muscle aches, and vomiting and may include jaundice (yellow skin and eyes), red eyes, abdominal pain, diarrhea or a rash. If the disease is not treated, the patient could develop kidney damage, meningitis (inflammation of the membrane around the brain and spinal cord), liver failure, and respiratory distress. In rare cases, death occurs. The risk of transmission of this disease through wastewater now seems minimal, but it has been considered a risk for collection system workers.

**AEROSOLS.** With the use of activated sludge as a standard treatment process, the wastewater worker walks above and around a cauldron of airborne aerosols. These aerosols may contain bacterial and viral infectious agents, and infections may result from contact with these aerosol mists. It is impossible to eliminate all sources of aerosol contamination in a wastewater treatment plant. Particle masks may help where there is heavy misting and could be part of PPE for those conditions. Aerosols settling on clothes are another infection vector. A good safety practice is to change clothes before leaving the facility so that possible infectious agents are not transported away from the site. Good personal hygiene practices and a health protection immunization program lower the risk of infection. Another practice that typically only larger plants can afford is to examine the processes used and analyze air samples to see if certain areas are higher risk areas, and then devise a PPE program and engineering practices to lower the risk. Otherwise, in areas where this is not affordable, the manager can consider possible risks and see which ones can be eliminated. Airborne infectious agents typically are not inhaled in a large enough number to be sufficient to cause infection (Kuchenrither et al., 2002).

**PERSONAL HYGIENE AND HEALTH PROTECTION.** This is an easily overlooked and highly effective method of protecting your health. Children are taught to wash their hands before putting anything into their mouths. The same applies to wastewater professionals. This very simple practice can eliminate some infectious agents before they come into contact with routes of infection. Washing hands thoroughly before eating, putting a cigarette in the mouth or eyedrops in the eyes, or blowing the nose lowers the chance of infectious agents reaching the mucous membrane. The following are personal hygiene and PPE guides that would lower workers' possible exposure:

- Keep hands and fingers away from eyes, nose, mouth, and ears.
- Wear rubber gloves when cleaning screenings, pumps, process sludges, grit, and wastewater or when doing any task that involves direct contact with wastewater or solids. Use gloves that are appropriate for the task (e.g., light nitrile or latex gloves—be careful of allergic reactions to latex—for sampling and heavier gloves for pump repairs).
- Always wear impervious gloves when the skin is chapped and burned or when the skin is broken or has open cuts and sores.
- Wash hands with warm water and soap thoroughly before eating, smoking, and after work. If there are no facilities available for this (small rural systems), use no-rinse antiseptic soaps. Remember, the steering wheel of an automobile can be a great place for the growth of microbes.
- Keep fingernails short and remove foreign material from underneath them with a stiff soapy brush.
- Store clean clothes in a locker and change before leaving the worksite. The locker should be separate from the one where work clothes are stored.
- Report all cuts and scratches and receive first aid for them. Clean them thoroughly.
- Shower before leaving work, if possible.
- When working in an area that will have a splashing hazard where microbial hazards may come into contact with the eyes, wear safety goggles or face shields.
- Always wear PPE where applicable. There should be some determination of what PPE is necessary by a safety committee or officer, and it is important to adhere to this. Again, there should be training for employees regarding microbial hazards in the workplace.
- Consider immunizations for common diseases. Table 5.2 shows a standard immunization program. Some employers pay for all of these, and others may pay for some of these.

As mentioned previously, there are many simple things that can be done to keep workers safe. While these hazards are present in the work environment, some prevention is common sense. If a person is constantly stressing the body by staying up late, then the immune response of the body decreases. Poor nutrition, cigarette smoking, and obesity will also decrease the immune response of the body. Lowering personal

**TABLE 5.2** A standard immunization program.

| Infection | Immunization |
|---|---|
| Hepatitis A | Immunoglobulin treatment |
| Hepatitis B | Three immunoglobulin vaccinations in a complete cycle |
| Influenza | Annual flu vaccines |
| Measles | Combined measles, mumps, and rubella (MMR) vaccines |
| Mumps | MMR vaccines |
| Rubella | MMR vaccines |
| Tetanus and diptheria | TD vaccine |
| Pnuemoccal disease | Pnuemococcal polysaccharide disease |

risk factors outside the workplace will help the body's immune response. "Workers with a weak immunity system continually get sick and leave the job. According to WRRC (2001), those who stay have a good immune system and continue working. Normal precautions may be sufficient for those workers who can adjust to the work environment." There does seem to be evidence that a person who works in wastewater treatment for fewer than 2 years is more likely to suffer gastrointestinal illness more often than seasoned veterans. Also, the sanitary conditions found at the plant and among workers do have an effect on health (Kuchenrither et al., 2002). Generally, however, cancer, airborne risks, and pathogen risks do not seem to be increased by the work environment after the initial work period.

## CONFINED SPACES

A confined space is a hazard that wastewater workers encounter with great regularity. Wastewater professionals are very familiar with confined spaces, such as manholes or storage tanks, but often overlook simple regular tasks in confined spaces, such as cleaning aerators in aeration basins, hosing down clarifiers, or even working in the subfloors of a building. A confined space is an accessible area with at least one of the following three conditions:

(1) Limited entrance and egress. There is typically one way in or out of the space.
(2) Unfavorable neutral ventilation (stale air). This is the most overlooked condition for a confined space.
(3) A design that allows for limited occupancy. The design for the area is one that really is not made for continual occupancy.

Many deaths, injuries, and job-related illnesses in the wastewater treatment industry result from occupancy in confined spaces or from hazards found in these spaces, such as toxic gases, oxygen deficiency, or engulfment. The following sections discuss some of the characteristics of confined spaces and the necessary precautions in more detail.

**LIMITED ENTRY AND EGRESS.** Confined spaces are limited by the entry and exit, by either size or location. Openings are typically small. Some openings are smaller than the standard 610-mm (24-in.) manhole openings. They are difficult to move through and pose a particular hazard if there is the need for an emergency retrieval. If a person must contort the body to get into the opening, getting an unmoving and uncooperative body out is even more difficult. It would not be unusual to have PPE, such as a respirator, hard hat, face shields, or SCBA, dislodged by the small opening. Keep in mind, not all confined spaces are defined by small openings, but could be aeration basins, digesters, or clarifiers that are difficult to enter or exit. Open top spaces could be excavations or the previously mentioned ones and would require a ladder or hoists to enter or exit. Being rescued from areas like this may be easier than from small openings, but require specialized equipment and at least one spotter or helper.

**VENTILATION HAZARDS.** A ventilation hazard may as simple as stale air or oxygen deficiency but can also be dangerously high levels of gases, such as hydrogen sulfide, carbon monoxide, or even carbon dioxide. There also can be explosive gases in the confined space. These can be particularly dangerous in areas such as manholes or lift stations, when a worker has no control or advance notice of the problem. The atmosphere may be so oxygen-rich that it also becomes a hazard. Later in this section, some of the more common gases will be defined and their hazards will be listed.

**LIMITED OCCUPANCY.** Most confined spaces not only have limited access, but are not designed to have continuous occupancy. They are designed, at best, to have short-term occupancy for maintenance or repair, cleanup, or similar tasks. These tasks are often dangerous because of the chemical or physical hazards they can pose, often in a limited-visibility situation.

***Hazards.*** The following are hazards that may be found in a confined space:

- Oxygen-deficient atmosphere. If the oxygen content is or drops below 19.5%, a worker should avoid entering a confined space or should leave that space immediately. Use of an oxygen sensor is necessary when entering a confined space and should be worn while in that space. Entry to an oxygen-deficient atmosphere should not be attempted without SCBA.

- Flammable atmosphere. There may be methane present from the decomposition of organic materials, or someone may pour flammables, such as gasoline, down the sewer. Any ignition source could cause an explosion or start a fire in such an atmosphere. There may be times when the upper explosive limit (UEL) of a flammable gas is exceeded before ventilation is applied. This means that the gas will not combust. When ventilation is applied, it could then be explosive. Monitoring while in the combined space by a gas monitor is a good practice for these reasons.
- Toxic atmosphere. Some gases and vapors, if present in a confined space, may be toxic, even if not flammable. The atmosphere inside a confined space may change while a person is working in it. There have been times when the atmosphere inside a confined space was above the UEL and the space was ventilated, which, in actuality, brought the working conditions in the space to a greater hazard limit as the concentration of the gas decreased to below the UEL. Fumes from equipment that is used may accumulate in the space (i.e., carbon monoxide from a combustion engine on a pump), or cleaning chemicals being used may give off vapors.
- Temperature extremes. Extremely hot or cold temperatures in a confined space can affect or injure workers. Again, temperatures may rise or fall while working in a confined because of equipment being used. Because of the chance of smaller air exchange through smaller entrances, temperatures can stay warmer or colder for long periods of time.
- Engulfment hazards. Materials stored in a bin or hopper can engulf or suffocate a worker. Chemicals, grain, and sand are typically stored in hoppers. If a worker is in an area such as the bottom of a clarifier or digester or pumping station, he or she should be sure that some sort of lockout/tagout (discussed later in this chapter) is used to ensure that fluid cannot engulf him or her.
- Noise. A confined space is a great echo chamber. When a person is in a confined space and the sound can do nothing other than assault a person's hearing, damage could occur. The sound can come from either inside or outside of the chamber and impair speech or drown out shouting.
- Slick or wet surface. Confined spaces often have a wet surface, which can be slippery and dangerous for maneuvering within it.
- Falling objects. It is important to be aware of the possibility of objects entering a confined space and falling on someone. This could be as simple as a person

walking by and unknowingly kicking a rock into the hole or unsafely throwing tools into and out of the hole. Use common sense and a rope and a bucket for such operations.

***Precautions.*** The following are safe work practices that are recommended before entering a confined space:

- Air monitoring. More precisely, this should be called *atmosphere monitoring*. First, the air is sampled with an oxygen sensor. There must be a minimum of 19.5% oxygen present before a person can go into the space. The worker should be sure the oxygen is measured at the levels he or she must pass through and the level where he or she will be working. There can be a stratification of gases caused by poor ventilation that could be oxygen-deficient. A gas monitor should always be used to check for the presence of explosive or deadly gases. Most four-gas detectors monitor for oxygen, combustibles, carbon monoxide, and hydrogen sulfide; however, with plug-in modules now available, some of these can be reconfigured to sample other gases. Air monitors must routinely be calibrated and tested with "bump gas" per manufacturer specifications. When working within a confined space, a worker should wear a portable detector to monitor atmospheric conditions and possible deadly changes. The oxygen sensor/broad-range sensor is best-suited for initial use in situations where the actual or potential contaminants have not been identified because broad-range sensors, unlike substance-specific sensors, enable employers to obtain an overall reading of the hydrocarbons (flammables) present in the space. However, such sensors only indicate that a hazardous threshold of a class of chemicals has been exceeded. They do not measure the levels of contamination of specific substances. Therefore, substance-specific devices, which measure the actual levels of specific substances, are best-suited for use where actual and potential contaminants have been identified.
- Ventilation. The confined space should have general ventilation or localized exhaust ventilation in place before entry and during occupancy. It is important to be sure that the air being used to ventilate the space is not pulling in toxic fumes from the outside (diesel fuel, gasoline vapors, etc.).
- Personal protective equipment. Respiratory equipment is essential for safe confined space entry. If a worker can be sure that the atmosphere is safe and that there is little chance it will change, PPE may not be necessary. If an oxygen-deficient atmosphere exists, workers will need to wear a SCBA. If the space cannot be entered while the worker is wearing the SCBA, equipment supplying forced

outside air or escape respirators may be used (such as a self-rescuer for mining). Workers need to be fit-tested for a respirator or an SCBA before using one, and workers with beards cannot wear one. It is important to remember that air-purifying respirators may only filter out or neutralize certain contaminants at specific concentrations. Atmospheres with oxygen concentrations that are either too high or too low cannot be made safe by using air-purifying respirators. Workers' questions should be directed to their supervisor or industrial hygienist.

- Labeling and posting. In a work environment, all confined-space entrances should be posted. The labels should clearly be marked so that all employees can understand them.
- Training. As with chemical hazards, all workers who must work around confined spaces need routine training, including knowing how to identify confined spaces, use air analyzers, and define permit-required spaces. They should be familiar with rescue procedures and communication in confined work spaces and complete lockout/tagout training and attendant training. Periodic follow-up training is recommended to reinforce safe work practices. This training should be evaluated to make sure that it is effective and current (a video that has been shown to workers multiple times will lose its effectiveness).
- Medical surveillance. Employees who must use a SCBA respirator to work in a confined space to perform their duties need to have their lung capacities checked annually and obtain medical clearances in addition to completing annual fitness tests.
- Isolation. The supervisor for any confined-space entry must be aware of any hazards that may exist. It is the supervisor's responsibility to lock up any lines and valves that may cause engulfment hazards. Also, electrical lockout/tagout procedures need to be followed. If an entrant notices any possible hazards that were not covered, he or she should communicate them to the supervisor.
- Attendant. Any time there is a confined-space entry, there should be an attendant whose sole duty is to communicate with entrants and make sure that they are not in danger. This may mean using a common sign language or radios (if the entrant gets out of the line of sight). It is also the attendant's responsibility to bring the entrant out using a harness and tripod if there are problems. This means the attendant cannot be performing traffic control, wandering off, or talking on the cell phone. The person below trusts the attendant with his or her life. No one should ever follow a collapsed fellow worker into a confined space. Often, there has been a rapid change of atmosphere that has caused the worker to collapse; when another worker enters the same atmosphere, he or she has just removed the victim's chance of rescue.

- Tripod and harness. The person entering a confined space should wear a harness. There are too many variations of harnesses to list; the best source for viewing them would be safety catalogs. An effective harness should allow the person being removed to be lifted straight up out of the confined space without forcing their head and neck to awkward angles. Harnesses should be comfortable to work in. One size does not fit all. Often, safety supply companies will allow customers to try on their stock before purchase to see if it works for the individuals. Tripods or lifting cranes are again varied and depend on the application. The tripod that fits over a manhole will not work to pull someone safely out of a deep clarifier or digester. There are even lifts designed to fit into mounts on bumpers of wastewater service vehicles for sewer lines. Maintenance of equipment and practice are essential to getting someone out of a dangerous situation safely.
- Procedures. There should be written procedures (typically in the form of a checklist) to be completed before entry to a confined space. When these procedures are followed and completed, the chances of injury decrease. A sample checklist will follow in this chapter. These procedures must be practiced. A life-or-death situation is no time to find out that something is not understood or that some procedures have been bypassed.

**PERMIT-REQUIRED VERSUS NON-PERMIT-REQUIRED CONFINED SPACES.** There is a distinct difference between these types of confined spaces. A permit-required confined space is one that meets the definition of a confined space and has one or more of the following characteristics:

- Contains or has the potential to contain a hazardous atmosphere,
- Contains a material that has the potential for engulfing an entrant,
- Has an internal configuration that might cause an entrant to be trapped or asphyxiated by inwardly converging walls or by a floor that slopes downward and tapers to a smaller cross section, and/or
- Contains any other recognized serious safety or health hazards.

It is the on-site supervisor's responsibility to determine whether the space is a permit-required confined space. To comply with OSHA Title 29 *CFR* Part 1910.146 (Permit Required Confined Spaces, 2005) (the rules that govern confined spaces), entry to a permit-required confined space requires that certain actions be performed by the employer:

- Follow the general requirements of paragraph (c) of the standard;
- Have a permit-required confined-pace program that complies with paragraph (d) of the standard;

- Follow the permit system requirements of paragraph (e) of the standard;
- Complete the entry permit as required in paragraph (f);
- Comply with employee training requirements of paragraph (g);
- Ensure that authorized entrants, attendants, and entry supervisors know their responsibilities as required in paragraphs (h), (i), and (j);
- Provide for rescue and emergency services as stated in paragraph (k); and
- Ensure that employees are allowed participation in these processes as required in paragraph (l) of the standard.

In general, employers must evaluate the workplace to determine if spaces are permit-required confined spaces (see Figure 5.5). If there are permit-required confined spaces in the workplace, the employer must inform exposed employees of the existence, location, and danger posed by the spaces. This can be accomplished by posting danger signs or by another equally effective means. If employees are not to enter and work in permit-required confined spaces, employers must take effective measures to prevent their employees from entering them. If employees are expected to enter these spaces, employers must comply with all requirements of the OSHA standard.

***Requirements of a Written Program.*** If the employer allows employee entry to a permit-required confined space, a written program must be developed and implemented. Among other things, the OSHA standard requires the employer's program to do the following:

- Identify and evaluate permit space hazards before allowing employee entry.
- Test conditions in the permit space before entry operations and monitor the space during entry.
- Perform, in the following sequence, appropriate testing for atmospheric hazards: oxygen, combustible gases or vapors, and toxic gases or vapors.
- Implement necessary measures to prevent unauthorized entry.
- Establish and implement the means, procedures, and practices (such as specifying acceptable entry conditions, isolating the permit space, providing barriers, verifying acceptable entry conditions, purging, making inert flushing, or ventilating the permit space) to eliminate or control hazards necessary for safe permit-space entry operations.
- Identify employee job duties.
- Provide, maintain, and require the use of PPE and any other equipment necessary for safe entry (e.g., testing monitoring, ventilation, communications, lighting equipment, barriers, shields, and ladders).

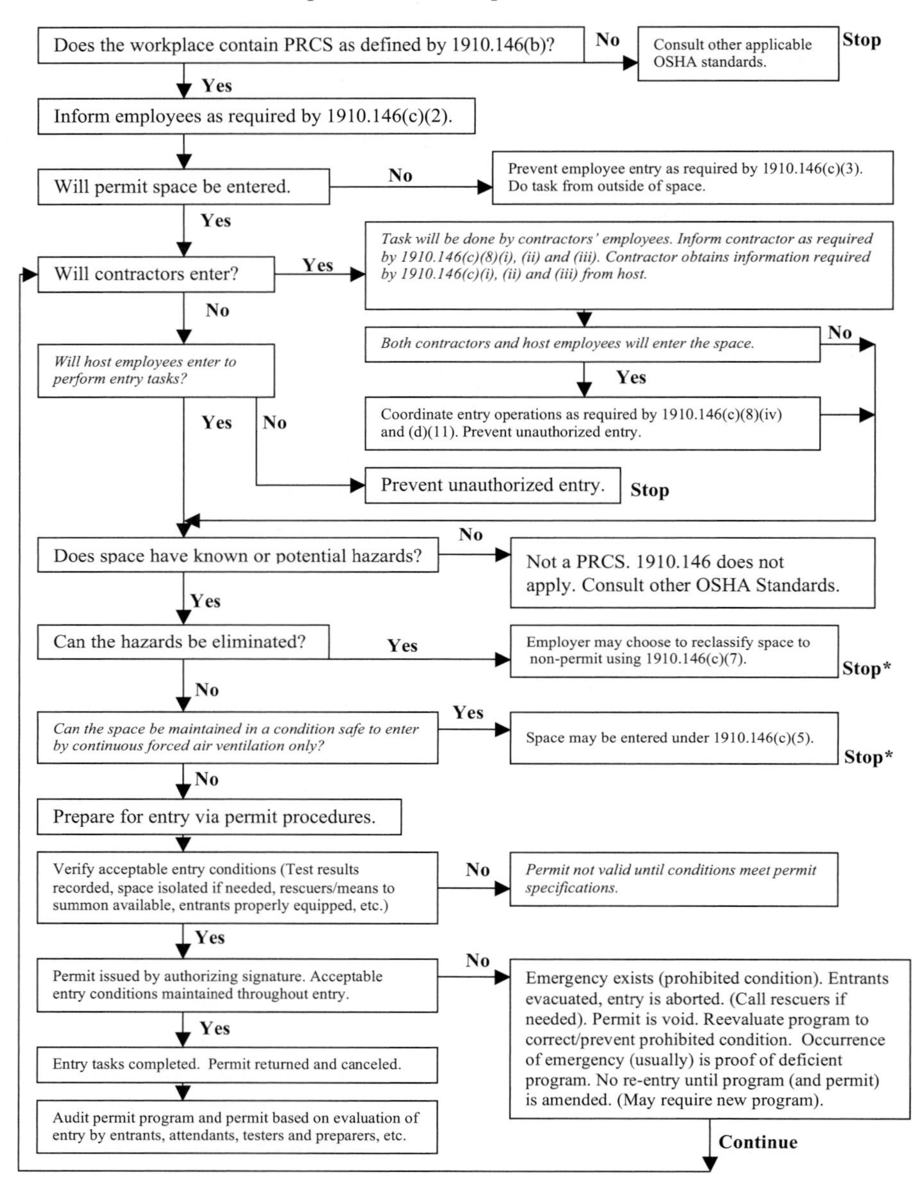


**FIGURE 5.5** Permit-required confined space (PRCS) decision flowchart (http://www.cehs.siu.edu/occupational/confined_space/flowchart.htm; accessed March 2006).

- Ensure that at least one attendant is stationed outside the permit space for the duration of entry operations.
- Coordinate entry operations when employees of more than one employer are to be working in the permit space.
- Implement appropriate procedures for summoning rescue and emergency services.
- Establish, in writing, and implement a system for the preparation, issuance, use, and cancellation of entry permits.
- Review established entry operations and annually revise the permit space entry program.
- When an attendant is required to monitor multiple spaces, implement the procedures to be followed during an emergency in one or more of the permit spaces being monitored.

If hazardous conditions are detected during entry, employees must immediately leave the space, and the employer must evaluate the space to determine the cause of the hazardous conditions.

When entry to permit spaces is prohibited, the employer must take effective measures to prevent unauthorized entry. Non-permit-required confined spaces must be reevaluated when there are changes in their use or configuration and, where appropriate, must be reclassified.

Contractors also must be informed of permit spaces and permit space entry requirements, any identified hazards, the employer's experience with the space (i.e., the knowledge of hazardous conditions), and precautions or procedures to be followed when in or near permit spaces.

When employees of more than one employer are conducting entry operations, the affected employers must coordinate entry operations to ensure that affected employees are appropriately protected from permit space hazards. Contractors also must be given any other pertinent information regarding hazards and operations in permit spaces and be debriefed at the conclusion of entry operations.

***Permit System.*** A permit, signed by the entry supervisor and verifying that pre-entry preparations have been completed and that the space is safe to enter, must be posted at entrances or otherwise made available to entrants before they enter a permit space.

The duration of entry permits must not exceed the time required to complete an assignment. Also, the entry supervisor must terminate entry and cancel permits when an assignment has been completed or when new conditions exist. New conditions must be noted on the canceled permit and used in revising the permit space program.

The standard also requires the employer to keep all canceled entry permits for at least one year.

***Entry Permits.*** Entry permits must include the following information (Figure 5.6):

- Test results.
- Tester's initials or signature.
- Name and signature of supervisor who authorizes entry.
- Name of permit space to be entered, authorized entrants, eligible attendants, and individuals authorized to be entry supervisors.
- Purpose of entry and known space hazards.
- Measures to be taken to isolate permit spaces and eliminate or control space hazards (i.e., locking out and tagging of equipment and procedures for purging, making inert, ventilation, and flushing permit spaces).
- Name and telephone numbers of rescue and emergency services.
- Date and authorized duration of entry.
- Acceptable entry conditions.
- Communication procedures and equipment to maintain contact during entry.
- Additional permits, such as for hot work, that have been issued to authorize work in the permit space.
- Special equipment and procedures, including PPE and alarm systems.
- Any other information needed to ensure employee safety.

***Training and Education.*** Before the initial work assignment begins, the employer must provide proper training for all workers who are required to work in permit spaces. Upon completing this training, employers must ensure that employees have acquired the understanding, knowledge, and skills necessary for the safe performance of their duties. Additional training is required when the following occur:

- Job duties change,
- There is a change in the permit space program or the permit space operation presents a new hazard, and
- An employee's job performance shows deficiencies.

Upon completion of training, employees must receive a certificate of training that includes the employee's name, signature or initials of trainer, and dates of training. The certification must be made available for inspection by employees and their authorized representatives. In addition, the employer also must ensure that employees are trained in their assigned duties.

## Confined Space Entry Permit

Date and Time Issued ____________________ Date and Time Expires ________________
Job Supervisor _______________________ Job Location _________________________
Equipment to be entered ________________________________________________________
Work to be performed __________________________________________________________

**Isolation and Lockout /Tagout**

•Pumps /Lines Blinded, Disconnected or Blocked ________Yes _______ No _______ N/A
•All Energy Sources Locked out/ Tagged out ________ Yes _______ No ______N/A

Ventilation__________ Mechanical __________ Natural __________ N/A _________

**Monitoring/Atmosphere Testing**

•Oxygen ____________________ 19.5% to 23.5%
•Flammable/Combustible ______________ % LEL
•Toxics - Hydrogen Sulfide _________ PPM
- Sulfur Dioxide____________ PPM
- Chlorine ________________ PPM
- Other(s) ________________

Meter Calibrated __________Yes ___________No __________ Date _____________ SN

____________________________________ __________

**Tester Signature** **Time**

Continuous Monitoring ____________ Yes ________ No_________N/A
Communication Procedure ___________________________________
Rescue Procedure ___________________________________
Emergency Phone Number ________________________________________________
Entry Watch/Standby Person ___________________________________

**Equipment /PPE**

•Safety Harness and Lifeline ________ Yes ________ No ________ N/A
•Hoisting/Retrieval Equipment ________ ________ Yes ________ No N/A
•Lighting (12 volt or GFI) ________ ________ Yes ________ No N/A
•Protective Clothing _______________________________________________
•Respiratory Protection _______________________________________________

**Period Monitoring/Atmosphere Testing**

| Oxygen | Flammable | Toxic(s) | | Initials |
|---|---|---|---|---|
| Time ______ | ______ | ______ | ______ | ______ |
| Time ______ | ______ | ______ | ______ | ______ |
| Time ______ | ______ | ______ | ______ | ______ |
| Time ______ | ______ | ______ | ______ | ______ |
| Time ______ | ______ | ______ | ______ | ______ |

Permit Approved/Issued By: ___________________________________

(a)

**FIGURE 5.6** (a) Sample permit and (b) entrants' log.

**Entrants Log**

| Name | In | Out | In | Out | In | Out |
|---|---|---|---|---|---|---|
| | | | | | | |
| | | | | | | |
| | | | | | | |
| | | | | | | |
| | | | | | | |
| | | | | | | |
| | | | | | | |
| | | | | | | |
| | | | | | | |
| | | | | | | |
| | | | | | | |
| | | | | | | |
| | | | | | | |

(a)

FIGURE 5.6 (a) Sample permit and (b) entrants' log.

***Authorized Entrant's Duties.*** The following are duties of the authorized entrant:

- Know space hazards, including information on the mode of exposure (e.g., inhalation or dermal absorption), signs or symptoms, and consequences of the exposure;
- Use appropriate PPE properly;
- As necessary, maintain communication with attendants to enable the attendant to monitor the entrant's status and to alert the entrant to evacuate;
- Exit from permit space as soon as possible when ordered by an authorized person, when the entrant recognizes that the warning signs or symptoms of exposure exists, when a prohibited condition exist, or when an automatic alarm is activated; and
- Alert the attendant when a prohibited condition exists or when warning signs or symptoms of exposure exist.

***Attendant's Duties.*** The following are duties of the attendant:

- Remain outside permit space during entry operations unless relieved by another authorized attendant.

- Perform non-entry rescues when specified by employer's rescue procedures.
- Know existing and potential hazards, including information on the mode of exposure, signs, or symptoms; consequences of the exposure; and their physiological effects.
- Maintain communication with and keep an accurate account of those workers entering the permit-required space.
- Order evacuation of the permit space when a prohibited condition exists, when a worker shows signs of physiological effects of hazard exposure, when an emergency outside the confined space exists, and when the attendant cannot effectively and safety perform required duties.
- Summon rescue and other services during an emergency.
- Ensure that unauthorized persons stay away from permit spaces or exit immediately if they have entered the permit space.
- Inform authorized entrants and entry supervisor of entry by unauthorized persons.
- Perform no other duties that interfere with the attendant's primary duties.

***Entry Supervisor's Duties.*** The following are duties of the entry supervisor:

- Know space hazards, including information on the mode of exposure, signs, or symptoms and consequences of exposure;
- Verify emergency plans and specified entry conditions, such as permits, tests, procedures, and equipment before allowing entry;
- Terminate entry and cancel permits when entry operations are completed or if a new condition exists;
- Take appropriate measures to remove unauthorized entrants; and
- Ensure that entry operations remain consistent with the entry permit and that acceptable entry conditions are maintained.

***Emergencies.*** The standard (OSHA Title 29 *CFR* Part 1910.146) requires the employer to ensure that rescue service personnel are provided with and trained in the proper use of personal protective and rescue equipment. All rescuers must be trained in first aid and cardiopulmonary resuscitation (CPR) and, at a minimum, one rescue team member must be currently certified in first aid and CPR. The employer also must ensure that practice rescue exercises are performed yearly and that rescue service personnel are provided access to permit spaces so that they can practice rescue operations. Rescuers also must be informed of the hazards of the permit space.

Also, where appropriate, authorized entrants who enter a permit space must wear a chest or full-body harness with a retrieval line attached to the center of their backs near shoulder level or above their heads. Also, the employer must ensure that the other end of the retrieval line is attached to a mechanical device or to a fixed point outside the permit space. A mechanical device must be available to retrieve personnel from vertical-type permit spaces more than 1.5 m (5 ft) deep.

In addition, if an injured entrant is exposed to a substance for which a MSDS or other similar written information is required to be kept at the worksite, that MSDS or other written information must be made available to the medical facility treating the exposed entrant.

This all may seem rather daunting, particularly in small systems, where there are one or two operators only. In situations like this, it is good practice to know the employees of surrounding towns and learn how to cooperate with them. With proper planning, two or three small communities may pool employees and do work that requires confined space entrance as a team, thereby staying within best safety practices, even in towns that have too few employees to be covered by OSHA regulations. This lowers any liability that may be faced by the employer.

***Confined Space Regulations for Collection System Workers.*** Sewer entry differs in the following three vital respects from other permit entries: (1) there rarely exists any way to completely isolate the space (a section of a continuous system) to be entered; (2) because isolation is not complete, the atmosphere may suddenly and unpredictably become lethally hazardous (toxic, flammable, or explosive) from causes beyond the control of the entrant or employer; and (3) experienced sewer workers are especially knowledgeable in entry and work in their permit spaces because of their frequent entries. Unlike other occupations where permit space entry is a rare and exceptional event, sewer workers' typical work environment is a permit space. The following is a synopsis of 1910.146 Appendix E for sewer system entry (Permit Required Confined Spaces, 2005).

- Adherence to procedure. The employer should designate as entrants only employees who are thoroughly trained in the employer's sewer entry procedures and who demonstrate that they follow these entry procedures exactly as prescribed when performing sewer entries.
- Atmospheric monitoring. Entrants should be trained in the use of and be equipped with atmospheric monitoring equipment that sounds an audible alarm, in addition to its visual readout, whenever one of the following conditions are encountered: oxygen concentration less than 19.5%; flammable gas or vapor at 10% or more of the lower flammable limit; or hydrogen sulfide or carbon monoxide at or above 10 or 35 mg/L (10 or 35 ppm), respectively, measured as an 8-hour,

time-weighted average. Atmospheric monitoring equipment should be calibrated according to the manufacturer's instructions

- Although OSHA considers the information and guidance provided above to be appropriate and useful in most sewer entry situations, the agency emphasizes that each employer must consider the unique circumstances, including the predictability of the atmosphere, of the sewer permit spaces in the employer's workplace when preparing for entry. Only the employer can decide, based on his or her knowledge of and experience with permit spaces in sewer systems, what the best type of testing instrument may be for any specific entry operation.
- The selected testing instrument should be carried and used by the entrant in sewer line work to monitor the atmosphere in the entrant's environment, and in advance of the entrant's direction of movement, to warn the entrant of any deterioration in atmospheric conditions. If several entrants are working together in the same immediate location, one instrument, used by the lead entrant, is acceptable.
- Surge flow and flooding. Sewer crews should develop and maintain liaison, to the extent possible, with the local weather bureau and fire and emergency services in their area so that sewer work may be delayed or interrupted and entrants withdrawn when sewer lines might be suddenly flooded by rain or fire suppression activities or when flammable or other hazardous materials are released into sewers during emergencies by industrial or transportation accidents.
- Special equipment. Entry to large bore sewers may require the use of special equipment. Such equipment might include such items as atmosphere-monitoring devices with automatic audible alarms, escape self-contained breathing apparatus with at least a 10-minute air supply (or other NIOSH approved self-rescuer), and waterproof flashlights, and may also include boats and rafts, radios, and rope stand-offs for pulling around bends and corners as needed.

## TRENCHING AND EXCAVATION SAFETY

This issue is one that is faced by collection system workers primarily and not by plant workers. However, because more than 60 deaths were caused in 2004 by trench collapses, it is worth mentioning. A trench or excavation is any digging that removes soil (a trench, by definition, is any excavation with a bottom less than 4.6 m [15 ft] wide.) Remember that soil is very heavy; 0.76 $m^3$ (27 cu ft) can weigh over 1400 kg (1.5 tons). As soil is removed from a ditch, the weight of the soil at the edge of the ditch presses against the ditch and can cause it to collapse. Any exposed trench face over 1.5 m (5 ft) in depth needs to be either shored or sloped back to prevent cave-in. Materials used for trench boxes, sheeting, sheet piling, bracing, shoring, and underpinning should be in good condition and should be installed so that they provide support that is effective to the bottom of the trench. Timber must be sound and free from large or loose knots. Vertical planks in the bracing system should be extended to an elevation no less than 0.3 m (1 ft) above the top of the trench face. Inspections of cribbing, bracing, shoring, or slopes in the surrounding area must be performed at least daily. Be mindful that changing weather conditions may drastically change the soil's stability. It is the supervisor's responsibility to ensure that all workers on the crew are provided with a safe work area.

Rainfall and accumulating water present another hazard to workers. Employees must exit trenches during rainstorms. The trench area should be reinspected after a rain event. Protective measures, such as surface ditches to limit runoff, can be useful tools. When there is standing water in the trench, it should be removed using pumps. These need to be monitored by a competent person. Safety harnesses and lifelines should be used (in accordance with 29 *CFR* 1926.104 [Safety Belts, Lifelines, and Lanyards, 2005]).

Another hazard that seems obvious and has not yet been mentioned is falling into the excavation. Workers must be careful around an excavation, making sure not to stand too close the edge. It is important to be aware of slipping or tripping hazards.

Pedestrians must be kept away from excavations and trenches so that they do not get injured by falling or being hit by equipment. All excavations must be protected to prevent pedestrian traffic near the excavation site. Examples of minimally acceptable protection may include cyclone fencing, complete barricading, or OSHA standard railings around the entire site. Protection should be placed well away from the excavation or trench, if at all possible. Plastic barricade tape at the edge of an excavation is *not* acceptable site protection. Accident prevention warning signs or flashing barricades should also be placed adjacent to excavations.

When employees are required to be in trenches that are 1.2 m (4 ft) or more in depth, an adequate means of exit, such as a ladder or steps, must be provided and located so that no more than 7.6 m (25 ft) of lateral travel is required for a person to reach the exit structure. The trench should be braced and shored during excavation and before personnel are allowed entry. Cross braces and trench jacks should be secured in true horizontal positions and spaced vertically to prevent trench wall material from sliding, falling, or otherwise moving into the trench. Portable trench boxes (also called *sliding trench shields*) or safety cages may be used to protect employees instead of shoring or bracing. When in use, these devices must be designed, constructed, and maintained in a manner that will provide at least as much protection as shoring or bracing, and extended to a height of no less than 150 mm (6 in.) above the vertical face of the trench.

During the backfill operation, it is important to backfill and remove trench supports together, beginning at the bottom of the trench. Jacks or braces should be released slowly and, in unstable soil materials, ropes should be used to pull them from above after employees have left the trench.

It is up to the supervisor or inspection engineer (legally, the competent person) to determine the soil classification. This is important because the soil classification determines the slope necessary for the trench or excavation. All previously disturbed soil is automatically classified as either Class B or Class C soil. If an excavation occurred to either replace or repair pipe, it is typically classified as Class C soil.

There are basic manual tests to be run by a competent person to determine the soil class. If a person pushes his or her thumb into the edge of the trench and it goes no further than the nail, it is probably Class B soil. If cracks are seen in the soil, it should be treated as class C soil. Bulging, sloughing, or water seepage, and excessive vibration from area traffic make the soil Class C also. If dry soils crumble freely, they are probably Class C soils; however, if they break into clumps, they have enough clay content to be Class B soils. If a moist soil sample can be rolled into a 0.318-mm (0.125-in.) thread approximately 50 mm (2 in.) long and it breaks, then the soils is probably Class B.

The soils classification defines how slopes at the edges of trenches and excavations may be made. If the soil is Class B, then a 1:1 slope on the edge of the trench or excavation is allowed. Class B soils may also be benched every 1.2 m (4 ft). A Class C soil cannot be benched and must have a 1:1.5 slope (34-degree angle from the bottom). This means much more excavation for safe working conditions. It is important to keep in mind that spoils must be kept back at least 0.6 m (2 ft) from the edge of the excavation.

One other factor not yet mentioned is the necessity of locating underground utilities before digging. It is common to see underground utilities torn up because they were either not located properly, or disregarded. In one case, a backhoe hit both an underground propane line and electrical line. What saved the operator was that the propane was so concentrated that it was above the UEL when the electrical lines were sparking. Locating utilities typically is included as part of the process of obtaining a permit for the excavation or trench. It is important to always check to see if a permit is needed, because some permits require specific safety requirements.

The atmosphere inside a trench should be routinely monitored for oxygen and explosive gases. Using a four-gas monitor as if the trench were a confined space is a good work practice. There are times a trench can become filled with exhaust fumes without the worker in the trench really noticing it because the change is so gradual.

## LADDER SAFETY

Closely related to excavation safety is ladder safety. Typically, there are not that many uses for ladders in the wastewater field, except as ingress and egress from trenches and excavations. Other uses are getting into storage areas, confined spaces, aeration basins, clarifiers, and digesters for maintenance and repair. Ladders leading to storage tanks may also be a common use. Falls are one of the most frequent causes of compensation injuries. The following is a list of safe practices to follow when using ladders; some of these standards are found in the OSHA Construction Standards (29 *CFR* 1926):

- Ensure that all ladders are equipped with approved safety shoes.
- Place the ladder so that the horizontal distance from its foot to the support it rests against equals one-fourth the length of the ladder.
- Do not allow working or standing upon the top two rungs of the ladder.
- **Never** splice two short ladders together.
- Never place a ladder against an unsafe support.
- Ensure that the ladder feet rest upon stable support. If the ladder feet are placed upon a platform, make sure that the ladder cannot slide or move off of the platform.
- Do not use ladders as scaffold platforms.
- Whenever possible, tie off the top of a straight ladder to a firm support.
- Ensure that a person holds a stepladder stable whenever someone is working at a height of 3 m (10 ft) or more.
- Make sure that any stepladder being use has its legs fully spread.
- When working near electrical lines, always use a nonconductive ladder.
- Avoid using stepladders as straight ladders.
- Extend stepladders at least 0.9 m (3 ft) above the top of the work platform to allow easy and safe ingress and egress.
- Routinely inspect ladders to make sure that they are in proper working order with no cracked or broken supports or steps.
- There should be a sticker on the ladder stating the weight limit on the ladder.
- Extension ladder sections must overlap. There should be a 0.9-m (3-ft) overlap in ladders up to 10 m (32 ft) long, 1.2 m (4 ft) for ladders 10 to 14 m (32 to 48 ft), and 1.5 m (5 ft) in ladders 14 to 18 m (48 to 60 ft).
- Always have three-point contact with the ladder (i.e., one hand and two feet).
- Never climb a ladder carrying tools. They should be put in a tool belt or lifted using a rope and bucket once the worker is at the desired height.
- If possible, use a fall protection system fastened to a secure place.
- Do not use a ladder on a windy day.
- Never leave a set-up ladder unattended.

## LIFTING

Far too many people have back problems as they get older. Many back problems can be prevented by proper lifting. The following are guidelines for lifting that may prevent a lifetime of nagging pain caused by improper lifting:

- Be careful when lifting. The size, shape, weight, and texture can actually change how much can be lifted. A person should lift what can be handled comfortably and get help if needed.
- Look at the surface to be lifted and see if the are any metal of wood slivers protruding from the edges. Make sure the surface is not too rough or too slippery.
- Use solid footing and keep the feet wide enough to provide stability and balance. A sudden jerk from lifting something that is out of balance or unstable can wrench the back and cause a lifetime of problems.
- Move as close to the load as possible and bend your legs at approximately 90 degrees at the knee. Keep the back as straight as possible. Grip the object firmly. Lift by straightening the legs. When lowering the load, reverse the procedure. Do not lift by bending the back over the object and trying to tilt the back into an upright position. This will stress the lower back muscles and discs.
- Do not carry loads that block vision.
- If a worker needs help to lift heavy or awkward objects, he or she should be sure to be coordinated with the other person or persons. Miscommunication when lifting heavy objects as a group can place all the stress of the lift on one person, putting that person's back at risk.
- Keep hands, fingers, and feet away from any points that could expose them to pinching or crushing, especially when setting the object down or going through doorways.
- Make sure the path through which a person will be walking will be clear of oil, grease, and water that may cause a slipping hazard.
- Make sure the package itself is clean and not covered in grease or oil or is wet so that it is less prone to slip out of the hands.

## LABORATORY SAFETY

There are many sources of injury and bacteriological contamination in a laboratory. While not all plants have much more than a rudimentary laboratory facility, the following are sound basic safety practices:

- Discard all chipped or cracked glassware. All glassware for disposal should be placed in appropriate containers.

- When using volatile chemicals that may cause an inhalation or atmospheric hazard, only use these under a ventilated hood.
- Store solvents or flammable liquids only in explosion proof cans or in a flammable storage locker. These are available from many suppliers of safety supplies or general contractor supply stores.
- When using acids that react violently with some organic materials (ammonia and nitric, acetic, and perchloric acids), take particular care to avoid possible fire or explosion.
- **Do not handle chemicals with bare hands.** Use gloves, tongs, spatulas, or whatever laboratory equipment may be best suited for the material and application being used.
- Provide an emergency eyewash station and shower in the laboratory. If in the field, make sure that if there is not immediate access to these items, there is an emergency eyewash bottle or two handy.
- Use suction bulbs on all pipettes. **Do not use the mouth.** It is much too easy for a worker to become distracted and ingest whatever substance he or she is trying to pipette, regardless of how practiced the worker may be.
- Wear a rubber apron when working with corrosive chemicals.
- Wear the appropriate face shield or chemical goggles when working with chemicals. Acids or bases in the eyes can cause a complete loss of vision in that eye. Safety glasses should be worn at all times in the laboratory.
- Clearly label all chemical containers, especially if transferring chemicals to smaller, more portable containers for use.
- Wear gloves when making rubber-to-glass connections.
- Properly ventilate to remove fumes and dust.
- Prohibit smoking, drinking, and eating in the laboratory.
- Workers must know where the fire extinguisher is and if it is usable. Routine inspections are necessary.
- Use tongs or insulated gloves to remove samples from hot places, such as hot plates, ovens, or muffle furnaces. Remember that ceramic wear remains hot to the touch for a very long time.
- Adequately shield centrifuges and caloric bombs.
- Make sure electrical outlets are properly grounded. Then make sure all equipment is plugged in properly. The male plug should have the appropriate grounding plug. If it has been removed, have it repaired.
- Thoroughly wash hands before eating, drinking, or smoking.
- Ensure that all laboratory faucets have either air gaps or vacuum breakers to prevent back-siphonage of toxic material to the water system. While this may

not deal particularly with laboratory safety, whenever a faucet is used in an application such as a polymer feeder or chemical feeder, it too should have an appropriate backflow preventer.

- Provide a spill kit for cleanup in case of spills.

## FIRE SAFETY

It may seem rather odd to cover fire safety in a wastewater treatment manual, particularly if one is working in the plains and the plant consists mainly of lagoons. There are many places in modern plants where fires can start and spread, and, with the chemicals that can often be found in activated sludge or trickling filter plants, there is a basic need to know some fire safety. Important questions an operator should ask include the following:

- Where are the extinguishers?
- What other fire-retardant systems are in place?
- Have the extinguishers been routinely inspected and are they operational? Old water-type extinguishers, even if improperly maintained, can become shrapnel if used when weakened.
- Has the operator been trained to use them?

Before examining extinguisher types and their uses, it is important to examine the basics of fire or combustibles. The most popular description of combustion, known as the fire triangle, is shown in Figure 5.7.

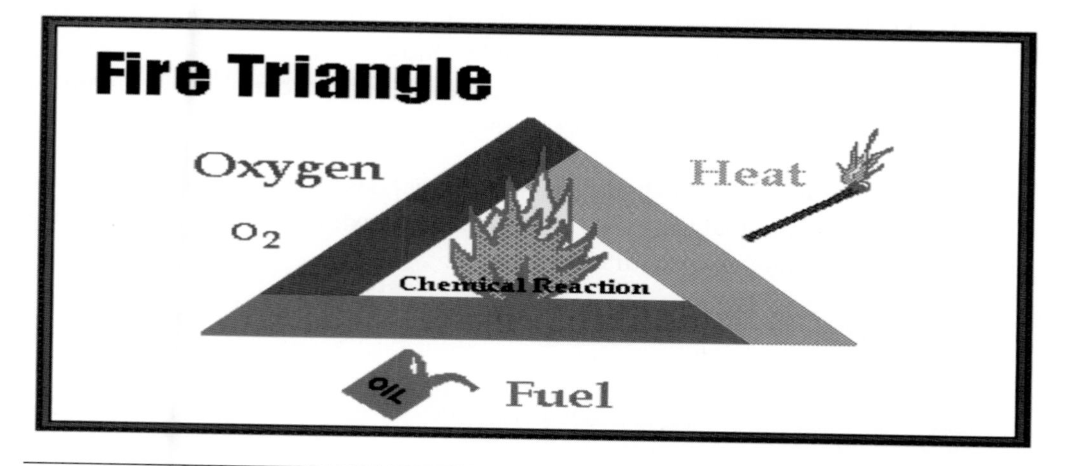


FIGURE 5.7 Fire triangle (http://www.pp.okstate.edu/ehs/MODULES/exting/Triangle.htm).

For an exothermic chemical reaction to occur (and fire is basically and exothermic reaction), there must be oxygen, fuel, and heat present. If any of the three components symbolized by the legs of the triangle (Figure 5.7) are taken away, then the fire will be extinguished. This is how fire extinguishers work. They take away one of the sides of the triangle. Dry chemical extinguishers take away the oxygen, smothering the combustion. Halon and carbon dioxide hopefully lower the temperature to below combustion parameters.

There are four basic designs of fire extinguishers. They are halon, carbon dioxide, water, and dry chemical extinguishers.

- Halon. These contain a gas that interrupts the chemical reaction that takes place when fuels burn. These types of extinguishers are often used to protect valuable electrical equipment because they leave no residue to clean up. Halon extinguishers have a limited range, typically 1.2 to 1.8 m (4 to 6 ft). The initial application of halon should be made at the base of the fire, even after the flames have been extinguished.
- Carbon dioxide. These are most effective on Class B and C (liquids and electrical) fires. Because the gas disperses quickly, these extinguishers are only effective from 0.9 to 2.4 m (3 to 8 ft). The carbon dioxide is stored as a compressed liquid in the extinguisher; as it expands, it cools the surrounding air. The cooling will often cause ice to form around the "horn", where the gas is expelled from the extinguisher. Because the fire could reignite, the agent should continue to be applied even after the fire appears to be out.
- Water. These extinguishers contain water and compressed gas and should only be used on Class A (ordinary combustibles) fires.
- Dry chemicals. These are typically rated for multipurpose use. They contain an extinguishing agent and use a compressed, nonflammable gas as a propellant.

The description of water-type extinguishers mentions Class A fires. Extinguishers are rated by the class of fire that they have the capability of extinguishing if they are used properly. The following are the four classes of fire.

- Class A, ordinary combustibles;
- Class B, combustible liquids and vapors;
- Class C, electrical fires; and
- Class D, combustible metals.

All extinguishers have the types of fire that they are capable of extinguishing posted on them. It is important to know what types of extinguishers are present before an emergency. An emergency is not the time to find out that only a Class D extinguisher is available and the fuel is gasoline. The person who has the responsibility to order or prepare the site for safety should assess the possible dangers and have the proper extinguishers in place before an emergency to the extent possible.

While nothing can replace hands-on training using an extinguisher, there is a basic mnemonic rule for using an extinguisher. That memory device is PASS. It stands for the following:

- *Pull* the safety pin (so the trigger can be pulled);
- *Aim* at the base of the fire from approximately 2.4 m (8 ft) away;
- *Squeeze* the trigger, aiming the extinguisher at the base of the fire; and
- *Sweep* the nozzle back and forth at the base of the fire.

While all this seems easy on paper, when there is an emergency, adrenaline starts flowing and people sometimes try to do more than they should. It is important to consider the situation. If there is a fire near the chlorine room, for example, a worker should not try to put it out alone if it looks like it is going to get out of control. The local fire department should be called for help. While one person may be able to put out the flames if it is small, if it gets out of control, that person is in danger and the surrounding populace is also in danger. That one person may be the only chance to warn them and get proper emergency equipment on the scene.

## MACHINE SAFETY

There are many different types of machinery that can be found in wastewater treatment plants, varying from pumps to belt filter presses. One thing that they have in common is moving parts. It is amazing how many moving parts there are and how damaging they can be to the human body.

The external moving arms that some primary sludge pumps use are mechanically related to the old piston rods and wrist pins of old-fashioned steamships. In past days, it was the job of an oiler to manually apply oil to the wrist pins and connecting rods (the connecting rods where often twice as tall as a man) amidst the heat of the steam engine. It was a job where one false move would cost the oiler his arm and sometimes his life. The working standards of the late 19th century should not apply to present-day workers.

Most pieces of machinery have guards that are designed to be in place over the moving parts. This is an engineering design feature that is meant to keep personnel from hazards. The guard encloses the danger. When a person comes across machinery that he or she knows can be dangerous, it is important to make sure that the guards are there. This conscious pattern of thought should extend to all tools used. Table saws and circular saws have safeties to prevent career-ending injuries to hands. **Use them.** If a worker encounters machinery that may need safeties beyond the original design, he or she should make sure that the proper chain of command is notified and action is taken. Belts and pulleys, chains and sprockets, rotating coupling, and racks and gears are among some of the common items that should have guards over them.

This is common sense. But what happens when these guards need to be taken off for maintenance? The correct answer is not for employees to work at their own risk. Most people are aware of lockout/tagout programs. These are normally accepted as standards when electrical work is being performed. There is energy, called *latent energy*, that may be stored in hydraulic systems or other forms of mechanical energy storage systems. This can be released if improperly locked out, with possibly dire consequences.

## LOCKOUT/TAGOUT PROGRAMS

In simple terms, a lockout/tagout program is a formalized program that has routine steps in it that will not allow anyone to turn on machinery that is being repaired or maintained. The main panel shutoff will have a lock placed on it by the person doing the work. In a place like a large wastewater plant, there could be electrical crews and maintenance and repair crews working on the same item at once. If, for example, there needed to be work on a clarifier drive, the supervisor may have the oil and flights worked on at the same time.

For this example, assume that there are three crews, each doing their particular jobs. The main shutoff for the clarifier would have a device that would fit through the hole in the main power shutoff, instead of the basic lock used if it were a one-man crew. Then locks from all three crews and the supervisor would fit into the holes of this device. There would be a tag stating "Do Not Use" if this is not printed on the device. When each crew has completed their project, their lock is removed from the device. When the job is complete, the supervisor would take the lock off of the device, remove the tag, and allow the device to be turned on. As a general rule, **never cut a lock off of a lockout device or a piece of equipment**. This could lead to serious injury. If someone were to quit without returning keys, suffer a long-term injury, or some other incident occurred that prevented that person from coming to work, then the decision to do something like cutting a lock becomes an administrative decision, where all consequences should be thoroughly considered. Illustrations of some common lockout devices are shown in Figure 5.8.

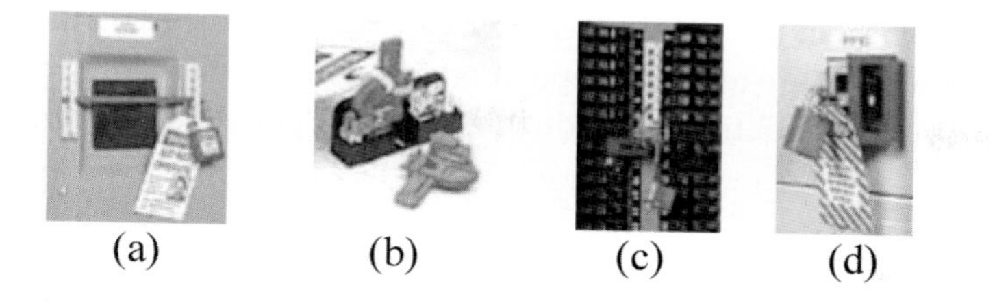

FIGURE 5.8 Illustrations of (a) 480v lockout device (courtesy of Lab Safety Supply), (b) circuit breaker lockout device (courtesy of Lab Safety Supply), (c) circuit breaker panel lockout device (courtesy of Lab Safety Supply), and (d) electrical panel lockout device (image courtesy of Brady® Worldwide Inc.) (http://www.labsafety.com/store/dept.asp?dept_id=5442).

In summary, a lockout/tagout program that meets OSHA specifications covers employee training in the three following aspects:

(1) The employer's energy control program,
(2) Elements of energy control programs that are relevant to the employee's duties, and
(3) Various requirements of the OSHA lockout/tagout standard.

The following are mandates that an employer must follow to protect employees:

- Develop, implement, and enforce an energy control program.
- Use lockout devices for equipment that can be locked out. Tagout devices can be substituted in lieu of lockout devices only if tagout devices can provide the same degree of protection.
- Ensure that new or overhauled devices are capable of being locked out
- If existing equipment cannot be locked out, develop a tagout program.
- Develop, document, implement, and enforce energy control procedures.
- Use only lockout/tagout devices authorized for the particular application and ensure that they are durable, standardized, and substantial.
- Ensure that lockout/tagout devices identify the individual users.
- Establish a policy that permits only the employee who applied the device to remove it (see 29 *CFR* 1910.147 (e)(3) [Occupational Safety and Health Hazards, 2005] for exceptions).
- Inspect and update energy control procedures annually.
- Provide effective training for all employees covered by the standard.

- Comply with the additional energy control procedures in OSHA standards when machines are tested or repositioned, when outside contractors work at the site, in lockout situations, and during shift and personnel changes.

## TRAFFIC SAFETY

Traffic safety can fall into the following two basic categories: (1) general driving practices and (2) signage and traffic patterns that need to be maintained when people are out working in streets.

**DEFENSIVE DRIVING.** General driving practices can be casually summed up as driving defensively. It is important to make sure vehicles are in working condition. All blinkers and safety lights, brakes, and wipers should work and windshield washer fluid should be full (especially in mountainous terrain). Seldom is getting to a job a few minutes earlier a life-or-death situation, so it is important to take time and be careful. The term "road rage" has become increasingly popular in the past few years. There are many reasons for this. The electronic age has brought a faster-paced lifestyle, with almost constant access and pressure. There often is little time to relax. Traffic has gotten worse. Deadlines are constantly looming. The road is often seen as place to make up time, and aggression is taken out anonymously. These conditions can lead to problems on the road. It is important to remember that there is no need to be aggressive, especially in a company vehicle. It is important to just get the job done and be careful. In the wastewater treatment industry, if a person loses his or her license, he or she often loses the job also.

**DRINKING AND DRIVING.** The use of illicit substances and driving do not mix. Many places of employment have drug testing before employment starts and can have random drug testing at any time. Anyone who has DOT licensing can be subject to random drug testing and, if there is an accident, the driver must undergo mandatory drug testing. Legal blood alcohol levels continue to drop and, in many states, the level is currently 0.08. It is important to remember that all drivers have different metabolisms and susceptibility to alcohol and take different medications. Many of these can cause unwanted side effects and lack of coordination. Drinking and driving remains one of the leading causes of death in the United States.

**CELL PHONE USAGE.** Cell phones have been a blessing for making communication easy. For driving, they have become a problem. Many studies have linked cell phone use when driving to accidents (Helperin, http://www.edmunds.com/ownership/

safety/articles/43812/article.html [accessed January 2006]). Originally, it was thought that using the hands to dial resulted in worse driving; however, the driving actually worsens because the driver is distracted. The driver's concentration is no longer on guiding the automobile down the road, but on what the driver is doing or saying. This endangers others. If it becomes necessary to talk while driving, please try to pull over. Also, some areas have actually made it illegal to use a phone and drive at the same time.

**SIGNAGE AND TRAFFIC PATTERNS.** One of the more dangerous things that wastewater workers can do is to work outside the plant performing collection system maintenance and repairs. The best defense and protection for wastewater treatment workers would be totally cordoning off the work area or worksite behind impenetrable concrete barriers. Often, that is impossible. The next best thing that can be done is protecting workers with a traffic control zone. The workers will need to set up a means of guiding traffic through the work zone so that they will not be harmed. This means proper placement of signs, cones, and possibly a flag person to guide the traffic. Before working in the roadway, remember to contact the local police or district to determine whether traffic control is necessary. In some areas, a traffic control plan must be submitted before any work can be started.

Typically, when working in a district where there is only one worker, the roads generally are not that heavily trafficked, so there will not be a need for more than the truck, cones, and basic "Road Work Ahead" or "Detour" signs (remember, the truck can be a safety device and just by its bulk can be used behind cones as a mobile barrier). Following are the recommended distances for simple signs:

- A "Road Work Ahead" sign is typically placed 46 m (150 ft) minimum in front of any other signage.
- A "Lane Closed" sign is typically placed 30 m (100 ft) minimum in front of the work zone.
- A "High Level Warning Device" is placed at the beginning of the work zone. This can be a multiple flag type of device or a flashing sign. It is important to check local regulations to make that the proper type of device is being used.
- Cones are placed to direct traffic around the work zone.

There are some basic rules that need to be followed. All flag persons should wear orange vests or shirts. Recently, some areas began allowing a fluorescent yellow-green color also. Reflective belts, suspenders, or edges on the normal orange vest are necessary for night work. Proper use of signs that say "Slow" on one side and "Stop" on the other are necessary. Flag persons should be positioned at least 30 m (100 ft) ahead of the work zone. The "Flagmen Ahead" sign should be approximately 152 m (500 ft) ahead of them. Flag persons are required where

(1) Workers or equipment will intermittently block traffic;
(2) One lane must be used for both directions of traffic (a flag person is required for both directions); or
(3) The safety of the public and/or workers requires it.

Local regulations should be checked for proper placement of signs. They may differ from area to area. Permits may be required at specific sites. Table 5.3 shows suggested distances between common warning devices.

The following are a few basic things to remember when working inside a cordoned-off area:

- Always use more than the minimum amount of warning signage.
- Anticipate unpredictable acts of drivers before setting up the traffic plan and then actualizing it. Cars and trucks are much larger than people, and, if poor driving is anticipated, it is much easier to protect workers from it, preventing accidents and injury.
- Keep all tools and materials behind the protected zones. If something were to fall into the flow of traffic, do not react and go after it without first looking and making sure that the area may be entered safely.
- The pedestrian, motorist, equipment, and work force need to be considered as traffic control plans are made.
- Finally, it is a good idea to review the safety plans with everyone on the work-site before work starts.

**TABLE 5.3** Suggested distances between common warning devices.

| | Speed (km/h [mph]) | Distance (m [ft]) |
|---|---|---|
| Traffic cones | 0 to 48 km/h (0 to 30 mph) | 3 to 6 m (10 to 20 ft) |
| | 48 to 64 km/h (30 to 40 mph) | 7.6 to 11 m (25 to 35 ft) |
| | 72 to 88 km/h (45 to 55 mph) | 12 to 15 m (40 to 50 ft) |
| High level warning devices | 0 to 40 km/h (0 to 25 mph) | 45 m (150 ft) before worksite |
| | 40 to 56 km/h (25 to 35 mph) | 76 m (250 ft) before worksite |
| | 56 to 72 km/h (35 to 45 mph) | 152 mm (500 ft) before worksite |
| | 72 to 88 km/h (45 to 55 mph) | 229 m (750 ft) before worksite |

# SAFETY CONSIDERATIONS IN PLANT DESIGN

When a new plant is designed or there are renovations made to an existing facility, possible safety hazards and design considerations must be taken into account by management to ensure a safe working environment. To encourage buy-in, management will invite the plant supervisor and operators to review the plans before the first piece of pipe is in the ground or before the concrete is poured. It is much easier and less expensive to change designs on paper than it is to change once the concrete has set. The following is a list of design considerations that should be considered before the construction process begins:

- Provide stairs instead of vertical ladders. If chemicals are to moved in that area, use a ramp instead of stairs.
- Provide nonskid floors and treads.
- Label all piping and color code it where possible.
- Maintain a minimum headroom of 2 m (7 ft).
- Provide guards on all accessible moving parts of machinery.
- Equip all stairs, openings, tanks, basins, ladder ways, and platforms with standard guard railings. If the railings and guards are at the base of an icy slope in mountainous terrain, make sure the guard is higher than minimum standard to prevent injury.
- Post appropriate warning signs in all hazardous areas.
- Provide adequate floor drains for washdown.
- Provide adequate lifting facilities.
- Isolate disinfection facilities from other buildings. Consider the use of alternative disinfection systems such as UV to lower the risk of using hazardous chemicals.
- If it is necessary to use gaseous chlorine and sulfur dioxide, provide leak detection systems and alarm systems.
- Provide combustible gas detectors, toxic gas detector, and oxygen-deficiency alarms and indicators where appropriate.
- Include SCBA in the plant equipment inventory. If the use of toxic chemicals can be lessened, then the maintenance costs and training may be eliminated from future operations budgets.
- Provide ship's ladders, stairs, or mechanical lifts for all installations deeper than 3.7 m (12 ft). A means of egress from aeration basin and aerobic digesters must be provided because the body loses buoyancy in aerated liquors.
- Provide adequate space for safe and efficient operation and maintenance of all equipment installed.
- Provide flush toilets, lavatories or wash fountains, hot and cold water, and showers.

- Provide adequate eating facilities with refrigerator, microwave, stove, and sink. This area should be out of the traffic pattern so that it may be maintained and adequately cleaned.
- Provide dressing rooms, benches, mirror, and metal lockers (two lockers is ideal so that work clothes may be kept separate from clean clothes).

## ACCIDENT REPORTING AND INVESTIGATION

It is mandatory that any injuries be reported within 48 hours of occurrence to management. This record will ensure that proper medical treatment is provided and workman's compensation and insurance claims can be filed and handled promptly. A worker who is injured will want fast action on claims. If the injury is serious enough, management is bound to notify the designated emergency contact. If there is the need for special treatment, such as treatment for allergic reaction to medications or diabetes, this should be noted on emergency contact forms so that accidents that could cause complications do not occur. Records of illness and accidents need to be kept so that individual workers and treatment processes can be monitored.

If accident reporting is to be effective, then follow-up to the initial accident report should be made. This may help management determine if there is a process that is too dangerous and should be redesigned. It also helps show if any worker is accident-prone and may need special help.

If management is serious about having a safe work facility, there is an appointed safety officer who will make detailed notes on each accident. The investigator's notes should include answers to the following questions:

- Who was present, who should have been present, and was someone out of place?
- Were any guards, kick plates, or other protective equipment out of place?
- Was access and egress open or blocked?
- What happened and what damage occurred (an accident can involve equipment damage and personal injury)?
- Is anyone hurt?
- Who was in charge at the time of the accident?

Diagrams of the scenario, videotape, and pictures may be helpful tools in investigating the incident.

Figures 5.9 to 5.11 are sample supervisor's report of accident forms and employee accident/incident report forms. Investigators may ask for recommendations to help prevent these accidents from repeating. Then, any changes that are necessary in process or equipment can be presented to management so that they may find a means

**Supervisor's Report of Accident**
**(Occupational Injury)**

INJURED EMPLOYEE EMPLOYEE # AREA

DATE OF ACCIDENT TIME AM/PM EMPLOYEE'S OCCUPATION

PLACE OF ACCIDENT

NATURE OF INJURY

NAME OF DOCTOR

NAME OF HOSPITAL

WITNESSES (NAME AND ADDRESSES)

**FIGURE 5.9** Supervisor's report of accident.

to pay for the necessary changes and prioritize changes so that they may happen in a decent work flow. If violations occurred in disregard of safety policies and procedures, these should be recorded so that appropriate disciplinary actions may be taken. If chemical spills occur, these need to be dealt with in a safe manner. A good report documents all that happened in a manner that is easily understood and suggests ways to prevent these incidents from happening again.

## OCCUPATIONAL SAFETY AND HEALTH ADMINISTRATION 300 LOGS

The OSHA 300 logs (Figures 5.12 to 5.14) are publicly posted records of accidents and lost work time that are mandated to be posted in the month of February where the labor force gathers. Typically, they are posted in either the lunch area or the time clock area. Please regularly check the log format on the OSHA Web site for changes, as they do change. Make sure the proper form is used and sent to the proper regulatory address and file.

## SAFETY PROGRAMS AND ADMINISTRATION

There has been much information presented in this chapter concerning safety, including mandatory safety items and procedures. The question then is how this becomes something useful and cohesive. Most of these are laws from the U.S. Congress and need to be followed. Others are industry standards and are accepted practice. Typically, a *written* safety and health program is needed. This manual should contain

**DESCRIPTION OF ACCIDENT**

**Information is to be used for preventing similar accidents. Answer questions specifically, not in generalities. The safety officer, as required, will conduct independent investigations of accidents.**

1. **What job was employee doing?**

   ________________________________________

2. **What tools, materials, and/or equipment were being used?**

   ________________________________________

3. **What specific action caused the accident?**

   ________________________________________

4. **How did the employee contribute to the accident?**

   ________________________________________

5. **Were the proper safety protection devices being used?**

   ________________________________________

6. **What materials, tools, etc. were defective or in unsafe condition? How?**

   ________________________________________

   ________________________________________

7. **What work methods or acts caused the accident?**

   ________________________________________

   ________________________________________

8. **What safeguards should have been used?**

   ________________________________________

   ________________________________________

9. **What steps will you take to prevent similar injuries?**

   ________________________________________

   ________________________________________

10. **What other steps should be taken to prevent a recurrence?**

   ________________________________________

**Figure 5.10** Description of accident.

safety and health policies, rules, emergency procedures, and emergency phone numbers. There is a need for ongoing safety training that will change and be updated as laws, standards, and the systems change. Wastewater treatment professionals working in a small plant may question the need for putting it all in writing, and then keeping it up-to-date. After all, there may only be one or two employees who know their jobs. However, what happens if someone leaves, retires or gets hurt? The replacement will not have the benefit of history and experience and could be overwhelmed. The town administration will need to know what to do.

11. Did you see the accident? ☐ Yes ☐ No

Date of Report ______________ ______________________________
Immediate Supervisor

Remarks Concerning Investigation ______________________________

Age of Injured ______________ Length of Service ______________

For Safety Department Use

Date Investigated______________ By ______________________
Signature

Remarks Concerning Investigation ______________________________

**Figure 5.10** Description of accident (continued).

In fact, the leadership of compiling a health and safety program can come from a worker or from higher levels of administration. It is best to work with administration closely, regardless of who is taking the lead in formulating such a program. If the town administration is taking the lead, the worker or manager can convey to them different things that are necessary so that the budgetary considerations can be made. If a wastewater treatment professional is taking the lead, he or she should take advantage of the opportunity to educate those with whom he or she is working.

Regardless of who is in charge and how the program is formulated, the safety and health program should start with a mission statement. The mission statement defines the need for such a program and should help define the working conditions expected and the follow-through required. It frames that attitude within which administrators and employees are expected to work. A good statement will define the goals and objectives of preventing job-related injuries and illness and the disruptions that they may

## EMPLOYEE ACCIDENT/INCIDENT REPORT

To be completed within 24 hours and sent to the Health and Safety Officer
Type or Print with Ball Point Pen

### EMPLOYEE IDENTIFICATION

1. Name ______________________ 2. Street Address ______________________
3. City, State, Zip Code ______________________ 4. Home Phone ______________________
5. Work Phone ______________________ 6. Length of Employment ______________________
7. SS# ______________________ 8. Birth date ______________________
9. Job Title ______________________

### INCIDENT INFORMATION (to be completed by Employee)

10. Date of Incident ______________________ 11. Time of Incident ______________________
12. Location of Incident:
    Inside Building (Building & Room #) ____________ Outdoors (Description) ____________
13. Was Supervisor Notified? _____Yes _____No
14. Date & Time Notified: ______________________
15. Name of Immediate Supervisor ______________________
16. Was injured person performing regular job duties at time of incident? _____Yes _____No
17. Did incident result in injury? _____Yes _____No
18. Did incident result in loss of property? _____Yes _____No
19. Complete description of incident ______________________
20. Circumstances that lead to incident, i.e., unfamiliar with task, lack of concentration, improper instruction, etc. ______________________
    ______________________
    ______________________
21. What measures could have been taken to prevent this incident?
    ______________________
    ______________________
22. If there was damage to property, describe the extent of loss to the best of your knowledge.
    ______________________
    ______________________
23. Witnesses _____Yes _____No

**FIGURE 5.11** Employee accident/incident report.

24. Name, address and phone number of witness:

______________________________________________

______________________________________________

25. Bodily injury - Body part injured:

| | Left | Right |
|---|---|---|
| hand elbow ankle | | |
| thumb shoulder foot | | |
| finger(s) thigh toe (s) | | |
| wrist knee eye | | |
| arm calf ear | | |
| face/teeth other | | |
| head | | |
| abdomen | | |
| back lower | | |
| back mid | | |
| back upper | | |
| groin | | |
| neck cervical | | |
| nose/throat/lungs | | |

26. Nature of Injury:

laceration sprain other puncture strain ______________________

insect/animal bite fracture/dislocation ______________________

burn inhalation ______________________

abrasion, scrape foreign matter ______________________

contusion, bruise skin irritation ______________________

exposure to body fluids (bbp) ______________________

Other:

______________________________________________

______________________________________________

Describe:

______________________________________________

______________________________________________

**FIGURE 5.11** Employee accident/incident report (continued).

27. Was this incident the result of a slip, trip, or fall? ____Yes ____No

28. Was this incident the result of lifting? ____Yes ____No
    Approximate weight of object: ____________ How high lifted? ____________
    Was kind of work performed regularly? ____Yes ____No

29. Did injury appear immediately? ____Yes ____No Explain: ____________
    ________________________________________

30. What was the length of time between the injury and your symptoms? ____________

31. Were you treated by a doctor? ____Yes ____No

    If yes, name of doctor ____________ Date treated: ____________

32. Did you go to the hospital? ____Yes ____No

    If yes, name of hospital ____________Date treated ____________

33. Was first aid given? ____Yes ____No

    By whom (self - using first aid kit, paramedic, other): ____________

34. Have you ever filed a Worker's Compensation claim? ____Yes ____No
    If yes, when and where:
    ________________________________________

35. Nature of previous claim(s): ____________

36. Is this injury an aggravation of an old injury? ____Yes ____No

I, the injured employee, herein certify that the information set forth above is true and correct to the best of my knowledge. By signing this form, I expressly waive all provisions of law which forbid any person or persons who heretofore did or who hereafter may medically attend, treat, or examine me or who may have information of any kind which may be used to render a decision in my claim for injury/disease of (Date)________ from disclosing such knowledge to my employer and/or any other agency contracted by employer to investigate this health claim. A copy of this form will serve as the original.

Employee Signature ____________Date: ____________
Print Name ____________

FIGURE 5.11 Employee accident/incident report (continued).

**OSHA's Form 300**

# Log of Work-Related Injuries and Illnesses

**Attention:** This form contains information relating to employee health and must be used in a manner that protects the confidentiality of employees to the extent possible while the information is being used for occupational safety and health purposes.

Year ________

**U.S. Department of Labor**
Occupational Safety and Health Administration

Form approved OMB no. 1218-0176

You must record information about every work-related injury or illness that involves loss of consciousness, restricted work activity or job transfer, days away from work, or medical treatment beyond first aid. You must also record significant work-related injuries and illnesses that are diagnosed by a physician or licensed health care professional. You must also record work-related injuries and illnesses that meet any of the specific recording criteria listed in 29 CFR 1904.8 through 1904.12. Feel free to use two lines for a single case if you need to. You must complete an injury and illness incident report (OSHA Form 301) or equivalent form for each injury or illness recorded on this form. If you're not sure whether a case is recordable, call your local OSHA office for help.

Establishment name ________

City ________ State ________

| Identify the person | | | Describe the case | | | Classify the case | | | | | | | | | | | |
|---|---|---|---|---|---|---|---|---|---|---|---|---|---|---|---|---|---|
| (A) Case No. | (B) Employee's Name | (C) Job Title (e.g., Welder) | (D) Date of injury or onset of illness (mo./day) | (E) Where the event occurred (e.g. Loading dock north end) | (F) Describe injury or illness, parts of body affected, and object/substance that directly injured or made person ill (e.g. Second degree burns on right forearm from acetylene torch) | Using these categories, check ONLY the most serious result for each case: | | | | Enter the number of days the injured or ill worker was: | | Check the "injury" column or choose one type of illness: (M) | | | | |
| | | | | | | Death | Days away from work | Remained at work | | On job transfer or restriction (days) | Away from work (days) | | | | | |
| | | | | | | | | Job transfer or restriction | Other record-able cases | | | Injury | Skin Disorder | Respiratory Condition | Poisoning | All other illnesses |
| | | | | | | (G) | (H) | (I) | (J) | (K) | (L) | (1) | (2) | (3) | (4) | (5) |
| | | | | | | | | | | | | | | | | |
| | | | | | | | | | | | | | | | | |
| | | | | | | | | | | | | | | | | |
| | | | | | | | | | | | | | | | | |
| | | | | | | | | | | | | | | | | |
| | | | | | | | | | | | | | | | | |
| | | | | | | | | | | | | | | | | |
| | | | | | **Page totals** | 0 | 0 | 0 | 0 | 0 | 0 | 0 | 0 | 0 | 0 | 0 |

Be sure to transfer these totals to the Summary page (Form 300A) before you post it.

| Injury | Skin Disorder | Respiratory Condition | Poisoning | All other illnesses |
|---|---|---|---|---|
| (1) | (2) | (3) | (4) | (5) |

Public reporting burden for this collection of information is estimated to average 14 minutes per response, including time to review the instruction, search and gather the data needed, and complete and review the collection of information. Persons are not required to respond to the collection of information unless it displays a currently valid OMB control number. If you have any comments about these estimates or any aspects of this data collection, contact: US Department of Labor, OSHA Office of Statistics, Room N-3644, 200 Constitution Ave, NW, Washington, DC 20210. Do not send the completed forms to this office.

Page 1 of 1

**FIGURE 5.12** Log of work-related injuries and illnesses (OSHA form 300; http://www.osha.gov/recordkeeping/new-osha300form1-1-04.xls; accessed January 2006).

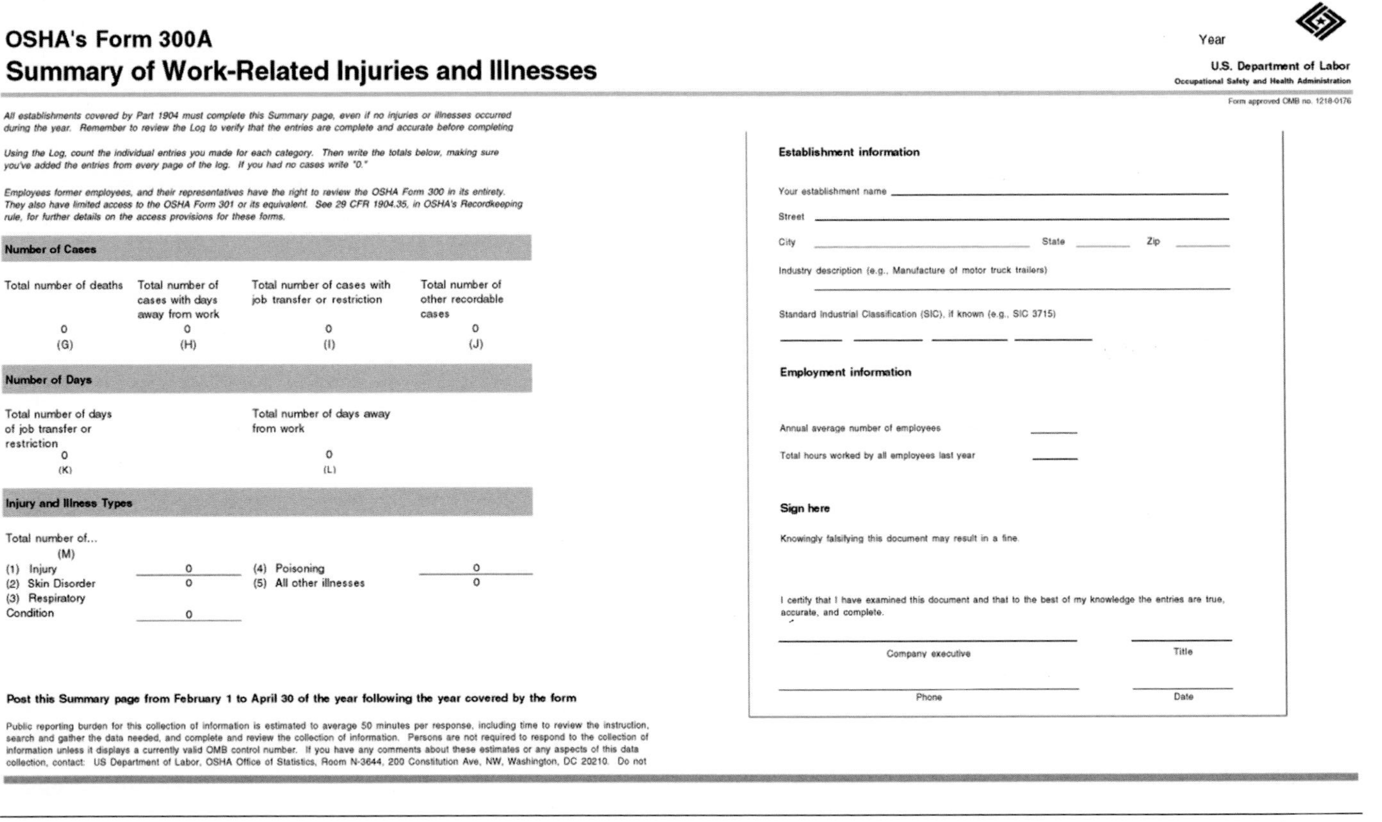

OSHA's Form 300A

Summary of Work-Related Injuries and Illnesses

Year

U.S. Department of Labor
Occupational Safety and Health Administration

Form approved OMB no. 1218-0176

*All establishments covered by Part 1904 must complete this Summary page, even if no injuries or illnesses occurred during the year. Remember to review the Log to verify that the entries are complete and accurate before completing*

*Using the Log, count the individual entries you made for each category. Then write the totals below, making sure you've added the entries from every page of the log. If you had no cases write "0."*

*Employees former employees, and their representatives have the right to review the OSHA Form 300 in its entirety. They also have limited access to the OSHA Form 301 or its equivalent. See 29 CFR 1904.35, in OSHA's Recordkeeping rule, for further details on the access provisions for these forms.*

**Number of Cases**

| Total number of deaths | Total number of cases with days away from work | Total number of cases with job transfer or restriction | Total number of other recordable cases |
|---|---|---|---|
| 0 | 0 | 0 | 0 |
| (G) | (H) | (I) | (J) |

**Number of Days**

| Total number of days of job transfer or restriction | Total number of days away from work |
|---|---|
| 0 | 0 |
| (K) | (L) |

**Injury and Illness Types**

Total number of...
(M)

| | | | |
|---|---|---|---|
| (1) Injury | 0 | (4) Poisoning | 0 |
| (2) Skin Disorder | 0 | (5) All other illnesses | 0 |
| (3) Respiratory Condition | 0 | | |

**Post this Summary page from February 1 to April 30 of the year following the year covered by the form**

Public reporting burden for this collection of information is estimated to average 50 minutes per response, including time to review the instruction, search and gather the data needed, and complete and review the collection of information. Persons are not required to respond to the collection of information unless it displays a currently valid OMB control number. If you have any comments about these estimates or any aspects of this data collection, contact: US Department of Labor, OSHA Office of Statistics, Room N-3644, 200 Constitution Ave, NW, Washington, DC 20210. Do not

**Establishment information**

Your establishment name ____________

Street ____________

City ____________ State ______ Zip ______

Industry description (e.g., Manufacture of motor truck trailers)

____________

Standard Industrial Classification (SIC), if known (e.g., SIC 3715)

____ ____ ____ ____

**Employment information**

Annual average number of employees ______

Total hours worked by all employees last year ______

**Sign here**

Knowingly falsifying this document may result in a fine.

I certify that I have examined this document and that to the best of my knowledge the entries are true, accurate, and complete.

____________ Company executive ______ Title

____________ Phone ______ Date

FIGURE 5.13 Summary of work-related injuries and illnesses (OSHA form 300A; http://www.osha.gov/recordkeeping/new-osha300form1-1-04.xls; accessed January 2006).

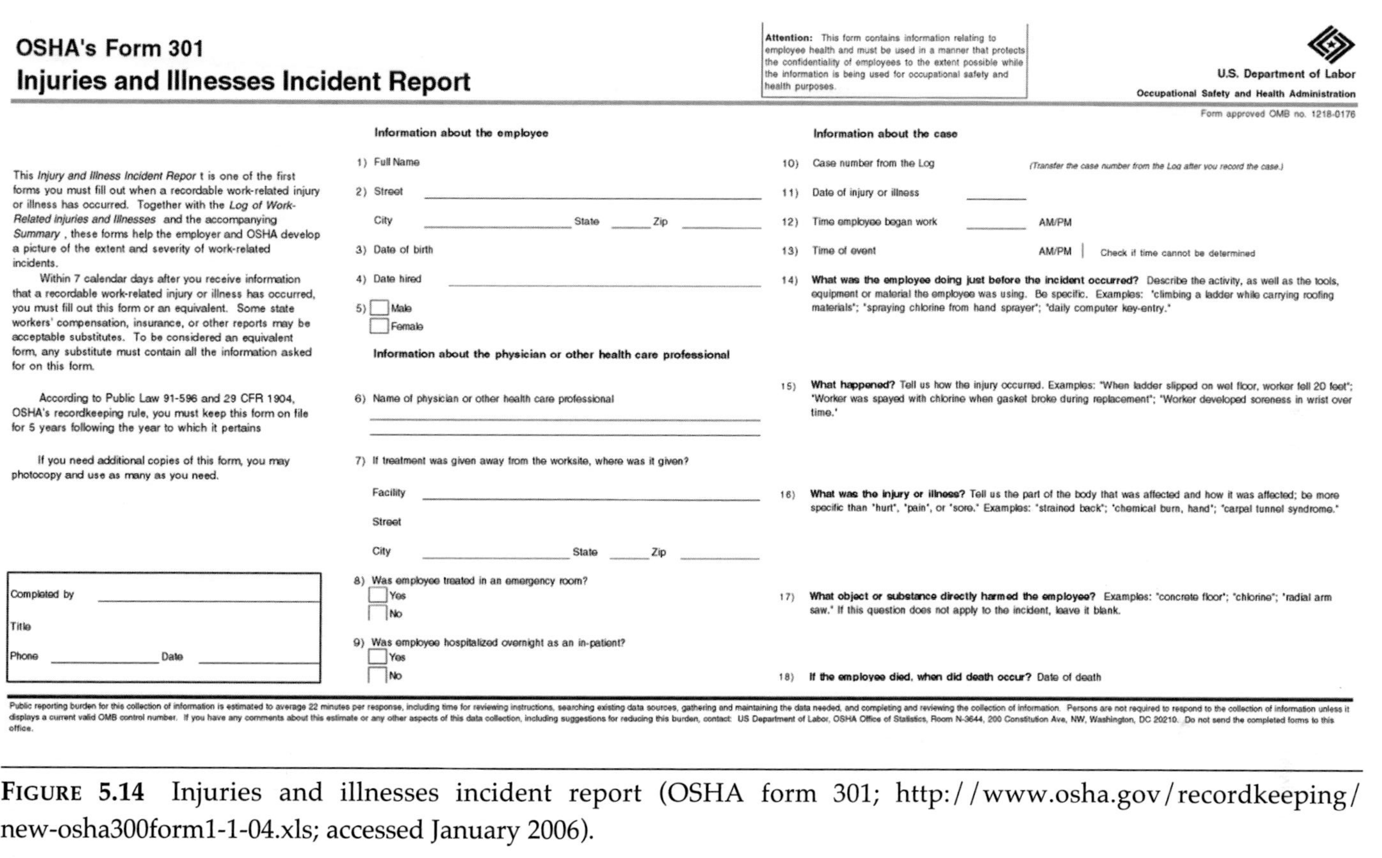

OSHA's Form 301
Injuries and Illnesses Incident Report

Attention: This form contains information relating to employee health and must be used in a manner that protects the confidentiality of employees to the extent possible while the information is being used for occupational safety and health purposes.

U.S. Department of Labor
Occupational Safety and Health Administration
Form approved OMB no. 1218-0176

This *Injury and Illness Incident Repor* t is one of the first forms you must fill out when a recordable work-related injury or illness has occurred. Together with the *Log of Work-Related Injuries and Illnesses* and the accompanying *Summary* , these forms help the employer and OSHA develop a picture of the extent and severity of work-related incidents.

Within 7 calendar days after you receive information that a recordable work-related injury or illness has occurred, you must fill out this form or an equivalent. Some state workers' compensation, insurance, or other reports may be acceptable substitutes. To be considered an equivalent form, any substitute must contain all the information asked for on this form.

According to Public Law 91-596 and 29 CFR 1904, OSHA's recordkeeping rule, you must keep this form on file for 5 years following the year to which it pertains

If you need additional copies of this form, you may photocopy and use as many as you need.

Completed by ______
Title
Phone ______ Date ______

**Information about the employee**

1) Full Name
2) Street ______
City ______ State ______ Zip ______
3) Date of birth
4) Date hired ______
5) ☐ Male
☐ Female

**Information about the physician or other health care professional**

6) Name of physician or other health care professional ______
7) If treatment was given away from the worksite, where was it given?
Facility ______
Street
City ______ State ______ Zip ______
8) Was employee treated in an emergency room?
☐ Yes
☐ No
9) Was employee hospitalized overnight as an in-patient?
☐ Yes
☐ No

**Information about the case**

10) Case number from the Log *(Transfer the case number from the Log after you record the case.)*
11) Date of injury or illness ______
12) Time employee began work ______ AM/PM
13) Time of event AM/PM | Check if time cannot be determined
14) **What was the employee doing just before the incident occurred?** Describe the activity, as well as the tools, equipment or material the employee was using. Be specific. Examples: "climbing a ladder while carrying roofing materials"; "spraying chlorine from hand sprayer"; "daily computer key-entry."
15) **What happened?** Tell us how the injury occurred. Examples: "When ladder slipped on wet floor, worker fell 20 feet"; "Worker was sprayed with chlorine when gasket broke during replacement"; "Worker developed soreness in wrist over time."
16) **What was the injury or illness?** Tell us the part of the body that was affected and how it was affected; be more specific than "hurt", "pain", or "sore." Examples: "strained back"; "chemical burn, hand"; "carpal tunnel syndrome."
17) **What object or substance directly harmed the employee?** Examples: "concrete floor"; "chlorine"; "radial arm saw." If this question does not apply to the incident, leave it blank.
18) **If the employee died, when did death occur?** Date of death

Public reporting burden for this collection of information is estimated to average 22 minutes per response, including time for reviewing instructions, searching existing data sources, gathering and maintaining the data needed, and completing and reviewing the collection of information. Persons are not required to respond to the collection of information unless it displays a current valid OMB control number. If you have any comments about this estimate or any other aspects of this data collection, including suggestions for reducing this burden, contact: US Department of Labor, OSHA Office of Statistics, Room N-3644, 200 Constitution Ave, NW, Washington, DC 20210. Do not send the completed forms to this office.

**FIGURE 5.14** Injuries and illnesses incident report (OSHA form 301; http://www.osha.gov/recordkeeping/new-osha300form1-1-04.xls; accessed January 2006).

cause. It should assign responsibility for carrying out policy. Details that would be contained in the program itself and not necessarily in the mission or policy statement would be identifying hazards, ongoing training, developing disciplinary, rewards procedures, and personnel training. For more details, see *Safety and Health in Wastewater Systems* (WEF, 1994) or *Supervisor's Guide to Safety and Health Programs* (WEF, 1992).

In larger systems, the responsibility for health and safety programs are also shared by administration and labor and are often formalized into committees. The committee is a shared responsibility where the program is defined and refined, ongoing inspections take place, and material is updated for the program. Regardless of who is ultimately responsible, records should be kept for training and accidents. As mentioned earlier, accident investigations must be conducted.

## MANAGEMENT RESPONSIBILITIES

As with any program, responsibility for creation, institution, and follow-through begins at the top. This can be the council, a commission, an administrator, or the plant superintendent. Leadership will set the policy and the attitude towards health and safety. The effectiveness of any program depends on the earnest concern of management. Management provides the control of the work environment and employee performance. Among management responsibilities are the following:

- Formulating a written safety and health policy,
- Enforcing the policy,
- Setting achievable goals,
- Providing adequate training, and
- Delegating authority so that the policy may be carried out.

These responsibilities typically are delegated to the line supervisor, crew chief, or foreman. It is his or her responsibility to make sure that labor has the proper tools to work safely and safe work practices are carried out, and that operators are using safe work practices and that proper training of labor has taken place. In larger cities, this responsibility is typically given to the safety director or equivalent position.

## EMPLOYEE RESPONSIBILITIES

The following are responsibilities of employees:

- Be familiar with safety policies and rules and follow them.
- Recognize job-related hazards and report to the supervisor any that seem unusual or new.

- Report all injuries and accidents.
- Before operating equipment, receive adequate instruction.
- Observe speed limits, traffic signs, and drive defensively, anticipating problems, whether on- or off-site.
- Keep hand tools and the work area clean and orderly at all times.
- Use the correct tool for the right task.
- Use proper PPE.
- Avoid wearing loose clothing that may be caught in moving equipment. If a worker has long hair, he or she should make sure it will not get caught in moving equipment.
- If wearing jewelry (watches and rings), make sure they do not get caught in machinery.
- Smoke only where allowed.
- Observe rules of personal hygiene to avoid infection.
- Never report to work under the influence of controlled substances or bring them on-site.
- Avoid practical jokes that could result in injury.
- Do not bring firearms to the worksite.
- Do not compromise safety for speed, regardless of circumstance.

## TRAINING

The first thing necessary for successful training, no matter how good the prospective program, is the proper attitude. The material could be great and the presentations awe-inspiring, but if management is not serious about its application, no improvements will occur. If there is not enthusiasm or financial backing for programs (including bonuses for safe work performance), then there will not be acceptance by the work force. They will know they are being trained in safety only because of liability issues. Leadership is one of management's primary responsibilities.

Safety and prevention of accidents do not occur by some random chance. These conditions prevail when they are topics accepted enthusiastically and, because of that enthusiasm, a systematic study of all hazards, illnesses, and accidents occurs. Then, ways to prevent these occurrences are applied to the particular facility.

All new employees must have basic safety training. They need to know where the health and safety manual is in case they are not presented with their own copy (in these days, with the availability of copiers and electronic storage, employees should be provided with their own copy). When new equipment is purchased, training must be held to uphold the standards that are being set in the general workplace. If there are continued injuries in one process, then a reassessment of the process needs to be made. Routine quality assurance/quality control inspections need to include safe working habits. Management observance of workers may also show unsafe working habits or careless attitudes. One of the worst things that can happen as a manager is to allow these attitudes to persist.

The following are some items that are included in a training program:

- Wastewater facility hazards, both general and site-specific;
- Employee health and industrial hygiene;
- Personal protective equipment, including respiratory protection;
- Materials handling and storage;
- Safe use of power and hand tools;
- Fire prevention and fire safety;
- Possibly first aid and CPR;
- Lockout/tagout;
- Confined space and entry;
- The MSDS and its use;
- Employee and community right-to-know;
- Emergency planning and response; and
- Chemical handling and storage.

Training serves as a preventative measure against accidents and job-related illness. It starts when a new employee comes on board and really never ends, when done correctly. Safety is a joint effort between management and labor and requires a commitment from both.

# REFERENCES

*American Heritage Dictionary* (1997) 3rd ed.; Houghton-Mifflin: Boston, Massachusetts.

Lue-Hing, C.; Tata, P.; Casson, L. (1999) *HIV in Wastewater: Presence, Survivability, and Risk to Wastewater Treatment Plant Workers;* Water Environment Federation: Alexandria, Virginia.

Casson, L. W.; Hoffman, K. D. (1999) HIV in Wastewater: Presence, Viability, and Risks to Treatment Plant Workers. In *HIV in Wastewater: Presence, Survivability and Risk to Wastewater Treatment Plant Workers;* Monograph; Water Environment Federation: Alexandria, Virginia.

Comprehensive Environmental Response, Compensation, and Liability Act (1980) *U.S. Code,* Chapter 103, Title 42; Washington, D.C.

Geldreich, E. (1972) Water-Borne Pathogens. In *Water Pollution Microbiology,* Mitchell, R., Ed.; Wiley & Sons: New York; Chapter 9; p 227.

Helperin, J. Driven to Distraction: Cell Phones in the Car (The Debate over DWY [Driving While Yakking]). http://www.edmunds.com/ownership/safety/articles/43812/article.html (accessed Jan 2006).

Kuchenrither, R. D.; Sharvelle, S.; Silverstein, J. (2002) Risk Exposure. *Water Environ. Technol.,* **14** (5), p. 37–40.

National Institute for Occupational Safety and Health (2005) *The NIOSH Pocket Guide to Chemical Hazards.* Department of Health and Human Services, Centers for Disease Control and Prevention, National Institute for Occupational Safety and Health: Washington, D.C.

National Fire Protection Association (1995) *Standard for Fire Protection Practice for Wastewater Treatment Plants and Collection Facilities,* NFPA 820. National Fire Protection Association: Quincy, Massachusetts.

National Fire Protection Association (2005) http://www.ilpi.com/msds/ref/nfpa.html (accessed Feb 2006). National Fire Protection Association: Quincy, Massachusetts.

National Fire Protection Association (2006) *Uniform Fire Code,* NFPA 1. National Fire Protection Association: Quincy, Massachusetts.

Occupational Safety and Health Hazards (2005) *Code of Federal Regulations,* Part 1910.

Permit-Required Confined Spaces (2005) *Code of Federal Regulations,* Part 1910.146.

Safety Belts, Lifelines, and Lanyards (2005) *Code of Federal Regulations,* Part 1926.104.

SARS Prevention: Hook Up Drains, Clean Up Waste (2003) Taiwan headlines. http://www.taiwanheadlines.gov.tw/20030506/20030506p5.html (accessed Feb 2006).

Thompson, S. S.; Jackson, J. L.; Suva-Castillo, M.; Yanko, W. A.; El Jack, Z.; Kuo, J.; Chen, C. L.; Williams, F. P.; Schnurr, D. P. (2003) Detection of Infectious Human Adenoviruses in Tertiary-Treated and Ultraviolet-Disinfected Wastewater. *Water Environ. Res.,* **75**, 163–170.

U.S. Environmental Protection Agency (2001) Consolidated List of Chemicals Subject to Emergency Planning and Community Right-to-Know Act (EPCRA) and Section 112(r) of the Clean Air Act, EPA-550/B-01-003. U.S. Environmental Protection Agency: Washington, D.C., http://www.epa.gov/ceppo/pubs/title3.pdf (accessed Feb 2006).

Water Environment Federation (1994) *Safety and Health in Wastewater Systems,* Manual of Practice No. 1; Water Environment Federation: Alexandria, Virginia.

Water Environment Federation (1992) *Supervisor's Guide to Safety and Health Programs,* Water Environment Federation: Alexandria, Virginia.

Water Resources Research Center (2001) Do Waterborne Pathogens Pose Risks to Wastewater Workers? *Ariz. Water Resour.,* **9** (6). http://www.ag.arizona.edu/AZWATER/awr/julyaug01/feature1.html (accessed Feb 2006).

World Health Organization (2003) *Severe Acute Respiratory Syndrome (SARS): Status of the Outbreak and Lessons for the Immediate Future.* Communicable Disease Surveillance and Response; World Health Organization: Geneva, Switzerland. http://www.who.int/csr/media/sars_wha.pdf (accessed Feb 2006).

# SUGGESTED READINGS

Brown, L. C.; Kramer, K. L.; Bean, T. L.; Lawrence, T. J.; Trenching and Excavation: Safety Principles, AEX-391-92; Ohio State University Extension Fact Sheet; The Ohio State University, Food, Agricultural and Biological Engineering, Columbus, Ohio. http://ohioline.osu.edu/aex-fact/0391.html (accessed Feb 2006).

Division of Occupational Safety, Rhode Island Department of Labor and Training; Introduction of Emergency Planning and Community Right-to-Know Act. http://www.dlt.state.ri.us/webdev/osha/sarasummary.htm (accessed Feb 2006).

Harford County, Maryland, Tier II Chemical Reporting Forms. http://www.co.ha.md.us/lepc/tier2form.html (accessed Feb 2006).

Kerri, K. D. (1995) *Small Water System Operation and Maintenance,* 3rd ed.; California State University: Sacramento, California.

Kneen, B.; Darling, S.; Lemley, A. (1996, updated 2004) Cryptosporidium; a Waterborne Pathogen. *Water Treatment Notes,* Fact Sheet 15, USDA Water Quality Program, Cornell Cooperative Extension; http://hosts.cce.cornell.edu/wq-fact-sheets/Fspdf/Factsheet15_RS.pdf (accessed Feb 2006).

Ministry of Health Singapore Home Page. http://www.moh.gov.sg (accessed Feb 2006).

Oklahoma State University Environmental Health and Safety (2004) Entering and Working in Confined Spaces; EHS Safety Manuals. http://www.pp.okstate.edu/ehs/manuals/CONFINED/Sec_1.htm (accessed Feb 2006).

Parents of Kids with Infectious Diseases; Hepatitis. http://www.pkids.org/hepatitis.htm (accessed Feb 2006).

University of Florida, Finance and Administration, Environmental Health and Safety (2002) EH&S Trenching and Excavation Safety Policy; UFEHS-SAFE1. http://www.ehs.ufl.edu/General/Trench02.pdf (accessed Feb 2006).

University of Wisconsin–Milwaukee; Excavation Safety. http://www.uwm.edu/Dept/EHSRM/EHS/SAFETY/excavations.html (accessed Feb 2006).

U.S. Department of Health and Human Services, National Institutes of Health, National Institute of Allergy and Infectious Diseases (2005) Parasitic Roundworm Diseases. *Health Matters.* http://www.niaid.nih.gov/factsheets/roundwor.htm (accessed Feb 2006).

U.S. Department of Labor, Occupational Safety and Health Administration (2002) Lockout/Tagout; OSHA Fact Sheet. http://www.osha-slc.gov/OshDoc/data_General_Facts/factsheet-lockout-tagout.pdf (accessed Feb 2006).

U.S. Environmental Protection Agency (1994) Fact Sheet: Emergency Planning and Community Right-to-Know Act of 1986 (EPCRA). http://es.epa.gov/techinfo/facts/pro-act6.html (accessed Feb 2006).

# Chapter 6

# Management Information Systems—Reports and Records

## INTRODUCTION

Data requirements at wastewater utilities have exploded in recent years and are expected to grow exponentially for the indefinite future. Operators need data to confirm their observations and judgments on how to control, adjust, and modify wastewater treatment processes. Regulators need periodic reports on utility operations so they can see whether public and environmental health are protected. Utility managers need data to determine whether to adjust sewer rates or upgrade the facilities. Lawyers need data to prove that the utility is operating in accordance with applicable laws and regulations. Outreach personnel need data to show the public why wastewater management is important.

Effectively collecting, querying, analyzing, reporting, distributing, storing, and archiving data—whether electronic, paper, audio, image, or video—have become key to effective and efficient operations. They are an important part of a utility manager's responsibilities. Today, most utilities could not operate for long without effective information technology (IT) support.

## MANAGEMENT INFORMATION SYSTEMS

Paper-based records require little equipment, but producing and updating them is labor-intensive. Accessibility and data sharing is limited, so people make copies of the paper and create duplicate files. Updates and additions are inconsistent, causing more work reconciling the differences. Data analysis is painstaking, time-consuming, and sometimes virtually impossible. Nevertheless, paper-based systems are still common.

Properly designed computer systems offer important advantages. They should help utility staff make better, faster decisions; automate routine tasks cost-effectively; and minimize risks. The utility's administrative and operations and maintenance (O&M) personnel should be involved in planning management information systems to ensure that they can get the information they need. With a good system, utility staff can

- Enter data once and use it for multiple purposes;
- Easily select, sort, manipulate, and print data in a variety of ways for various purposes;
- Easily display data graphically for analysis;
- Attain better data accuracy via range checking and other editing techniques;
- Automate routine tasks (e.g., automatically re-ordering materials and supplies when the stock falls below a certain level) to make them more consistent;
- Track data changes;
- Organize and store data securely;
- Improve customer service;
- Know the plant's operating status at a glance; and
- Effectively and rapidly communicate with colleagues and external stakeholders.

The information technology industry changes rapidly, making it a challenge for utility managers and staff to stay current. Even the name of the utility's "computer department" is evolving: staff may call it Data Processing (now generally considered outdated), Management Information Systems (MIS), Information Services, or other names. (In this chapter, the term *information technology* refers to information systems, hardware, software, and the associated business processes required to select, implement, and manage the technology.)

Basically, management information systems can be divided into four categories: hardware, software, data, and supporting business processes. *Hardware* refers to the physical devices used to input, store, process, and exchange data [e.g., mainframes; servers; routers, switches, and other network devices; and personal computers (PCs), personal digital assistants (PDAs), printers, copy machines, private branch exchanges (PBXs), video cameras, and programmable logic controllers (PLCs)]. Hardware's lifespan varies but typically ranges from 3 to 7 years. *Software* is any program that runs on the hardware. Its typical lifespan ranges from 3 to 10 years or longer. Software upgrades (often called "maintenance") extend this lifespan. *Data* are the content manipulated by the computer system. The lifespan of data is often very long—sometimes 100 years or more—spanning multiple generations of hardware and software. *Integration* is the concept of a unified system of hardware, software, and data (from a user's perspective).

Many wastewater utilities still have fairly primitive information management systems, and many of those with sophisticated systems have yet to truly leverage the benefits. The utilities that effectively manage their information and supporting technologies will be better positioned for future success (AWWA, 2001).

For management information systems to be effective, a utility's business processes (the way that work gets done) and its software must be aligned in the following eight areas (AWWARF, 2004):

- Overall/common (practices not unique to MIS, such as developing and training staff, managing and measuring performance, and handling customer feedback);
- Applications and integration;
- Data (practices related to data definition, standards, analysis and reporting, management, and quality);
- IT infrastructure management (IT assets, architecture, and help-desk practices).
- Service delivery and sourcing (service delivery alternatives, service level management, and software life-cycle management);
- IT organization, direction, oversight, and planning (governance and policy definition, budgeting, total cost of ownership, and business continuity);
- Program and project management (program management, project management, quality control, and testing strategies); and
- Security [risk-management and security issues (e.g., personnel security, production controls, and audit trails)].

**TYPICAL SOFTWARE APPLICATIONS.** Today, wastewater utilities typically use a number of software applications [e.g., e-mail, word processors, spreadsheets, graphics applications, Laboratory Information Management Systems (LIMSs), and financial management systems]. To reduce the costs of integration, these applications should be based on common standards.

Following are some typical software applications found at wastewater utilities.

*Computerized Maintenance Management System.* A computerized maintenance management system (CMMS) [also called a computerized work management system (CWMS) or a work management system (WMS)] enables utilities to manage maintenance work and minimize equipment downtime cost-effectively. The system is designed to plan, schedule, and manage maintenance activities; control parts inventories; coordinate purchasing activities; and help prioritize long-term asset investment needs.

A CMMS typically consists of the following six components:

- Work management (corrective, preventive, and predictive maintenance scheduling, activities, and procedures);
- Equipment inventory (an inventory and description of equipment and support systems requiring maintenance, along with other technical or accounting information);
- Inventory control, tools, and materials management (materials, tools, and spare parts management, scheduling, and forecasting);
- Purchasing or procurement (maintenance-related requisition, procurement, and accounting);
- Reporting and analysis (standard and ad hoc reports); and
- Personnel management (staff skills, wages, and availability).

Effective CMMSs range from single-user PC systems to comprehensive networked client–server-based systems for utilities with several hundred maintenance personnel in multiple locations. While a manual maintenance system may seem adequate for small treatment plants, they should consider converting to a PC-based CMMS to enhance record accessibility, sharing, and security.

Relatively new to the wastewater treatment industry, CMMSs are used extensively in other industries. The rising popularity of CMMSs at treatment plants is the result of concerns about deteriorating infrastructure, the need for better asset management, and related new regulations [e.g., Governmental Accounting Standards Board (GASB) 34 and the capacity, management, operations, and maintenance (CMOM) rule].

The information from a CMMS, especially that pertaining to major equipment and work history, is an integral part of an asset management plan.

Some utilities connect key assets directly to the CMMS, so it can automatically create work orders based on vibration analysis, lubricant analysis, thermography, and other real-time data. Some connect the process control system (PCS) to the CMMS, so it can automatically generate work orders based on equipment run time.

***Geographic Information System.*** According to Huxhold (1991), at least 80% of data collected and used for utility and environmental information systems are related to geography. Geographic information systems (GISs) allow users to link databases and maps to better see topographical-related patterns and trends. These systems are often at the center of a utility's data integration strategy because they are data-intensive, can use data from many sources, and can be specialized to meet specific utility needs.

***Laboratory Information Management System.*** According to *Good Automated Laboratory Practices* (GALPs) (U.S. EPA, 1995), a LIMS is an automated system that collects and manages laboratory data. It typically connects the lab's analytical instruments to one or more PCs that are part of the utility's overall network. (Sometimes the interface software is custom and complex.) A full-featured LIMS can track chain-of-custody forms, print bar codes for sample containers, hold data in the queue until all validations are completed, generate various trend charts and reports, perform on-the-fly sample cost analyses, bill customers, monitor quality assurance and quality control activities, track instrument calibration and maintenance, and control inventory. Various levels of security and access are available.

There is virtually a limitless range of possible configurations for LIMSs. However, not all automated laboratory systems are LIMSs; those that record data but do not allow changes (e.g., analytical balances) are not LIMSs (Figure 6.1). The agency's GALPs are intended for comprehensive LIMS configurations in which data are entered, recorded, manipulated, modified, and retrieved.

When implementing a LIMS, laboratory managers should ensure that

- Personnel, resources, and facilities are adequate and available as scheduled;
- Personnel clearly understand the function(s) they will perform on the LIMS;
- Each applicable GALP is followed;
- There are standard operating procedures (SOPs) to protect raw LIMS data integrity;
- A quality assurance unit (QAU) monitors LIMS activities;
- Any deviations from SOPs and applicable GALPs are corrected and both the deviation and correction are appropriately documented;

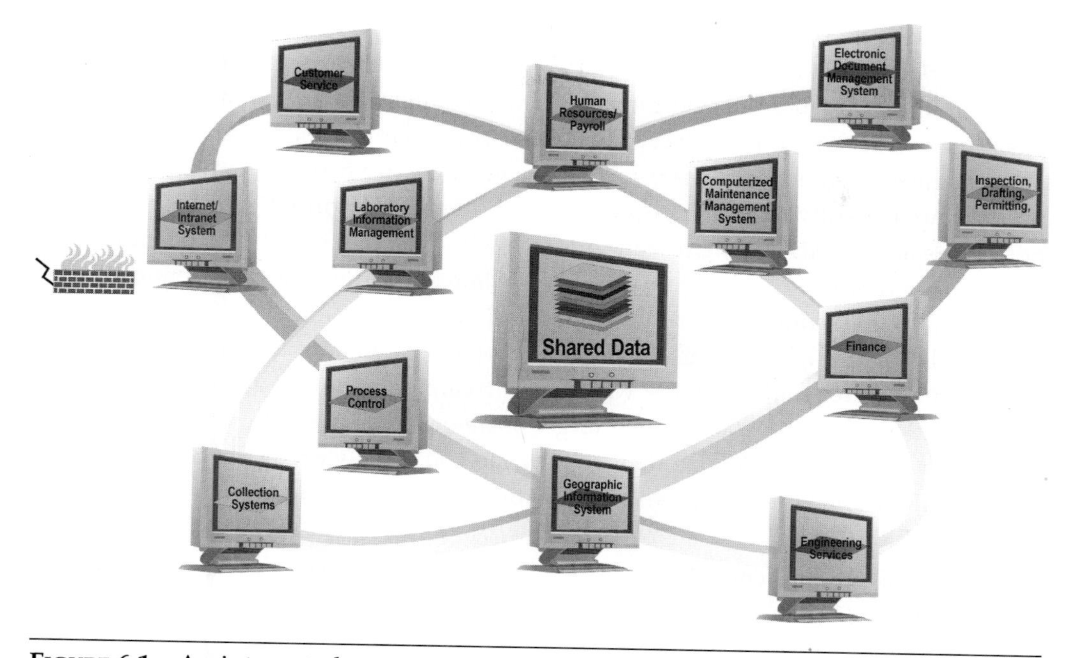


FIGURE 6.1 An integrated system makes data available across systems.

- Appropriate changes are subsequently made to SOPs; and
- Any data deficiencies noted in LIMS QAU inspection reports or audits of raw LIMS data are promptly corrected.

***Human Resources Management System.*** A comprehensive human resources management system (HRMS) can provide flexible, accurate control over payroll accounting and all employee personnel records, as well as in-depth inquiry and reporting capabilities. It also includes payroll, benefits administration, and pension administration. Advanced HRMSs often support career development, training, and succession planning. They are sometimes supplemented by a training information management system (TIMS), which tracks personnel training.

***Financial Information System.*** A financial information system (FIS) typically includes modules on general ledger, budgeting, accounts receivable, accounts payable, project management, investment tracking, and job costing. (Many wastewater utilities do not use the accounts receivable module because their needs are too simple for it to be practical.) Ideally, the FIS should be interfaced with the CMMS and other applications involving inventory management, fixed-asset management, and other cost-related issues.

***Process Control System and Supervisory Control and Data Acquisition System.*** Process control systems and supervisory control and data acquisition (SCADA) systems automatically monitor and can control wastewater collection, treatment, and disposal processes in real time. Except for the instrumentation used to collect field data and control equipment (e.g., start or stop a pump), their hardware and software components are basically the same as those used in business information systems. In fact, both PCSs and SCADA systems can interface with business systems to promote cost-effective operations (e.g., enable plants to calculate up-to-date energy, chemical, labor, and other costs so control algorithms can be adjusted, if possible).

The hardware has become relatively inexpensive and simple to install, program, and maintain, so even the smallest wastewater treatment plants can afford to automate their processes (WERF, 2002). Some large wastewater utilities connect their PCSs or SCADA systems to dynamic models, so they can continually predict how influent conditions will affect performance and automatically adjust key operating parameters as needed (WERF, 2002).

The same security considerations that apply to other business systems also apply to PCSs and SCADA systems.

***Personnel Productivity Applications.*** There are many applications available to enhance utility personnel's productivity (e.g., word processors, spreadsheets, graphics software, e-mail, and calendars).

***Collaboration Software.*** Web-based software (e.g., wikis, blogs, and virtual meeting software) allows staff, vendors, consultants, and regulators to share ideas, solve problems, and jointly develop or review reports from any location at any time.

***Other Support Systems.*** Many utilities have electronic controls for heating, ventilation, and cooling (HVAC); interior lighting; and emergency and security systems (e.g., fire alarms, electronic key access, motion sensors, and closed-circuit television cameras). To help reduce energy costs, the heating and cooling controls should be optimized based on work schedules.

In addition to these core applications, there are a number of other software applications that wastewater utilities require, depending on their size and complexity. Wastewater models, for example, can predict inflow and precisely describe combined sewer overflows (CSOs) from collection systems into receiving waters. Capital projects planning and management systems help utility staff manage construction budgets and schedules. Many utilities use an electronic document management system to manage all the documents generated as part of conducting business. These systems, which can convert paper documents into electronic format, allow multiple users in multiple locations to "share" the same document. Their capabilities typically include check-in and check-out, version control, categorization, search, and archive. An electronic document management system is similar to a records management system, which provides a centralized location to store, access, and retrieve critical information sources in compliance with legal requirements. All of these applications are commercially available, and most wastewater utilities have adopted or are migrating to a strategy of procuring them from vendors rather than developing custom applications.

In addition, many utilities now have developed custom Web-enabled systems that may be used via intranet (internal communications) or Internet (external communications). These Web-enabled systems consist of various technologies (e.g., relational databases, user interfaces, and messaging systems) and allow varying levels of integration with other utility systems. They may provide a portal to other systems.

Most wastewater utilities have implemented a local area network (LAN), which allows users at different computers to share data, and have connected to a wide area network (WAN), which links multiple LANs. Utilities also typically are connected to the Internet and can electronically communicate with others, regardless of their location. The clear trend is for utilities to exchange data electronically with individuals (via e-mail and a utility's Web site) and other businesses (e.g., electronic banking between the utility and its bank). As utility Web sites become richer and offer more capabilities (e.g., allowing significant users to enter their own data), the utility's internal business processes must become more streamlined and reliable. This typically involved redesigning the business processes.

The U.S. Environment Protection Agency (U.S. EPA) is moving toward accepting electronic reports via the Cross Media Electronic Reporting and Recordkeeping Rule [40 *CFR* 3 (New) and 40 *CFR* 9 (Revision)], which provides a legal framework for electronic reporting and recordkeeping under the agency's environmental regulations. The intent is to provide a uniform, technology-neutral framework for electronic reporting and recordkeeping across all U.S. EPA programs. The final rule, which was published in the *Federal Register* in October 2005, does not mandate electronically submitted reports or records but does

- Modify requirements in the *Code of Federal Regulations* to remove any obstacles to electronic reporting and recordkeeping;
- Allow regulated entities to submit any report or maintain any record electronically once U.S. EPA announces that electronic reporting or recordkeeping is available for that document;
- Require electronic reports to be submitted via U.S. EPA's central data exchange (CDX) or another designated report-submittal system;

- Require validation of electronic signatures on reports submitted electronically (valid electronic signatures have the same legal force as their "wet-ink" counterparts); and
- Set forth electronic-reporting requirements that U.S. EPA-authorized programs must satisfy and provide a streamlined process for the agency to approve of electronic-reporting implementations.

A utility's software applications may reside at a variety of locations. For example, the FIS may be in the municipal administration building downtown, the GIS may be distributed across multiple locations, and some parts of the CMMS may be implemented via wireless devices used by mobile maintenance staff.

The software applications may be supported and maintained via utility staff, staff augmented by contractors or consultants, a third-party vendor, or another outsourced arrangement.

Integrating these applications is a major management issue for many utilities. Integration has many advantages but can be expensive (Figure 6.1). There are a number of strategies that can be adopted to integrate systems, each with advantages and disadvantages.

Cost-effectively implementing, using, and supporting all these systems includes making sure business processes are aligned with technology, ensuring the technology addresses overall business requirements (not individual needs), adequate training for end-users and support personnel, sufficient support resources, and standardization. Successful technology deployment requires addressing not only technology issues, but also organizational issues and business processes.

Particular challenges often include issues related to integration, data management, governance, and staff skills.

## INFORMATION TECHNOLOGY GOVERNANCE

Effective governance ensures that a utility's IT investment meets its business requirements and produces expected results. A good governance method should ensure that the

- Goals are clearly defined,
- Business and IT strategies are aligned,
- IT issues are understood,
- Strategies can be implemented efficiently,
- Related risks are managed, and
- Progress is tracked and reported.

Effective governance means that technology decisions and investments are based on good business practices. In other words, the "customer's" needs are addressed, managers receive regular feedback, technology investments are planned, the desired results are clearly identified before final signoff, and the resulting process is audited after the technology has been implemented.

**PLANNING AND ORGANIZING.** Every wastewater utility should have an IT strategy. Developed by a cross-functional team with strong management support, the strategy should set the overall direction, establish key standards, identify and prioritize major projects (including each project's purpose, schedule, and estimated costs), and account for utility-support issues (e.g., training, required resources, and reporting requirements). The strategy also should specify how various applications will work together so users can easily enter, access, and use the utility's data. It should be updated annually and completely overhauled every 3 to 5 years (or whenever a major shift occurs in the utility's strategies).

In addition to technical and financial issues, the IT strategy should address organizational issues (e.g., centralized, distributed, matrixed, or some combination) and personnel needs. The utility's IT staff must have the skills to implement the strategy, or else the strategy should provide a mechanism for them to acquire the necessary skills. Also, staff retention is becoming important as the baby boom generation begins retiring.

Each year, the utility should create an IT operations plan that specifies which IT strategy elements staff will accomplish. Then at the end of the year, utility staff should review the related accomplishments and lessons learned, and then generate a new, appropriate IT plan for the upcoming year.

When working on the strategy, team members should be judicious in requesting custom software. Historically, all IT departments wrote custom software for their organizations, but now many more applications are commercially available. Software should only be custom-developed when no viable commercial off-the-shelf (COTS) package meets the utility's need.

As a result of the shift from custom to COTS applications, IT jobs are now primarily about project management rather than software development. So, the strategy should account for the training that existing staff need to become effective project managers. (Managing IT projects is similar to managing any other project; it requires clear charters, teams, timelines, and costs, as well as involvement by key stakeholders.)

**ACQUISITION AND IMPLEMENTATION.** To ensure that any new information technology will meet the utility's business needs and deliver the required results, the project team should first define the goals and requirements. Team members also should become familiar with the possibilities and limitations of the available COTS technology, so they can make the best decisions.

In addition, the team should examine the business process that will be affected by the new technology. Often, this business process needs to be redesigned when the new technology is deployed or else the workload will increase because employees are struggling to do things the old way, while the system requires things to be done in a new way. The struggle typically results in a failed technology, or at best, its anticipated benefits fail to materialize.

The project team also must decide which procurement option is preferred. Software may be "purchased" (i.e., utilities may purchase licenses to use the software), leased from a vendor, or run on a third-party's Web site [i.e., rented, using an application service provider (ASP) model]. Software financing and licensing approaches vary significantly and change frequently, so the team should evaluate the options carefully when selecting a new technology.

Commissioning the system requires as much focus as selecting it. There are several key steps:

- Documenting the new system and how it will be used,
- Training technical staff and others to use it,
- "Cleansing" the data,
- Developing the necessary reports, and
- Conducting a postimplementation audit (i.e., assessing whether the desired results were achieved and whether more benefits can be gained).

*Data cleansing* is the process of improving data quality when migrating data from an old application to a new one. The team should evaluate the existing data for quality problems (e.g., missing, incorrect, incomplete, or redundant data) and take the necessary steps to solve them. Often overlooked, data cleansing ensures that the new system has correct and complete data expected by users.

**ONGOING SUPPORT.** Once implemented, the new technology will require ongoing support from both the vendor and the utility's IT staff to function properly and optimize its effectiveness.

***Support Services.*** Utilities should establish the minimum level of support services expected from the technology vendor or on-site IT staff. If the utility signs a Service Level Agreement (SLA) with the vendor, the minimum service levels should be well-defined to ensure that the utility gets what it pays for. Aspects to consider include vendor response time, escalation procedures, the problem tracking mechanism, frequency of upgrades, upgrade documentation, training requirements, and duration of the support contract. Some utilities avoid vendor SLAs because they do not think the services provided are worth the costs involved. They prefer to develop an in-house SLA, which enables technology users and IT staff to jointly define expectations and discover the cost implications of a particular level of support.

Poorly executed SLAs can result in antagonistic relationships between the service provider and the service receivers. A well-executed SLA requires significant management involvement. For example, expected and actual service levels should be regularly compared, reported, and adjusted as necessary.

***Training.*** Training should be provided on a "just-in-time" basis. Staff new to a technology cannot absorb expert-level training because they are simply trying to understand the basics. However, the training can be very valuable later, when staff gain more experience. Follow-up training helps reinforce new concepts, resolve misunderstandings, and promote acceptance of the new technology.

***Ongoing Technology Management.*** Technology management involves adapting reports or creating new ones as users gain experience, changing configurations, addressing problems and complaints, and other user-driven support needs.

***Maintenance.*** Technology maintenance includes backing up files (and occasionally restoring some part of the system to ensure that the correct files have been backed up and that the backup/restore system operates as expected); applying software patches and updates; tuning the databases, operating system, and hardware; and ensuring license compliance as users are added.

***Security.*** Information security should be a part of a utility's overall security strategy. Appropriate security measures should be defined and implemented in tandem with the new technology itself, and everyone must follow proper IT security practices (e.g., password maintenance; no shared passwords; and appropriate security updates as people leave the utility, take on new roles and responsibilities, etc.). Also, the security system should be audited periodically to confirm that it is appropriate and effective. For mission-critical systems, this may include probes and attacks to prove that the protection works.

**MONITORING.** Monitoring ensures that the new technology is delivering intended results, that risks are mitigated, and that issues are managed appropriately. Effective monitoring is a matter of measuring and regularly reporting on appropriate performance indicators—perhaps in the form of scorecards (e.g., the Balanced Scorecard, customer-satisfaction assessments, management reporting, and external assessments). Monitoring reports should illustrate the technology's current effectiveness and recommend improvements, as needed.

## TYPES OF RECORDS

What is a record? According to 44 U.S.C. Chapter 33, Section 3301, *records* include "all books, papers, maps, photographs, machine-readable materials, and other documentary materials made or received by a U.S. government agency under federal law or in connection with the transaction of public business and preserved or appropriate for preservation by that agency or its legitimate successor as evidence of the organization, functions, policies, decisions, procedures, operations, or other activities of the government or because of the informational value in them." So, records can take many forms (e.g., printed or "hard" copy, electronic or "soft" copy, and even e-mail and voicemail).

Records are the backbone of all management systems. Capturing and managing records are not only regulatory requirements, but also vital to well-run operations. Each record houses important information for a multitude of "lookup" uses. When linked with other records and organized chronologically, spatially, or in some other pattern, they enable utility staff to analyze the age, condition, and many other attributes of the facilities and make highly informed decisions. Therefore, records should be accurate, accessible, and yet secure to ensure that utility have the best information available.

Following are the types of records that most wastewater utilities need to manage.

**PHYSICAL PLANT RECORDS.** Utility staff should always have easy access to current records of the physical treatment plant. Such records typically include construction specifications; the engineer's design report; as-built drawings of the wastewater treatment, pumping, and collection facilities; a map of the collection system; an O&M manual; equipment descriptions; shop drawings; manufacturers' literature; and manufacturers' O&M instructions.

Records of existing facilities and equipment are not only invaluable to O&M staff handling day-to-day operations, but also to managers scheduling services, regulators evaluating permit compliance, and engineers and contractors designing and constructing facility improvements or additions. The format and content of these records should remain as consistent as reasonable, while allowing for useful improvements. For example, paper-based O&M manuals typically did not include record drawings because they were too large. Instead, the drawings were stored in bins, file drawers, or plan racks. However, electronic record drawings [produced via computer-aided design (CAD) software] can be linked to electronic O&M manuals and continually updated by engineers.

***Record Drawings.*** Record (as-built) drawings of a wastewater collection and treatment system provide a concise, illustrated record of each item of equipment, its location, and its relationship to other pieces of equipment. They typically include a hydraulic profile, flow schematics, valve tables, and diagrams. The drawings are categorized as mechanical, electrical, civil or structural, architectural, instrumentation and controls, and landscaping. A large wastewater utility may have to manage several thousand of them.

Record drawings are used by engineers, O&M staff, consultants, and contractors for planning; project design, development, and management; procurement of equipment and materials; construction management; and O&M support. They typically are produced toward the end of a new construction project and should be updated whenever the project is rehabilitated or undergoes major repairs. The drawings or updates should be completed and checked before project closeout.

Plant engineers may establish utility-specific standards for submitting, updating, indexing, or layering record drawings. Typically, engineers note significant modifications or additions on the appropriate tracings (or electronic CAD files) and revise the record drawings accordingly. Doing this simplifies future repairs, modifications, or additions considerably.

***Operations and Maintenance Manual.*** The O&M manual describes each basic piece of equipment and explains its function, capabilities, and effect on other units, as well as the factors that affect its operation (Table 6.1). It also should include technical information (e.g., vendor-provided specifications), model and serial numbers, and graphics or photographs (especially if the equipment is submerged or buried).

The manual should serve as an instruction manual for inexperienced personnel and a useful reference for experienced ones, so the information must be clear and easy to find. A properly written O&M manual also should help operators establish SOPs—documented, utility-specific methods for performing various O&M activities. The procedures also should note all related safety measures (e.g., emergency procedures, hazardous materials and waste-handling procedures, and lockout or tagout procedures) and reflect the degree to which the PCS and the CMMS are integrated.

Operations and maintenance manuals should be electronic whenever possible so they can be updated easily. When paper-based O&M manuals are converted into an online interactive document, they should include search engines to help staff find information quickly. Forms available electronically should have links to the related

**TABLE 6.1** Topics an operations and maintenance manual should cover.

1. Managerial responsibility, effluent quality requirements, and plant description.
2. Copies of water quality standards and permits.
3. Operation and control of wastewater treatment facilities.
4. Operation and control of sludge handling facilities.
5. Personnel qualifications and requirements and number of people to be employed.
6. Laboratory testing.
7. Records.
8. Safety.
9. Maintenance.
10. Emergency operating response program.
11. Utilities.
12. Electrical system.

SOPs so staff can print them or access them on portable devices (e.g., PDAs or notebook computers) that can be used in the field, thereby encouraging better plant performance and asset preservation.

***Manufacturers' Literature.*** Manufacturers' literature is the best source of detailed information on a particular piece of equipment. It typically includes shop drawings with fabrication and construction details, installation directions, recommended operating guidelines, and suggested lubricants and maintenance instructions. It also may include exploded views and parts lists; shop drawings; recommended O&M schedules; data on each part (e.g., names, model numbers, types, sizes, certifications, and warranties); and suppliers' names, addresses, and phone numbers.

This information is often accessible via the CMMS for ongoing maintenance and asset management purposes, and via the PCS to support ongoing operations.

***Equipment Descriptions.*** An equipment description typically is a 1-page form in the O&M manual that provides key information on a piece of equipment, a checklist of routine maintenance activities, and a record of repair and overhaul services done. It also may include a list of tools, parts, and consumable items (e.g., filters and lamps) needed for routine maintenance and emergency repairs.

***Specifications.*** *Specifications* provide a detailed description or assessment of the requirements, dimensions, materials, and so on needed for a construction project. The specifications accompanying utility construction drawings typically include the following data:

- Descriptions of equipment and their capacities (e.g., pumping capacity);
- The proper materials (e.g., polyvinyl chloride pipes for chemical solution lines and ductile iron pipe for pressure lines) needed;

- Details about the quality of materials, workmanship, and fabrication required;
- Details about related appurtenances and chemicals;
- Descriptions of the construction procedures (e.g., proper pipe bedding and backfilling procedures) to be used;
- A program for keeping existing services in operation during construction;
- A shutdown schedule (if service must be interrupted);
- An outline of the performance tests needed to ensure that the completed work meets design standards; and
- Other data not included in the construction drawings.

Utility staff should keep copies of the specifications for each construction project. Addenda often supersede parts of the specifications, so they should be filed in the same place.

***Design Engineer's Report.*** Essentially, the design engineer's report should justify the project team's design selections. It typically includes brief descriptions of the project and the existing conditions, discussions of design parameters and testing results, justifications of the design recommendations, and the engineer's cost estimate. It should also note the project's service areas, their projected populations, the design capacities and parameters used, and the basis on which the facility was designed.

## COMPLIANCE REPORTS

***Discharge Monitoring Reports.*** Under the 1972 Clean Water Act (CWA), wastewater treatment plants must have National Pollutant Discharge Elimination System (NPDES) permits. These 5-year permits limit the volume of effluent that may be discharged and the concentration of pollutants that it may contain. They also dictate how often the wastewater must be tested (e.g., daily, once a week, or twice a week) and what types of samples (e.g., grab, 8-hour composite, or 24-hour composite) must be analyzed.

In addition, the CWA requires wastewater utilities to submit monthly NPDES monitoring reports [called *discharge monitoring reports* (DMRs)] to U.S. EPA and state regulators. (Some small treatment plants may submit DMRs quarterly as required by permit.) The report forms, which note all effluent limits, are prepared by the U.S. EPA in accordance with each treatment plant's permit.

There are applications that enable operators to generate DMRs quickly. If the treatment plant's monitoring information is already stored electronically, then DMRs can be generated via templates, which contain the permit limits and calculation logic needed to generate DMR results. These results can be reviewed, refined, and approved as necessary before utility personnel print the final DMR for submission. Some utilities now submit computer-generated versions of DMRs—or just the data itself—electronically.

Occasionally, questions arise about permit limits, sampling requirements, reporting frequency, or contact information. So, NPDES permit limits and related information should be kept current and made readily available and searchable online.

***Wasteload Management and Projection Reports.*** Wasteload management and projection reports help utilities predict influent organic and hydraulic loadings 5 years into the future to determine whether they will exceed those of a treatment plant, pumping station, or collection system. The information enables utility managers to make practical decisions about extending collection systems and expanding treatment plants. The reports typically are filed with regulators and updated annually.

Utility staff begin by calculating the annual average flow, three consecutive maximum monthly average flows, and the biochemical oxygen demand (BOD) contribution per person. This information can be determined from reliable influent BOD, flow, and service population data. Statistics show that the per-person BOD contribution is approximately 0.07 to 0.09 kg/cap (0.15 to 0.2 lb/cap), and the average wastewater flow per capita ranges from 0.3 to 0.5 $m^3/d$ (75 to 125 gpd). Comparing the utility's data to established norms will show staff whether there are significant differences that should be investigated.

Staff can obtain the capacities of pumping stations, treatment plants, and collection systems from the design engineer's reports or the NPDES permit applications. They then should predict the future population based on census data and known growth patterns. (Between censuses, staff should use approved developer plans and issued building permits to predict population changes.)

Staff then make wasteload projections based on population projections and estimated per-capita hydraulic and organic loadings. If the 5-year wasteload projections exceed the facilities' capacities, staff should take steps to plan appropriate facility expansions or restrict growth in the relevant area(s).

***Pretreatment Program Reports and Records.*** Pretreatment programs are designed to control discharges of "known" priority pollutants to wastewater utilities so they do not endanger utility employees, damage biological treatment processes, or accumulate in biosolids or receiving waters (U.S. EPA, 1983). (For more information on pretreatment programs, see Chapter 4.)

Pretreatment program records typically include the following information:

- The program development report, which contains a pretreatment ordinance based on U.S. EPA's model pretreatment ordinance;

- Industrial wastewater permit applications;
- Copies of permits issued, including special conditions;
- Administrative records on sampling and testing costs;
- Inspection records;
- Enforcement records;
- Compliance schedules; and
- The annual pretreatment program report required by U.S. EPA.

A separate file must be maintained for each industrial discharger. Utilities should seriously consider an electronic information management system for this program because of the large amount of data and recordkeeping involved. Over time, the accumulated data can help utility staff troubleshoot malfunctioning processes and improve process-control efficiency.

There are applications available to manage industrial discharge permits (e.g., calculate draft limits, track permit revisions, generate permit documents, and enable the contents of these documents to be searched). They also can generate self-monitoring report (SMR) forms for pretreatment program participants to use. Once the sampling results or completed SMRs have been added to the database (manually or imported from a LIMS or Internet-based SMR module), the applications can automatically detect noncompliance events and notify both utility and industrial staff via e-mail. Another important application feature to consider is automated evaluations of significant noncompliance, according to a U.S. EPA guidance document (U.S. EPA, 2002). Whenever this feature detects a violation, it would record the incident for further action and issue a Notice of Violation.

***Toxics Reduction Evaluation Records.*** The U.S. Environmental Protection Agency requires that the toxics (also called *priority pollutants*) discharged to wastewater utilities be regulated. Whenever NPDES permits are being renewed, utilities with industrial customers (e.g., manufacturers, hospitals, medical centers, and water purveyors) must test treatment plant influent and effluent for toxics. (For a list of identified toxics, see CWA §307.) If toxics are found, the NPDES permitting authority will set effluent toxics limits for the treatment plant.

To achieve these limits, the utility must begin with a *toxics reduction evaluation* (TRE)—sampling and analysis of potential on-site and off-site sources to determine how the toxics are entering the system. These evaluations should be coordinated by a multidisciplinary team of toxicologists, chemists, engineers, and plant personnel. The overall goals of a TRE are to

- Sample potential sources of toxics,
- Characterize any toxics found in a sample,

- Pinpoint the source(s) by evaluating changes in toxicity after various chemical and physical processes, and
- Implement measures to reduce and eliminate the toxics.

In the process, the team must compile extensive records of each sample; these records include field collection sample forms, chain of custody forms, and quality assurance project plans. Team members examine these testing records to determine the significant contributors of toxics. Depending on the source, the toxics then can be abated via an in-house pollution prevention program, new pretreatment facilities, or household-chemical regulations (implemented by U.S. EPA). The identified toxics also are continuously monitored via NPDES reporting.

***Risk Management Program Reports.*** Clean Air Act §112(r)(7) requires U.S. EPA to develop a Risk Management Program (RMP) to help prevent accidental releases of regulated substances (e.g., chlorine gas, methane, and various flammable compounds) and reduce the severity of a release if one occurs. So, utilities and other organizations must develop and implement RMPs in accordance with the agency's accidental release prevention regulations (40 *CFR* 68).

For more information on RMP guidance and related records such as air-dispersion modeling results and storage records see http://yosemite.epa.gov/oswer/ceppoweb.nsf/content/EPAguidance.htm#Wastewater and http://yosemite.epa.gov/oswer/ceppoweb.nsf/content/RMPS.htm.

***Capacity, Management, Operations, and Maintenance Compliance Reports.*** Under U.S. EPA's draft proposed Sanitary Sewer Overflow (SSO) Rule, collection system utilities must meet five performance standards:

- Properly manage, operate, and maintain all parts of the collection system;
- Provide adequate conveyance capacity;
- Reduce the effect of any SSOs;
- Notify parties who may be exposed to an SSO; and
- Document the CMOM program in a written plan.

The CMOM program is a structured framework designed to help utility staff properly manage the collection system to optimize system performance, maintain facilities and capacity, and respond appropriately to any overflows. Currently, most wastewater utilities are developing strategies to meet anticipated CMOM requirements, and some are implementing them.

Training and recordkeeping are important aspects of a CMOM program. One best management practice is ensuring that O&M staff have the necessary knowledge and skills to keep the collection system functioning optimally. Each employee should be trained as is appropriate for his or her roles and responsibilities, and the O&M staff should keep staff-training records.

Another important aspect is an overflow response plan (a preparedness plan to deal with sewer overflows or line breaks). *Preparing Sewer Overflow Response Plans: A Guidebook for Local Governments* (APWA, 1999) is designed to help municipalities prepare overflow response plans. The guidebook's model plan includes objectives and organization, overflow response procedures, public advisory procedures, regulator notification procedures, media notification procedures, and steps to distribute and maintain the plan.

The overall goal of a CMOM program is to create a regular and consistent cycle, a more proactive approach to collection system maintenance. To ensure future CMOM compliance, utilities should optimize their collection systems, focus on eliminating preventable SSOs, and train their O&M staff appropriately. To determine how well the CMOM program is working, utilities should identify performance measures that correspond to program goals and track them. The following are commonly used performance measures:

- Number of customer complaints,
- Number of overflows,
- Number of blockages per mile,
- Number of pump station failures,
- Ratio of base flows to peak wet weather flows,
- Number of manholes inspected, and
- Miles of sewer cleaned.

The information needed to create a CMOM plan and monitor compliance can be found in the following documents:

- Collection system maps,
- Collection system design and construction standards,
- Complaint logs,
- Work orders,
- Sewer assessment reports,
- Standard maintenance procedures,
- SOPs,
- Typical forms used by field and office personnel,
- Organization charts, and
- Safety and training program manuals.

***Air Pollution Control Permit Reports.*** Local or regional air-pollution-control districts may regulate ambient air emissions from some wastewater utilities. The permits they issue may require utilities to report odor-control equipment outages, unusual plant conditions, unusual wastewater or biosolids characteristics, and complaints from the public or other agencies. Traditionally, the related recordkeeping was paper-based, but SCADA systems, CMMSs, and Customer Information Systems provide automated alternatives.

**OPERATING RECORDS.** Operating records consist of daily logs, weekly operating reports, monthly operating reports, and laboratory records. All of the data involved should be pertinent and useful because recordkeeping—especially manual recordkeeping—can be expensive and time-consuming. Traditionally, most recordkeeping was paper-based, but now most treatment plants are implementing computerized recordkeeping systems.

An operations management system (OMS) collects both plant operating data and laboratory test results. It also enables users to analyze the data and produce regulatory and other routine reports. There are many types of OMSs available. Some integrate real-time PCS or SCADA system data with water quality, flow, and related data. Some also address safety, quality control, and environmental management requirements. Many are intranet-based, so utility employees can access it directly via a WAN and remotely via dial-in facilities. All are updated regularly—so users typically have access to current information—and include links to SOPs, regulators' home pages, and O&M, health, safety, and other reference documents.

All of the records in this section can be directly integrated into an OMS. Operators should only keep those that are pertinent to wastewater processing, management, and administration. Any data that are not needed to control treatment processes or determine their efficiencies should not be entered into the system. [For more information on process control testing, see *Activated Sludge* (WEF, 2002).]

Because only laboratory test results can illustrate final effluent quality, all results—both good and bad—must be secure and reported accurately, even if the permitted parameters are monitored more frequently than regulators require. The NPDES permit requires that all results obtained via NPDES testing must be reported.

***Daily Operating Records.*** There are typically three forms of daily records—plant logs, daily bench sheets, and routine reports—and they may be paper- or computer-based. The data in these records include wastewater and treatment process measurements, power used, weather (e.g., temperature, rainfall, and other hydrological data), and other physical information. These data may be collected every few seconds, minutes, or hours, or only when the parameter has changed by at least a preset interval. They may be entered manually or via a PCS or SCADA system.

Daily administrative data include routine reports on personnel, payroll, small purchases, and related information that may or may not be part of an OMS.

In addition, a plant log may contain data on the progress of construction or maintenance work, equipment failure(s), employee accidents, floods or unusual storms, process bypasses, complaints registered, the names and affiliations of all visitors, and other matters. These data are valuable references.

These daily operating records can be the first step in a performance improvement program.

***Weekly and Monthly Operating Reports.*** In addition to the data in daily operating records, weekly and monthly operating reports contain pertinent information on wastewater treatment operations (e.g., sedimentation, aeration, disinfection, digestion, and solids management). Such process-control data include air application, centrifuge test, chemical dosages, clarifier loading, digester gas production, influent flow, mixed-liquor volatile suspended solids, organic loading, sludge age, mean cell residence time, microscopic analyses, return sludge flow, sludge blanket levels (via sludge judge or instrumentation), sludge settleability (via settleometer), sludge volume index, wastewater flow, and waste sludge quantity (yield).

Monthly operating reports should contain a summary of all data collected daily or weekly. Calculating monthly averages of useful process-control parameters can help reveal trends (e.g., consistent performance or changing conditions).

The reports should be printed on good-quality paper so they are suitable for submission to local regulators. (For samples of reports formatting and the types of data required, see Figures 6.2 through 6.7.)

Analyses of weekly and monthly report data will show any departures from previously established norms—especially when graphing both test results and operating parameters. For example, comparing a plot of mixed-liquor suspended solids versus time with a 7-day moving average versus daily data can reveal whether the amount of solids under aeration is adequate for treatment and if the wasting rate is correct. [For more information, see *Activated Sludge* (WEF, 2002).]

***Laboratory Records.*** Laboratory data must be recorded—at the very least, on printed bench sheets or in lab notebooks. Because there are no standard forms, most wastewater utilities develop their own data sheets for recording lab test results and calcula-

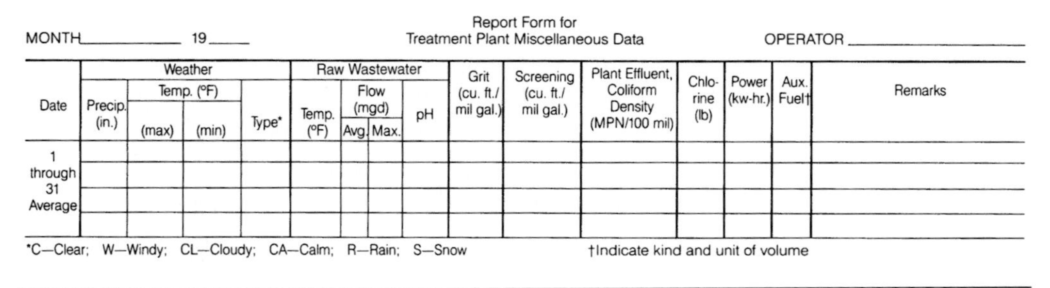

MONTH________ 19____ Report Form for Treatment Plant Miscellaneous Data OPERATOR____________

| Date | Weather | | | | Raw Wastewater | | | | Grit (cu. ft./mil gal.) | Screening (cu. ft./mil gal.) | Plant Effluent, Coliform Density (MPN/100 mil) | Chlorine (lb) | Power (kw-hr.) | Aux. Fuel† | Remarks |
|---|---|---|---|---|---|---|---|---|---|---|---|---|---|---|---|
| | Precip. (in.) | Temp. (°F) (max) | Temp. (°F) (min) | Type* | Temp. (°F) | Flow (mgd) Avg. | Flow (mgd) Max. | pH | | | | | | | |
| 1 through 31 Average | | | | | | | | | | | | | | | |

*C—Clear; W—Windy; CL—Cloudy; CA—Calm; R—Rain; S—Snow †Indicate kind and unit of volume

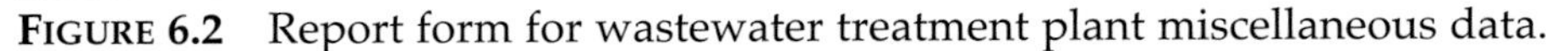

**FIGURE 6.2** Report form for wastewater treatment plant miscellaneous data.

MONTH________ 19____ Report Form for Treatment Plant Primary Treatment Data OPERATOR____________

| | 5-Day BOD | | | | | Suspended Solids | | | | | Suspended Volatile Solids | | | | Remarks |
|---|---|---|---|---|---|---|---|---|---|---|---|---|---|---|---|
| | Influent | | Effluent | | Removal | Influent | | Effluent | | Removal | Influent | | Effluent | | |
| | (mg/L) | (lb) | (mg/L) | (lb) | (%) | (mg/L) | (lb) | (mg/L) | (lb) | (%) | (mg/L) | (lb) | (mg/L) | (lb) | |
| 1 through 31 Average | | | | | | | | | | | | | | | |

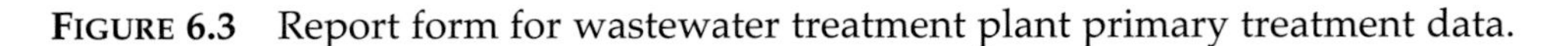

**FIGURE 6.3** Report form for wastewater treatment plant primary treatment data.

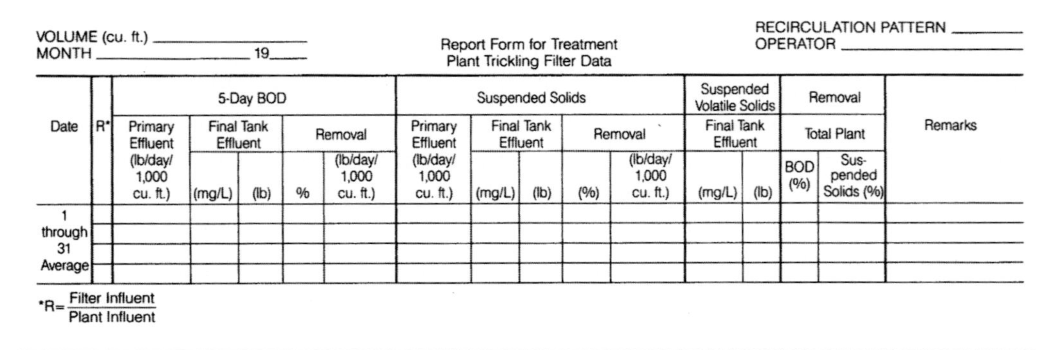

VOLUME (cu. ft.) ____________ RECIRCULATION PATTERN ________
MONTH ____________ 19____ Report Form for Treatment Plant Trickling Filter Data OPERATOR ____________

| Date | R* | 5-Day BOD | | | | | Suspended Solids | | | | | Suspended Volatile Solids | | Removal | | Remarks |
|---|---|---|---|---|---|---|---|---|---|---|---|---|---|---|---|---|
| | | Primary Effluent | Final Tank Effluent | | Removal | | Primary Effluent | Final Tank Effluent | | Removal | | Final Tank Effluent | | Total Plant | | |
| | | (lb/day/ 1,000 cu. ft.) | (mg/L) | (lb) | % | (lb/day/ 1,000 cu. ft.) | (lb/day/ 1,000 cu. ft.) | (mg/L) | (lb) | (%) | (lb/day/ 1,000 cu. ft.) | (mg/L) | (lb) | BOD (%) | Suspended Solids (%) | |
| 1 through 31 Average | | | | | | | | | | | | | | | | |

$*R = \frac{\text{Filter Influent}}{\text{Plant Influent}}$

**FIGURE 6.4** Report form for wastewater treatment plant trickling filter data.

MONTH__________ 19____

Report Form for Treatment Plant Activated Sludge Data

OPERATOR __________

| Date | Air Applied | | | 5-Day BOD | | | Suspended Solids | | Suspended Volatile Solids | | Removal | | DO Final Effluent (mg/L) | Mixed Liquor | | | | Return Sludge | | Waste Sludge (1,000 gal.) | Remarks |
|---|---|---|---|---|---|---|---|---|---|---|---|---|---|---|---|---|---|---|---|---|---|
| | (hr) | (cfm) | (cu. ft/lb BOD Rem.) | Primary Effluent (lb/day/ 1,000 cu. ft.) | Final Effluent | | Final Effluent | | Final Effluent | | Total Plant | | | MLSS (%) | Settled Sludge Volume @ 30 min ($SSV_{30}$) | Sludge Volume Index | DO (mg/L) | $Q_{RAS}$ | SS (mg/L) | | |
| | | | | | (mg/L) | (lb) | (mg/L) | (lb) | (mg/L) | (lb) | BOD (%) | SS (%) | | | | | | | | | |
| 1 | | | | | | | | | | | | | | | | | | | | | |
| through | | | | | | | | | | | | | | | | | | | | | |
| 31 | | | | | | | | | | | | | | | | | | | | | |
| Average | | | | | | | | | | | | | | | | | | | | | |

FIGURE 6.5 Report form for wastewater treatment plant activated sludge data.

MONTH__________ 19____

Report Form for Treatment Plant Anaerobic Digester and Sludge Data

OPERATOR __________

| Date | Raw Sludge | | | | | Supernatant | | | | | | Sludge near Bottom | | | | | Temp. (°F) | Gas | | Remarks (include gal. of sludge to beds) |
|---|---|---|---|---|---|---|---|---|---|---|---|---|---|---|---|---|---|---|---|---|
| | (gal) | Loading (lb/1,000 cu. ft.) | pH | Total Solids (%) | Volatile Solids (%) | (gal) | pH | Total Solids (%) | Volatile Solids (mg/L) | Suspended Solids (mg/L) | 5-Day BOD (mg/L) | Total Solids (%) | Volatile Solids (%) | Volatile Acids (mg/L) | pH | Alkalinity | | Produced (cu. ft.) | Wasted (cu. ft.) | |
| 1 | | | | | | | | | | | | | | | | | | | | |
| through | | | | | | | | | | | | | | | | | | | | |
| 31 | | | | | | | | | | | | | | | | | | | | |
| Average | | | | | | | | | | | | | | | | | | | | |

FIGURE 6.6 Report form for wastewater treatment plant anaerobic digester and sludge data.

MONTH__________ 19 ____

Report Form for Treatment Plant Aerobic Digestion and Sludge Data

OPERATOR __________

| Date | Raw Sludge | | | | | | | Digesting Sludge | | | | | | | Digested Sludge | | | Supernatant | |
|---|---|---|---|---|---|---|---|---|---|---|---|---|---|---|---|---|---|---|---|
| | Volume (gal.) | Leading (lb/1,000 cu. ft.) | pH | TS (%) | VS (%) | COD (mg/L) | N (mg/L) | pH | TS (%) | VS (%) | COD (mg/L) | N (mg/L) | DO (mg/L) | pH | TS (%) | VS Reduction (%) | N (mg/L) | TS (%) | N (mg/L) |
| 1 | | | | | | | | | | | | | | | | | | | |
| through | | | | | | | | | | | | | | | | | | | |
| 31 | | | | | | | | | | | | | | | | | | | |
| Average | | | | | | | | | | | | | | | | | | | |

FIGURE 6.7 Report form for wastewater treatment plant aerobic digester and sludge data.

tions. The sheets should be designed to simplify the process of recording, reviewing, or recovering data.

A more sophisticated option would be LIMS software, which can automatically acquire data from instruments, manage lab work data, and store and print test results. [For more information on LIMS, as well as sound data management advice, see *Good Automated Laboratory Practices* (U.S. EPA, 1995).]

Laboratory records contain information related to the sample chain of custody, the test(s) performed on the sample, and the test results.

***Industrial Waste Management Records.*** When industrial and domestic wastewater have similar characteristics, they typically are treated together because it is considerably less expensive to treat them jointly than separately. Municipalities that operate wastewater utilities typically have ordinances specifying the parameters that industrial wastewater must meet to be accepted by the utilities. Otherwise, the utilities' biological treatment processes could be damaged.

To ensure that the ordinances' requirements are met, municipalities periodically sample and test wastewater from nondomestic sources. They also should confirm that the industrial wastewater can be biologically treated.

So, industrial waste management records should include the following information:

- The industrial wastewater permit application and permit;
- The required sampling, analyses, and chain-of-custody records;
- Test results (preferably certificates of analyses issued by certified testing laboratories);
- Notices of ordinance violations, consent decrees, and other enforcement actions;
- Surcharge calculations for industrial wastewater discharges that were biologically treatable but stronger than domestic wastewater (e.g., higher concentrations of total dissolved solids, soluble BOD, BOD, total suspended solids, phosphorus, and ammonia-nitrogen);
- Compliance schedules (with milestone dates); and
- A record of all costs associated with the sampling and testing (suitable for billing purposes).

Industrial sources subject to pretreatment-program reporting requirements must maintain readily available, monitoring-related records for at least 3 years (longer if in the midst of litigation). All pretreatment activities should be documented and maintained in the manner prescribed by the utility.

The industrial pretreatment program authority should maintain the following records:

- Industrial waste questionnaire;
- Permit applications, permits, and fact sheets;
- Inspection reports;
- Industrial source reports;
- Monitoring data (e.g., laboratory reports);
- Required plans (e.g., slug control, sludge management, and pollution prevention);
- Enforcement activities;
- All correspondence to and from the industrial source;
- Phone logs and meeting summaries;
- Program procedures; and
- Program approval and modifications of the utility's NPDES permit(s).

A computerized data management system can make it easier for pretreatment program staff to track due dates, submissions, deficiencies, notifications, and so on, and calculate noncompliance with effluent limits. Many also use standard forms (e.g., inspection questionnaires, chains-of-custody, and field measurement records) and procedures (e.g., sampling and periodic compliance report reviews) to help organize data and ensure that they are consistent.

Program staff also must annually publish—in its area's largest daily newspaper—a list of industrial sources that were in significant noncompliance with their requirements at any time during the previous 12 months. In addition, 40 *CFR* §403.12(I) requires them to submit annual reports to the NPDES permitting authority. The reports must document the program's status and describe all activities performed during the previous calendar year.

All laboratory records must be based on U.S. EPA-approved analytical tests. Many of these tests can be found in the most recent edition of *Standard Methods for the Examination of Water and Wastewater* (APHA et al., 1998).

Some pretreatment programs also develop local limits for pollutants (e.g., metals, cyanide, 5-day BOD, total suspended solids, oil and grease, and organics) to prevent interference, pass-through, sludge contamination, or worker health and safety problems at the wastewater treatment plant. Typically, these limits are imposed at the "end-of-pipe" discharge from an industry (i.e., where the company's discharge pipe connects to the plant's collection system).

***Relating Recordkeeping to Operations.*** The analysis results that a utility chooses to record should reflect the treatment plant's operating conditions (e.g., influent flowrate, influent wasteload, the wastewater's characteristics and treatability, and effluent concentrations). With these data, operators should be able to determine how efficient each treatment process is and predict how the plant's effluent will affect receiving waters.

If the data are entered into a computerized OMS database, utility staff can analyze various trends more easily. Varying results, for example, would alert staff to expect difficulties in plant operations. Uniformly increasing loads could be evaluated to determine when wastewater loads will reach the plant's capacity. Sudden variations in results may reveal accidental discharges, collection-system damage, or other incidents.

***Energy Management.*** Energy (e.g., electricity, gas, and fuel oil) is a significant line item in a treatment plant's operating budget. Many utilities' operating reports include a standard report or graph on energy use.

To manage energy costs, utilities should monitor the price and use of each form of energy via a PCS, SCADA system, or special monitoring equipment, so they can plot monthly trends and note changes. A formal energy audit, based on monitoring data and historical records, would show staff precisely how power is used on-site and how use varies throughout the day, week, and seasons. Utility staff also can determine the energy efficiency of each treatment process if they collect energy use, energy cost, and process volume statistics and plot daily or monthly trends (expressed as energy use or cost per million gallons treated). Once personnel have baseline information about each process' energy consumption, they can make changes (e.g., automate controls) to improve energy efficiency and measure resulting improvements.

Electricity's demand charges and other characteristics (e.g., lower-than-desired power factor) can be significant, so staff should be familiar with the electricity rate schedule. Operators could even have a list of equipment—chosen based on monitoring and energy audit data—that they could shut down during peak demand times without impairing treatment results.

**PREVENTIVE AND CORRECTIVE MAINTENANCE RECORDS.** Maintenance has become more important as permit limits become stricter and utilities become more automated. The former emphasis on *corrective maintenance*—fixing equipment when it breaks—is shifting to *preventive maintenance*—keeping equipment in good working order—and ultimately to comprehensive asset management to maximize the equipment's lifespan.

As a result of these changes, utilities need a dedicated, well-trained maintenance staff to keep their facilities in satisfactory repair. They also need a comprehensive maintenance management system to keep track of their activities.

Whether the maintenance system is computerized or manual, it should include equipment-, storeroom-, and inventory-management records; work-schedule records; budget records; and maintenance-cost records. The samples shown in Figures 6.8 through 6.11 illustrate the information typically needed to ensure that the equipment maintenance history is up to date.

***Predictive Maintenance.*** A method for protecting valuable items, *predictive maintenance* involves testing equipment to identify incipient problems and scheduling whatever maintenance is needed to avoid them. The most common predictive maintenance methods are infrared thermography, vibration analysis, oil analysis, and surge testing. The equipment can sometimes be tested while in operation. (For more information on predictive maintenance methods, see Chapter 12.)

Work Order No. ____________________ Date ________

| Location | Requested by: (Phone) | Priority: |
|---|---|---|

| Equipment Name | No. | | | |
|---|---|---|---|---|
| | | ☐ Inspect | ☐ Replace | ☐ Service |
| | | ☐ Repair | ☐ Overhaul | ☐ Paint |
| | | Work Description | | |
| | | | | |
| | | | | |
| | | | | |
| | | Work Performed/Comments | | |
| | | | | |

Job Estimate

Labor $ ________

Material $ ________

________________ Shift Operator

________________ Maintenance Superintendent

WORK ORDER

| Personnel Assigned | Manhours | Date | Work Done | Parts & Materials |
|---|---|---|---|---|
| | | | | |
| | | | | |
| | | | | |
| | | | | |
| | | | | |
| | | | | |
| | | | | |
| | | | | |
| Total | | | | |

Work Completed By ______________________________ Date ________

Work Accepted By ______________________________ Date ________

FIGURE 6.8 Sample work order form.

Plant No. ____ MACHINE NO. ____ INVENTORY NO. ____ MOTOR NO. ____
H.P. ____ MANUFACTURED BY ____

| SERIES | SHUNT | COMPOUND | SYNCHRONOUS | INDUCTION |
|---|---|---|---|---|
| TYPE | FRAME | SPEED | VOLTS | AMPERES |
| PHASE | CYCLES | TEMP. RISE | EXCITATION AMPS. | ROTOR OR ARM, SER. NO. |
| MODEL NO. | FORM NO. | STYLE OR S.O. NO. | SERIAL NO. | |
| MFGRS. ORDER NO. | | OUR ORDER NO. | | |

CONNECTION DIAGRAM—ROTOR OR ARMATURE

| SPECIFICATION | BEARINGS | SHAFT EXTENSION | PULLEY | GEAR | V BELT DRIVE |
|---|---|---|---|---|---|
| OPEN ☐ | SLEEVE ☐ | DIA. ____ | DIA. ____ | TEETH ____ | NO. GROOVES ____ |
| EXP. PROOF ☐ | BALL ☐ | LENGTH ____ | FACE ____ | PITCH ____ | PITCH DIA ____ |
| DRIP PROOF ☐ | ROLLER ☐ | KEYWAY ____ | | FACE ____ | BELT SECTION |
| TOTALLY ENCL. ☐ | | | | | ☐ A-1/2×11/32 |
| VERTICAL ☐ | | | | | ☐ B-21/32×7/16 |
| ____ ☐ | | | | | ☐ C-7/8×17/32 |
| | | | | | ☐ D-1-1/4×3/4 |

MOTOR SERVICE RECORD

| DATE INSTALLED | LOCATION | APPLICATION |
|---|---|---|
| | | |
| | | |
| | | |

| DATE REPAIRED | REPAIRS OR PARTS REPLACED | CAUSE | REPAIRED BY | TOTAL COST |
|---|---|---|---|---|
| | | | | |
| | | | | |
| | | | | |
| | | | | |
| | | | | |
| | | | | |

**FIGURE 6.9** Sample motor service record—form 1.

| MOTOR H.I. | A.C. | D.C. | TYPE | FRAME | R.P.M. | PHASE | CYCLES | VOLTS | DIVISION | SECTION |
|---|---|---|---|---|---|---|---|---|---|---|

| WINDING | TOOL NO. | STYLE NO. | SERIAL NO. | LOCATION-SECTION | COL. NO. |
|---|---|---|---|---|---|

| DATE INSTALLED | DATE IN STORAGE | DATE INSTALLED | DATE IN STORAGE | DATE INSTALLED | DATE IN STORAGE |
|---|---|---|---|---|---|
| DRIVES MACHINE | TOOL NUMBER | DRIVES MACHINE | TOOL NUMBER | DRIVES MACHINE | TOOL NUMBER |

| DATE | BLOWN OUT | BEARINGS | END PLAY | BRUSHES & HOLDER | COMMU. | MEGGER | GENERAL CONDITION | INSPECTED BY |
|---|---|---|---|---|---|---|---|---|
| | | | | | | | | |
| | | | | | | | | |
| | | | | | | | | |
| | | | | | | | | |
| | | | | | | | | |
| | | | | | | | | |
| | | | | | | | | |
| | | | | | | | | |
| | | | | | | | | |
| | | | | | | | | |
| | | | | | | | | |
| | | | | | | | | |

DATE TAKEN OUT OF SERVICE ____ DATE SCRAPPED ____

**FIGURE 6.10** Sample motor service record—form 2.

| JAN. | FEB. | MAR. | APR. | MAY | JUNE | JULY | AUG. | SEPT. | OCT. | NOV. | DEC. |
|---|---|---|---|---|---|---|---|---|---|---|---|
| 1 2 3 4 | 1 2 3 4 | 1 2 3 4 | 1 2 3 4 | 1 2 3 4 | 1 2 3 4 | 1 2 3 4 | 1 2 3 4 | 1 2 3 4 | 1 2 3 4 | 1 2 3 4 | 1 2 3 4 |

| Preventive Maintenance Program | Equipment Record Number ______ |
|---|---|
| EQUIPMENT DESCRIPTION | ELECTRICAL OR MECHANICAL DATA |
| Name | Size |
| Serial No. | Model |
| Vendor | Type |
| Vendor Address | |
| Vendor Rep. Phone | |
| Initial Cost Date | |

| WORK TO BE DONE | FREQUENCY | TIME |
|---|---|---|
| | | |
| | | |
| | | |
| | | |
| | | |
| | | |
| | | |

| DATE | WORK DONE | SIGNED | DATE | WORK DONE | SIGNED | DATE | WORK DONE | SIGNED |
|---|---|---|---|---|---|---|---|---|
| | | | | | | | | |

FIGURE 6.11 Sample equipment record card.

Predictive maintenance records are simple (e.g., a vibration analysis printout of a bearing check, an infrared image of electrical facilities, or a surge-testing photograph of motor windings). They should be linked with the corresponding equipment record and kept as long as the asset is in service.

***Equipment Management.*** Before creating an equipment management system, utility staff should decide how to number each piece of equipment (for identification purposes, quick digital access, and links among related records). The simplest method is to "count off" each area or building by thousands (e.g., the headworks is 1000, the primary treatment area is 2000, and the primary clarification area is 3000) and then number each piece of equipment within the area or building appropriately (e.g., the first bar screen is 1001, the second is 1002, and the third is 1003). There should be plenty of unassigned numbers left for equipment added later. Complex wastewater treatment facilities may require a more sophisticated approach to identification, in which case some CMMSs can help generate and organize unique equipment identifiers.

In addition to alphanumeric identifiers, each equipment record should include the

- Equipment's name, description, and location;
- Name and address of its manufacturer, supplier, or builder;
- Cost and installation date;
- Type, style, and model;
- Capacity and size rating;
- Mechanical and electrical data;
- Serial code;
- Preventive maintenance requirements (procedures and frequency);
- Proper lubricants and coatings; and
- Link(s) to the list of spare parts.

Over time, each record also should accumulate a history of the corrective and preventive maintenance work performed on the equipment. This should include related problems, scheduled tasks, completed tasks, the initials of the person(s) who did each task, and the costs, materials, and hours involved.

***Maintenance Schedule.*** Utilities schedule maintenance work to reduce equipment failure, enhance asset performance, comply with manufacturer warrantees, prioritize tasks, minimize idle time, and reduce overtime. Preventive and corrective maintenance scheduling also should take into account the maintenance group's size, each employee's capabilities, and all of the equipment's daily, weekly, monthly, quarterly, semiannual, and annual maintenance requirements. Seasonal weather patterns also should be taken into account when scheduling certain tasks (e.g., painting and tank cleaning) to improve the likelihood of prompt completion.

At a small treatment plant (i.e., less than 1890 $m^3/d$ [0.5 mgd]), depending on treatment system complexity, one employee may be responsible for scheduling, tracking, and perhaps performing much of the work. Small plants may use paper-based work order forms to assign maintenance and repair work (Figure 6.8). These forms should note the task, the personnel assigned, the parts and tools needed, the steps involved, and any safety issues. For example, an equipment-lubrication work order should note the manufacturer's recommendations, points of lubrication, frequency of lubrication, and the need to follow lockout/tagout procedures before beginning the task.

Once completed, labor and materials costs also should be noted on the form. Utility staff can use this information to better prepare maintenance budgets and determine whether a particular piece of equipment is cost-effective.

Staff at medium and large treatment plants (i.e., greater than 1890 $m^3/d$ [0.5 mgd]) typically rely on CMMSs, which integrate work orders, scheduling, performance, cost, equipment history, workforce requirements, budgets, parts inventory, and maintenance work backlog into one comprehensive equipment-management process. Staff input all the equipment's preventive maintenance schedules, along with time and materials estimates, into the CMMS. The system then establishes a calendar of tasks and issues work orders. Data on completed work orders are entered into the CMMS by either clerical staff or the staff who did the work.

Computerized maintenance management systems can automatically provide reports on work schedules, labor and parts costs, maintenance history, and workforce requirements. They also can provide information on work order backlogs and available resources.

Managers should analyze maintenance reports periodically to ensure that the system is operating efficiently. For example, they should review the daily list of work order requests. At the beginning of each month, two work order reports should be produced: one on work completed the previous month and one on the work order backlog [noting the reason(s) for noncompletion and schedule deviation]. Both reports should include statistics (e.g., type, priority, and age of the work order). Managers should evaluate these reports, compare them with those for previous months, and note any trends or problems with receiving parts or supplies.

***Storeroom and Inventory Management.*** As a minimum, inventory records should include the part or material description, number, quantity, reorder level, purchase date, cost, and vendor(s). They also should note when items were used and for what purpose so staff can properly monitor inventory and schedule re-orders. Staff also should maintain a complete catalog of parts and equipment in stock, and always keep storeroom and inventory system records up to date.

A CMMS typically includes a module for automating all inventory transactions. To ensure that this inventory module remains accurate and orders replacements appropriately, staff should promptly note when each part or material is used or received. They also should manually count parts and materials periodically and compare their results with the inventory module's records.

Utility staff should use a purchase-order system to track orders (e.g., item ordered, quantity, cost, vendor, date ordered, date received, and date invoice paid). This system may be part of a CMMS or a financial software suite. If part of the financial software, there should be a bi-directional interface between it and the CMMS purchase-order module.

***Budgets and Maintenance Costs.*** Complete maintenance records can help staff prepare accurate budgets. To do this, the records should distinguish between preventive-maintenance and corrective-maintenance costs. They also should note the costs of both personnel's and contractor's efforts.

## ADMINISTRATIVE REPORTS

***Annual Operating Reports.*** When projects are financed via municipal bond issues, financial institutions require the utility to publish an annual report that summarizes the current status of the facilities and developments that occurred during the previous year.

The report should have four sections: general, operations, maintenance, and improvements. The general section should outline the scope of the report and list the names and addresses of persons associated with the facility. It also should note the facts about the project financing, the facility's financial status (e.g., O&M costs, construction costs, and remaining debts), and estimated completion dates for construction projects.

The operations section should briefly describe the wastewater treatment system and how it works. It should contain flow data and graphs comparing hydraulic capacities and loadings (past, present, and 5-year projections). It also should discuss the past year's flow data and operating efficiency (with a supporting table), compare past and present water quality (noting any efficiency trends), and outline process problems and their solutions.

The maintenance section should outline the maintenance projects completed during the year; describe preventive, predictive, and corrective maintenance activities; and list equipment repair and replacement costs. It also should describe the major projects and projected costs planned for the upcoming year.

The improvements section should describe capital additions to storage facilities, pumping stations, distribution system, and wastewater treatment plants. It also should list their costs and completion dates (for future reference).

***Insurance Coverage.*** The insurance coverage report should include the types of policies, names of insurance companies, policy numbers, effective and expiration dates, and amounts of coverage. A wastewater utility's insurance coverage should be based on replacement costs and updated annually. Coverage based on property-value appraisals should be updated when appropriate.

Utility staff should hire a responsible insurance broker to recommend coverage. Some larger agencies may be self-insured.

***Annual Budget.*** The annual budget may be included in the annual report or maintained as a separate report. It should list the previous year's budget, the current year's budget and actual expenditures and receipts, and the upcoming year's budget.

***Annual Audit Report.*** Most utilities are legally required to be annually audited by a certified public accountant. Even if audits are not required, annual audits are a sound business practice. After examining a sample of receipts and expenditures, the accountant should be able to certify that the utility has been managed according to acceptable accounting practices. The utility should keep copies of the audit reports for as long as it exists.

***Personnel Training Records.*** Every employee should have a training record. The U.S. Occupational Safety and Health Administration and other regulators require evidence of minimum training. The record should contain the employee's name, job description, work experience, education, and training needs. Comparing job descriptions with employees' knowledge and skills may reveal training needs. Also, maintenance work could be planned based on staff skills (often via a CMMS linked to employee training and certification records).

There are many types of training, all of which should be recorded in the participants' records. On-the-job training, for example, involves pairing an inexperienced worker with an experienced one until the inexperienced worker has learned enough to work independently. The training time and skills achieved should be recorded. Also, an in-plant library of technical materials and manuals could be made available to personnel for self-study.

In-house training sessions can be useful for learning safety procedures, process changes, and first aid, as well as preparing for licensing exams. Workshops, short courses, and seminars address a wide variety of O&M topics, and participants who satisfactorily complete such training typically receive certificates noting the topic and number of hours involved. Copies of these certificates should be included in the training records.

Training is not just for new employees; experienced staff may need training to upgrade skills and learn how to use new materials and equipment efficiently. Also, given today's concerns about terrorism and security, all staff should be trained on the following:

- The characteristics and uses of potential weapons,
- Vulnerability assessments of local water and wastewater utilities,

- Emergency response plans,
- Responding to and recovering from incidents, and
- Interacting with the media and residents during and after an act of terrorism.

For more information on the contents of a personnel or human resources management system, see Chapter 3.

## ACCOUNTING RECORDS

***Bill Payment.*** A requisition system can make bill paying simpler and more prompt. A purchasing system or CMMS can automatically create and convert requisitions to purchase orders (POs). Purchase orders are sent electronically or mailed to vendors and suppliers. Then, when products and/or services are received, invoices are received, and the appropriate personnel should approve and sign them. Each requisition, PO, invoice, and bill should be assigned a code number so all financial information can be categorized and tracked appropriately (Table 6.2).

Requisition systems can be a useful management tool at all wastewater utilities. They can help staff maintain financial records more easily. Staff also can use the system to compare options and control collection, treatment, and administrative costs. Requisition system data also can serve as a basis for developing future budgets.

***Financial Reports.*** The reporting frequency typically coincides with the timing of the operating or permit holder, from the perspective of the regulators.

Periodically comparing expenditures and receipts with budgets enables staff to adjust their plans as necessary to avoid problems at the end of the fiscal year (Table 6.3). This process can be automated so weekly, monthly, or quarterly financial status reports will be readily available. Then, staff can compare their budgets and expenses in each category. Staff also can compare budgeted and actual revenues, as well as revenues and expenditures. The automated process also can be designed to project the full year's expenses based on receipts logged to date, so staff can note trends and prevent budgetary overruns.

An established, regular financial reporting system will make annual-budget and rate-increase approvals less controversial because staff will be more familiar with the plant's finances and prepared to act responsibly.

**EMERGENCY RESPONSE RECORDS.** Treatment plant staff should document emergencies and their responses to them because this information can be used to update the emergency response plan and track problem areas for future construction projects. Whenever a significant emergency threatens or occurs, staff should compile an emer-

**TABLE 6.2** Example of expenditure categories.

| Code | Function |
|---|---|
| Operations and maintenance—wastewater treatment plant | |
| 101 | Labor cost—regular operating personnel. |
| 102 | Labor cost—part-time personnel. |
| 103 | Electrical power purchased. |
| 104 | Fuel. |
| 105 | Process chemicals (lime, ferric chloride, aluminum sulfate, polymer, and chlorine). |
| 106 | Equipment and supplies. |
| 107 | Equipment and maintenance. |
| 108 | Buildings and grounds maintenance (cleaning and grounds care supplies). |
| 109 | Emergency repairs (by outside repair service). |
| 110 | Sludge dewatering and disposal. |
| 111 | Telephone, telemarketing, and metering. |
| 112 | Miscellaneous. |
| 113 | Capital improvement. |
| Operations and maintenance—wastewater collection system | |
| 201 | Labor cost—regular operating personnel. |
| 202 | Labor cost—part-time personnel. |
| 203 | Electrical power. |
| 204 | Fuel. |
| 207 | Equipment maintenance. |
| 208 | Buildings and grounds maintenance. |
| 209 | Emergency repairs (by outside repair service). |
| 210 | Meter calibration repair and replacement. |
| 211 | Line flushing and cleaning. |
| 212 | Telephone. |
| 213 | Miscellaneous. |
| 214 | Capital improvements. |
| Administrative and clerical | |
| 301 | Labor cost |
| 302 | Billing service. |
| 303 | Legal fees. |
| 304 | Audit fees. |
| 305 | Engineering fees. |
| 306 | Insurance. |
| 307 | Social security. |
| 308 | Meter reading. |
| 309 | Miscellaneous. |

TABLE 6.3 Sample format for comparing revenues and expenditures to budgeted amounts.

| Receipt or expenditure code number | Current budgeted amount | Expended or received this month | Expended or received year-to-date | Percent expended or received year-to-date |
|---|---|---|---|---|
| | | | | |

gency conditions report detailing the event. If the plant is flooded, for example, the report should note the following:

- When plant staff were notified of impending flooding,
- When flood water entered the site,
- Where water first entered the plant,
- The highest water level on-site (measured in relation to the plant's physical structures),
- The equipment or structures damaged by the flood,
- Reports of the receiving stream's maximum flood stage,
- The protective actions that staff undertook,
- Other organizations or agencies contacted and actions taken,
- How long and how much the water quality was affected,
- Descriptions of the repairs and replacements required to restore the plant to pre-flood conditions,
- Any job-related staff injuries,
- Any contractors, repair services, or equipment vendors involved in the repairs and replacements (including the names of their representatives),
- The costs of repairs and replacements,
- Actions taken to prevent the recurrence of flooding, and
- Photographs or videos of the flood and resulting damage.

Much of this information also would be included in the utility's annual operating report and may be necessary if insurance claims must be submitted [especially if submitting to the Federal Emergency Management Agency (FEMA)]. For more information on FEMA's reporting requirements—which have been extensively revised—see the agency's Web site (www.fema.gov). For more guidance on disaster management, see *Natural Disaster Management for Wastewater Treatment Facilities* (WEF, 1999). This manual explains what to expect during a disaster (e.g., floods, earthquakes, and hurricanes) and focuses on disaster-recovery planning and management.

**RECORD PRESERVATION.** Typically, paper correspondence and reports should be stored in folders in filing cabinets. Equipment records should be stored and kept at least as long as the equipment itself is in service. (It is better to keep records too long than dispose of them too soon.)

Digital records (those stored on computers) should be backed up routinely, and copies of the backups should be maintained off-site. A critical computer system may be

**TABLE 6.4** Distribution of records.

| | Treatment plant | Operator's vehicle | Engineer | Federal | State | Trustee |
|---|---|---|---|---|---|---|
| Collection system drawings | ■ | ■ | ■ | ■ | | |
| Treatment plant drawings | ■ | | ■ | ■ | ■ | |
| Operations and maintenance manual | ■ | | ■ | ■ | ■ | |
| Manufacturer's literature | ■ | | ■ | ■ | ■ | |
| Weekly operating reports | ■ | | ■ | | ■ | |
| Annual operating reports | ■ | | ■ | | | |
| Expenditure reports (requisition) | ■ | | | | | ■ |
| Monthly reports | ■ | | ■ | ■ | | |
| Insurance certificates | ■ | | | | | |
| Correspondence with regulatory agencies | ■ | | ■ | | | |
| Audit | ■ | | ■ | | | |
| Personnel records | ■ | | | | | |

backed up hourly or more frequently and have a fully redundant system available in a secondary, remote location when the primary system is unavailable.

To prevent records from being destroyed in a flood, fire, or other disaster, they should be copied and distributed among multiple sites (Table 6.4). Some large utilities are going a step further and having their employees access records remotely via mobile data terminals or wireless notebook computers.

## REFERENCES

American Public Health Association; American Water Works Association; Water Environment Federation (2005) *Standard Methods for the Examination of Water and Wastewater,* 21st ed.; American Public Health Association: Washington, D.C.

American Public Works Association (1999) *Preparing Sewer Overflow Response Plans: A Guidebook for Local Governments;* American Public Works Association: Washington, D.C.

American Water Works Association (2001) *Excellence in Action: Water Utility Management in the 21st Century,* Lauer, W. C., Ed.; American Water Works Association: Denver, Colorado.

American Water Works Association Research Foundation (2004) *Creating Effective Information Technology Solutions.* American Water Works Association: Denver, Colorado.

Huxhold, W. E. (1991) *Introduction to Urban Geographic Information Systems.* Oxford University Press: New York.

U.S. Environmental Protection Agency (1995) *Good Automated Laboratory Practices,* Report 2185; U.S. Environmental Protection Agency: Washington, D.C.

U.S. Environmental Protection Agency (1983) *Guidance Manual for POTW Pretreatment Program Development*; U.S. Environmental Protection Agency: Washington, D.C.

U.S. Environmental Protection Agency (2002) *Guidance on Environmental Data Verification and Data Validation;* U.S. Environmental Protection Agency: Washington, D.C.

Water Environment Federation (2002) *Activated Sludge,* 2nd ed.; Manual of Practice No. OM-9; Water Environment Federation: Alexandria, Virginia.

Water Environment Federation (1999) *Natural Disaster Management for Wastewater Treatment Facilities*; Special Publication; Water Environment Federation: Alexandria, Virginia.

Water Environment Research Foundation (2002) *Sensing and Control Systems: A Review of Municipal and Industrial Experiences.* Water Environment Research Foundation: Alexandria, Virginia.

## SUGGESTED READINGS

Water Environment Federation (2006) *Automation of Wastewater Treatment Facilities,* 3rd ed.; Manual of Practice No. 21; Water Environment Federation: Alexandria, Virginia.

Water Environment Research Foundation (1999) *Improving Wastewater Treatment Plant Operations Efficiency and Effectiveness;* Water Environment Research Foundation: Alexandria, Virginia.

# Chapter 7

# Process Instrumentation

# INTRODUCTION

With each renewal of a plant's discharge permit, operations are required to meet additional and more stringent requirements for the quality of a plant's effluent. As a result, plant operations now control a growing number of more complicated processes. Often, plant staff levels are not increased commensurately with the increased numbers and complexities of the processes. However, if the processes are to operate efficiently and meet specific standards, proper monitoring and control are essential.

Traditionally, many types of analyses (e.g., chlorine residual, dissolved oxygen, pH, turbidity, and nutrient concentrations) have been performed in the laboratory on grab or composite samples. However, today, instrumentation allows continuous, online monitoring of these parameters. Real-time automatic monitoring and control of costly resources can routinely achieve significant savings in energy and chemical consumption without sacrificing effluent quality.

# INSTRUMENTATION

Instrumentation in wastewater treatment plants includes a wide array of pneumatic, hydraulic, mechanical, and electrical or electronic equipment, ranging from simple gauges, flow meters, and control panel indicators to sophisticated plant-wide control systems with multiple computers, video displays, and printers. Instrumentation can do the following:

- Provide a real-time window on process operations by pinpointing transient conditions and allowing correlation of interprocess operations,
- Improve facility maintenance and equipment availability,
- Carry out automatic shutdown to reduce equipment damage resulting from mechanical failures by documenting and quantifying run status,
- Simplify implementation of preventive maintenance programs, and
- Facilitate information management systems.

Many process measurements may not be important enough to economically justify online instruments. The individual parameters to be measured online will vary from plant to plant and even within the same plant. Justification of an online instrument or control system includes identifying and quantifying both tangible and intangible benefits and liabilities. Tangible benefits include direct savings of funds by decreasing resource consumption or staffing levels and avoiding discharge permit violations and subsequent fines. Although sometimes considered an intangible benefit, instrumentation can significantly improve personnel protection from hazards found in treatment plants and collection systems. The cost of appropriate safety instrumentation pales compared with the cost of a human life or limb, or a lawsuit.

Intangible benefits are those for which a funding level is difficult to quantify. They include detailed process performance accountability, reduction of operator busywork, and peace of mind. Each plant's management staff and governing bodies most often determine the importance of intangible benefits.

Benefits are balanced by liabilities. Liabilities associated with instrumentation and control systems include significant capital expenditures and ongoing maintenance requirements. Sophisticated instruments and control systems require new staff skills and may warrant significant changes in a facility's organizational structure. Poorly designed or applied instrumentation and control systems can cost much more than the most idealistically projected benefits. Too much instrumentation or automation can be just as bad as (or worse than) having none.

**SENSORS.** A sensor is a device that measures a process variable (e.g., flow) and converts the information about the variable into a form operators can use. The conversion may result in a signal that permits economical transmission of the measurement signal to another device remote from the sensor. Here, the term *transmitter* is more appropriately used. Although the terms *sensor, instrument, transducer, converter,* and *transmitter* have broad and often overlapping meanings, the correct terminology depends on the function of the specific device, installation, or system.

Sensors that are in direct physical contact with the treatment process flow streams or the process equipment are *input devices*. They do not control equipment; they make up the input basis for control. Properly functioning sensors make process monitoring and control possible. The sensor, the eyes and ears of the system, is the most important device in the successful application of instrumentation systems. If properly installed and diligently maintained, sensors benefit plant operations.

Sensors monitor three categories of parameters: status, physical, and analytical. Typical status sensors include valve or gate status (opened or closed), equipment operating status (on or off), and remote switch position (automatic or manual). Status sensors are also used to generate alarms to notify operators of process upsets or dangerous conditions. They can interlock process equipment to shut down automatically, thus avoiding or limiting damage to essential or expensive equipment.

Physical parameters include flow, level, pressure, position, temperature, speed, and vibration. Typically the more significant factors in process and equipment control, physical parameters are used to monitor equipment performance, establish treatment requirements, generate historical data, and, when combined with analytical data, calculate process efficiencies.

Analytical parameters include pH, residual chlorine, turbidity (suspended solids), sludge density, ammonia, dissolved oxygen, nitrous oxides, and hydrocarbons. These parameters are used as input for automatic control functions or regulatory monitoring reports. Typically, analytical sensors are more expensive and complicated than those associated with status and physical parameters. Because the laboratory routinely tests for many of these parameters on grab and composite samples, the plant laboratory is closely involved with ensuring that these online measurements are correctly maintained and calibrated.

**FINAL CONTROL ELEMENTS.** The final control element is an electric, electronic, pneumatic, or hydraulic device or equipment adjusted to achieve a desired process variable measurement. A sensor provides input for the control; the final control element accepts outputs made to implement a control action. In wastewater treatment plants, the final control element is typically a valve, gate, or pump. Included with each final control element are secondary conversion devices that accept control-signal transmissions and convert them to a mechanical or physical action. The converters include valve actuators, positioners, and a variety of variable-speed, pump-control equipment (e.g., magnetic clutches, motorized pulleys, and variable frequency drives). If properly applied, installed, and diligently maintained, the final control elements and their supporting control and conversion equipment yield successful process control.

**CONTROL PANEL INSTRUMENTATION.** Control panel instrumentation includes a broad array of devices that provide operators with process information and points of control. Examples of control panel instrumentation are panel meters, indicators, alarm annunciators, status indicator lamps, recorders, controllers, pushbuttons, and selector switches. Sensors, final control elements, and control panel instrumentation allow operators to inspect and control a process from one convenient location. The devices mounted in a control panel that operators do not directly see include electro-mechanical relays, signal-transmission conditioners and converters, and power supplies.

**MICROPROCESSORS AND COMPUTERS.** Microprocessors and computers are now commonly used in many applications, including individual instruments and entire digital-based control systems. Housed in small spaces, microprocessors can perform extremely complex functions at high speeds with low power consumption. For example, before microprocessors, the calculation required to convert the measured level in a flume to a flow signal required a hardwired electrical circuit that consumed much of the space in a flow meter cabinet. Today's microprocessors occupy less space, are more reliable, and permit calculations to be changed or reprogrammed easily, thereby allowing flexibility in use.

Industrial applications of programmable logic controllers (PLCs) have replaced the large banks of electromechanical control relays that used to be required for equipment control (Figure 7.1). The PLC, adapted to the industrial environment, allows plant personnel to easily change the logic used for equipment control.

Perhaps the most significant example of the utility of the microprocessor is the use of the personal computer (PC). These microprocessors are inexpensive, stand-alone computer systems that help a plant operator or manager enter, store, retrieve, and manipulate enormous amounts of plant data quickly and easily. Before PCs, even simple calculations and recordkeeping consumed hundreds of working hours annually. Now, PCs perform these activities quickly and accurately. In addition, available software packages make computer power affordable for even the smallest facility.

**CONTROL SYSTEMS.** Because of the nature and variety of wastewater treatment facilities, the processes are typically spread over a large area. Instrumentation and automation can centralize data reporting and control commands. Centralization saves operators time, because they no longer need to go to the piece of equipment to perform routine monitoring or control. However, centralization does not replace the need for operators to periodically inspect a piece of equipment. No control system can replace that function.

Control systems consist of individual electrical, electronic, and pneumatic components connected in a network hierarchy. Redundant control equipment components may be warranted to ensure that failure of a single control component does not result in critical process failure, equipment damage, or hazards to personnel. No matter how extensive or sophisticated the control system, there must always be a provision for local manual control of the equipment.

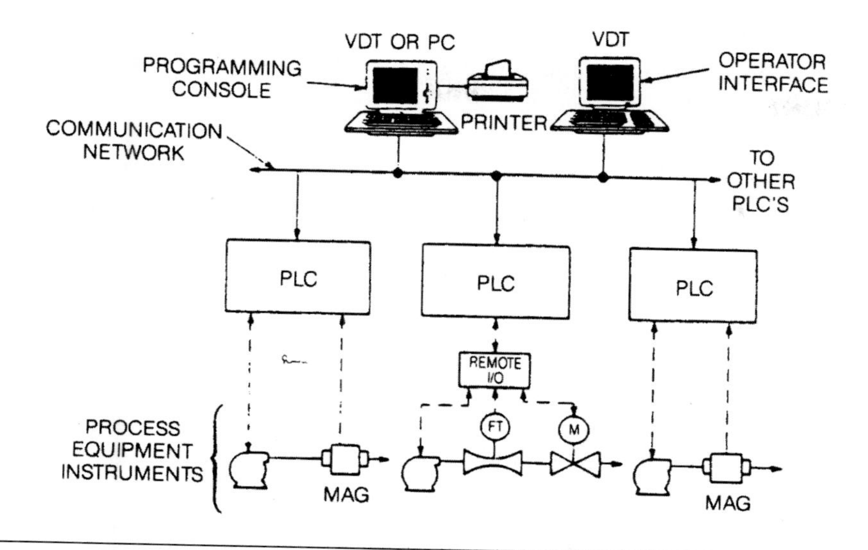


FIGURE 7.1 Distributed control system based on programmable logic controllers.

## CONTROL SENSOR APPLICATION GUIDELINES

**INSTALLATION.** Operators should follow the manufacturer's recommendations when installing an instrument and should discuss any deviations or problems with the manufacturer. If the equipment is installed incorrectly, irreparable damage or malfunction may result. Careless installation will undermine even the most carefully selected and applied equipment. Successful installation incorporates the operator's knowledge and experience of the liquids and gases being measured and controlled. For example, all wastewater treatment plant operators are familiar with the tendency of wastewater and solids to plug small openings. Operators should consider this and other qualities of wastewater and solids when installing instruments. For inline installations, a means of bypass or a spool piece should be provided so the measuring element may be removed from service for replacement, maintenance, or calibration that must be done offsite.

**ENVIRONMENT.** Operators should minimize or eliminate dust, moisture, extreme temperature, and corrosive gases. Although the presence of low levels of corrosive gas (e.g., hydrogen sulfide) may not be detectable or considered hazardous to personnel, continued direct exposure of electronic parts to minute levels of corrosive gas will eventually result in damaged or inoperative equipment. To prevent the intrusion of moisture and gases, operators should seal all the instrumentation's electrical conduits and covers. Multiple instrument arrays should be protected in cabinets by producing a slight positive pressure in the cabinet using clean, dry, compressed air with a dewpoint less than 4.4 °C (40 °F).

**PLACEMENT.** Placing the instrument at the proper location in a process is one of the more difficult problems in applying instruments in a wastewater treatment plant. With some instruments (e.g., flow meters), test data have scientifically documented improper placement. With other instruments (e.g., pH or dissolved oxygen meters), the best location is based on experience in similar applications or trial and error. With ultrasonic devices, care must be taken to install them where sidewalls or piping obstructions do not interfere with the instruments' measurements. Often, an instrument has been deemed defective or unreliable when, in fact, it was only placed improperly. Before abandoning an instrument, the manager should confirm that the sensor location is correct and, where possible, that alternatives have been tried. Manufacturers typically will provide considerable guidance, assistance, and suggestions for sensor placement.

**MEASUREMENT RANGE.** The expected measurement range of an instrument is based on current, rather than future, requirements. Range selection is based on the instrument turndown (i.e., the maximum expected measurement divided by the minimum expected measurement). When ultimate design values are used to determine the instrument range, the maximum expected measurement is large, resulting in an excessively large range. Over-ranging a device leads to instrument insensitivity and inaccuracy, because resulting measurements use only the lower end of the instrument's capacity.

**ACCURACY AND REPEATABILITY.** *Accuracy* expresses how close an instrument is to the true value of the variable being measured. *Instrument error* is the difference between the instrument reading and the true value. Because accuracy is specified in many ways, the operator should verify the accuracy criteria being used in each specific case. The most common specification of instrument accuracy is as a percentage of either the span of the instrument or the actual reading (Figure 7.2). For example, if a magnetic flow meter has a span of 0 to 5451 $m^3/d$ (0 to 1000 gpm) and a specified accuracy of 1% of span, the reading would be expected to be within 55 $m^3/d$ (10 gpm) of the actual flow. If the same instrument is specified as a percentage of the actual reading, the expected variation from the true flow will change, depending on the actual reading. At 545 $m^3/d$ (100 gpm), the reading would be expected to be within 5 $m^3/d$ (1 gpm). At 4910 $m^3/d$ (900 gpm), the reading would be expected to be within 49 $m^3/d$ (9 gpm).

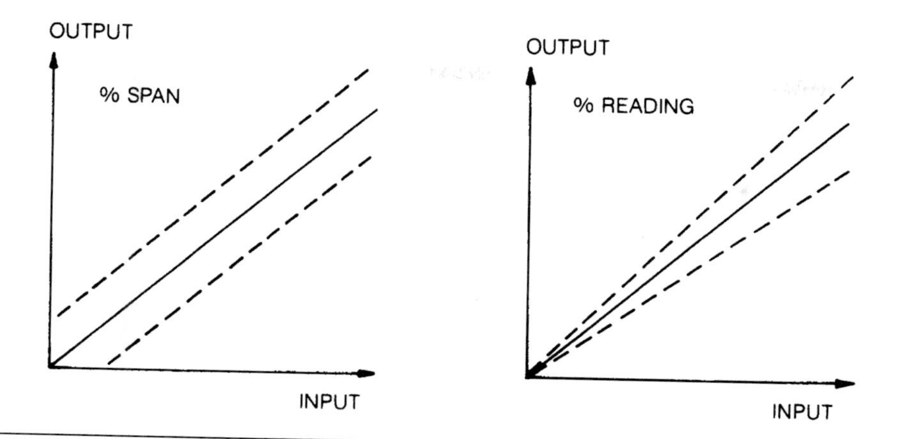


**FIGURE 7.2** Accuracy: the solid line represents true values; the reading is within the band bordered by the dashed lines.

*Repeatability* is the closeness of agreement among a number of consecutive output measurements for the same input value under identical operating conditions and approaching from the same direction, for full-range traverses. If an instrument is repeatable and operators know the inherent errors, this type of input for control is often useful.

Other important instrument characteristics that are examined to learn if the instrument meets the needs of the application include hysteresis, sensitivity, linearity, dead-band, drift, and response time.

## FLOW MEASUREMENT

Flow is one of the most important wastewater parameters to be measured. The many types of flow-measuring devices have three basic criteria that determine their performance: area, velocity, and device characteristics. The two basic types of flow measurement are open-channel and closed-pipe. For good measuring-device performance, both types require approach conditions free of obstructions and abrupt changes in size and direction. Obstructions and abrupt changes produce velocity-profile distortions that lead to inaccuracies.

**OPEN-CHANNEL FLOW MEASUREMENT.** A weir or flume measures the liquid flowing in open channels or partially filled pipes under atmospheric pressure (Figures 7.3 and 7.4). This device causes the flow to take on certain characteristics (e.g.,

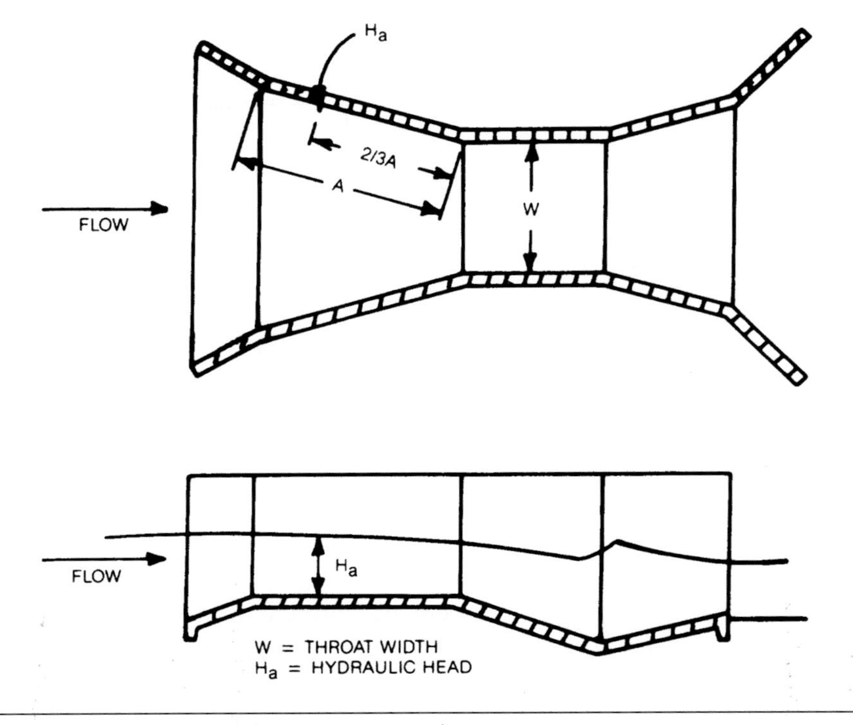


**FIGURE 7.3** Parshall flume (Skrentner, 1989).

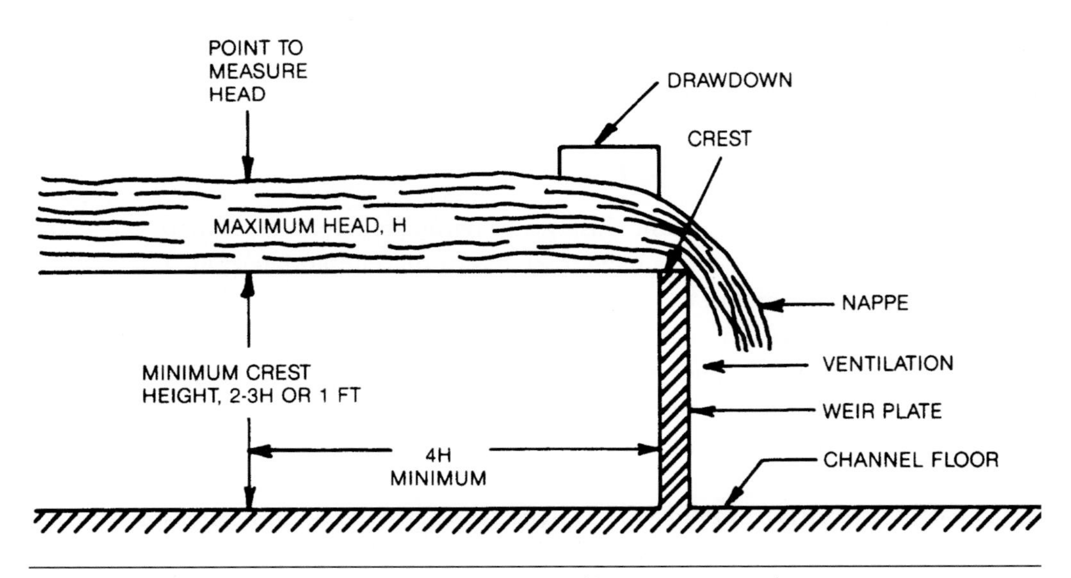


**FIGURE 7.4** Typical weir: generated elevation (ft × 0.3048 = m).

shape and size), depending on the device used. Changes in flowrate produce a measurable change in the liquid level near or at the device. This level is related to flowrate via an appropriate mathematical formula. The specific device determines the location and accuracy of level measurements and is extremely important for accurate performance. Many texts deal extensively with open-channel flow metering. For more information, see the *Open Channel Flow Measurement Handbook* (Isco, Inc., 1989).

**CLOSED-PIPE FLOW MEASUREMENT.** When the pipe runs full in closed-pipe flow measurement, the cross-sectional area can be accurately determined. However, measuring velocity is more difficult, because friction from the pipe wall, fluid viscosity, and other factors affect the flow velocity. There are five classes of closed-pipe flow meters typically used in wastewater treatment: differential head, mechanical, magnetic, ultrasonic, and mass flow.

***Differential-Head Flow Meters.*** Differential-head flow meters include orifice plates, venturi tubes (Figure 7.5), nozzles, pitot tubes, and variable-area rotameters. When properly selected and applied, they measure clean liquids (e.g., potable water and filtered water) and relatively clean gases (e.g., low-moisture digester gas and compressed air).

Differential-head flow meters use an in-pipe constriction that produces a temporary and measurable pressure drop across it. This pressure drop is related mathematically by the principles of mass and energy conservation to the velocity and flowrate of the fluid being metered. The meter's range is typically limited to a 4:1 ratio of maximum to minimum flow. For proper operation, differential-head meters require clear pipe taps both upstream and downstream of the constriction. Routine maintenance includes clearing these taps by blowing them out and removing any accumulated moisture in gas applications.

***Mechanical Flow Meters.*** The propeller meter, the most common mechanical flow meter, functions on the principle that liquid impinging on a propeller causes rotation at a speed proportional to the flowrate (Figure 7.6). The propeller is linked mechanically to an indicator and often to a totalizer or transmitter. Proper operation requires low suspended solids and debris.

***Magnetic Flow Meters.*** Magnetic flow meters are used extensively in applications ranging from filtered effluent to thickened or digested solids (Figure 7.7). They function via electromagnetic induction, in which the induced voltage generated by a conductor moving through a magnetic field is linearly proportional to the conductor's velocity. As the liquid (the conductor) moves through the meter (generating the magnetic

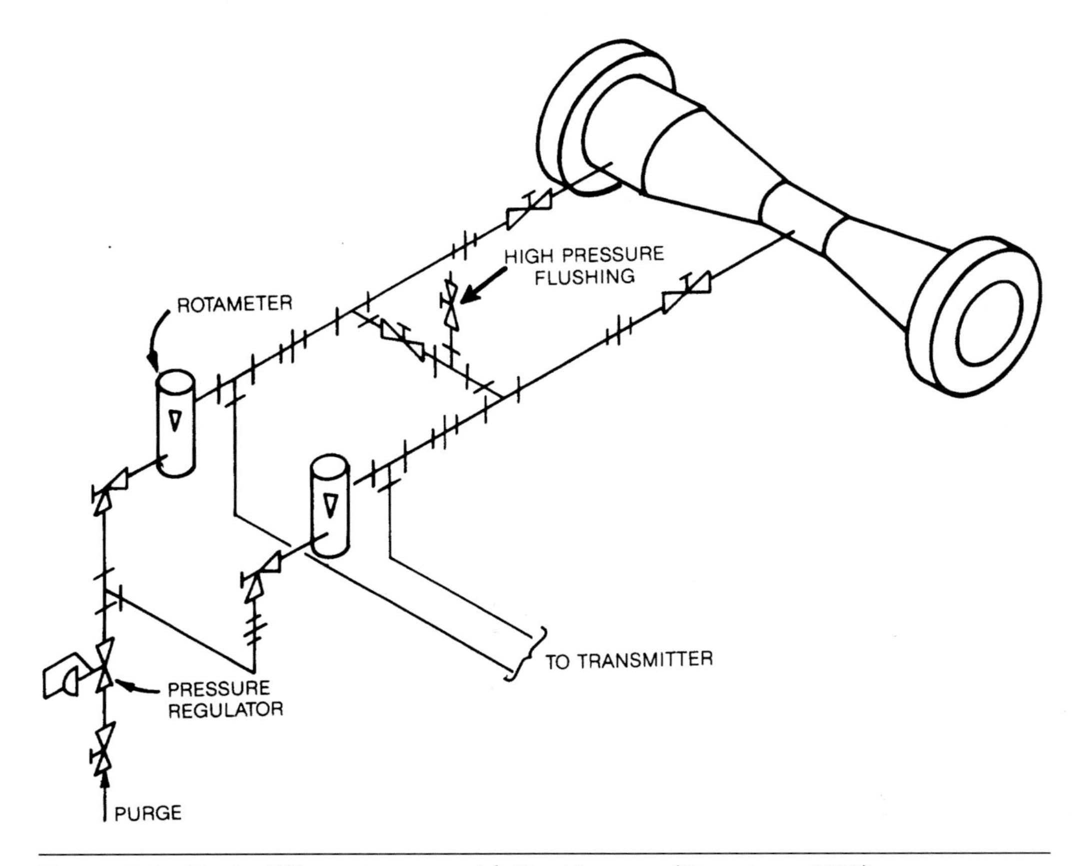


**FIGURE 7.5** Typical Venturi meter with liquid purge (Skrentner, 1989).

field), the voltage produced is measured and converted to a velocity and, thus, a flowrate. Magnetic meters require a full pipe for proper operation. Proper grounding is important for certain brands. In applications where greasing of electrodes is likely, additional equipment for degreasing the electrode may be required. Magnetic flow meters provide no obstructions and are manufactured with abrasion- and corrosion-resistant liners, which is why they are frequently used in solids metering. Their major disadvantage is their relatively high price.

***Ultrasonic Flow Meters.*** Ultrasonic flow meters are based on the measurement of ultrasonic wave-transit time or frequency shift caused by the flowing fluid. An instrument that measures wave-transit time is called a time-of-flight or counterpropagation ultrasonic flow meter (Figures 7.8 and 7.9). Ultrasonic waves of known frequency and

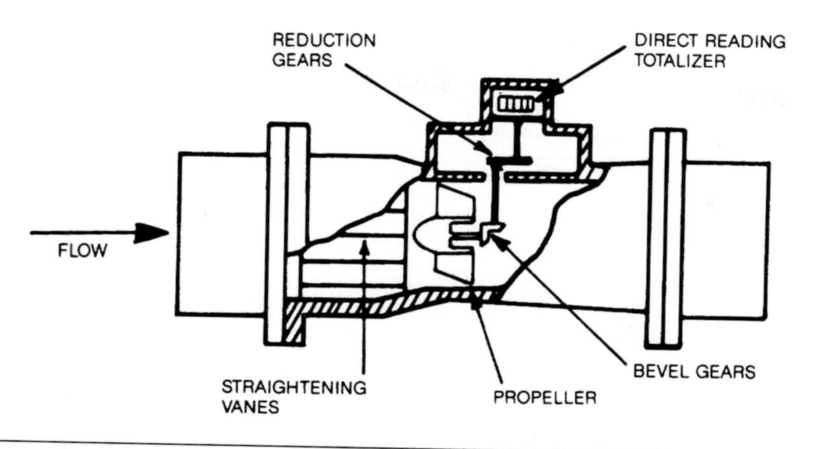


**FIGURE 7.6** Propeller flow meter.

duration are beamed across the pipe at known angles. The waves are sensed either directly by an opposing receiver or indirectly as reflected waves. The changes in wave transit time or frequency caused by the flowing liquid are linearly proportional to the liquid velocity. This velocity is converted from flow and output to a display by the conversion electronics. The presence or absence of air bubbles and density of solids in the

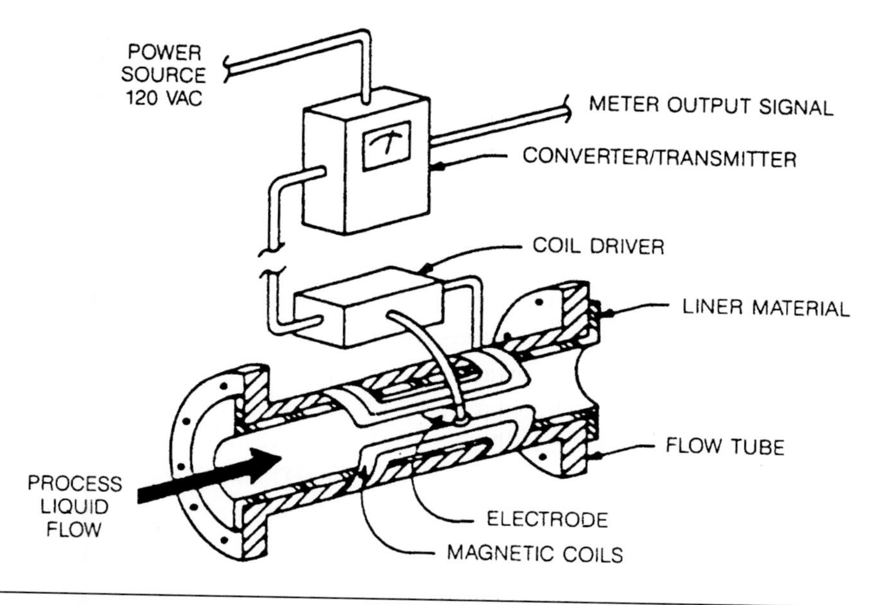


**FIGURE 7.7** Magnetic flow meter.

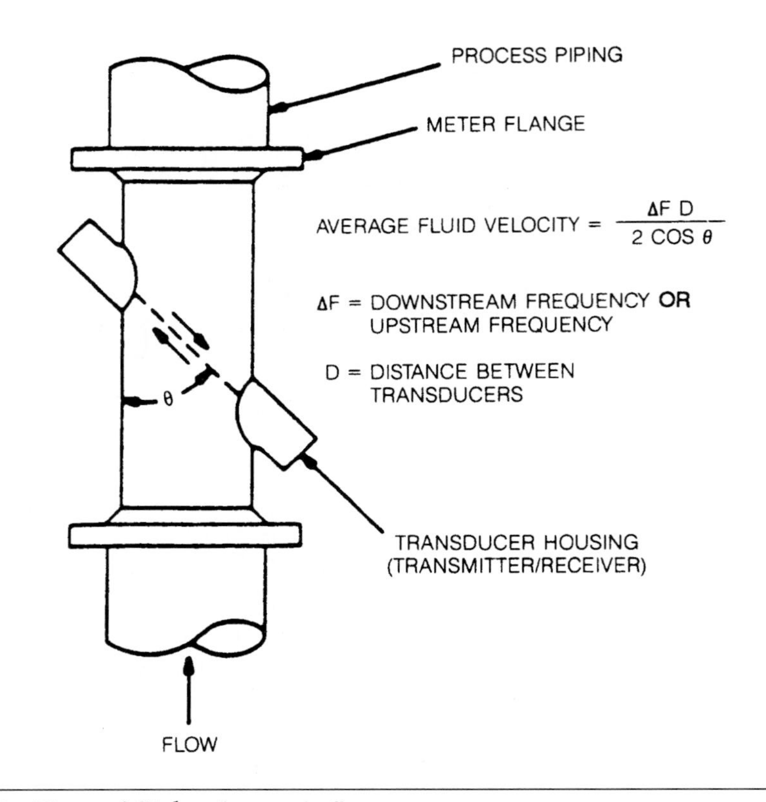


FIGURE 7.8 Time-of-flight ultrasonic flow meter.

fluid being metered affect the meters. Operators should follow the manufacturer's specifications and carefully match the meters to the application.

***Mass Flow Meters.*** Mass flow meters measure the mass of material delivered in a unit of time. They use the density of the flowing fluid or gas and either the Coriolis effect or thermal dispersion to translate the measurement to mass flow.

In the Coriolis flow meter, either straight or bent tube(s) may be used. As the fluid flows through the tube, translational and rotational Coriolis forces simultaneously cause angular deformation or movement of the tubes. The amplitude of the deformation is measured; it depends directly on the velocity and density of the moving mass (mass flow). Another meter uses an oscillation instead of a constant angular velocity. Two parallel tubes are oscillated in anti-phase, much like a tuning fork. When a fluid flows through the tubes, a phase shift is generated. As fluid flow varies, electrody-

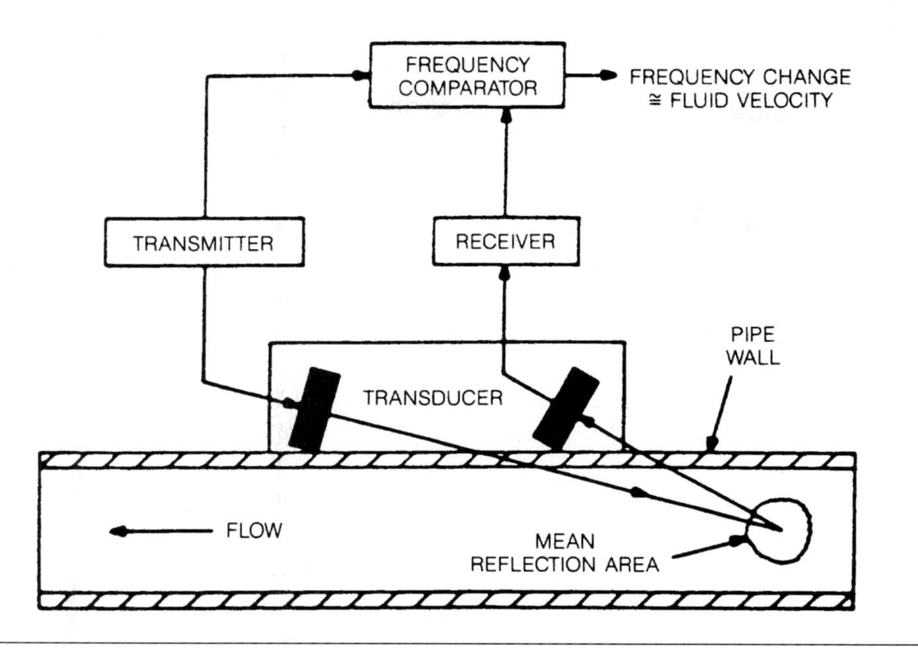


FIGURE 7.9 Reflecting ultrasonic flow meter.

namic sensors at the inlet and outlet measure the phase shift, which is directly proportional to mass flow.

Thermal-dispersion mass meters use heat transfer to determine the mass of material moved per unit of time. It measures the rise in temperature of a fluid after a known amount of heat has been added. From the thermodynamic characteristics of the flowing media, the volumetric and mass flow can be determined.

These meters are subject to fouling and should not be used on fluids with any appreciable suspended solids. They have been used to measure polymer, air, and digester-gas flows.

## PRESSURE MEASUREMENT

Pressure is a parameter used extensively in wastewater treatment plants. Pressure measurements are used directly for the following:

- Monitoring most types of pumping equipment,
- Regulating pressure in various types of pressurized processes, and
- Maintaining proper pressures in lubrication and seal systems.

Some types of devices for measuring the full range of pressures include purely mechanical devices with local indicators (e.g., Bourdon tube gauges and spiral, helical, and bellows gauges) (Figure 7.10). Pneumatic or electronic pressure transmitters measure pressure and transmit the measurement to a remote location. Differential-pressure transmitters are also widely used for level- and flow-measurement applications. When properly selected and applied, all of these devices have proved reliable.

**MECHANICAL PRESSURE GAUGES.** Mechanical gauges work on the principle of physical displacement caused by changes in pressure (Figure 7.11). A link to an

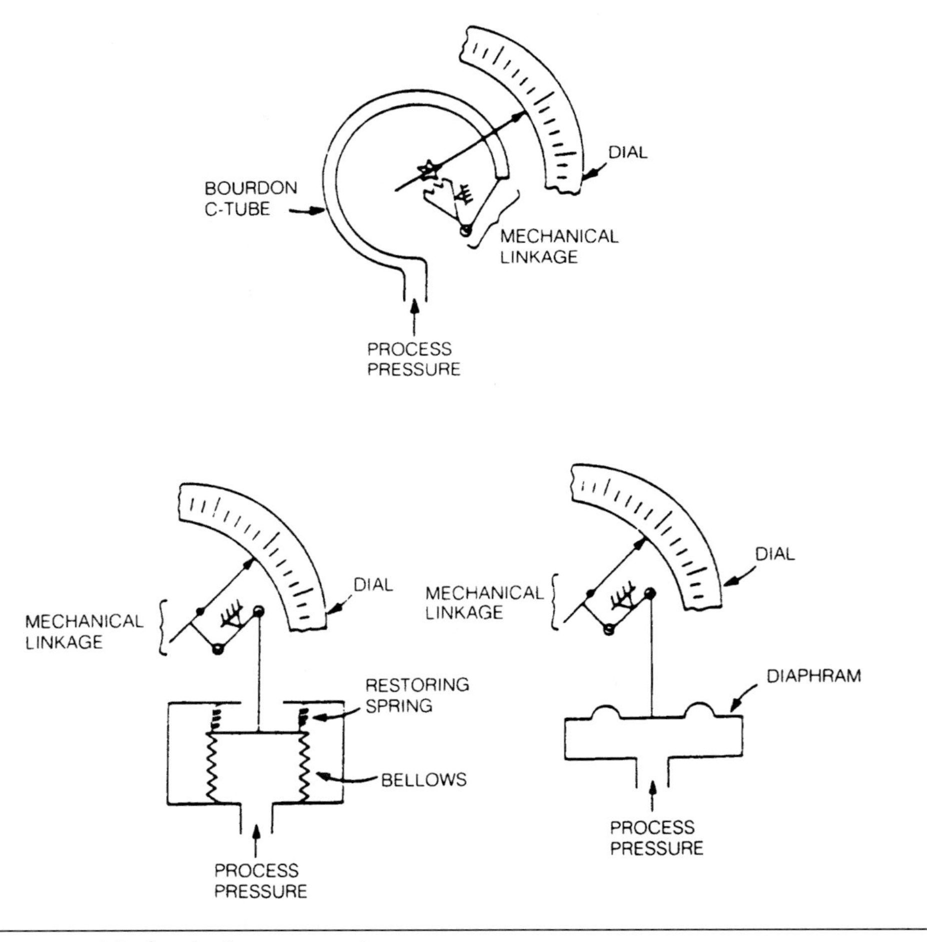


**FIGURE 7.10** Mechanical pressure elements.

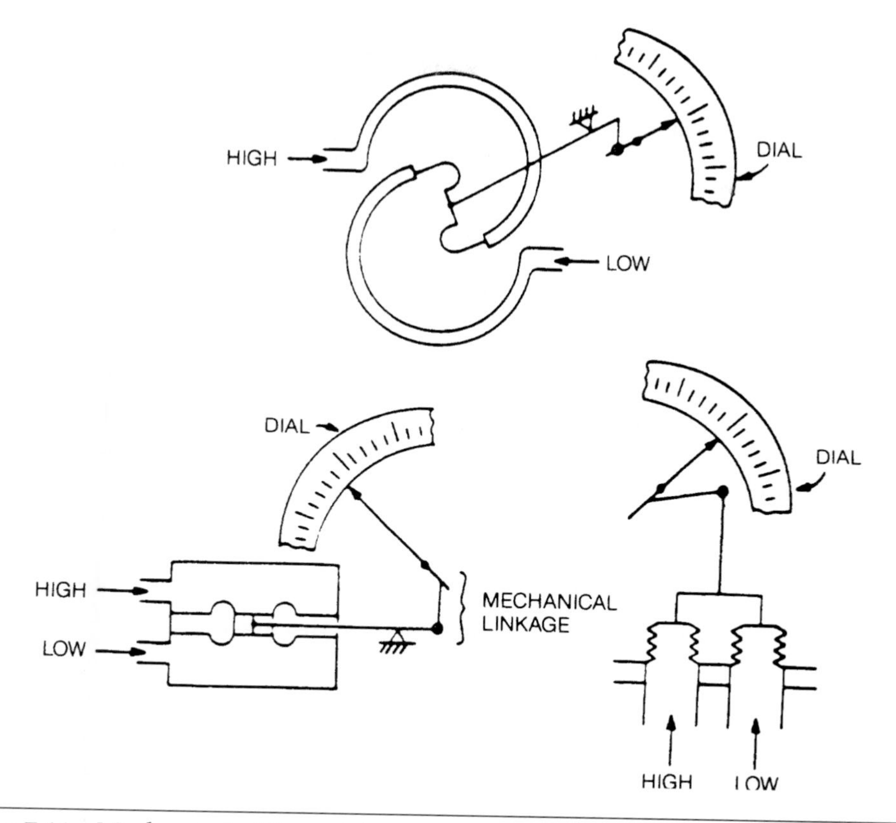


FIGURE 7.11 Mechanical differential-pressure elements.

indicator or pointer on a scale shows operators the pressure measurement. A wide array of pressure ranges is available for mechanical pressure gauges. For proper operation, the process tap for installation of the pressure gauge is made so the tap will not plug with solids or debris. Operators should isolate the gauges from highly pulsating pressures that will severely vibrate the instrument's delicate mechanical linkages. For acceptable accuracy, gauges should be selected to operate at the midpoint of the specified measurement range. If high accuracy is critical, gauges not in active use should be removed.

**PNEUMATIC PRESSURE TRANSMITTERS.** Pneumatic pressure transmitters using a compressed-air transmission system allow transmission of sensed pressure to a remote location. These transmitters use the principle of balancing pressures across a metal diaphragm using the process fluid on the "wet" side and instrument-supplied air (regulated by a nozzle–baffle system) on the "dry" side. Any tendency of the diaphragm to move because of changes in process pressure causes a restriction of the nozzle-flapper system, which increases air pressure to bring the system into balance. Operators should apply these devices carefully in much the same way as mechanical pressure gauges to protect them against plugging and excessive vibration. They need to be calibrated periodically to maintain the relationship established between the process pressure and the change induced in the instrument-supplied pressure.

**ELECTRONIC PRESSURE TRANSMITTERS.** An electronic component attached to a mechanical pressure device (typically a diaphragm) generates an electrical signal proportional to the applied pressure. A change in the device changes the electrical property of the electronic component (capacitor or strain gauge). These devices are becoming widely used for remote pressure transmission with electronic display and control systems. They need to be calibrated periodically to maintain the relationship between the actual process pressure and the magnitude of the electronic transmission signal.

**DIFFERENTIAL-PRESSURE TRANSMITTERS.** Differential-pressure transmitters use one of the pressure-measurement devices discussed previously, along with a special housing or fittings. The housing or fittings provide two pressure connections that balance the pressure. Differential-pressure transmitters are used for level measurement on pressurized tanks and with differential-head flow measurement devices on pressurized pipes.

## LEVEL MEASUREMENT

Level measurement is primarily used to control pumps to maintain levels in wet wells, tanks, and grit chambers. Level measurement is also used with other devices (e.g., flumes or weirs) for open-channel flow measurement. The three main methods of determining level are physical measurement of the liquid surface from above, measurement of head pressure from below, and electrical methods.

**FLOATS AND ULTRASONIC DEVICES.** A float or ultrasonic device measures a liquid level from above. Floats, one of the oldest and simplest methods of level measurement, are used extensively in wet wells or sludge vaults that require a discrete high- or low-level indication. They are also used for local indication of level in tanks and open channels. The primary maintenance concern of a float system is accumulation of solids, debris, or ice on the float or in the stilling well, when used (Figures 7.12 and 7.13).

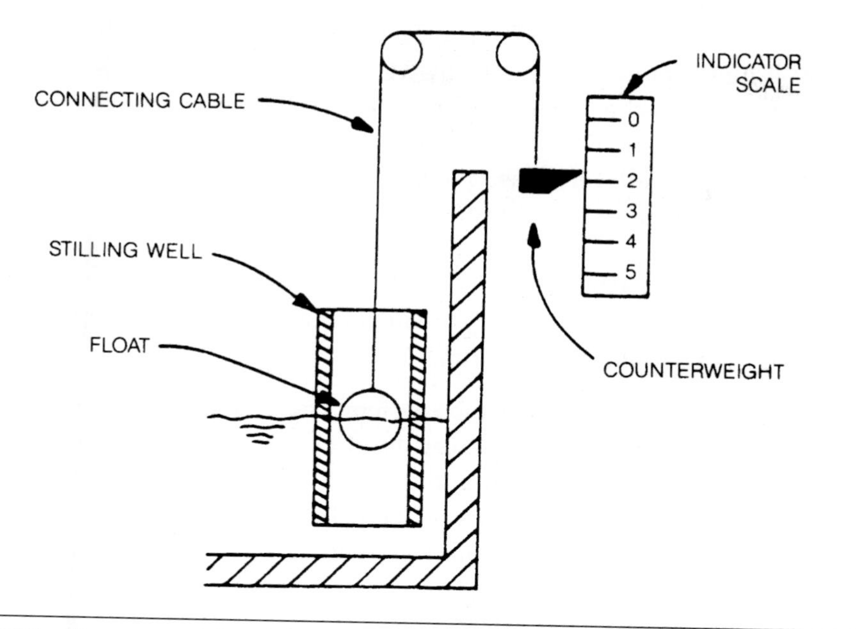


FIGURE 7.12 Counterweighted float-level indicator.

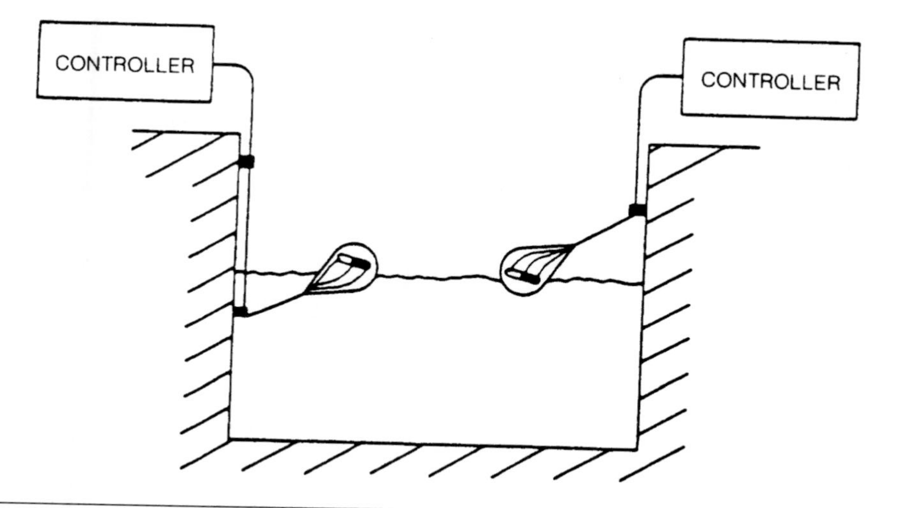


FIGURE 7.13 Float switches.

Ultrasonic level measuring devices installed above the liquid surface measure the level by generating a pulse of ultrasonic waves that bounce off the liquid surface. The instrument detects the echo, calculates the echo's travel time, and converts it to a level measurement (Figure 7.14). Changes in air temperature, surface foams, and fog can adversely affect ultrasonic level measurements. Operators should shield the sensitive electronic components from moisture and corrosive gases.

**HEAD-PRESSURE LEVEL MEASUREMENT.** Bubbler tubes and diaphragm bulbs measure the liquid level's head pressure and are often used in open-channel or nonpressurized tank applications (Figure 7.15).

***Bubbler Tube System.*** The bubbler tube system uses a small, regulated airflow that constantly bubbles into the liquid. Because the airflow is small, the system produces a backpressure equal to the static head of the liquid. A conventional pressure gauge or transmitter measures this backpressure as the height of an equivalent water column. Corrections are required when the liquid's specific gravity differs significantly from

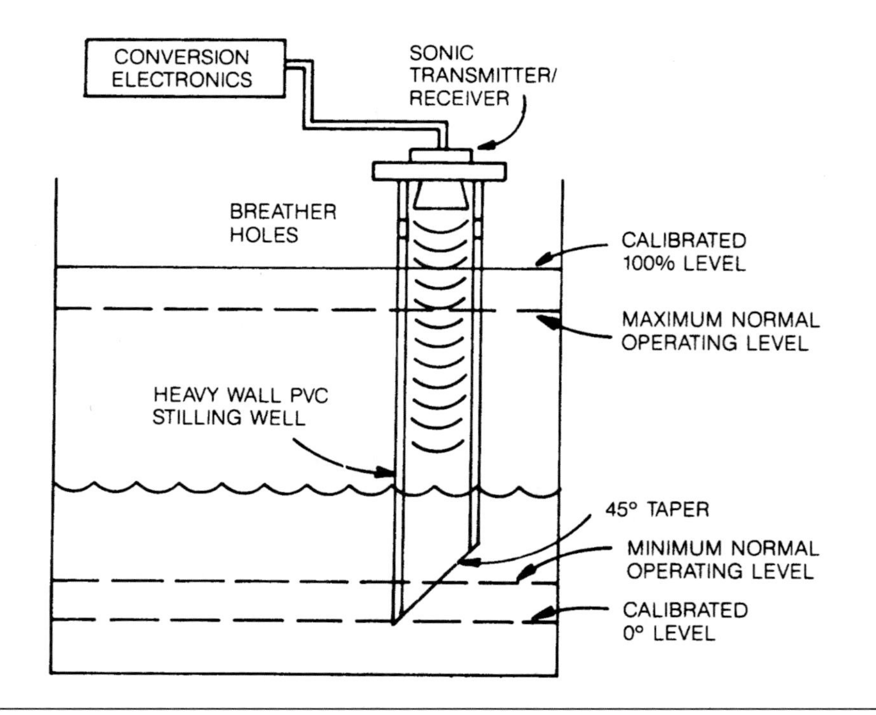


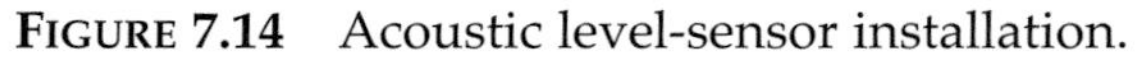
FIGURE 7.14 Acoustic level-sensor installation.

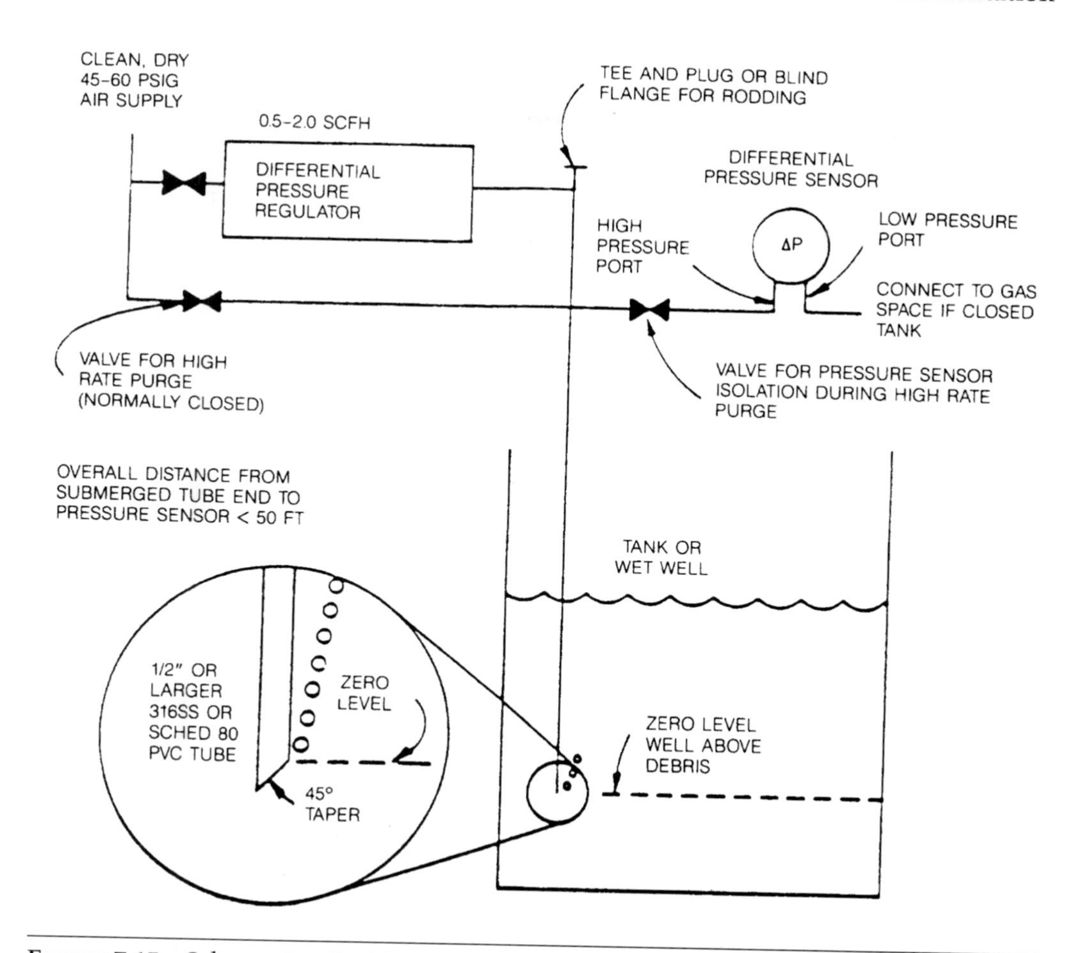


FIGURE 7.15 Schematic of bubbler-level system (ft × 0.3048 = m and in. × 25.4 = mm).

water. Because air is constantly bubbling out of the bubbler tube, the system is typically self-purging. Valving may be arranged to isolate the pressure-measuring device while providing high purge flow through the tube for preventive-maintenance blowdown if fouling occurs. Stilling wells are often used to protect the bubbler tube from turbulence and damage. To protect the pneumatic instruments and regulators, operators should clean the air supply of excessive moisture and oils.

***Diaphragm Bulb System.*** The diaphragm bulb system operates on the principle that air sealed between the dry side of the diaphragm (in the capillary tube) and the receiver compresses or expands with the movement of the diaphragm. A change in the static head of the liquid being measured moves the diaphragm, so the pressure of the trapped air is the same as the head pressure. Temperature changes because of sunlight or building heat, particularly along the capillary tube, can cause measurement errors as a result of expansion of the trapped air. To reduce the effect of temperature, the capillary can be filled with a fluid unaffected by operating temperature; however, this often affects the measurement response time.

**ELECTRICAL METHODS.** Electrical methods of level measurement consist primarily of conductance and capacitance probes (Figure 7.16). Conductivity probes are discrete devices used much like floats (e.g., turning pumps on and off and alarming).

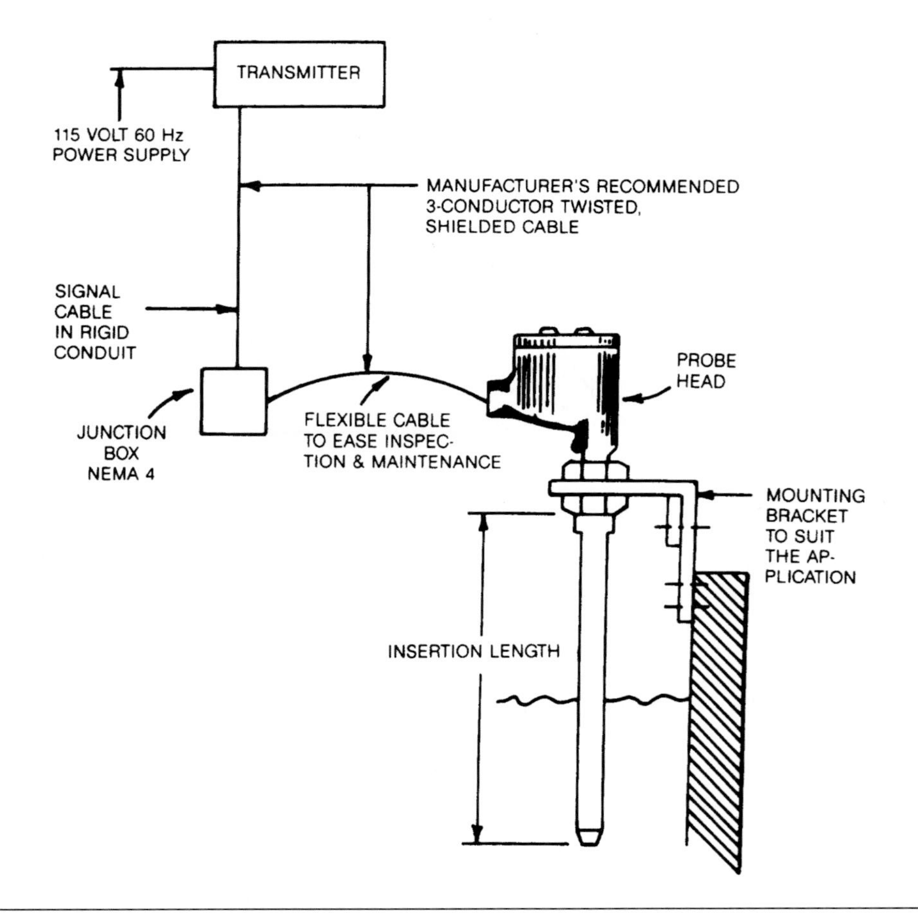


FIGURE 7.16 Typical capacitance probe installation (Skrentner, 1989).

Conductivity probes use the liquid to complete a circuit between two probes or between two sections of one probe. The devices should be used in relatively clean liquids because buildup on the probe can either complete the circuit or act as an insulator.

A capacitor consists of two electrically conductive plates separated by nonconductive material. In wastewater treatment applications, a capacitance probe acts as one conductive plate and the nonconductive material. The liquid acts as the other conductive plate. As the liquid level changes, the system's capacitance changes proportionally. The instrument measures this capacitance and displays the level. Problems occur if the probe accumulates solids, but many manufacturers have designed the instrumentation to compensate.

## TEMPERATURE MEASUREMENT

Even though most of the major wastewater treatment processes are not temperature-controlled, many temperature measurements are required. Obvious applications for temperature measurement are anaerobic digesters, chlorine evaporators, incinerators, and equipment protection. Less obvious are temperature controls for analyzers and flow meters. Temperature-measurement devices include liquid thermometers, bimetal thermometers, pressure on liquid or gas expansion bulbs, thermistors, resistance temperature detectors (RTDs), infrared detectors, and crystal window tapes. The RTD is typically used on lower, ambient-range temperatures, while thermocouples provide better reliability in higher ranges. Also, gas- and liquid-filled temperature sensors and thermistors are frequently used for equipment-protection and cooling systems. For continued accurate service, operators should periodically calibrate the instruments using a standard temperature-measurement device with high accuracy.

**THERMAL BULB.** The most commonly used thermal bulb is gas-filled and operates on the physical law that the absolute pressure of a confined gas is proportional to the absolute temperature (Figure 7.17). Installed in a thermal well for protection and ease of inspection and calibration, the thermal bulb is available in several temperature ranges and accuracies. They are typically used for local control and indication only.

**THERMOCOUPLE.** The thermocouple operates on the principle that current flows in a circuit made of two different metals when the two electrical junctions between the metals are at different temperatures (Figure 7.18). The various combinations of metals used are tabulated in most engineering handbooks, and the selection of metals is based on the maximum temperature to be measured. Thermocouples measure as high as 980 °C (approximately 1800 °F), with an accuracy of 1% of full scale.

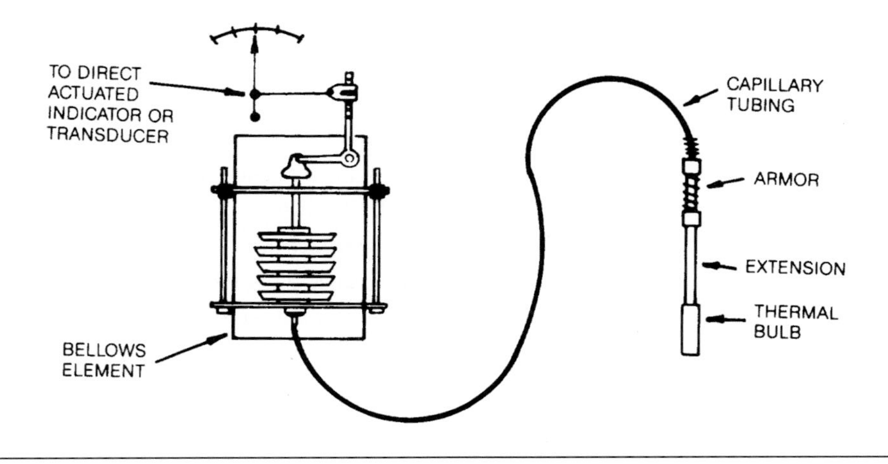


FIGURE 7.17 Thermal bulb system.

**RESISTANCE TEMPERATURE DETECTORS.** A resistance temperature detector has a temperature-sensitive element in which electrical resistance increases repeatedly and predictably with increasing temperature. The sensing element is typically made of small-diameter platinum, nickel, or copper wire wound on a special bobbin or otherwise supported in a virtually strain-free configuration. The detector is typically selected for high accuracy and stability. A common RTD application is the measurement of bearing and winding temperatures in electrical machinery.

**THERMISTORS.** A thermistor (the name comes from "thermally sensitive resistor") is a solid-state semiconductor whose resistance varies inversely as a function of

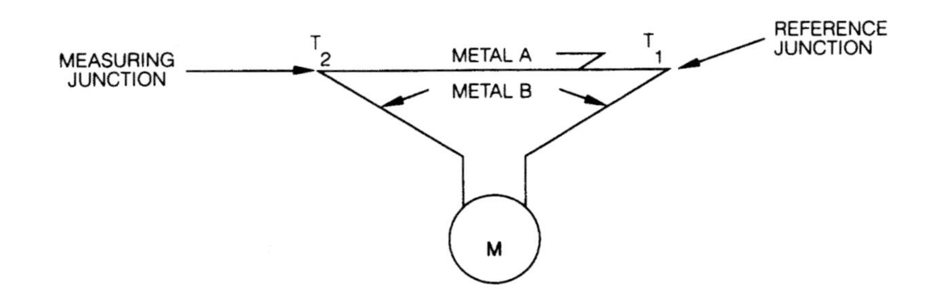


FIGURE 7.18 Typical thermocouple circuit.

temperature. Some thermistors are available with a sharp change in slope at some characteristic temperature. In some situations, a thermistor can replace an RTD as a temperature sensor. One advantage of the thermistor over the conventional wire RTD is that the thermistor has a greater resistance change for a given temperature change. A disadvantage is that the accuracy available, although good, is inferior to that of the conventional RTD.

**TEMPERATURE SENSOR INSTALLATION.** The temperature sensor's lifespan depends on its installation. Thermo-wells, made of metal or ceramic materials and threaded into the wall of a pipe or tank, shield the temperature sensors from the process and enable them to be removed without disturbing the process. In wastewater treatment applications, particularly where solids are involved, operators should not obstruct the full flow of fluid in the pipe. Of particular concern is the collection of debris (strings, plastics, and large chunks) on the thermo-well. Typically, if the thermo-well protrudes less than 25% of the pipe diameter and is at an elbow or T, the flow will wash away the debris.

## WEIGHT MEASUREMENT

Weight measurement, an important part of operations, provides accurate data about chemical use or solids concentration before reduction or disposal. Weight measurements account for the dewatering and hauling costs. Several proven mechanical and electronic methods for measuring weight are weigh beam, hydraulic load cell, and strain gauge.

**WEIGH BEAM.** The standard weigh beam, the simplest mechanical method, involves a basic lever mechanism with a fulcrum between the two forces (Figure 7.19). For the beam indicator to match up with its reference, the load multiplied by its distance to the fulcrum must equal the counterbalance weights multiplied by their distance to the fulcrum. Scales of this type range from fractions of a kilogram (pound) to megagrams (tons) and may have accuracy of 0.1% of the actual weight.

**HYDRAULIC LOAD CELL.** This is a closed system in which the hydraulic oil in the cell is loaded by external force and transmits a corresponding pressure via a diaphragm to a pressure gauge (Figure 7.20). Multiple load cells may be used and the sum of their forces applied to a conventional scale beam. Hydraulic load cells are most effective on high-capacity applications and have an accuracy of 1% of full-scale.

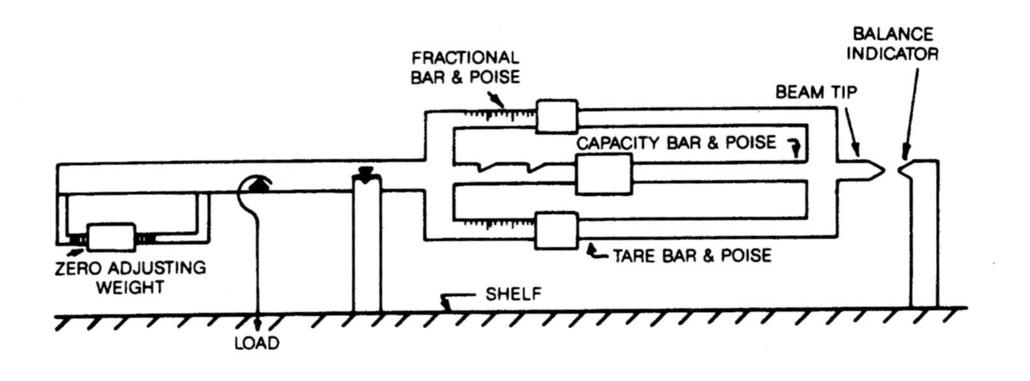


FIGURE 7.19 Weigh beam with sliding poises.

**STRAIN GAUGE.** A strain gauge is an electronic method of weighing. The load button transfers its load force to a preselected steel load column to which the strain gauges are attached. The strain or deflection of the steel column alters the gauges' resistance. Multiple units may be tied together electronically and produce standard output signals via electronic amplifiers. The accuracy of this system is typically 1% of full-scale.

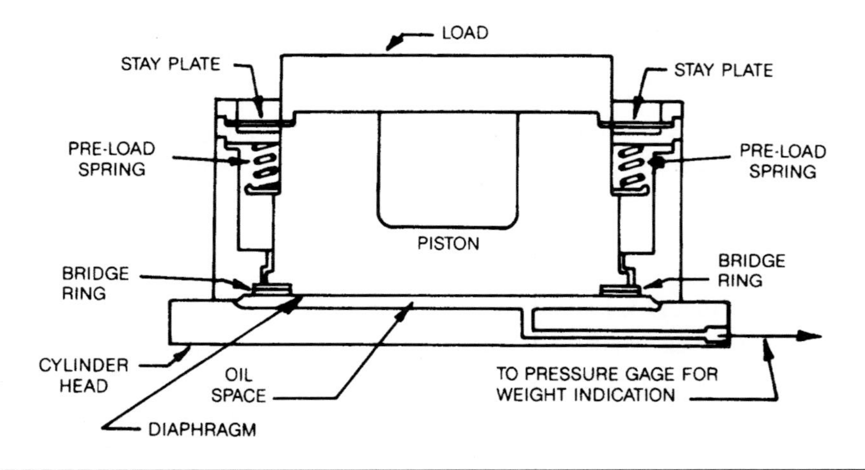


FIGURE 7.20 Hydraulic load cell.

## SPEED MEASUREMENT

Speed measurement is used as a report-back signal or local indication for various types of mechanical equipment (e.g., variable speed drives on pumps, blowers, and some mechanical mixers). Speed-measurement methods include direct mechanical connection to a rotating shaft or gear and noncontact approaches that use magnetic or optical approaches.

**TACHOMETER GENERATORS.** The tachometer generator consists of a stator and a rotor. The stator (nonmoving member) typically consists of laminated steel plates with inward projections for the pickup coils. The rotor (rotating member) is a permanent magnet mounted to the shaft. The shaft is coupled to the member whose speed is to be measured. The output is read on an indicator, or via conversion devices, amplified into one of several standard output signals. Tachometers have an accuracy of 1% of full-scale.

**NONCONTACT FREQUENCY GENERATORS.** Noncontact frequency generators use a shaft-mounted gear and a noncontact electronic or optical pickup mounted close to the gear. The pickup detects the gaps between the tear teeth and the frequency with which they pass by the pickup. This frequency is then converted to a useful display given in revolutions per minute or percent of speed. The advantages of this type of speed sensor include accuracy and lack of moving parts.

## PROXIMITY SENSORS

Proximity sensors—inductive, capacitive, or optical—measure equipment speed, alignment or position, and indexing without touching the equipment.

**MAGNETIC PROXIMITY SENSORS.** The three basic elements of proximity sensors are a sensing oscillator, a solid-state amplifier, and a switching device. Proximity sensing is achieved when a metallic object (target) enters the sensing field. The presence or absence of a metal target is sensed by a change in the magnetic field. This, in turn, changes the internal impedance of the oscillator, which provides a signal output. The output activates a solid-state amplifier and switching device.

**CAPACITIVE PROXIMITY SENSORS.** Capacitive proximity sensors, which sense the difference in capacitance, can sense nonmetallic materials (e.g., water or gasoline). The sensors are used as liquid-level sensors. Sensing is done directly through nonmetallic tank wells, ports, or metal tank walls, or else from inside a capped nonmetallic pipe submerged in the tank.

**OPTICAL PROXIMITY SENSORS.** Optical proximity sensors use a visible or infrared light beam to sense the presence or absence of an equipment component or process fluid. In wastewater treatment plant applications, optical sensing is used to sense the position of a valve or level in a storage bin.

## PHYSICAL–CHEMICAL ANALYZERS

Physical and chemical analyzers measure the physical and chemical properties of process streams (e.g., plant influent, effluent, process sludges, centrate, process gases, and chemicals). Given the increasing regulatory requirements for excellence in overall treatment efficiencies, inline physical–chemical analyzers are playing an ever-increasing role in plant operations. Some physical–chemical analyzers are specific (i.e., they measure one chemical ion), while others measure a group of chemicals, a type of substance, or an effect:

- Specific chemical ions—dissolved oxygen, pH, ammonia, and oxidation–reduction potential (ORP);
- Group of chemicals—total chlorine residual and nitrogen oxides (commonly known as $NO_X$);
- Type of substance—total suspended solids and suspended solids; and
- Effect—biochemical and chemical oxygen demand.

**ION-SELECTIVE ELECTRODES.** A wide variety of ion-selection electrodes, some of which are specific-ion probes, are currently marketed. These electrodes have good monitoring capabilities. The most widely used electrodes measure ammonia, cyanide, dissolved oxygen, chloride, calcium, pH, fluoride, and sulfide. Typically, they are accurate in the laboratory but require careful and periodic maintenance in field applications. However, the reliability of many of these instruments has improved considerably in recent years. Some of the more noteworthy probes used for online monitoring or control in the wastewater treatment industry are described in the following paragraphs.

**DISSOLVED OXYGEN.** Aerobic microbial life processes need dissolved oxygen. Monitoring dissolved oxygen is a prime requirement in the activated sludge process (Figure 7.21). For optimum treatment, a certain level of oxygen must be maintained in the aeration tanks, particularly if nitrification is required. Controlling the quantity of air supplied to the activated sludge process can result in significant cost savings. Temperature, pressure, and dissolved solids all affect how much oxygen can be dissolved

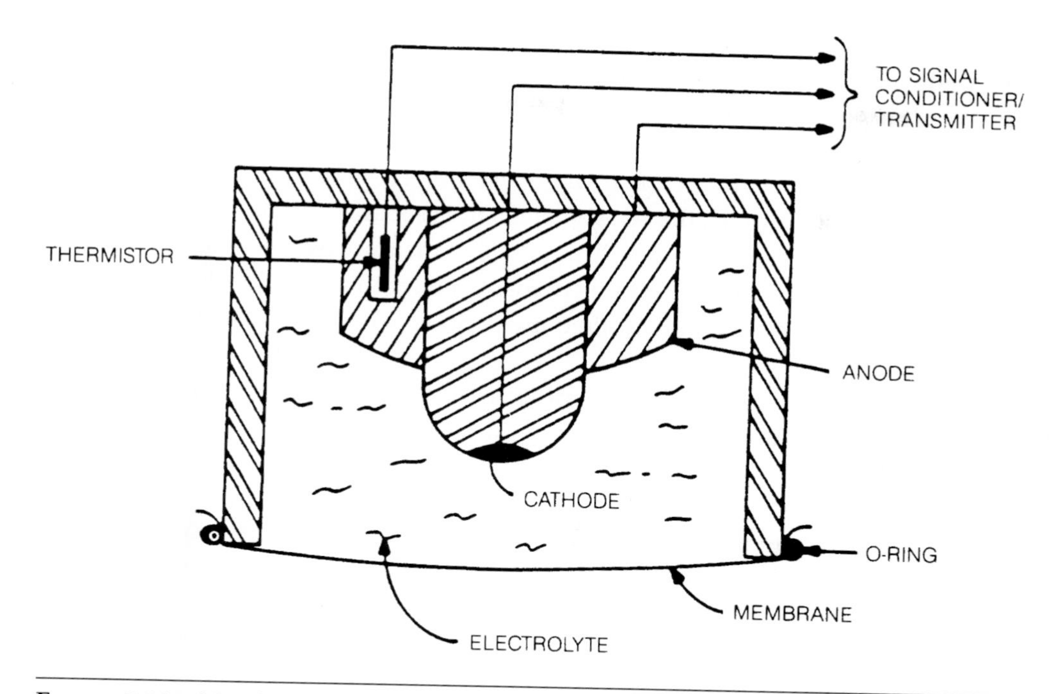


**FIGURE 7.21** Membrane dissolved-oxygen probe.

in a solution. Most dissolved oxygen meters automatically compensate for temperature and pressure.

Volumetric (potentiometric) methods of oxygen measurement depend on the electrolytic reduction of molecular oxygen at a negatively charged electrode (cathode) to generate an electric current directly proportional to the oxygen content. The two volumetric electrode systems in popular use are the active and the passive systems.

***Active System.*** The active system's electrodes measure the oxygen concentration by reducing molecular oxygen at the cathode. The electrons needed to reduce molecular oxygen are supplied by the active cell that uses dissimilar metals and an electrolyte. Its ability to generate voltage can be compared with the lead–acid storage battery in an automobile.

***Passive System.*** The passive system's electrode operates on the same principle as the active cell, which is reduction of oxygen at a polarized cathode. The active cell's source of electrons is the cell itself, from the anode to the cathode. The passive cell's source of electrons is an external power source (e.g., a battery). The voltage is applied across the anode and cathode, and any oxygen passing through the permeable membrane is consumed (reduced) at the cathode, causing current to flow. The magnitude of the current produced is proportional to the amount of oxygen in the sample and is, therefore, a measure of the oxygen present.

It is possible to measure the dissolved oxygen in a solution containing ionic and organic contaminants because the sensing element is separated from the test solution by a selectively permeable membrane. This ability to measure oxygen without interference from other material allows for measurement of oxygen *in situ*. Weekly calibration checks are typically required, and most require periodic wiping of the membrane to remove accumulations.

Manufacturers have realized the importance for easily removable probes and now supply excellent mounting hardware.

**pH.** The pH is a measurement of the acidity of a process liquid. Continuous measurements of the pH of incoming wastewater are frequently used, particularly in plants where drastic changes in pH (as a result of industrial discharges) cause treatment problems.

The glass electrode, which is sensitive to hydrogen ion activity, measures the pH of an aqueous solution (Figure 7.22). The electrode produces a voltage related to hydrogen ion activity and to pH. The pH is determined by measuring the voltage against a reference electrode. While it is generally assumed that no other ions seriously affect the pH electrode in an aqueous system, sodium ions can have an effect. Temperature corrections are also necessary but are typically done automatically by the meter.

**AMMONIA.** Ammonia is a chemical compound that undergoes oxidation and reduction to various forms of nitrogen compounds. To reduce the oxygen uptake in a receiving stream, many plants are required to oxidize ammonia. Many wastewater treatment plants have discharge limits expressed in terms of ammonia-nitrogen ($NH_3$-N). Alternatively, the amount of nitrate and nitrite (oxidized ammonia) can be used to determine if there is adequate dissolved oxygen in an aeration basin or if there are proper conditions in an aerobic digester. Ammonia analyzers use the selective ion probe approach to determine the amount of ammonia present. Operators should calibrate the ammonia analyzers periodically and maintain them to standard solution concentrations. Sampling systems can provide representative samples and filter sample streams to improve accuracy.

**OXIDATION–REDUCTION POTENTIAL.** Oxidation–reduction potential is a measure of the easily oxidizable or reducible substances in a wastewater sample. An operator can control the process better by knowing if there is a large quantity of reduc-

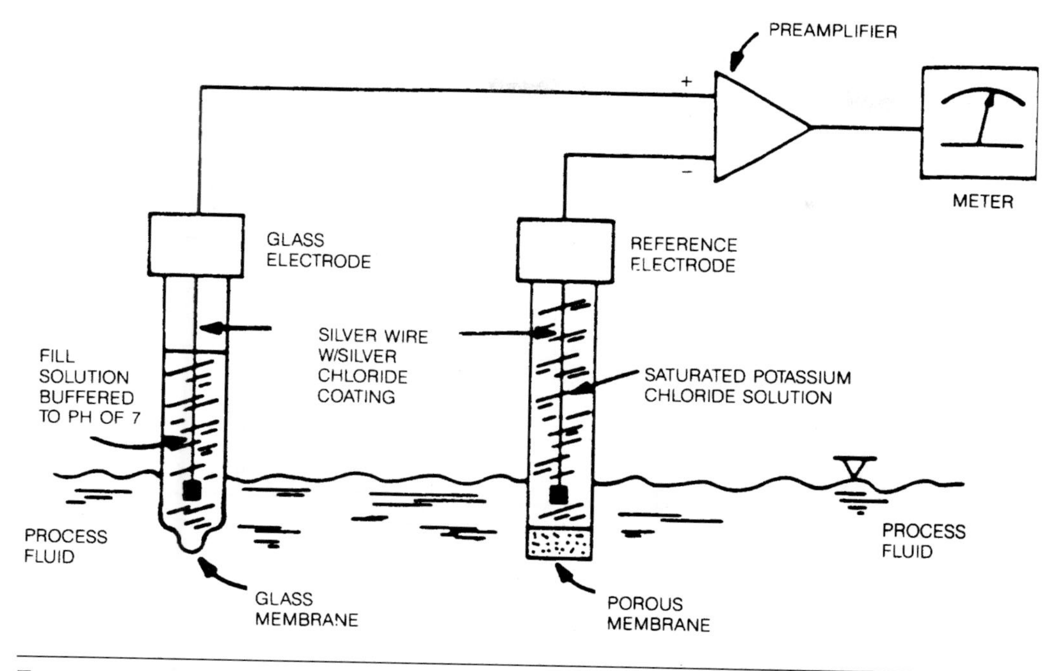


**FIGURE 7.22** Typical pH sensor.

ing substances (e.g., sulfide and sulfite) that may have an immediate, high oxygen demand and may result in an inadequate supply of oxygen for the microorganisms in the secondary process. Although not specific, the ORP measurement is instantaneous (an electrode is used) and can be used to help maintain dissolved oxygen in the aeration tank. Another application is to evaluate the progress of digestion and process stability in the anaerobic digesters.

**CHLORINE RESIDUAL.** The continuous measurement of residual chlorine is important in final effluent disinfection. In addition, some areas of the United States require dechlorination to reduce the toxic effects of chorine residual on aquatic life. Dechlorination requires the use of chlorine residual analyzers to prove that virtually no chlorine remains. Clorination controls become much more important when dechlorination is required. Logically, the more constant the chlorine residual after chlorination, the easier dechlorination is.

The most commonly used chlorine analyzers are designed to measure chlorine amperometrically (Figure 7.23). The analyzer has a flow-through cellblock that contains the platinum measuring electrode and the copper reference electrode. As the sample

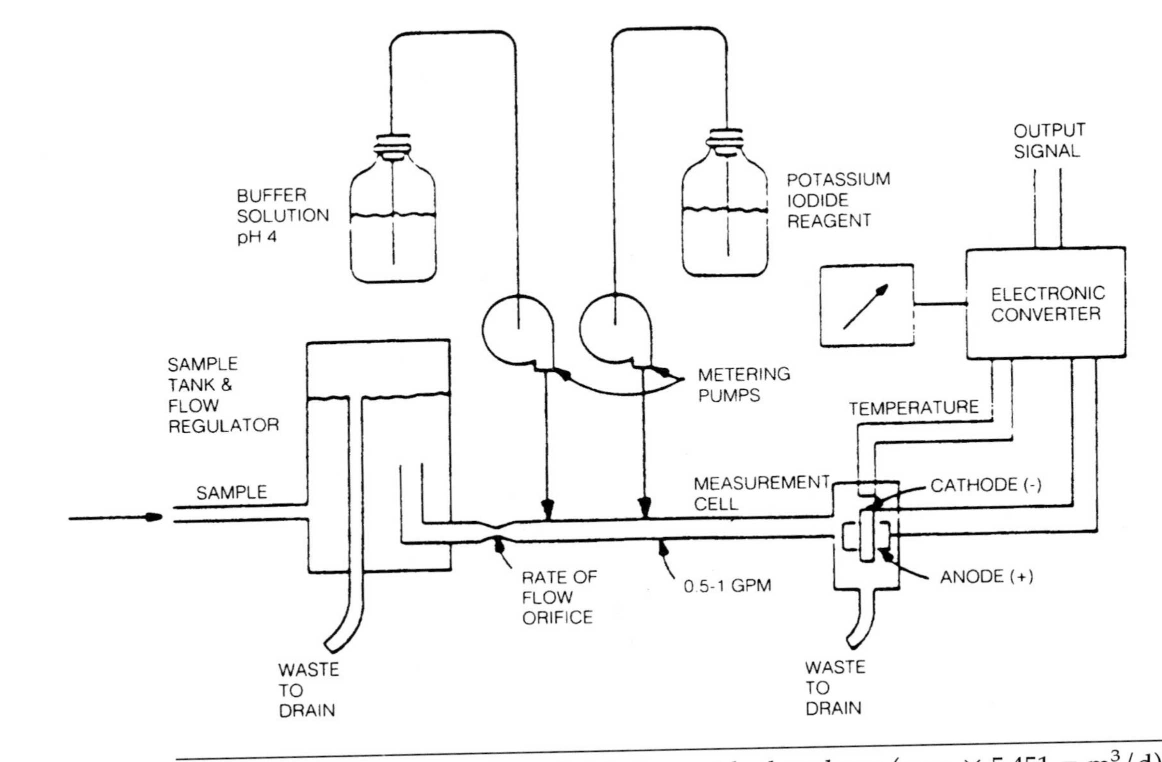


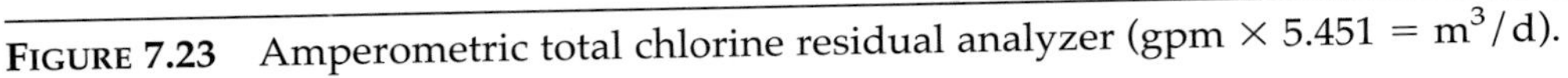
**FIGURE 7.23** Amperometric total chlorine residual analyzer (gpm × 5.451 = $m^3/d$).

flows past the electrodes, a direct current is generated that is proportional to the chlorine residual in the sample. Other units use optical methods, in which a color is developed and then analyzed via light absorption. Manual amperometric titration is widely used as a standard to calibrate the analyzers.

There are several forms of chlorine. Typically, either free or total chlorine residuals are of interest (Chapter 26). Analyzers for a variety of these groups are available.

Chlorine residual meters require periodic maintenance and calibration. Annual maintenance and supplies needed to maintain the instrument may exceed its capital cost. However, in a large plant, improved stability and reduced chemical cost can easily outweigh the capital cost of the instruments.

**NITRATE AND NITRITE.** Nitrate is typically used to monitor and control biological nutrient removal (BNR) plants. Although present in BNR plants, nitrite quickly oxidizes to nitrate or reduces to nitrogen. Probes are available for its measurement, but it is of less interest than nitrate. Online nitrate instruments are used in BNR plants to control anoxic-zone recycle rates and intermittent operation of aeration systems, as well as monitor effluent.

Nitrate and nitrite ions absorb UV light up to 240 nm. In online measurement, a beam of UV light is directed through the sample, and the amount that passes through to the detector is measured. The amount absorbed is converted to a nitrate measurement. The instruments are provided with a means to compensate for any turbidity interference.

Because the value of nitrite is typically insignificant, the value is widely accepted as $NO_3$ for plant control purposes.

Nitrate can also be monitored using the specific-ion approach, but it suffers from interferences and is rarely applied to wastewater.

## SOLIDS CONCENTRATION

Generally, the complexity of wastewater solids prevents any attempt to measure a particular type of solid. In the laboratory, these solids include suspended, dissolved, and settleable solids. At times, the term *density* has been mistakenly used as an equivalent to solids concentration. Density describes the weight of fluid rather than weight of solid per unit volume. Most sludges are denser than pure water. The largest causes for variations in the density of many sludges are temperature, the amount of entrained gases, or the amount of low specific-gravity particles in the water. An example of a low-density sludge with a high solids concentration is a rising or floating sludge in a clarifier. To avoid confusion, operators should use the terms *solids* or *solids concentration*, not density.

Because a continuous online analyzer cannot dry the sample, all solids concentration meters use indirect methods (e.g., optical, ultrasonic, and nuclear). Indirect methods correlate the solids concentration with a measurable factor. The limitation of not relating perfectly to the quantity of suspended matter does not seriously affect the analyzers' ability to produce a repeatable signal of great value in process control.

**NEPHELOMETERS.** When a light beam is directed into a liquid containing suspended particulates, the suspended particulates scatter some of the light. The nephelometer helps observers measure the amount of light that the particulate matter scatters (Figure 7.24). The amount of scattered light relates approximately to the amount of particulate matter, particle size, and surface optical properties. The nephelometer is a photoelectric device that uses an incandescent light source (lamp), which produces light in wavelengths from blue to red. The light is directed to a liquid and, if the liquid contains particles, some of the light strikes the particles and scatters. By placing a photocell

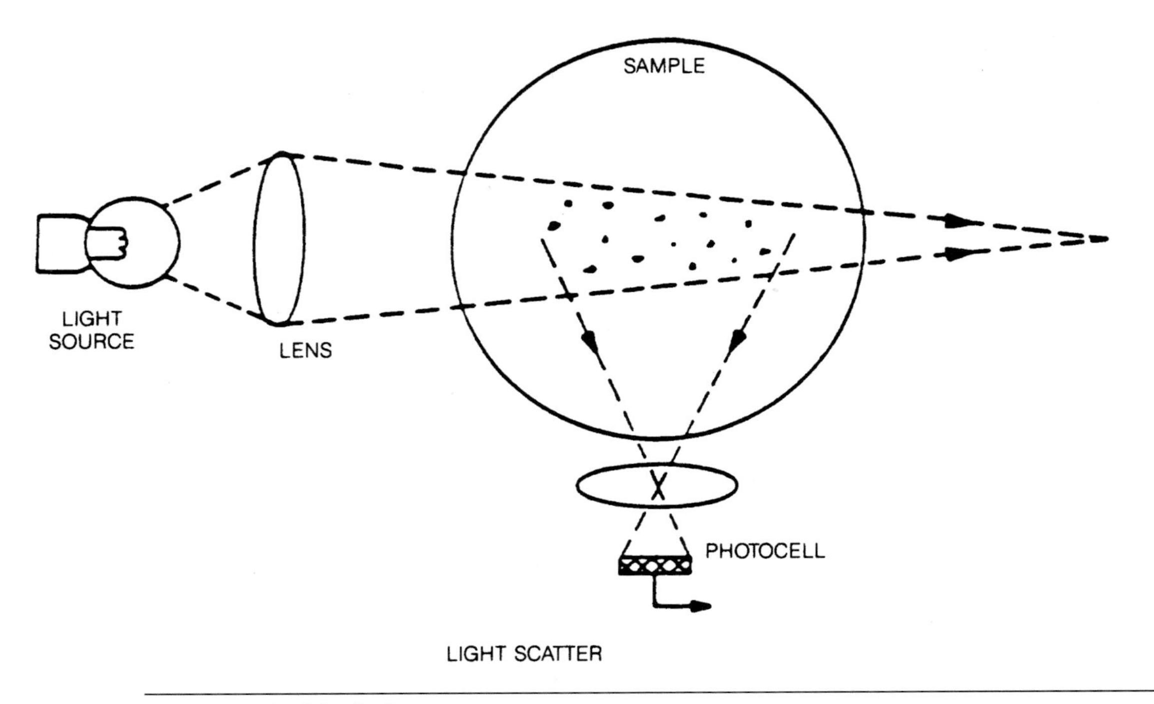


**FIGURE 7.24** Nephelometer.

or light detector at an angle to the light beam rather than directly in front of it, the detector receives only light scattered by the suspended particulate matter. Most nephelometers have the photo detector placed at a 90-degree angle to the incandescent light source.

**TURBIDIMETERS.** In process control, one way of determining the solids concentration is to relate the solids concentration to the turbidity caused by the presence of suspended matter. Turbidity is the optical property of a water sample that causes light to be scattered and absorbed rather than transmitted in straight lines through the sample (Figure 7.25). Typically, a beam from a standard light source is directed through a clear glass window into the fluid stream to be measured. A percentage of this light is reflected (backscattered) by the suspended matter in the liquid, and a percentage is absorbed. A photosensitive detector uses an attenuation beam of reflected light as a measure of the solids concentration. Some turbidity meters use a second (reference) detector to compensate for the changes in temperature or light source.

To reduce the interferences caused by accumulated dirt, slime, or bubbles in the sampling chamber, one manufacturer has developed a method in which a reciprocat-

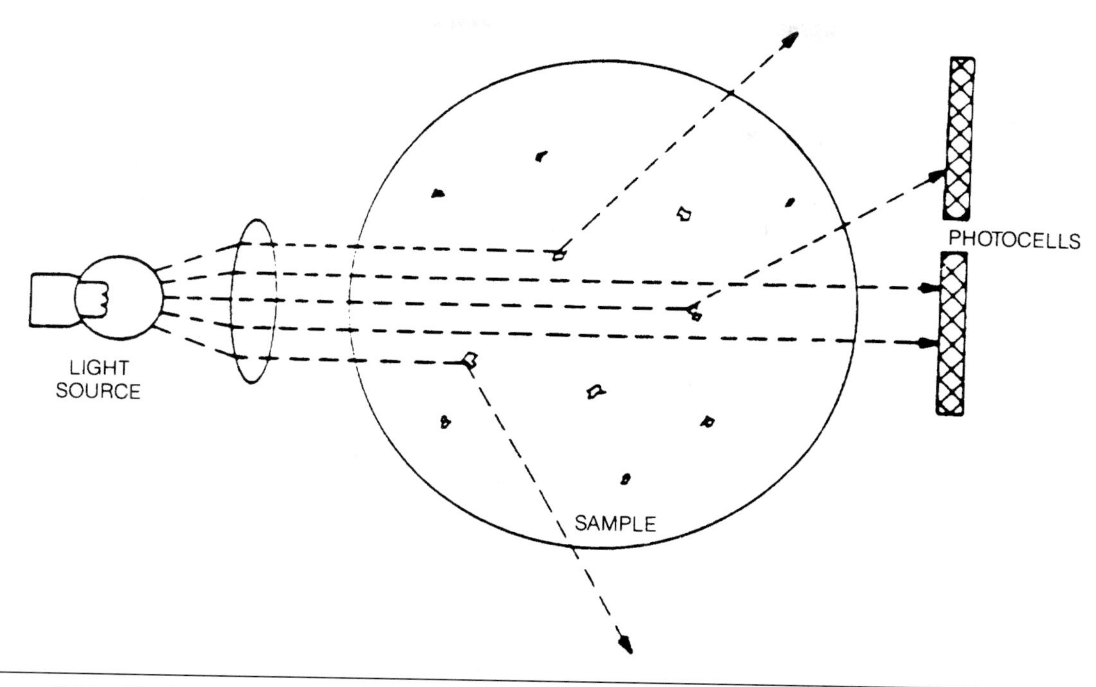


FIGURE 7.25 Turbidimeter.

ing motor-driven piston draws and expels the sample every few seconds while wiping the optical surface of the sampling tube on each cycle. Other manufacturers use either special optical surfaces or the scouring action of the process flow to keep the optical surfaces clean. Certain sensors use surface scatter to keep both the light source and the detector away from the fouling sludge.

**ULTRASONIC SOLIDS METERS.** The operating principles of ultrasonic solids meters are based on the attenuation of mechanical vibrations because of solids. Attenuation meters consist of transmitting and receiving transducers. The transmitter generates vibration signals that travel through the liquid, which are attenuated (dampened) by scattering caused by the suspended matter. The receiving transducer picks up the attenuated signal. The reduction of signal strength relates to the solids concentration. The speed of sound through sludge is used to measure the solids concentration. This change in speed is quite small but can be measured with extreme accuracy. The meters may need to be corrected to compensate for the effects of temperature and pressure on sound velocity.

**NUCLEAR-DENSITY METERS.** Gamma radiation (nuclear-density) detectors depend on a source of radioactive material that emits energy in the gamma region and a detector for this energy (Figure 7.26). The radioactive source is placed on one side of a pipe and the detector on the opposite side. Both water and solids act as gamma-radiation absorbers. For thicker sludges, the solids concentration may vary proportionally with the density of the sludge, thus allowing the use of a density meter as a solids meter. The benefits of nuclear-density meters include the following:

- Reasonable sensitivity,
- No moving parts, and
- Reduced maintenance because they do not touch the material.

On the other hand, the difficulties of nuclear-density meters include the following:

- The reading is not accurate if the pipe is not full of sludge or the material stratifies;
- Periodic calibration or compensation is required if the sludge's temperature changes, because temperature affects density;
- Entrained air renders the method useless for many applications;
- The density of greases and oils differs from that of sludge; and
- A U.S. Nuclear Regulatory Commission (Washington, D.C.) license is required to use, calibrate, and dispose of the radioactive materials.

**SLUDGE BLANKET SENSORS.** Many solids-concentration measurement techniques are used to detect sludge blankets. The instruments can be fixed at a certain depth to show the presence or absence of a sludge blanket. The instruments can also be movable, allowing the actual blanket interface to be located. Automatic detection of changes in blanket level is used to increase process-control stability.

## SIGNAL-TRANSMISSION TECHNIQUES

For maximum use and effectiveness, signals generated by various sensors and instruments are transmitted from the sensor to a receiver at another location. Often, the sensor output is transmitted to a control panel or computer system, which allows operators to inspect many process variables simultaneously.

The three components of a signal-transmission system are the transmitter, receiver, and transmission medium (the connection between the transmitter and receiver). The transmitter converts a mechanical or electrical signal from the sensor into a form that

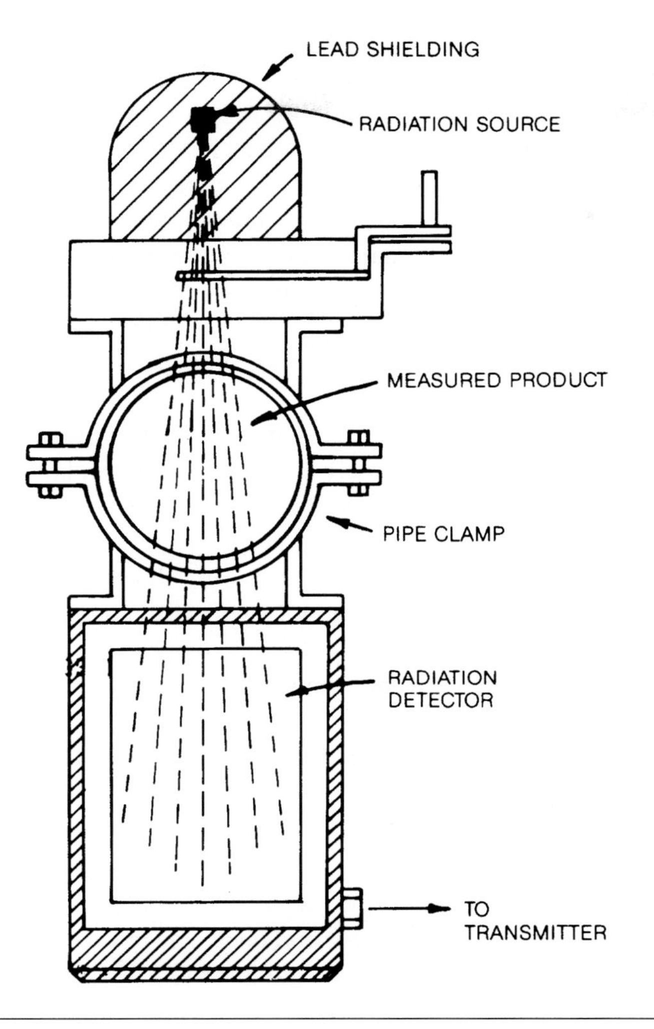


FIGURE 7.26 Nuclear solids analyzer (Skrentner, 1989).

the transmission medium can use. The transmission medium contains the signal and transfers it to the receiver. The receiver subsequently converts it into a form that the receiving system can use.

**ANALOG SIGNAL TRANSMISSION.** Analog signal-transmission systems continuously and proportionally convert a continuous sensor output to another form. For example, an analog pneumatic transmission system converts the sensor output from a mechanical displacement to a proportional pressure, typically 20 to 100 kPa (3 to 15 psig). The key characteristic of analog systems is that the transmission signal is continuous and proportional to the sensor output.

Pre-electronic analog transmission systems included pneumatic (compressed air) and hydraulic (compressed fluid) systems. In each case, a mechanical sensor displacement is converted and conveyed by the system without the use of electricity. These systems used to be the workhorse transmission systems in wastewater and industrial treatment plants. Their advantages include intrinsic safety (no sparking potential in hazardous areas) and ease of understanding by mechanical maintenance personnel. However, response time is sluggish in these systems, particularly where long transmission distances are involved.

Electronic analog transmission systems use relatively low voltage and current electricity. They offer the instantaneous response and low-cost characteristics of electricity. Standard electronic transmission systems use 4 to 20 mA-DC or 1 to 5 V-DC signals. As with pneumatic and hydraulic systems, they provide a continuous signal proportional to the sensor output. Electronic signal wiring should be shielded to protect it from high voltages and the other electronic noise encountered in an industrial environment. While they are more responsive and less expensive to install, electronic systems require the skills of an electronics technician for adequate maintenance.

**DIGITAL SIGNAL TRANSMISSION.** Like computers, digital signal transmitters use the binary number system, which expresses all decimal numbers as a combination of zeros and ones. Electrically, this is manifested as either the presence or absence of electrical voltage. To transmit an analog signal via a digital system, an analog-to digital converter is used. This device accepts the sensor's analog signal and electronically converts it to an appropriate series of zeros and ones. The receiver may use a digital-to-analog converter if the digital signal is not used directly.

The advantages of a digital signal-transmission system are that large amounts of data may be inexpensively and accurately transmitted at high speeds. Digital transmission systems use either an electronic transmission technique or, through further conversion, optical techniques involving optical fibers instead of wires.

**SIGNAL-TRANSMISSION NETWORKS.** Many sensor and control signals can be combined to form an electronic or fiber-optic transmission network. Various electronic components and computers linked to the network consolidate large amounts of plant data to be used at various locations in the facility. Networks can drastically reduce the total number of wires required to transmit data from the field sensors to an operator station. The many networks vary according to the type of medium and the network components required. With the advent of digital sensors, it is now possible to connect a series of field instruments or final control elements into networks (Fieldbus).

**TELEMETRY.** Both analog and digital signal-transmission techniques link remote facilities (e.g., lift stations) with a central monitoring point (Figure 7.27). These specialized systems, often called *telemetry systems*, use radio, telephone, microwaves, or lasers as communication media to transmit information over long distances or difficult terrain. When combined with computers, telemetry systems easily consolidate the monitoring and control of many remote facilities. Because telemetry systems are purely electronic, they require the skills of an electronics technician for adequate preventive maintenance.

## CONTROL CONCEPTS

Control of a process variable would not be required if processes were always steady-state (i.e., operated without change). However, process conditions in a treatment plant are always changing based on the flow and strength of the wastewater being treated. These changes create deviations from ideal process adjustments or setpoints, resulting in process inefficiencies, including over- and under-treatment of wastewater and the

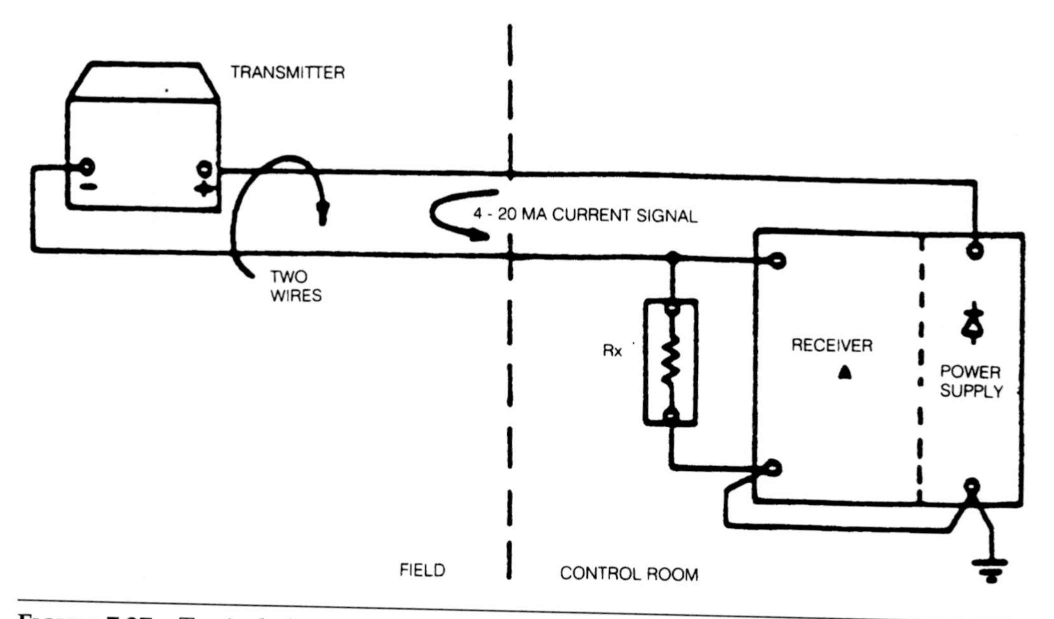


**FIGURE 7.27** Typical electronic analog signal-transmission system.

removed solids. Control of a process variable (e.g., flow or level) is accomplished by moving valves or changing pump speed, commonly called the *manipulated variable*.

**MANUAL CONTROL.** An operator performs manual control. For example, if an operator observes a rising level in a wet well and slightly increases the speed of a pump to match the flow into the well, the new level will be maintained. To take the level back to its original point, the operator will have to increase the pump speed to lower the level sufficiently and then slow it again to match the inlet flow. With experience, the operator may guess the speed necessary to maintain the wet well at an acceptable level, but as the flow changes, the level will again deviate from the optimum level. This kind of control is also called *open-loop control* because there is no direct connection between the desired value of the process variable (setpoint), the instantaneous value of the process variable, and the action of the controlled variable, unless the plant operator constantly observes and changes the manipulated variable.

**FEEDBACK CONTROL.** In the above example, if the operator is continuously monitoring the level, the operator is providing the feedback. However, doing so is not an efficient use of resources. Feedback control is the easiest way to automate the control of a process. A feedback controller (Figure 7.28) receives sensor output on either the process variable or the desired result and compares it with an operator-entered set-

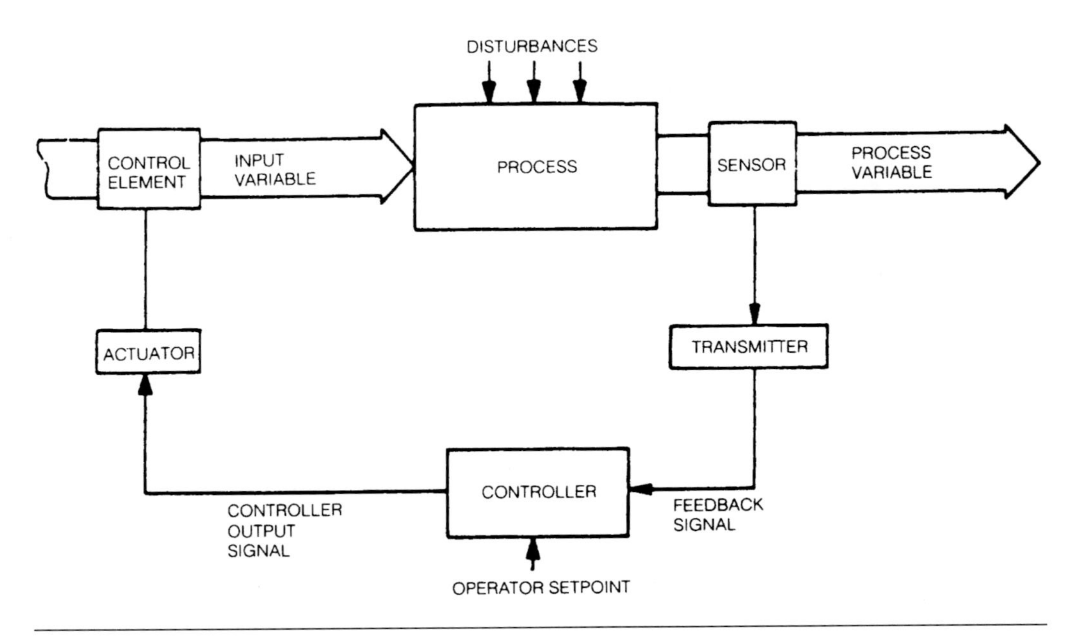


FIGURE 7.28 Feedback control.

point. The controller automatically calculates the difference (error) and determines an appropriate control action for the input variable. Because the feedback controller is always on duty, the level in the wet well can be maintained close to the optimum required (the operator's setpoint).

Feedback control accounts for all existing possible errors, process disturbances, and process inputs. The individual errors and disturbances need not be specifically quantified, because the feedback mechanism gives the controller a cumulative effect.

One disadvantage of feedback control is the time lag between a process adjustment and the evidence of that adjustment. The time lag is typically not a problem when controlling the flow or level in relatively small vessels. However, if the time lag is 30 minutes or longer, feedback control alone may be inadequate.

**FEEDFORWARD CONTROL.** Feedforward control uses information about the inputs to a process to prevent process errors (Figure 7.29). A good example of feedforward control is ratio control. In a chlorination system, the rate of chorine application is based on the amount of plant flow entering the chlorination facility. An automatic ratio-control system allows the operator to establish a mass-to-flow chlorine-dose setpoint. Then, as the plant flow increases or decreases, the chlorine application rate adapts automatically and proportionally.

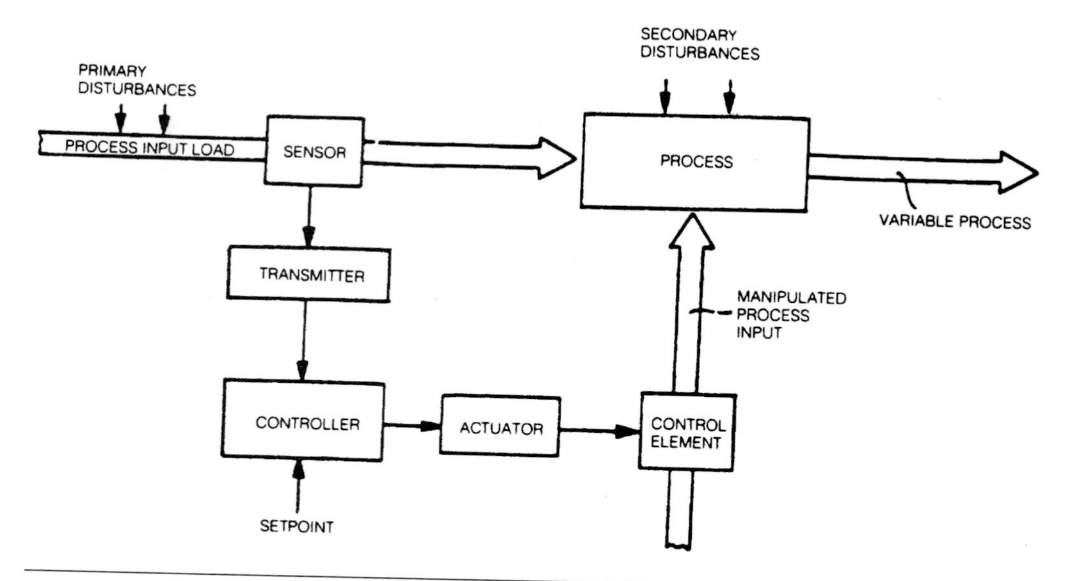


FIGURE 7.29 Feedforward control.

**COMPOUND CONTROL.** Combining feedback and feedforward controls provides the ultimate in automatic control for certain processes. As in the chlorination example, the operator determines a ratio of chlorine to be applied as the plant inflow changes. The feedforward controller faithfully meters the chlorine according to this requirement. However, the ultimate process goal is maintaining a residual level of chorine. Suspended solids, pH, and other process characteristics not detected by the feedforward control equipment will produce considerable variation in the chlorine residual, despite the constant adjustment provided by the chlorine-application ratio.

In a compound control loop for chlorination, feedback control adjusts the ratio setpoint of the feedforward control loop. A chlorine residual analyzer continuously monitors the chlorine residual in chlorination effluent. As the process variable deviates from the desired setpoint, a new setpoint is sent to the ratio controller, which readjusts the ratio of chlorine to plant inflow accordingly.

**ADVANCED CONTROL.** Computers, microprocessors, and other programmable devices allow unique control concepts to be applied to a specific process or group of processes (e.g., the concept of sequential control or logic statements that compare process conditions with preprogrammed conditions allowing specific outputs to the process equipment). The control system outputs precisely imitate an operator's actions in response to the process changes. Pure feedback or feedforward controls may be only partially adequate for the control situation encountered. Advanced control is typically best applied where multiple, parallel treatment units are used or where changing process conditions cannot be anticipated by the change in one process input or output.

Advanced control programs take into account dozens of process sensors while manipulating selected pieces of equipment to achieve the desired process control. Advanced mathematical calculations, models, or condition matrices are used to maintain correct setpoints or properly correct for extreme process conditions. Besides variable-speed pump or valve adjustment, equipment may be started or stopped. These programs are often generated by trial-and-error techniques until the best match is found. The feedback and feedforward control concepts are often used as portions of the overall control logic.

## AUTOMATIC CONTROLLERS

Automatic controllers are prepackaged devices that not only accept input signals from sensors and operators but also output calculated control signals to a final control element to maintain an operator-entered or otherwise calculated setpoint. The controllers in use today are primarily electronic in principle, although pneumatic and mechanical controllers are still used successfully, depending on the application. The following modes of control are available and applied based on control requirements.

**ON–OFF CONTROL.** On–off (differential) control is the most widely used automatic control in industrial and domestic control applications. On–off control is used when extremely tight control of a specified setpoint is not required (Figure 7.30). The process variable is controlled based on a specific band of values rather than just one value. By nature, on–off control is discontinuous.

A common domestic example of on–off control is a thermostat. A common wastewater treatment example of on–off control is that used in a small single-pump lift station. Two floats (sensors) indicate the high and low levels desired in the wet well. When the high float is tripped, the pump starts and runs until the low float is tripped, shutting off the pump. Thus, the liquid level in the wet well is maintained between the levels of the floats (the acceptable band of operation).

**PROPORTIONAL CONTROL.** Proportional control is often used when continuous control is required and a wide operating band for the process variable is not acceptable. Proportional control is based on a linear mathematical relationship in which the controller output is equal to the error between the setpoint and actual value of the process variable multiplied by a constant adjustable factor known as *gain* (Figure 7.31).

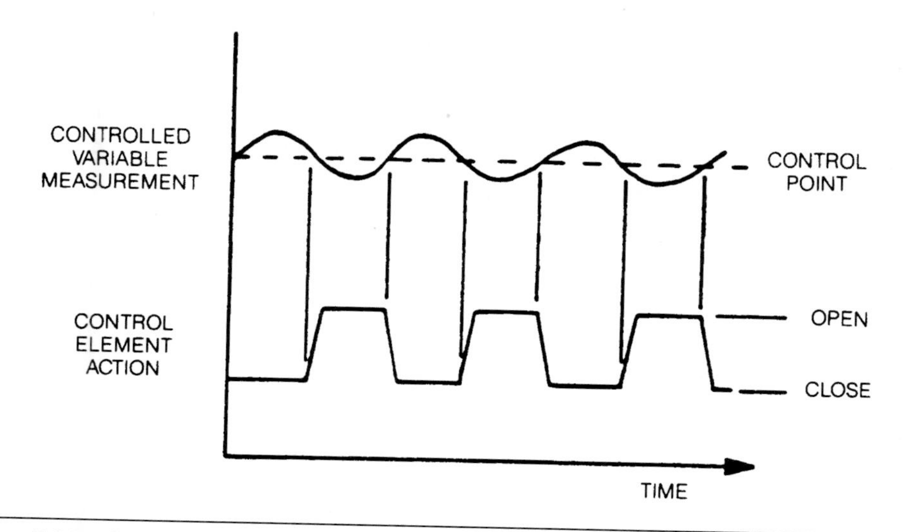


FIGURE 7.30 On–off control.

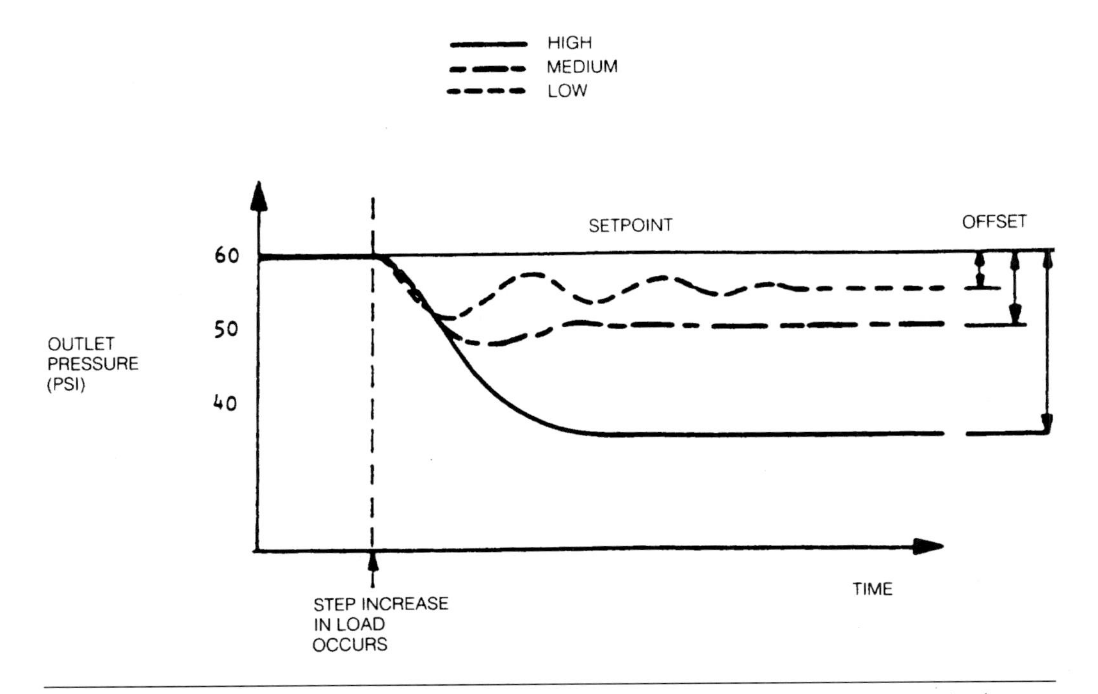


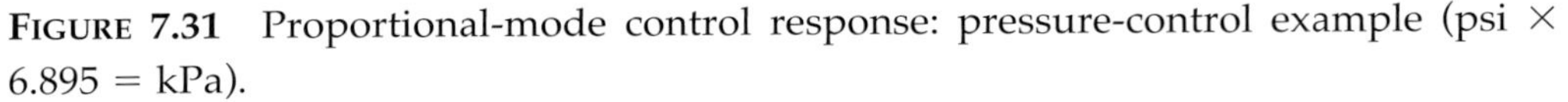

**FIGURE 7.31** Proportional-mode control response: pressure-control example (psi × 6.895 = kPa).

As the error increases, the controller output increases proportionally to compensate. The higher the gain, the larger the output increases for each unit change in the error.

Proportional control has one major disadvantage. At steady state, the controller exhibits an offset between the ultimate desired setpoint and that achieved with the controller. Often, this offset is considered insignificant, so proportional control is more than adequate.

**RESET CONTROL ACTION.** In cases where the inherent offset of proportional control is unacceptable, reset control action may be required. Reset control action uses a mathematical integration of the error signal, allowing the controller output to change at a rate matching the change of the error signal over time. This action, in combination with proportional control, corrects for the offset condition inherent with proportional control alone. This type of controller is often referred to as having proportional–integral action (Figure 7.32). The disadvantage of this controller is that it is inherently less sta-

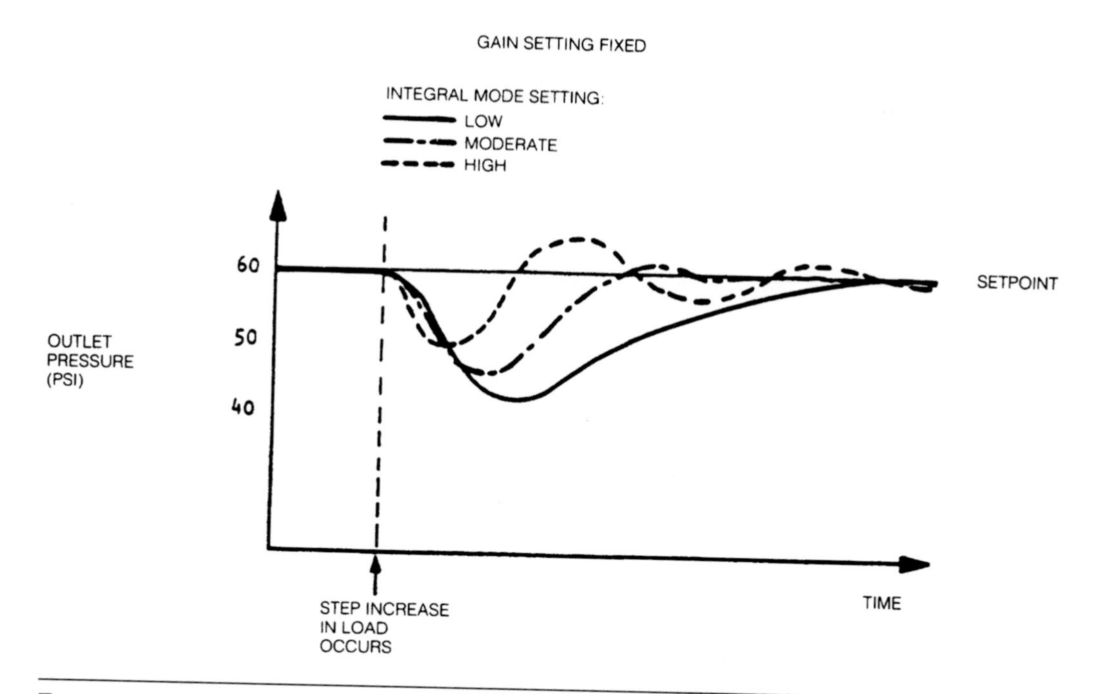


**FIGURE 7.32** Proportional plus integral mode control response: pressure-control example (psi × 6.895 = kPa).

ble than a proportional-only controller and so may be difficult to tune during initial application.

**THREE-MODE CONTROL.** A three-mode controller (commonly known as a *proportional–integral–derivative controller*) is the most sophisticated standard controller available for feedback control loops (Figure 7.33).

This controller uses the features of the proportional–integral controller and adds a third action (derivative control) for even more responsive control. Derivative action provides a mechanism for adding lead action to the control loop, compensating for time lags in the control loop. It cannot be used alone because a large continuous, unchanging error would result in no controller response. Derivative controller applications are rare in wastewater treatment control applications. This type of controller is covered here because controllers may often be found installed with all three modes available for use. Operators should be aware that all three modes are not required for adequate control.

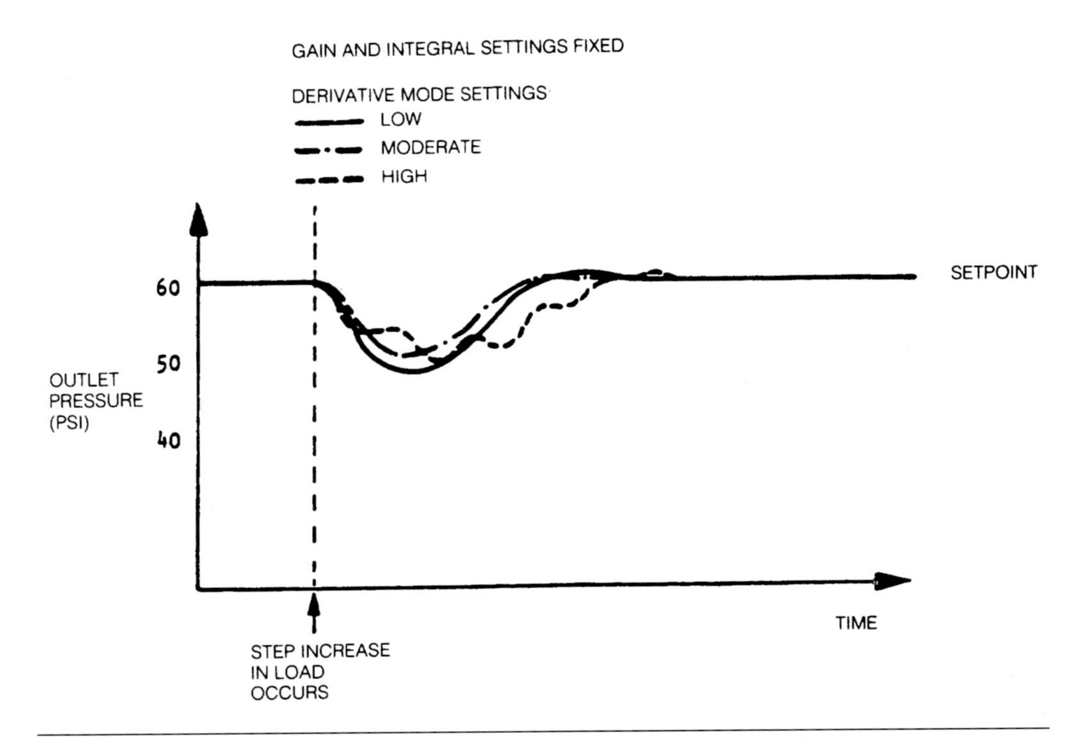


FIGURE 7.33 Proportional plus integral plus derivative control response: pressure-control example (psi × 6.895 = kPa).

# PERSONNEL PROTECTION INSTRUMENTATION

Specialized gas-detection instrumentation is an essential component of a modern personnel protection program. Instrumentation systems sample ambient-air conditions in a plant building or manhole and provide early warning of hazardous conditions. These systems may also automatically activate ventilation equipment and other alarm systems. Hazardous atmospheric conditions typically found in wastewater collection and treatment facilities include oxygen deficiency; explosive gases; and toxic gases (e.g., hydrogen sulfide, chlorine, sulfur dioxide, and carbon monoxide). *Safety and Health in Wastewater Systems* (Water Environment Federation, 1994) provides more information on personnel protection programs.

Personnel-protection instrumentation systems are packaged for both portable and stationary applications. Portable applications include confined-space entry (manholes and wet wells) for inspections and maintenance. Stationary applications include equipment structures below grade (e.g., lift stations or pipe galleries) or any enclosed area where gases may escape from wastewater or sludges in a routine way (e.g., a screening room, grit chamber, or dewatering areas). Managers of any wastewater treatment or collection operation, no matter how small, should consider purchasing a minimum complement of hazardous-condition instrumentation, because this equipment is reasonably priced and reliable.

**PORTABLE EQUIPMENT.** Portable, battery-powered alarm systems that detect multiple gases are essential equipment for any personnel entering a confined space, even if it pretested as safe. Worn by personnel, the units will sound an alarm at preset gas levels. Health and safety organizations and federal, state, and local regulations establish the alarm limits. Essential features of portable multigas testing equipment include the following:

- Visible and audible alarms for each gas according to real-time exposure limits, 15-minute short-term exposure limits, and 8-hour time-weighted-average limits;
- A display for the actual numerical concentration of each gas;
- An operating battery life of at least 10 hours, with audible low-battery charge alarm;
- Certification as intrinsically safe in hazardous environments;
- Ability to function over a range of temperatures, from below freezing to approximately 49 °C (120 °F);
- Accessories for calibration and remote sampling; and
- Documentation that clearly describes operation, calibration techniques, and equipment and provides a list of gases that could interfere with a particular sensor.

Operators should use a portable gas measurement system that meets the established plant standard safety procedures.

Operators should keep written records of routine calibration, repairs, sensor replacement, and all alarm incidents. Sensor life varies depending on the environment where it is stored and used. Operators should replace sensors routinely, based on the factory-recommended schedule or on the sensor failure history, whichever is more frequent.

**STATIONARY EQUIPMENT.** Stationary systems include gas sensors and conversion electronics permanently mounted in a given location. Chemical storage and feed areas, pumping stations, the digester-gas compression room, and other areas often use stationary systems connected to plant alarm systems that trigger the operation of ventilation equipment. Because sensor installation is permanent, proper placement is essential to allow accurate sampling of the atmosphere in an area. For gases heavier than air, sensors should be no more than 0.3 m (1 ft) above the floor. Pockets of toxic gas may collect near the floor, even on a floor that is above grade. If auxiliary equipment (e.g., a sampling pump) is used with the system, its failure should be treated as a hazardous condition.

Because most of the areas where stationary systems are used are often moist with varying concentrations of corrosive gases, operators should protect the electronics and power-supply components by putting them in a clean environment or suitable enclosure. Many toxic gas hazards can occur during electrical power failure, so plant staff should consider providing a source of uninterruptible electrical power (e.g., a battery) for a stationary system. The essential features of a stationary system and a portable system are the same. As with portable gas-detection equipment, thorough records of calibration, repair, and alarm conditions are required.

**OXYGEN DEFICIENCY.** An oxygen-deficiency sensor measures and shows the percentage of oxygen in the air being sampled. It is important to test various levels of the airspace being sampled because gases lighter or heavier than air will form layers. The fact that oxygen is present at safe levels does not necessarily mean that the area is not potentially dangerous because other toxic gases may be present.

**EXPLOSIVE GASES.** Explosive gases are dangerous combinations of oxygen and hydrocarbon gases (e.g., methane, ethane and gasoline and paint solvent vapors). Tests for these gases must be conducted with extreme caution, avoiding sparks that could ignite an explosive gas mixture.

**HYDROGEN SULFIDE.** Hydrogen sulfide gas, commonly found in wastewater treatment and collection systems, is malodorous (smells like rotten eggs at lower concentrations). High concentrations of the gas quickly numb the sense of smell and may mislead an operator into thinking that conditions are safe. Besides being toxic and potentially fatal to personnel, hydrogen sulfide is corrosive.

## FACILITY PERSONNEL PROTECTION AND SECURITY

Facility-wide instrumentation systems are often provided to protect the facility's personnel, structures, and equipment. These systems may include fire detection, card access, page party, and closed-circuit television. The fire detection system is used to warn personnel of a fire for evacuation purposes, notify the local fire response agencies, and to locate the fire. Card access systems are used to permit site access, to limit access within plant areas, for personnel recordkeeping, and for security emergency response. The closed-circuit television can be used to monitor public areas surrounding the plant and help protect against unauthorized access to areas within the plant. A page-party system is used to alert personnel during an emergency, provide communication between personnel on the plant site, and help locate personnel by voice announcement and response.

**FIRE DETECTION SYSTEM.** Fire detection system standards for wastewater treatment facilities have been established by the National Fire Protection Association (Quincy, Massachusetts) and are often augmented by state and local building codes to define minimum requirements for alarm systems and fire-suppression systems. Typical suppression systems include water sprinklers or sprays, foam, halon, carbon dioxide, dry chemicals, and preaction sprinkler systems. A fire-detection system supervises and monitors these systems and monitors for fire or smoke via photoelectric, ionization, and temperature detectors. In sprinkled areas, the system would also monitor the manual pull stations.

The fire-detection system, via audible and visual indication devices, provides early warning to building occupants so a safe and orderly evacuation can be made. Local fire panels annunciate the detector's location so fire respondents can quickly address the problem. The panel can also be interfaced with heating, ventilation, and air conditioning systems to help evacuate smoke, elevators to limit access and return to the proper floor, page-party systems to provide personnel evacuation instructions, closed-circuit television systems to monitor the affected area, and the telephone system to contact the local municipal fire station for appropriate response.

On large digital systems, local panels are connected via data highways and monitored via cathode ray tube screens. Each device can be monitored for current status, and alarm limits can be modified for differing area working conditions (reducing false alarms). Operating and alarm history records are maintained in the system database.

Fire-detection systems must be tested quarterly to ensure that the monitoring devices are properly cleaned and calibrated and that the system performs as required.

**CARD-ACCESS SECURITY SYSTEM.** Card-access systems are used to limit access to an area and can be provided as standalone devices interfacing electrically to a magnetic door latch (or gate motor) to permit entry and an infrared detector (or pressure-sensing device) to permit egress. With more sophisticated card readers or scanners, these devices can be connected to a computerized system, which can be programmed to permit facility access or restrict access to certain areas within the plant. Individual access can be programmed and limited by time of the day or day of the week. Schedules can be set up to cover work shift assignments and more. Information stored on the system can be used to identify personnel access and maintain time and attendance records.

**INTRUSION DETECTION.** Like the card-access system, intrusion detectors are used to monitor unstaffed areas of the facility. Depending on the area to be protected, limit switches, conductive tape, and magnetic intrusion detectors can be mounted on doors, hatches, or windows, with passive infrared motion detectors installed to cover specified areas. Remote pumping stations often use this hardware to notify plant personnel via radio or telephone telemetry that the station is being accessed. To provide facility perimeter detection, taut wire systems within security fencing and bistatic and monostatic microwave systems can be used. The latter devices require an overlapping layout to prevent blind spots.

**CLOSED-CIRCUIT TELEVISION SYSTEM.** A closed-circuit television system uses a camera and cathode ray tube to monitor an area. These devices are hardwired or networked back to a local control or security office to provide visual input to the operator of existing field conditions. Cameras can be provided with sun screens; lenses for low-light operation; pan, tilt, and zoom capabilities to observe a wider area or get a close-up view; heaters for winter applications; and wipers to combat rain or snow. Controls can be implemented with the card-access system or intrusion-detection devices to focus on a certain area if unauthorized access is attempted or access is permitted after normal working hours. The cameras can be connected to video recorders so a record of previous activity is maintained. These same functions can be interfaced with the fire alarm system to provide a video record of the event.

**PAGE-PARTY SYSTEM.** The page-party function supports two types of voice communication: paging, which allows announcements to be made; and party lines, which allow "telephone" communications between persons connected to the same party line. A page-party system may use integrated page-party stations providing both types of voice communication in a single enclosure and using separate wiring to connect the integrated page-party stations together. A page-party system may also be an independent paging system using amplifiers, wiring, and speakers to provide paging and separate handsets to provide point-to-point voice communication. A hybrid solution using both types of page-party systems may also be used to provide the page-party function. The benefit of an integrated page-party system is simple and efficient communication in both the process and administrative areas of the plant.

Each page-party station, including the handset, speaker, and speaker amplifier, can be provided with internal monitoring and diagnostic capability. These functions can be centrally monitored for use and maintenance purposes. This capability has proven useful when personnel use the system in an unauthorized manner.

## INSTRUMENT MAINTENANCE

**RECORDKEEPING.** There is no substitute for a good recordkeeping program. A record of instrument performance and repairs allow operations or maintenance personnel to properly evaluate an instrument's effectiveness and determine if the instrument meets the objectives used to justify its purchase and installation. As a minimum, the following basic information should be maintained for each instrument in the wastewater treatment plant:

- Plant equipment identification number,
- Manufacturer,
- Model number and serial number,
- Type,
- Dates placed into and removed from service,
- Reasons for removal,
- Location when installed,
- Calibration data and procedures,
- Hours required to perform maintenance,
- Cost of replacement parts, and
- Operations and maintenance manual references and their locations.

**MAINTENANCE.** Preventive maintenance requirements are best met by combining the recommendations of the equipment manufacturer with the experience and knowledge acquired over time by operations and maintenance personnel. Typically, preventive maintenance for instrumentation is divided into four areas: reasonability checks, cleaning, detailed inspection, and calibration.

A *reasonability check* is the act of observing the indication of the instrument with process changes and in comparison with other related instrumentation. For example, acceptable operation of an influent flow meter can be verified by observing its response to step-flow changes in the main flow to the plant as main lift pumps are cycled off and on. The same meter can be compared with the sum of multiple, parallel downstream meters. For a dissolved oxygen sensor, a second portable dissolved oxygen meter can be used to make a quick comparison. Reasonability checks are not a replacement for a detailed inspection and calibration. Instrumentation technicians can do reasonability checks, but they are mandatory for operations personnel.

Cleaning involves everything from wiping the cover of a recorder to removing inline pH or dissolved oxygen probes for removal of precipitates or solids deposits.

Detailed inspections may be part of the cleaning activity or can be scheduled on a calendar basis. To prevent instrument damage, a trained technician should inspect the instrument. To maintain consistency and ensure that the instrument is not damaged, specific cleaning procedures need to be followed. The specific environment in which the instrument is installed determines the frequency and methods of cleaning.

*Calibration*—the process of fixing, checking, or correcting the graduation of an instrument—is the comparison of the instrument's performance against some accepted standard followed by adjustment to match the standard. Excessive calibration may indicate that the instrument is failing. If an instrument appears to be malfunctioning, operators should first determine that it is not showing transient process conditions or other non-instrument-related conditions. Calibration adjustment should not be made casually to force an instrument to match what is perceived to be correct based on observed process performance, because that would defeat the purpose of instrumentation as a reliable tool for verifying process performance. Operators should replace parts when it is confirmed that a device has malfunctioned or when it is called for in preventive maintenance procedures. The plant should stock adequate spare parts, because delivery times can range up to several months, depending on the device. Certain parts are best replaced by the manufacturer, while others can be easily replaced by nearly anyone who can follow a written procedure. Electronic components must be handled carefully, because they are subject to severe damage from static electricity and even weak magnetic fields. Personnel who replace parts need proper training for the specific activity.

**PERSONNEL TRAINING.** Despite plant or staff size, each plant develops its specific instrumentation training program. At a minimum, operators must be trained in the principles of instrument operation that allow detection of malfunctions. Maintenance personnel should be trained to use the specialized equipment required for instrument maintenance. Training can include short courses, seminars, or formal classes conducted by instrument manufacturers or consultants. It may include the use of videotapes offered by professional organizations [e.g., the Instrumentation, Systems, and Automation Society (Research Triangle Park, North Carolina)]. Regardless of the training source, the format can include lectures and, more importantly, hands-on activities. Whenever possible, hands-on training in the plant should use the actual instruments to be maintained. Written documentation and audio–visual techniques should be used to record the training.

Annual allocations for personnel training should be included in the budget. Because advances in the field of instrumentation and control systems occur more rapidly than other areas of plant maintenance (e.g., repair of mechanical process equipment) instrument repair personnel need more frequent training to stay abreast of the industry.

**MAINTENANCE CONTRACTS.** An effective alternative to having a full-time maintenance technician on staff is to contract with the instrument or control-system supplier or a qualified third-party maintenance organization. This is particularly true for smaller treatment facilities. Larger facilities should carefully evaluate the relative benefits of this approach. A contract with a reputable maintenance organization typically provides access to a skilled technician and specialized test equipment. Various types of service packages are available, ranging from complete service (e.g., parts, routine service calls, and emergency repairs) to simple quarterly visits during normal working hours using parts supplied by the treatment plant.

The annual cost of maintenance contracts may be 10% of the equipment cost or more, depending on the response provisions. A negative aspect of maintenance contracts is the outside technician's relative unfamiliarity with the plant's equipment compared with that of employees. Additionally, the service organization may not appreciate the relative importance of one device over another. Thus, plant managers must provide direction and arrange priorities for the service people. To ensure that the work performed meets the contract terms, service personnel should be monitored.

Some wastewater treatment plants have created hybrid operator–instrument technician positions to allow the needed intimacy with the equipment without creating a full-time, dedicated position. This approach yields additional benefits because the operational knowledge allows the employee to fully understand the process function and importance of a particular device.

## REFERENCES

Isco Inc. (1989) *Open Channel Flow Measurement Handbook;* Isco Inc. Environmental Division: Lincoln, Nebraska.

Skrentner, R. G. (1989) *Instrumentation Handbook for Water and Wastewater Treatment Plants;* Lewis Publishers: Chelsea, Michigan.

Water Environment Federation (1994) *Safety and Health in Wastewater Systems,* Manual of Practice No. 1; Water Environment Federation: Alexandria, Virginia.

## SUGGESTED READING

American Water Works Association (2001) *Instrumentation and Control (M2),* 3rd ed.; American Water Works Association: Denver, Colorado.

# Chapter 8

# Pumping of Wastewater and Sludge

# INTRODUCTION

Pumping systems (which include pumps, motors, valves, and controls) are integral to wastewater treatment. Collection and treatment system operations depend on pumping systems to help transport wastewater and solids to the treatment plant and between various treatment processes. Pumping systems transfer fluid from one point to another or one elevation to another so it can then flow via gravity through the collection or treatment system to the next pumping system. [In this chapter, *fluid* is any material being pumped, *liquid* is any fluid that is predominantly water, and *solids* is a fluid with significant solids content. Raw wastewater, for example, is a liquid; it typically contains 200 to 300 mg/L of total suspended solids (0.02 to 0.03%)—too little to affect flow characteristics. Return activated sludge could be a liquid or solids; it typically contains 10 times more solids, which could affect flow characteristics. Thickened, dewatered sludge is a solids; it can contain more than 20% solids, which profoundly affects flow characteristics.]

Operators should be able to recognize pumping system components, know how they function, understand how basic hydraulic conditions affect performance, and learn how to optimize pumping systems for long, productive service lives. They also should be familiar with wastewater-related hazards. For example, gritty and stringy solids can plug pumping systems or wear them out more rapidly. Both wastewater and solids contain high concentrations of pathogens known to cause illness in people. They also are susceptible to biological reactions that can increase pumping system pressures and generate poisonous gases that could both build up to dangerously high levels if not vented properly.

Pumping strategies (e.g., increasing off-peak pumping and maintaining high wet well levels) should be designed to lower energy costs and balance flow through treatment processes. However, these strategies will alter the design and operation of various process units and could create the following problems:

- Increasing evening flows through the treatment plant will increase chemical-addition and solids-pumping monitoring needs during off hours, when the requisite staff may not be present.
- Maintaining continuously higher wet well levels will lower velocities in the collection system, promoting deposition of solids.
- Detaining flow in the collection system can lead to more odors.
- Holding flow in the collection system will shrink the reserve capacity available for emergency or wet-weather conditions.

So, utility staff should thoroughly investigate potential consequences before implementing any pumping changes.

(For even more information, see the "Hydraulics Institute Standards", which provides detailed information on pump types, nomenclature, ratings, test standards, and applications, as well as instructions for installing, operating, and maintaining centrifugal, rotary, and reciprocating pumps.)

## PUMP AND PUMPING FUNDAMENTALS

A pumping system typically includes a pump, suction and discharge reservoirs, interconnecting piping, and various appurtenances (e.g., valves, flow-monitoring devices, and controls). To deliver a given volume of fluid through the system, a pump must transfer enough energy to the fluid to overcome all system energy losses.

In the following subsections, various energy needs (headlosses) are identified and defined, and the energy (head) that the pump imparts to the fluid is described. Head is typically measured as meters of water (m $H_2O$) or feet of water (ft $H_2O$).

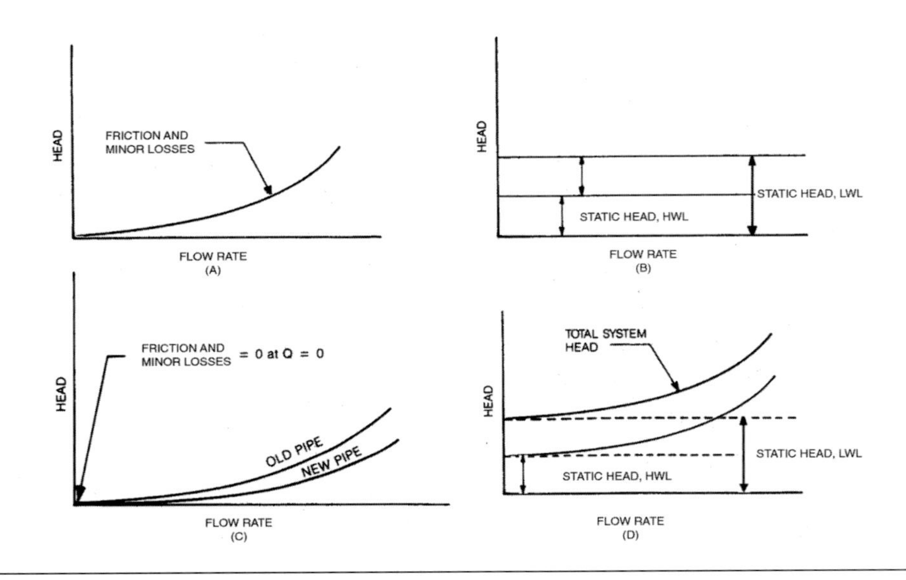


**FIGURE 8.1** Pump system head curves: (A) friction head loss versus flow, (B) velocity head and static head versus flow, (C) friction loss curves, and (D) total system head versus flow.

**SYSTEM HEAD.** A pumping system requires a certain amount of energy to convey a given amount of flow at a specific rate. Its "system curve" is a graph charting the total energy requirement at various flow rates (Figure 8.1). The *total energy requirement* includes both the energy needed to overcome velocity losses as the fluid moves through the pumping system and that needed to overcome elevation changes from the pumping system's inlet to its outlet.

***Static Head.*** *Static head* is the difference between the fluid's surface level (pressure head) at the pumping system's entrance (suction point) and that at its exit (discharge point) (Figure 8.1, Part B). In liquid pumping systems, the fluid level in the influent wet well varies, while that in the discharge tank is relatively constant. So, the static head is a larger value when the influent level is low and a smaller one when it is high. The static head can be zero or less than zero, depending on the downstream fluid surface elevation and pressure head.

***Velocity Head.*** *Velocity head* is the distance required to accelerate the fluid from a standstill to a particular velocity. It depends on the desired fluid velocity and is calculated as follows:

$$\text{Velocity head} = v^2/2g$$

where

$v$ = the fluid velocity in meters per second (m/s) or feet per second (ft/sec),

1 ft/sec = 0.3048 m/s, and

g = 9.80 $m/s^2$ (32.2 $ft/sec^2$).

Typically, the velocity head is added to a pumping system's headlosses.

***Headlosses.*** A pumping system's headlosses include both friction and minor losses.

*Friction Losses.* *Friction loss* is the energy that flow loses because of frictional resistance in pipes. It can be calculated via various formulas (e.g., Darcy, Darcy–Weisbach, and Hazen–Williams). Over time, the friction head at a specified flow within a particular system may not equal the design head because

- The "as built" system differs from the design;
- The mechanical equipment's manufactured tolerances vary;
- A fluid's viscosity changes when its temperature changes [applies to fluids with a viscosity much greater than water (e.g., solids)];
- Solids have accumulated or air is entrained in the system, causing a partial restriction;
- The piping system is deteriorating.

These changes could either increase or decrease friction loss, so utility staff should calculate friction losses for both newer and older piping.

*Minor Losses.* Whenever a fluid changes direction and speed to travel through a pumping system's valves, bends, and other appurtenances, it loses energy. Such losses are called *minor losses* because each individual loss is not a large value. They are calculated as follows:

$$\text{Minor loss} = kv^2/2g$$

where $k$ = an appurtenance-specific loss (Brater et al., 1996; Street et al., 1995).

While each individual loss is minor, the cumulative effect of these losses can be substantial. It can even exceed the pumping system's total friction losses.

*Total Headlosses.* A pumping system's *total headloss* is the sum of both friction losses and minor losses. It depends on the square of the flow rate (Figure 8.1, Part A). Typically, new piping has fewer headlosses than old piping (Figure 8.1, Part C).

***Total System Head or Total Dynamic Head.*** Total system head, which is typically synonymous with total dynamic head at a given flow rate, is the sum of the static head, velocity head, minor losses (including entrance and exit losses), and total friction losses through all piping (Figure 8.1, Part D). It serves the following two purposes:

- Defines the total energy losses realized at a given flow rate through a pumping system, and
- Determines the total energy required to convey and lift fluid at a given rate through a pumping system.

Typically, the velocity head is ignored because it is inconsequential compared to total minor and friction losses. Total minor losses are estimated based on $k$ factors, which are only accurate to within 10 or 20% (unless the pumping system has been calibrated to field measurements). Depending on the pumping system's configuration, the total system head could be zero at a given flow rate, but this is unusual.

Most pumping systems are not defined by one system head curve, but rather by a band of total system headlosses or head requirements lying between the maximum and minimum total system head curves. This band may or may not be significant, depending on the pumping system.

**PUMP PERFORMANCE.** Every pumping system needs a pump that can provide the necessary energy to overcome the total system head for a particular operating point. Pumps are defined by their performance curves, efficiency curves, and net positive suction head (NPSH) requirements. Manufacturers typically provide this information in one graph, which plots each characteristic for varying flow rates.

***Pump Performance Curves.*** Pump performance curves (also called "pump curves") show how much energy (head) a pump will impart to the fluid. This typically is illustrated as one line per impeller diameter operating in similar casings at one constant speed (Figure 8.2a). It can also be shown as several curves illustrating multiple speeds for one impeller diameter and casing (Figure 8.2b) (Karassik, 1982).

***Pump and Motor Power and Efficiency.*** *Power.* The power generated or used by a pump can be expressed in three forms:

- *Water power* ($P_W$), which is the energy needed to pump fluid from one location to another, as well as that fluid's energy on leaving the pump (it equals all of the energy losses in a pumping system).
- *Brake power* ($P_B$), which is the energy provided to the pump by the motor (it equals the power applied to the pump shaft) ($P_B > P_W$).

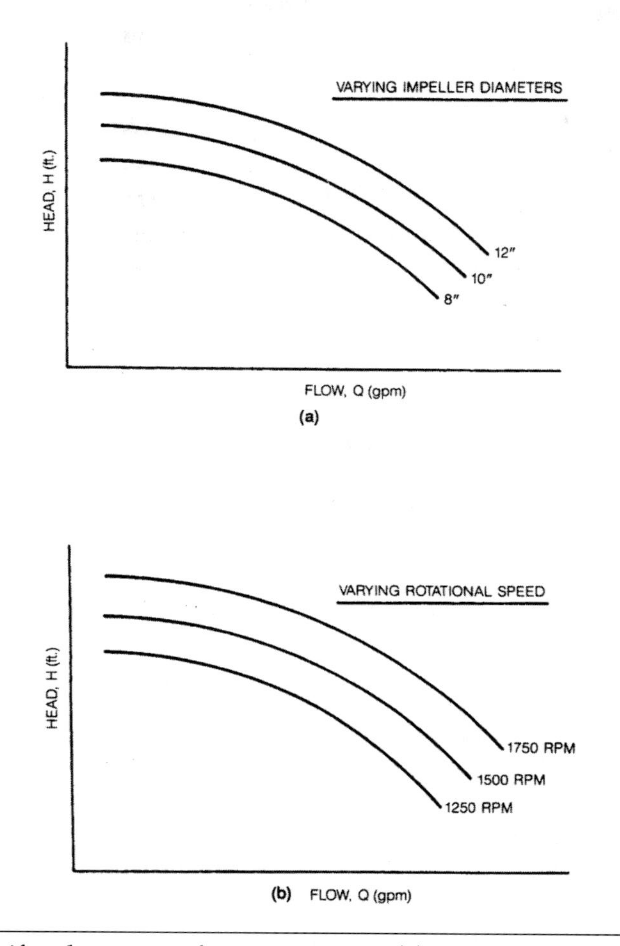


**FIGURE 8.2** Centrifugal pump performance curves: (a) curves for varying impeller diameters at constant rotational speed and (b) curves for varying rotational speed with a constant impeller diameter (ft × 0.304 8 = m; gpm × 5.451 = $m^3/d$; and in. × 25.4 = mm).

- *Wire or motor power* ($P_M$), which is the electrical power required by the motor ($P_M > P_B$).

*Efficiency.* A pump's *efficiency* is the difference between brake power input and water power output. It is always less than 1.0. A motor's efficiency is the difference between motor power input and brake power output. The percent efficiency of a pump is:

$$100 \times P_W/P_B$$

and the percent efficiency of a motor is:

$$100 \times P_M / P_B$$

where 1 hp = 550 ft-lb/s of work, 33 000 ft-lb/min of work, or 0.7457 kW.

Pump efficiency curves illustrate how efficiently a pump impeller transfers energy into the fluid (Figure 8.3). Pump manufacturers provide graphs depicting pump efficiency versus flow rate for a given pump, typically on the same graph that shows the pump curve.

***Pump Affinity Laws.*** *Affinity laws* are the relationships that determine pump performance based on rotational speed or impeller diameter. Utility personnel use them to determine how to alter a pumping system to meet a new design or operating condition. For example, if flows have increased, a pump's speed or impeller diameter could be increased to compensate. The affinity laws quantify the alteration required to meet a new condition.

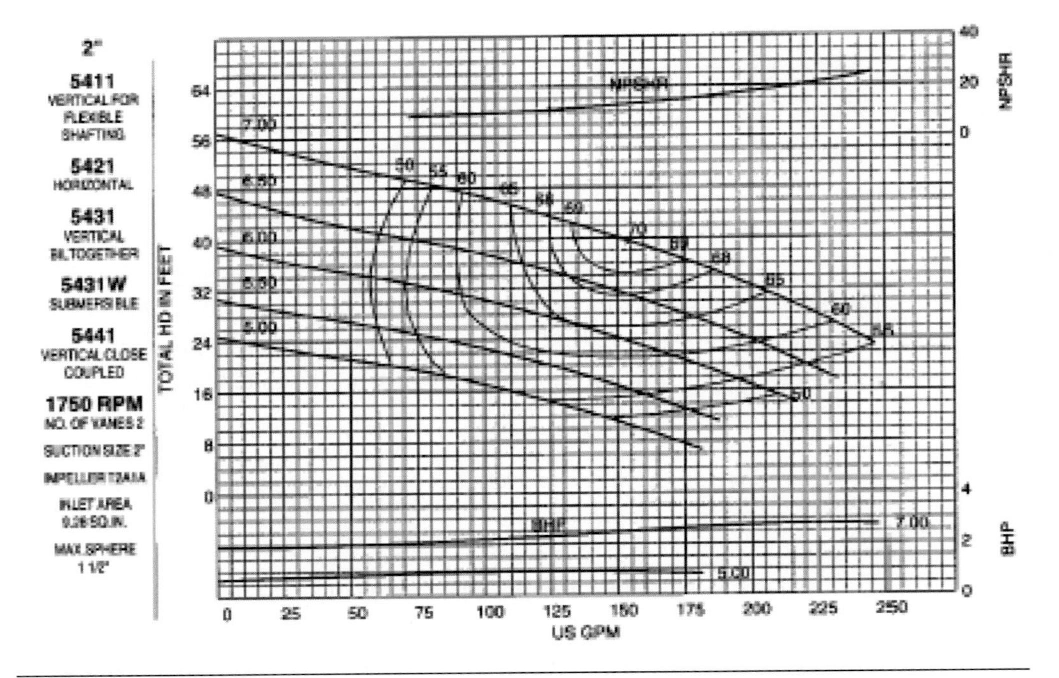


FIGURE 8.3 Example of pump efficiency curves and $NPSH_R$ curve (Courtesy of Fairbanks Morse).

*Pump Rotational Speed.* When a pump's rotational speed ($\omega$) is changed, the following three relationships apply:

- The flow capacity ($Q$) varies directly with the speed:

$$Q_2/Q_1 = \omega_2/\omega_1.$$

- The head ($H$) varies with the square of the speed:

$$H_2/H_1 = (\omega_2/\omega_1)^2.$$

- The brake horsepower ($P$) varies with the cube of the speed:

$$P_2/P_1 = (\omega_2/\omega_1)^3.$$

Subscript 1 indicates known (existing) characteristics. Subscript 2 indicates the new characteristics.

*Pump Impeller Diameter.* When a pump impeller's diameter ($D$) changes, the following three relationships apply:

- The flow capacity ($Q$) varies directly with the cube of the diameter:

$$Q_2/Q_1 = (D_2/D_1)^3.$$

- The head ($H$) varies with the square of the diameter:

$$H_2/H_1 = (D_2/D_1)^2.$$

- The brake horsepower ($P$) varies with the quintic of the diameter:

$$P_2/P_1 = (D_2/D_1)^5.$$

Subscript 1 indicates known (existing) characteristics. Subscript 2 indicates the new characteristics.

*Example.* For a given pump, utility staff can develop a new pump curve by doing the following:

- Select a point on the existing pump curve and note its coordinate ($Q_1$, $H_1$).
- Choose a new operating speed.
- Calculate a point on the new pump curve using the first two affinity laws for speed; this establishes a new coordinate ($Q_2$, $H_2$), which lies on the new pump curve.

- Repeat this process with enough points to establish a smooth pump curve for the new operating speed.
- Determine whether the new pump curve can meet the changed conditions.
- Select a new operating speed and repeat this process, if needed.

***Specific Speed.*** An impeller's *specific speed* is the number of revolutions per minute at which a geometrically similar impeller would rotate if it were sized to discharge against a head of 1 ft. It classifies pump impellers based on their geometric characteristics by correlating pump capacity, head, and speed at optimum efficiency. Specific speed is typically expressed as:

$$\omega_S = \frac{\omega\sqrt{Q}}{H^{0.75}}$$

where

$\omega_S$ = pump specific speed,
$\omega$ = rotational speed (rpm),
$Q$ = flow rate at optimum efficiency (L/s or gpm), and
$H$ = total head (m or ft).

Specific speed indicates an impeller's shape and characteristics because the ratios of major dimensions vary uniformly with specific speed (Figure 8.4). So, it helps designers predict the proportions required and helps application engineers check a pump's suction limitations.

Although pumps are traditionally divided into three classes (radial-flow, mixed-flow, and axial-flow), each has a wide range of specific speeds:

- In radial-flow pumps, fluid enters the hub, flows radially to the periphery, and discharges at about 90 degrees from its point of entry. Those with single-inlet impellers have a specific speed less than 4200; those with double-inlet impellers have a specific speed less than 6000.
- Mixed-flow pumps have a single-inlet impeller with the flow entering axially and discharging in both axial and radial directions (at about 45 degrees from the point of entry). Their specific speed ranges from 4200 to 9000.
- Axial-flow (propeller) pumps have a single-inlet impeller with the flow entering axially and discharging nearly axially (at nearly 180 degrees from the point of entry). Their specific speed is typically more than 9000.
- Pumps that fall between the radial- and mixed-flow types typically have a specific speed between 1500 and 4000.

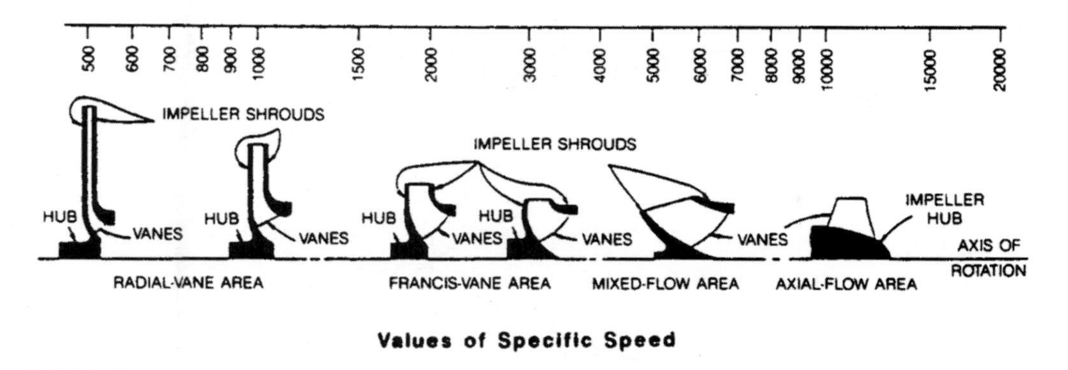


FIGURE 8.4 Standard, specific scale and impeller types.

The distinctions among pumps are not rigid; many are combinations of two types. Also, in the specific speed range of about 1000 to 6000, double-suction impellers are used as often as single-suction impellers.

***System Operating Point.*** The *system operating point* is the intersection of a pump performance curve and a total system head curve at a particular speed (Figure 8.5). It indicates the flow rate at which a given pumping system will operate when using a specific pump. (If the pumping system is properly designed, this flow rate should exceed the design flow rate.) The goal of pump design and operations is for this operating point to equal or be very near the best efficiency point (BEP). The *best efficiency point* is the point on a pump curve where a pump would operate at maximum efficiency.

***Net Positive Suction Head.*** The *net positive suction head* is the remaining energy (head) on the suction side of the pump. If headlosses in the suction piping are too large, then energy and pressure will drop. If the pressure becomes too low (reaches the vapor pressure of the fluid), the fluid will convert into the gas phase and form vapor cavities, a process defined as *cavitation* and referred to as *pump cavitation* in this case. Cavitation can be damaging to the pump when the vapor cavities collapse.

The minimum head required [required NPSH ($NPSH_R$)] is pump-specific. To determine whether a pump will operate properly, engineers compare the required NPSH to the available NPSH ($NPSH_A$), which depends on elevation and pumping-system characteristics (it is independent of the pump, however). Typically, NPSH and its associated calculations and analyses are a design issue, but if system changes are implemented, then NPSH should be checked.

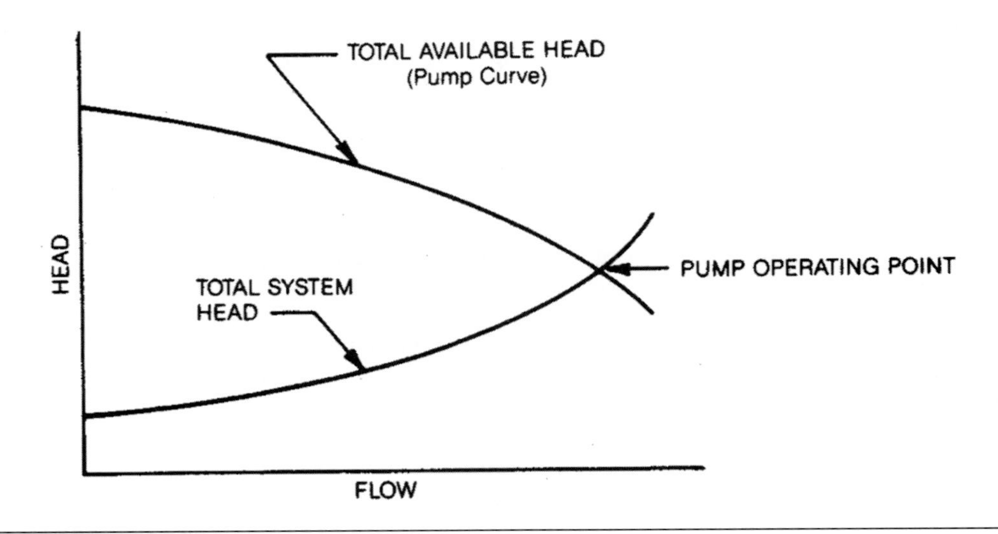


**FIGURE 8.5** Centrifugal pumping system head curves.

*Required Net Positive Suction Head.* The *required NPSH* is the minimum head needed on the suction side of the pump to avoid cavitation as the fluid is conveyed through the pumping system. Pump manufacturers provide graphs depicting a pump's required NPSH versus flow rate, typically on the same graph that shows the pump curve (Figure 8.3).

*Available Net Positive Suction Head.* The available NPSH is calculated as follows:

$$NPSH_A = \frac{P_{atm} - P_{vp}}{\gamma} \pm Z_W - H_L$$

where

$P_{atm}$ = the absolute atmospheric pressure exerted on the free fluid surface on the suction side of the pump, where the atmospheric pressure is based on absolute elevation (above sea level) of the free fluid surface;

$P_{vp}$ = the fluid's vapor pressure, where the vapor pressure is based on the fluid's temperature at the pump suction;

$Z_W$ = the vertical distance between the fluid surface and the pump centerline (it will be negative if the fluid surface is below the centerline, and positive if it is above the centerline); and

$H_L$ = the sum of all headlosses in the suction piping (Figure 8.6).

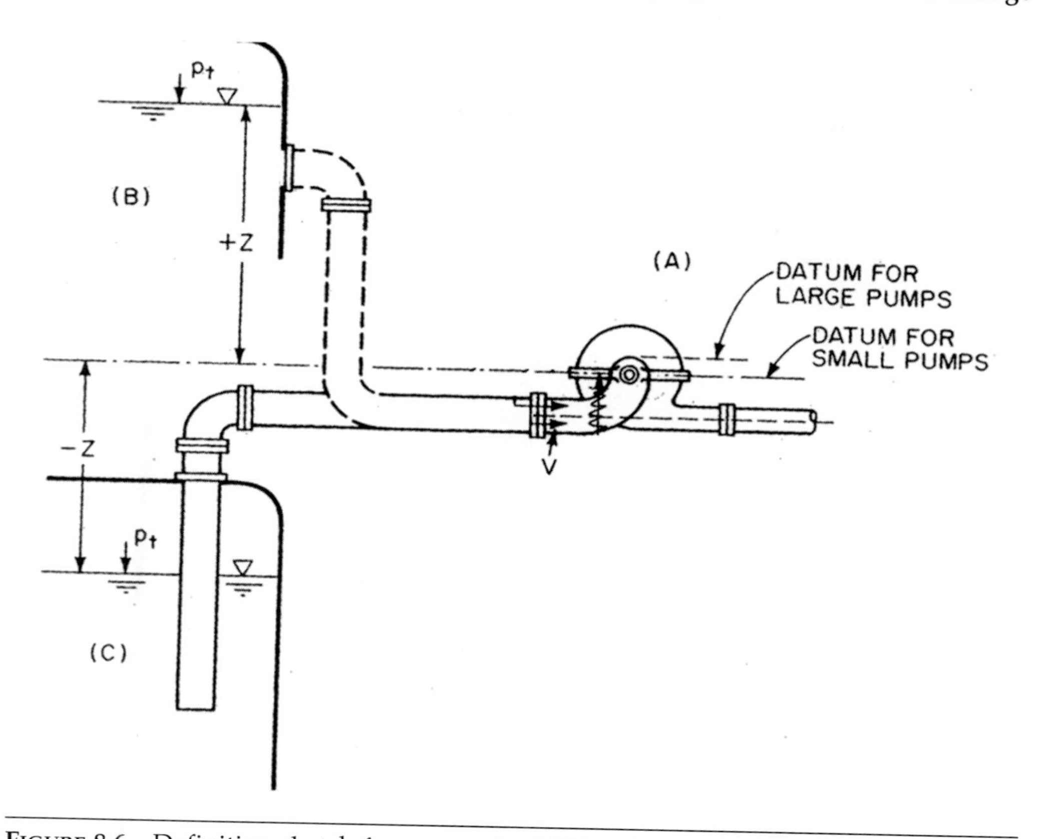


**FIGURE 8.6** Definition sketch for computing NPSH. (Karrasik et al., 1976)

*Net Positive Suction Head Evaluation.* To avoid pump cavitation, the following must be true:

$$NPSH_A > NPSH_R$$

*Determination of Pumping Energy Requirements and Costs.* The following example illustrates pumping energy requirements and associated costs. Calculate the horsepower delivered by a pump, the size (wire hp) of the pump's electrical motor, the power consumption of the motor, and the annual cost of the required electrical energy, given the following data:

- Flow = 300 gpm,
- Total head = 175 ft,

- Pump efficiency = 85%,
- Motor efficiency = 90%,
- Pump run time = 12 hours/day and 4 days/week, and
- Energy cost = $0.07/kWh.

**Step 1.** Calculate the pump's water power:

$$
\begin{aligned}
P_W &= \frac{Q \times 8.34 \text{ lb/gal} \times TDH}{33{,}000 \text{ ft-lb/min}} \\
&= \frac{300 \times 8.34 \times 175}{33{,}000} \\
&= 13.3 \text{ hp}
\end{aligned}
$$

**Step 2.** Calculate the motor's brake power:

$$
\begin{aligned}
P_B &= \frac{P_B}{\text{PumpEfficiency}(\%/100)} \\
&= \frac{13.3}{(85/100)} \\
&= 15.6 \text{ hp}
\end{aligned}
$$

**Step 3.** Calculate the motor's wire power:

$$
\begin{aligned}
P_M &= \frac{P_B}{\text{MotorEfficiency}(\%/100)} \\
&= \frac{15.6}{(90/100)} \\
&= 17.3 \text{ hp}
\end{aligned}
$$

**Step 4.** Calculate the motor's power consumption:

$$
\begin{aligned}
W_M &= P_M \times 746 \text{ W/hp} \\
&= 17.3 \times 746 \\
&= 12{,}900 \text{ W} \\
&= 12.9 \text{ kW}
\end{aligned}
$$

**Step 5.** Calculate the pump's annual electrical costs:

Annual energy cost = Cost/kWh × annual running time (hours/year) × kW
Annual running time hours = 12 (hours/day) × 4 (days/week) × 52 (weeks/year)
= 12 × 4 × 52
= 2,496 (hours/year)
Annual energy cost = \$0.07 kWh × 2,496 hours/year × 12.9 kW
= \$2,253.89

**PUMP CLASSIFICATION.** In this manual, pumps are classified based on how they add energy to the fluid being pumped (Figure 8.7). They are either kinetic or positive-displacement.

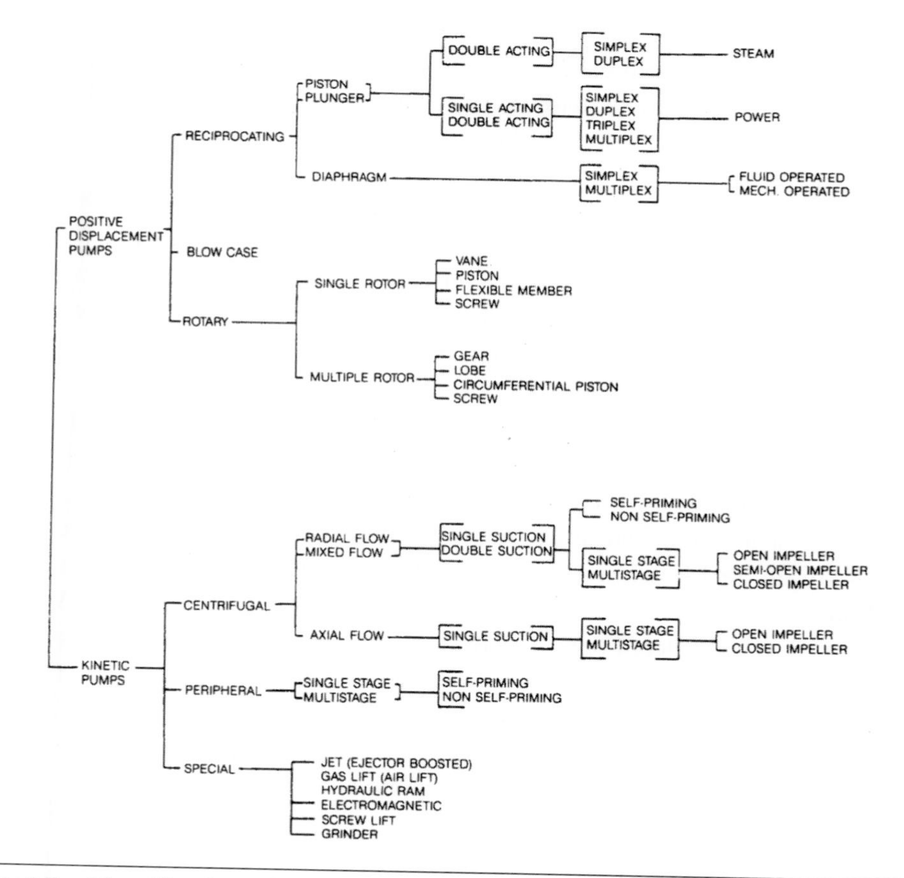


FIGURE 8.7 Classification of pumps.

***Kinetic.*** Kinetic (dynamic) pumps continuously add energy to the fluid as it is conveyed through the pumping system. The most common is the centrifugal pump. Kinetic pumps are mainly used for liquids (e.g., raw or treated wastewater).

***Positive-Displacement.*** Positive-displacement pumps periodically add energy to a defined volume of fluid via a movable boundary. They are primarily subcategorized as either reciprocating or rotary, depending on how the movable boundary travels. Positive-displacement pumps are primarily used for solids.

**PUMP SEALS.** Most pumps transfer mechanical energy to the fluid via a rotating mechanical component (e.g., a shaft) that is in contact with both the fluid and the motor while passing through the stationary pump housing. The places where this component enters and exits the pump housing must be sealed so the fluid being pumped does not leak from the housing. Seal friction and wear also must be minimized.

Three types of seals are used: packing, seal water, and mechanical seals. Proper operations and maintenance (O&M) ensures proper sealing and pump operation. Seal failure can lead to either excessive leakage or critical damage to a pump.

***Packing.*** *Packing* is the term for a braided "rope" wedged tightly into the space between the moving and stationary components. Because the rope is made of carbon and other materials that do not adhere to the moving component, it reduces leakage while allowing the component to move relatively freely and not overheat. However, packing degrades over time and must be replaced.

***Seal Water.*** Seal water is not a mechanical means to stop or reduce leakage. Instead, seal water is supplied under higher pressure than the pumped fluid and keeps the pumped fluid from leaking out of the pump between the moving and stationary components. The seal water then flows out of the pump and must be collected and conveyed to a drain. Seal water must be provided in a continuous stream (typically, a couple gallons per minute on average) whenever the pump is operating. Wastewater utilities typically use filtered or screened secondary effluent for this purpose.

The drawbacks of this option are the need for distribution, control, monitoring, and drainage systems for the seal water and the need to deal with a steady stream of seal water whenever operating, maintaining, or simply observing the pumps (i.e., it can be messy because the water dribbles down the pump and across the floor to the drain).

***Mechanical Seals.*** A mechanical seal is a device affixed to both the rotating and stationary components to prevent fluid leakage while allowing the rotating component to move freely (i.e., a more precise packing). All mechanical seals have three basic sets of parts (Figure 8.8):

- A set of primary seal faces: one rotary and one stationary (e.g., a seal ring and insert);
- A set of secondary seals called *shaft packings* and *insert mountings* (e.g., O-rings, wedges, and V-rings); and
- Hardware (e.g., gland rings, collars, compression rings, pins, springs, and bellows).

The two flat, lapped faces of the primary seal are perpendicular to the shaft and rub together, minimizing leakage. One is typically made of a non-galling material (e.g., carbon-graphite), while the other is a relatively hard material (e.g., silicon-carbide) so they will not adhere to each other. The softer face is typically smaller and is called the *wear nose*.

For example, an end face mechanical seal has four main sealing points: A, B, C, and D (Figure 8.9). The primary seal is at the seal face (A). The leakage path at B is blocked by an O-ring, V-ring, or wedge. The leakage paths at C and D are blocked by gaskets or O-rings.

The seal faces are lubricated via a boundary layer of gas or liquid between the faces. Design engineers must decide which lubricant will provide the desired leakage,

FIGURE 8.8 Mechanical seal section (Courtesy of Goulds Pumps).

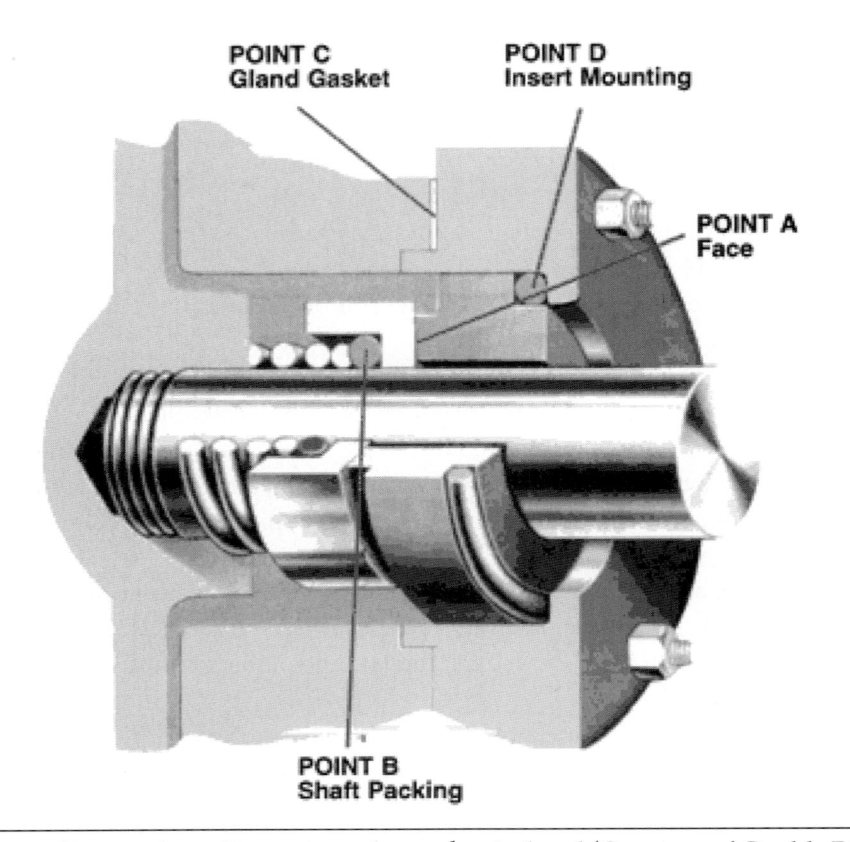

FIGURE 8.9 Four main sealing points of a mechanical seal (Courtesy of Goulds Pumps).

seal life, and energy consumption. Operators must provide this lubrication, monitor the seals, and rebuild them when needed.

Although mechanical seals are similar to packing, they provide tighter seals with less friction (thereby reducing a pump's power consumption). They also last longer, reducing maintenance costs. The drawbacks include greater initial costs than packing and greater overall costs than seal water.

## GENERAL OPERATIONS AND MAINTENANCE REQUIREMENTS

All pumping systems need similar O&M activities to maximize their operating lives. In addition to the activities recommended below, utility staff should follow the manufacturer's O&M recommendations because of their equipment-specific expertise and warranty requirements.

Whenever mechanical problems occur, utility staff should review them to determine whether they are the result of normal operations, inadequate equipment, or an abnormal situation. If caused by one of the two latter conditions, process or equipment modifications may remedy or reduce the problem.

***Operations.*** The following operations activities directly affect pump performance.

*Appurtenances.* Major pumping system appurtenances include the motor or drive system (which can be constant-speed, variable-speed, or variable-frequency); standby or redundant capacity; pipes and valves; and surge protection or mitigation. Monitors and controllers (e.g., time clocks and timing controls, pressure gauges, flow meters, solids analyzers, and level detectors) are also important.

Although appurtenances are typically chosen and laid out during design, these decisions can profoundly affect pump O&M. Improper appurtenances or inaccurate monitoring will reduce pump effectiveness, no matter how well-maintained or precise the appurtenance or monitoring system is. To ensure personnel safety and avoid major disruptions, operators need to know what the pumping system's appurtenances are and how they work. Taking a pump out of service for repair or replacement, for example, is a lot simpler when there are valves on both sides of the pump that can be closed to isolate it from other pumps and associated piping. Without them, removing a pump from service would require more extensive efforts (e.g., shutting down the entire pumping station).

When choosing appurtenances, design engineers must anticipate potential process and mechanical problems, as well as provide enough flexibility for the process to achieve the necessary operating range. Also, pumping systems typically must be able to accommodate ever-increasing flows (i.e., handle both smaller startup flows and larger design flows). In addition, they must be flexible enough to allow for bypassing, standby, or redundant pumping when needed.

*Testing Procedures.* Pumps can be either factory-tested or field-tested. Factory tests occur in a controlled environment with prescribed hydraulic, mechanical, and electrical equipment. They are typically performed in accordance with Hydraulic Institute (Parsippany, N.J.) guidelines to ensure that the pump will meet the specified performance conditions.

Field tests involve all pumping-system components (e.g., the pump, driver, electrical equipment, piping, and overall installation). Although precise data measurements can be more difficult to obtain because of installation limitations (e.g., buried piping and tight site constraints), field tests evaluate the entire pump installation under actual service conditions. It is typical for pump efficiencies measured via field tests to be lower than those measured via factory tests.

Specific testing procedures depend on the pump involved, but typically a pump's capacity, total head, speed, and input horsepower are tested before the pump is put into service and periodically thereafter.

*Startup.* Before starting up a pump, operators should do the following:

1. Review the manufacturer's literature on all systems and subsystems before initial use.
2. Review all of the manufacturer's step-by-step procedures for starting up the equipment.
3. Inspect and clean all debris from the tanks, piping, and suction well.
4. Check all hand-operated valves for smooth operation, proper seating, and proper orientation.
5. Confirm that the inlet and discharge valves are open.
6. Ensure that all equipment is properly installed and lubricated, and if applicable, that lubricating fluid is in the pump.
7. Check that all the belts or couplings are aligned properly.
8. Ensure that all guards and other safety devices have been installed properly.
9. Check that the seal has lubrication or water at the proper pressure (this depends on the type of pump seal involved).
10. Confirm that all electrical connections are complete (incorrectly wired motors can run in reverse) and the required voltage is available.
11. Bump motor to see that it rotates freely.

*Shutdown.* To shut down a pump, operators should do the following:

1. Disconnect or turn off the power to the driver. Then, lock out the appropriate breaker and tag it "out of service" to prevent accidental or inadvertent operation.
2. Close the suction and discharge valves, and isolate any external service connections. In solids and other applications where pressure could build in the out-of-service pump, leave one valve cracked open or install an appropriately sized pressure-relief valve.
3. If the pump will be out of service for a long time (typically on the order of several months or more—please consult with the manufacturer to confirm the specific period because different types of pumps require different levels of care), refer to the O&M manual for long-term storage instructions.

*Monitoring.* Monitoring various pumping system components confirms that pumps are operating properly or indicates the need for maintenance. Below are typical monitoring devices and their uses:

- Pump-motor indicators note whether the motor is on (i.e., whether the pump's switch is set to "Auto" or "On").
- Flowmeters indicate the rate at which fluid is being conveyed through the pumping system.
- Pressure sensors note the pressure in the piping. Low pressure in the suction piping can indicate blockages or unexpectedly low fluid levels in the wet well. Low pressure in the discharge piping can indicate impending cavitation or total system head alterations because of line breaks, major leaks, or other piping malfunctions. High pressure in the discharge piping can indicate blockages. Pressure and flow rate combined indicate the pump's water power (i.e., how much the pump is working). Increasing water power can indicate blockages or restrictions (e.g., improperly opened valves) in the line.
- Voltage and amperage meters show how much electricity the equipment is drawing, so operators can optimize pump operations. Increasing energy draw can indicate a developing problem (e.g., if the motor needs more energy and the pump's water power is the same, then the system's efficiency is decreasing because of impeller wear, bearing wear, motor wear, etc.).
- Temperature sensors note the temperature of a component (e.g., bearings). An incremental change can point to a potential problem.
- Vibration sensors note how much the pump or piping is vibrating. Increasing vibration may be the result of internal problems (e.g., bearing failures or impeller damage) or deteriorating mountings or anchors. In large installations, critical equipment has online sensors that can monitor vibration in real time, signaling problems before failure and more serious damage occurs.

Wastewater and solids are inherently corrosive and can hinder the proper performance of instruments in contact with them. Monitoring devices installed in such harsh environments may need more maintenance, more frequent replacement, or protection from the fluid. A magnetic flowmeter, for example, may use externally mounted magnetic sensors to measure velocity or flow. If a sensor must be directly installed in the fluid, a diaphragm seal can be placed between it and the fluid.

*Controls.* Pumping-system controls include on–off switches that are activated by pressure sensors, temperature sensors, or the depth of fluid in the wet well (as measured by float switches, bubbler systems, pressure transducers, or ultrasonic devices). They also include variable-speed or variable-frequency drives, which change speed to match inflow rates. In addition, solids pumps may be controlled via time clocks, secondary treatment process flow rates, etc.

Although the choice of controls is typically considered a design issue, controls can profoundly affect pump O&M. Knowing how the control system functions will help operators better maintain the utility's pumping systems. Suppose, for example, that a pumping station's pumps activate automatically based on wet-well fluid levels. If operators isolate one of the pumps for service, removal, or repair without modifying the controls first, the results could be disastrous. The solution may be simple (e.g., flip a local switch from "Auto" to "Off" and lock- and tag-out the pump at the breaker), but operators need to know what it is to ensure safe operations.

Sophisticated control systems monitor both the equipment and the treatment process itself, continually monitoring various pump or appurtenance conditions. Knowing this enables operators to note the current equipment and pumping system status and track changes that indicate potential failures or substandard operating conditions.

The current trend is to control pumps and processes via programmable logic controllers or supervisory control and data acquisition (SCADA) systems and integrate them into an overall control scheme (e.g., the Wonderware system). However, operators still need to know how to operate this equipment manually when the automated control system is offline or when unusual loadings or abnormal conditions occur.

***Maintenance.*** The following maintenance activities may not directly affect pump performance, but they can affect pump longevity. For example, monitoring vibrations does not affect system operations because a vibrating pump can still deliver design-level flows. However, excessive vibration can lead to failure of the pump or related equipment.

Periodic cleaning and a simple "look–feel" inspection of each pump are prudent, systematic procedures for detecting early signs of trouble. They only require a few minutes and may save a lot of money and inconvenience.

When performing a typical "look–feel" inspection, utility staff should

- Observe the pumps, motors, and drives for unusual noise, vibration, heating, or leakage;
- Check the pump's discharge lines for leaks and confirm that the valves are in the proper positions;
- Check the pump seal water and adjust if necessary (if the pump has mechanical seals, a seal water system is unnecessary);
- Confirm that the control panel switches are in the proper positions;
- Monitor discharge flow rates and pump revolutions per minute; and
- Monitor pump suction and discharge pressures.

These activities can be automatically monitored and archived via the pump's control system, so operators must be familiar with the control system.

*Wear.* All rotating equipment will wear over time, but the amount and degree of wear depends on the abrasiveness of the fluid being pumped, the pump's run time, and the operating conditions. Operating pumps at faster than optimal speeds, for example, can increase wear no matter what fluid is being pumped.

Detecting early signs of wear enables operators to repair pumps at minimal cost and downtime. A gradual drop in flow or pressure, for example, indicates that the impeller or wear rings may be worn. Excessive leaking from the gland suggests that the seal or packing may be worn. Vibration, noise, or hot spots on the pump are signs that the bearings are worn.

*Standard Maintenance.* Pumps are maintenance-intensive—perhaps more so than any other equipment at a wastewater treatment plant. Solids pumping and grinding operations are particularly wearing, often causing pumping equipment to deteriorate rapidly and need replacements (of either parts or the entire system). So, manufacturers recommend specific maintenance procedures to maximize the useful life of pumping systems.

Preventive maintenance (e.g., lubrication, cleaning, and component inspections) can reduce the likelihood of major-equipment damage, failure, downtime, and replacement, thereby improving equipment reliability and lowering equipment costs. This is important for wastewater pumping systems, where redundancy may be required but downtime can be disastrous. Major pump overhauls should be included in the O&M budget and planned for times when the treatment plant or pumping system is not being used to its design capacity.

Preventive maintenance can be tracked manually or electronically. A simple card or paper filing system can record all the equipment's run-time intervals, maintenance history, problems, repair frequency, and the materials and costs involved. A computerized maintenance management system takes this a step farther; not only can it maintain equipment management histories, but it also can issue and monitor work orders, and enable utility staff to analyze all O&M data. Maintenance information is valuable for planning capital and O&M needs, estimating related budgets, and detecting and diagnosing problems early.

Equipment maintenance schedules must conform to manufacturers' O&M manuals to ensure that warranty requirements are met. These manuals can be kept in bound notebooks in the operators' office and other appropriate locations, but making the data even more accessible to field staff further facilitates maintenance activities by making them easier to accomplish. Some wastewater utilities are converting to electronic documentation, which can be searched, referenced, and viewed from any computer connected to the database, and issuing staff personal data assistants (PDAs) that contain maintenance schedules and checklists.

Sometimes lubricant companies will review the plant's equipment and prepare a master lubrication schedule. Alternatively, utility staff can put together this schedule themselves based on reviews of all the equipment O&M manuals. The recommendations should be commensurate with the damage that could occur because of inadequate lubrication. For example, equipment that needs continuous or nearly continuous lubrication should have automated oilers or grease-fitting systems whose reservoirs permit routine inspections and refills.

Whenever applying lubricants, operators should first take unfiltered samples of existing lubricants and have them analyzed for pH, wear-related debris, water content, and viscosity. The results will show whether the bearings or seals are excessively worn and are allowing process fluid to pass through the seal. Lubricant analysis will detect bearing and seal wear up to 6 weeks before vibration analysis detects them. (Most lubrication companies will analyze lubricant samples for free, but if the utility's supplier cannot—or its analysis is self-serving—the plant should consider an independent lab.)

If a utility is following the manufacturer's guidelines and the equipment seems to need excessive maintenance, staff should consult the manufacturer. Most are willing to review their recommendations and make adjustments that will alleviate the problem. However, if the excessive maintenance is the result of improper or atypical use, then equipment or process modifications may be required.

*Predictive Maintenance.* Predictive maintenance involves monitoring operating conditions and tracking performance trends to detect when equipment is beginning to operate outside its long-term, normal range. Then, operators can maintain the equipment before it fails.

Typical predictive maintenance activities include vibration monitoring, infrared analysis, alignment checks, motor current analyses, ultrasonic testing, and oil analyses. Vibration monitoring is used to determine equipment deterioration. Vibration readings can be taken by a specialist or an operator (via an inexpensive vibration pen), depending on the utility's financial resources. Critical equipment should have monthly readings; noncritical equipment should have quarterly readings. (If vibrations are monitored automatically, the data should be reviewed monthly or quarterly, as appropriate.) The readings should be taken in the horizontal, vertical, and axial directions on the inner and outer bearings of the pump, motor, and couplings. Each site should be plotted separately to develop a trend line; the horizontal (X) axis notes the dates of each reading, while the vertical (Y) axis lists vibration readings in centimeters per second (cm/sec) or inches per second (in/sec). The equipment should be closely monitored when vibration levels reach 0.3 in/sec. It should be shut down when vibration reaches 0.5 in/sec. After each overhaul, a complete vibration analysis should be done to establish a baseline for comparison.

Used to inspect equipment while it is operating, infrared analyzers note whether there are temperature variations in pumping-system components (e.g., bearings). They can indicate potential problems before failure occurs.

If the pump, coupling, and driver are misaligned, the pumping system will vibrate more and bearing life will shorten. Equipment alignment may gradually change under normal operating conditions, so it should be checked periodically (according to manufacturer recommendations). Alignment tools are available to check the alignment of integral pump components.

Motor current analyses are performed to check on the current draw of an electrical motor. If a motor is not operating correctly, it will not draw the correct current and can overheat or burn out.

Ultrasonic testing can be used to track and monitor bearing wear by observing changes in a pump's ultrasonic signature and can be used to expose electrical tracking and corona effects.

Oil analyses can be performed, as noted above with respect to lubrication, to check if seals are beginning to fail.

*Spare Parts.* Even with a well-planned and -executed O&M program, all equipment eventually wears and then fails. Treatment plants typically are designed with some redundancy, but once a backup system is activated, there is no backup for it. So, an inventory of critical spare parts must be maintained onsite.

Operators should begin by determining what spare parts are needed for each pumping-system component. This information is available in the manufacturer-provided O&M manuals. When calculating how many of each part to keep onsite, operators should consider the pumping system's critical needs, its redundancy, the part's failure rate, and its availability.

An area should be set aside for an orderly inventory of spare parts. They should be stored in a clean, dry area away from vibration and in accordance with manufacturer instructions. They also should be inspected for corrosion every 6 months. Whenever a spare part is removed, a replacement should be automatically reordered to keep the inventory up-to-date. Otherwise, the spare parts program will be ineffective.

Some spare parts may need to be kept at the pumping system site if it is critical, operating conditions are severe, and repairs can be performed in the field.

When specifying or purchasing a pump, utilities should ask the manufacturer to provide a suggested spare parts list that includes an estimate of lead time between ordering and delivery. If the pump is critical, it and its spare parts should be purchased simultaneously to avoid unexpected downtime. Utilities also should ensure that the manufacturer keeps detailed records (by serial number) of all their pumps.

*Troubleshooting.* In addition to monitoring leaks, excessive noise, vibration, and high temperatures, operators can track pump hydraulics to find signs of impending trouble. Hydraulics monitoring can be the best and simplest trouble alert because a correctly functioning pump should be capable of pumping its design flow rate, while a pump that is failing to provide its design flowrate obviously has a problem. This approach can provide a first clue to the advent of a problem; however, it cannot diagnose the cause. (Although the pumping system is chosen during plant design, operators need to know how to capitalize on using hydraulic monitoring to alert them to trouble.) Other simple tools for troubleshooting pumps include:

- Pressure gauges, which indicate whether $NPSH_A$ is adequate and the pump is operating in an acceptable range on the pump curve (the gauges should be installed as close to the suction and discharge ends of the pump as possible);
- Flow meters, which help further determine the pump's operating point;
- Voltage and amperage meters, which determine if the incoming power is sufficient or if the motor has problems; and
- Tachometers, which check the speed of dry-pit pumps.

For pump-specific troubleshooting guidance, see the manufacturer's O&M manual.

# LIQUIDS PUMPING

Most wastewater and some solids are pumped via kinetic pumps. In wastewater applications, centrifugal pumps are primarily used.

**TYPES OF KINETIC PUMPS.** There are three types of kinetic pumps: centrifugal, peripheral or regenerative turbine, and special (Figure 8.7).

***Centrifugal.*** A centrifugal pump has one moving part: an impeller rotating within a casing (Figure 8.10). The impeller is mounted on a shaft that is supported by bearings and connected to a drive. Leakage along the shaft is limited by seals.

***Peripheral or Regenerative Turbine.*** The peripheral or regenerative turbine pump is a low-flow, high-head pump using peripheral or side channel vanes on a rotating

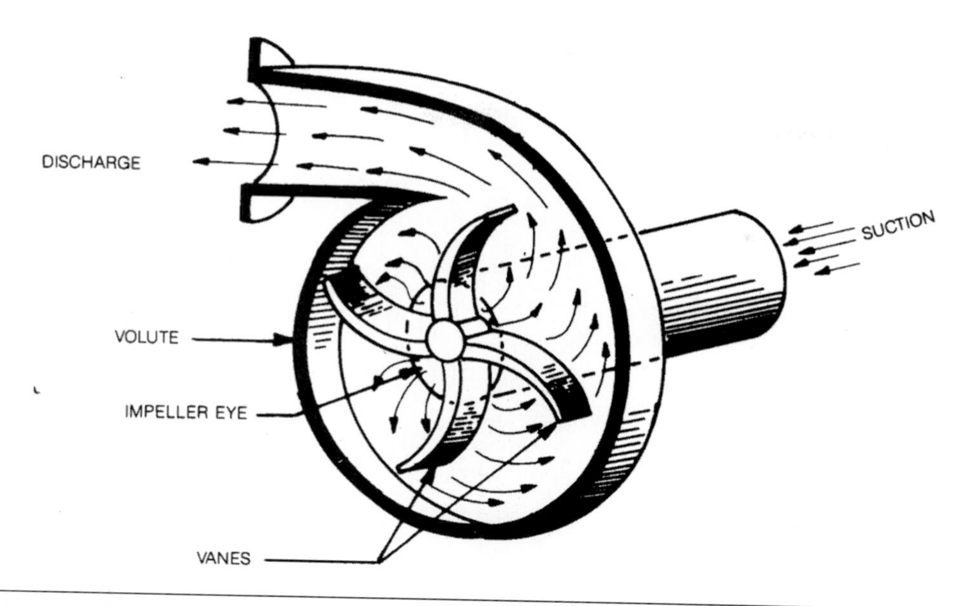


**FIGURE 8.10** Diagram of centrifugal pump.

impeller to impart energy to the pumped fluid. It differs from a centrifugal pump, because fluid enters near the impeller outside diameter and exits the pump discharge near the same radius. Pressure is generated in the pump because the fluid is accelerated tangentially in the direction of rotation and radially outward into the casing channel because of centrifugal force. As the fluid hits the casing wall it is redirected back onto an adjacent blade where more energy is imparted.

The regenerative turbine pump can achieve more head at a given impeller diameter than any centrifugal pump. Unfortunately to achieve this high head, the pump has very tight clearances between the impeller vanes and the casing walls, so pumping solids is impractical.

***Special.*** Special pumps include propellers, screw lifts, and grinders.

*Propeller.* Propeller pumps are similar to centrifugal pumps, except the impeller resembles a common boat propeller. They can efficiently lift up to 5000 L/s (50 000 gpm) of fluid at up to 10 m (30 ft) of head. Flow is conveyed through the pump axially (Figure 8.11).

*Screw-Lift.* A screw-lift pump consists of a spiral screw—a tube with spiral flights set in an inclined trough or enclosed tube—and an upper bearing, a lower bearing, and a drive arrangement. Used in many applications (e.g., raw wastewater, return activated

**FIGURE 8.11** Propeller pump section (Courtesy of ITT Industries, Flygt).

sludge, and stormwater), screw-lift pumps efficiently lift large volumes of fluid at low heads, automatically adjusting their pumping rate and power consumption to match the inflow depth while operating at a constant speed. They operate economically down to 30% of maximum design capacity. They need little maintenance and no prescreening, but require a lot of space and can aerate the pumped fluid.

*Grinder.* Grinder pumps typically are centrifugal pumps with a mechanical grinder placed across the suction side of the pump. The grinding mechanism chews incoming solids into finer solids, so the impeller only has to handle the finer solids.

**CATEGORIZATION.** Liquid pumps typically are categorized based on impeller type, suction, location, and orientation.

***Impellers.*** A kinetic pump imparts energy to fluids via a rotating element (called an *impeller*) that is shaped to force the fluid along or through the pump and increase its pressure. The fluid can be forced outward perpendicular from the impeller's axis (*radial flow*), along the impeller's axis (*axial flow*), or in both directions (*mixed flow*).

Radial- and mixed-flow pumps typically are called *centrifugal pumps* because the impeller develops its head via the centrifugal force imparted to the fluid. Flow enters the eye of the impeller and exits perpendicular to the shaft at high velocity. In the process, most of the fluid's velocity head is converted to pressure head by the pump's casing or volute. Mixed-flow impellers develop head mostly via centrifugal force and partly via the axial thrust imparted to the wastewater.

Induced vortex pumps have impellers recessed in the volute chamber (Figure 8.12). The impellers induce a vortex to pull the fluid through the pump with minimal impeller contact (Krebs, 1990).

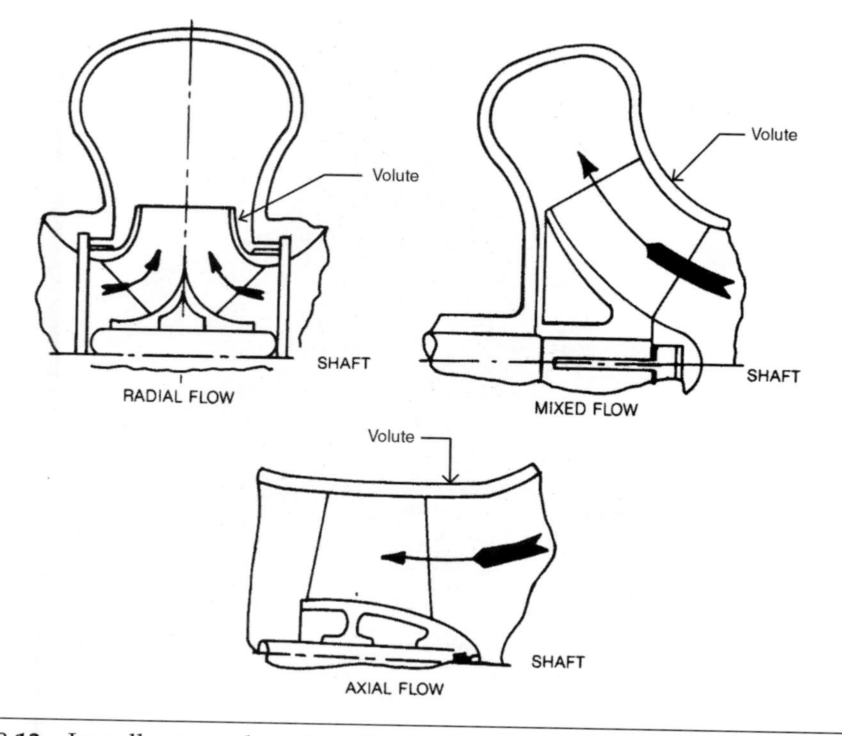


**FIGURE 8.12** Impeller types based on flow pattern.

The choice of pump depends on the hydraulic conditions involved. Radial-flow pumps are best for high heads, axial-flow pumps are best for low heads, and mixed-flow pumps are best for moderate heads. Mixed-flow impellers also are used when the flow is relatively high but the head is relatively low.

*Open and Semi-open Impellers.* Open and semi-open impellers are typically used to pump raw wastewater and sludge with a high solids content or large solids particles. They operate under high volume and low pressure. Their solids-pumping capabilities depend on the gap between the impeller and the suction side of the volute case. This gap can vary from 0.38 mm (0.015 in.) to several inches.

Open impellers have no front or back shroud; they are typically used to pump wastewater containing large solids. (A shroud looks like a duck's foot webbing.) The impeller in Figure 8.13 rotates counterclockwise.

Semi-open impellers have a back shroud and several vanes (Figure 8.14). They are typically used to pump return activated sludge or raw wastewater containing medium-size solids (Karassik et al., 1976).

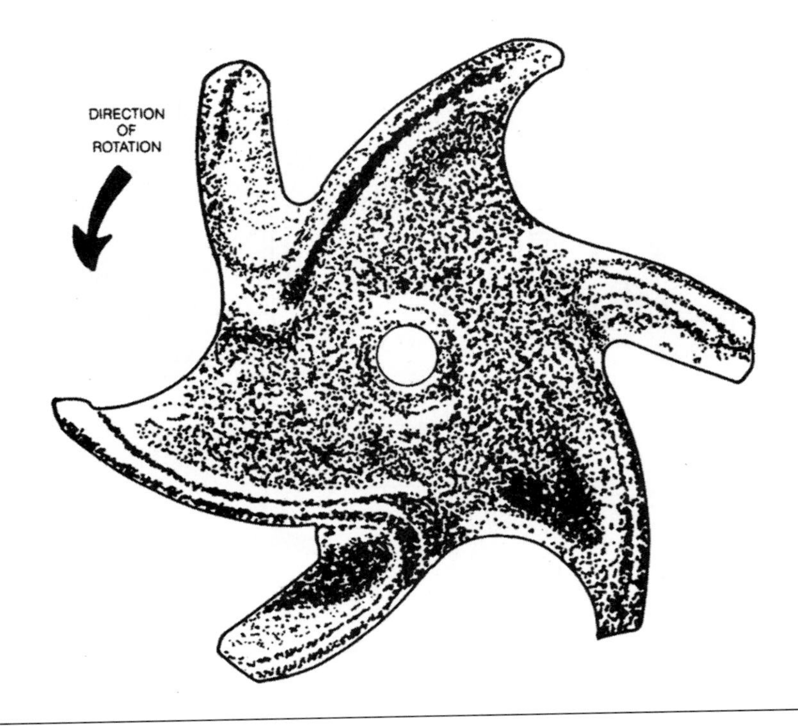


FIGURE 8.13 Open impeller.

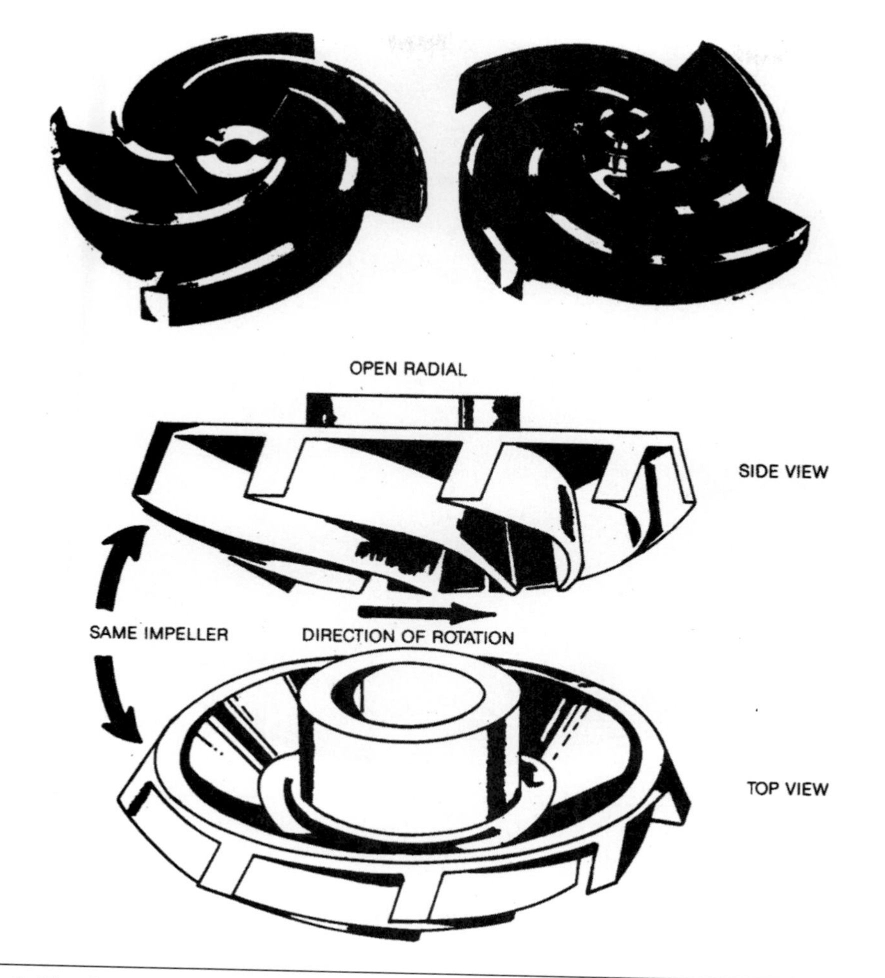


**FIGURE 8.14** Semi-open impeller.

*Closed Impellers.* Enclosed, nonclog impellers with two vanes handle 80% of wastewater with low solids concentrations [e.g., raw and settled wastewater, return sludge (from all secondary processes), waste sludge, filter feed, and plant effluent]. Fully closed impellers, which rotate counterclockwise, transfer energy most efficiently (Figure 8.15). The largest solids particle that can be passed by a closed impeller depends on the distance between shrouds.

*Propeller.* The axial-flow (propeller) impeller develops its head via the axial thrust imparted to the fluid by the propeller blades. It is used at low heads and operates at rela-

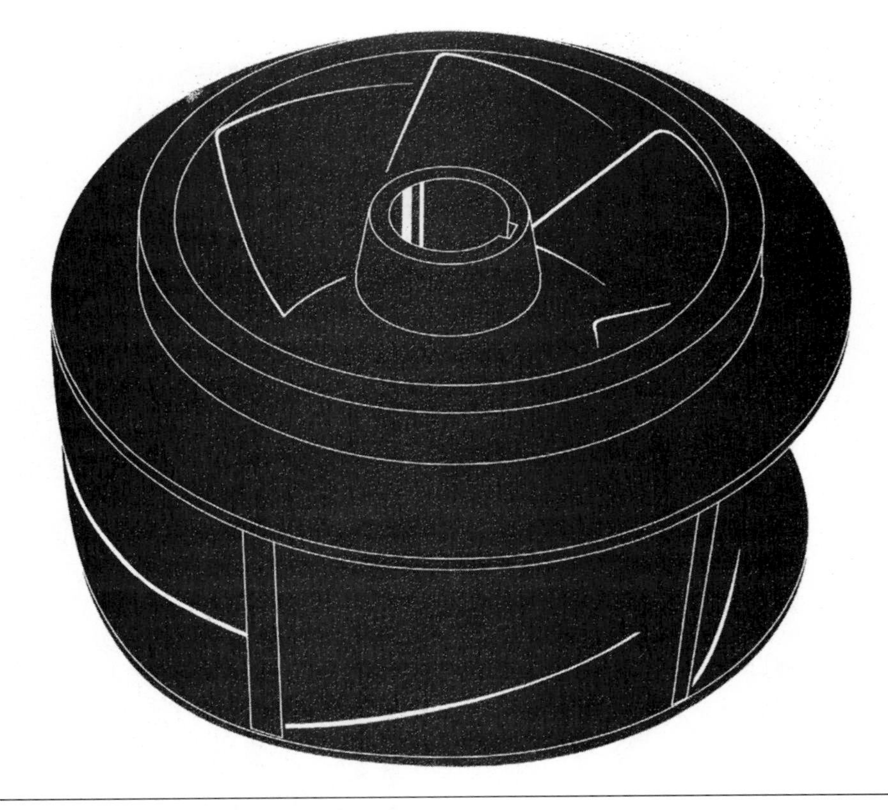

FIGURE 8.15 Closed-radial impeller.

tively low revolutions per minute. Some have large capacities but must operate with positive suction pressures.

*Grinders.* Although a grinder is not an impeller, it is intimately associated with impellers. Grinders can be fitted across the suction end of centrifugal or positive-displacement pumps. In centrifugal pumps, the grinding mechanism and pump impeller are typically connected to a common shaft.

***Suction.*** Pumps may be designed as single-suction (end-suction) or double-suction units. Generally, all of the impellers above are available for any type of kinetic pump suction. An axial-flow pump requires a single-suction setup and an open or closed impeller; it cannot support double-suction setups or semi-open impellers (Figure 8.7). Radial- and mixed-flow pumps, on the other hand, can support both types of suction and all styles of impellers.

*Single-Suction.* Single-suction pumps typically handle raw wastewater or other solids-bearing liquids. In these pumps, fluid only enters one side of the impeller, which is designed with relatively large clearances, fewer blades, and no shroud connecting the blades. Such open-impeller (nonclog) pumps handle solids of various sizes (ask the manufacturer to specify the maximum solids size that can safely pass through a particular pump). However, strings and rags may entangle the impeller blades and clog the pump, so they and excessively large solids must be removed from the liquid before it is pumped.

*Double-Suction.* In a double-suction pump, internal passages split the influent equally between the two sides of the impeller. Such pumps, which have small internal clearances, typically are designed to handle clear liquids (e.g., relatively "clean" plant water for seal water, washdown water, and fire hydrants, or to boost potable water).

***Location.*** Pumps can be installed inside or outside the fluid being pumped.

*In Fluid.* Pumps installed within the fluid are called *submerged pumps* or *vertical wet-well pumps*. They can be mounted horizontally or vertically. The impeller is submerged or nearly submerged in the fluid, and no additional piping is required to convey the fluid to the impeller.

*Outside of Fluid.* Pumps installed outside the fluid are called *dry-pit pumps*; they can be mounted horizontally or vertically in a dry well. Fluid is piped to the suction of these pumps.

***Orientation.*** Pumps may be oriented (mounted) vertically or horizontally.

*Vertical.* Vertical wet-pit pumps are mounted vertically with the impeller submerged in the fluid and its driver above both the suction and any "high" fluid level. On the discharge side of the impeller, the *bowl*—gradually expanding passages with guide vanes (baffles)—converts velocity head to pressure (Figure 8.16). Vertical wet-pit pumps with radial flow or low–speed, mixed-flow impellers are called *turbine pumps*. Because turbine pumps convert energy via a bowl rather than a volute, they can be constructed with small diameters (similar to those in water-supply wells). Turbine pumps are readily adaptable because more bowls and impellers can be attached to the turbine shaft.

*Horizontal.* Horizontal pumps are typically installed in dry pits with the motor next to the pump. They are typically centrifugal pumps (e.g., standard, non-clog, recessed impeller, and closed coupled horizontal).

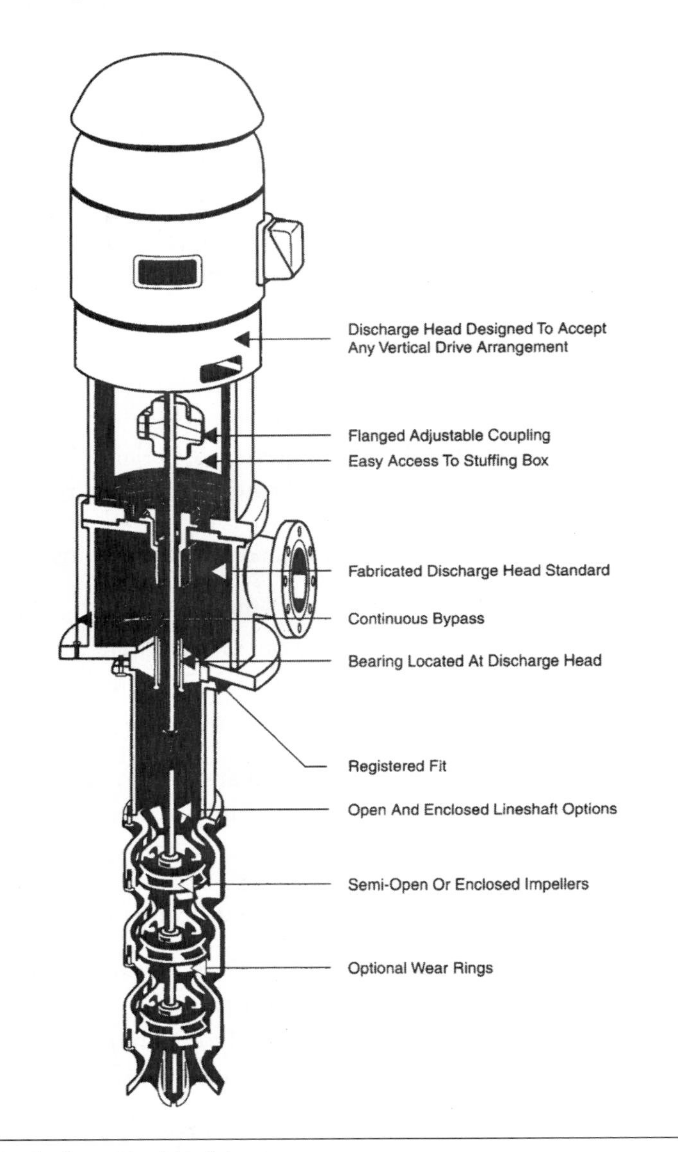


**FIGURE 8.16** Typical vertical turbine pump.

**APPLICATIONS.** ***Horizontal or Vertical Nonclog Centrifugal Pumps.*** Horizontal or vertical nonclog centrifugal pumps (Figures 8.17 and 8.18) handle many types of wastewater (e.g., raw wastewater, return activated sludge, and recirculated trickling filter effluent). These pumps—the "workhorses" of centrifugal pumps—have an extensive history in wastewater pumping.

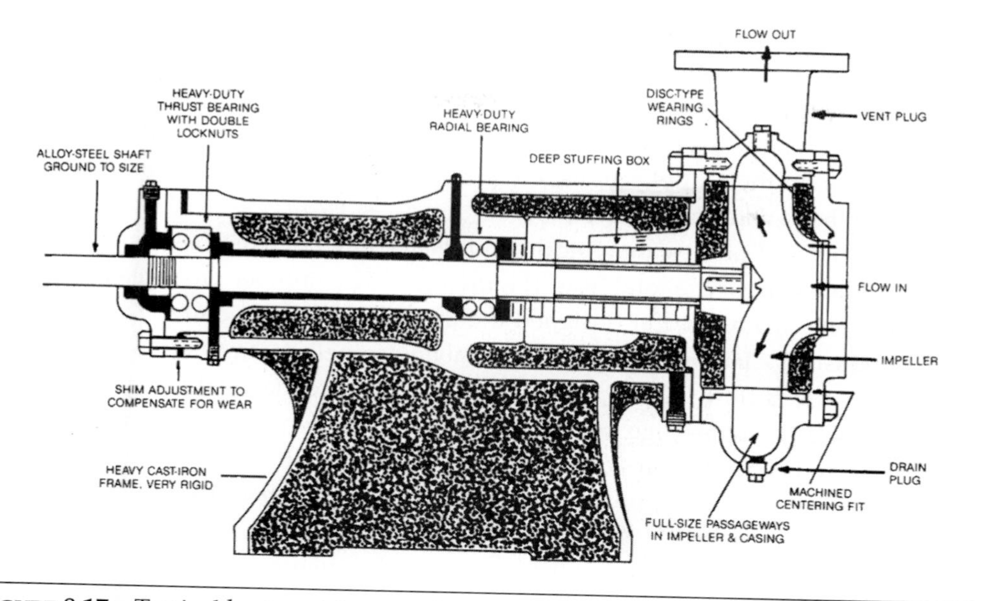


**FIGURE 8.17** Typical horizontal pump for raw wastewater and return activated sludge.

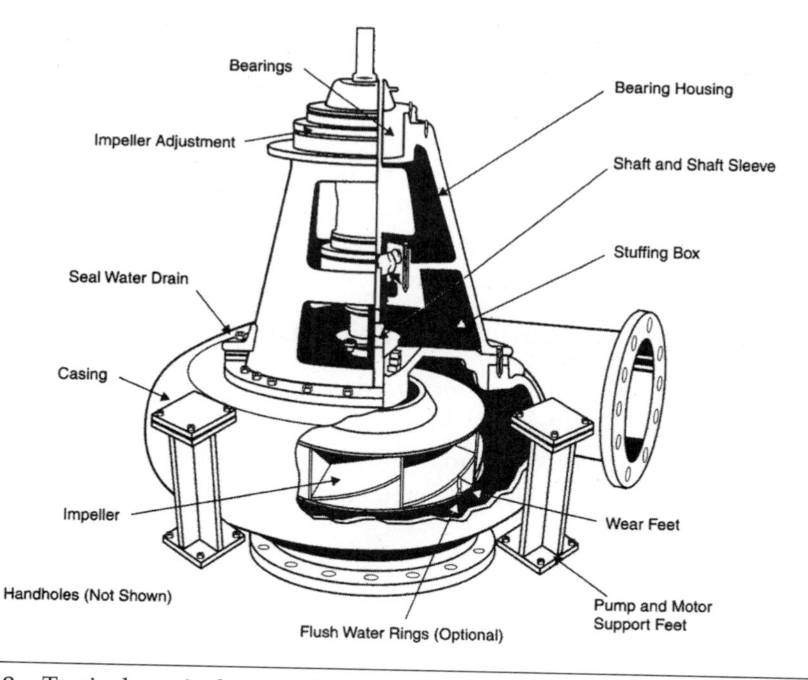


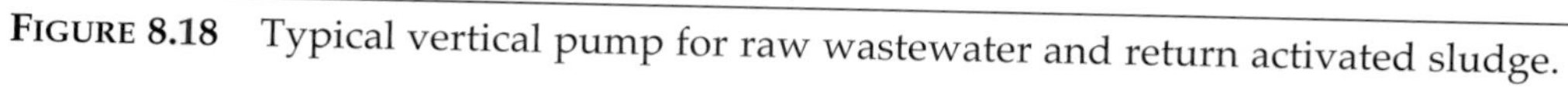
**FIGURE 8.18** Typical vertical pump for raw wastewater and return activated sludge.

***Vertical Centrifugal Pumps.*** Vertical turbine pumps typically handle "cleaner" wastewaters (e.g, secondary effluent, drainage, and washwater). For example, the City of Corona (California) Wastewater Treatment Plant uses vertical turbine pumps to transfer equalized secondary effluent to filters, and filtered, disinfected effluent to plant water systems. In these applications, the pumps are efficient (70% or more) and can achieve high head and flow rates.

Vertical wet-pit pumps are typically multistage units in which each stage handles a nearly equal share of the total head (Figure 8.16). The number of stages needed depends on how a manufacturer designed the impeller, the shaft, and the pump bowls (impeller housings). Turbine pumps with radial flow impellers may have up to 25 stages. Pumps with mixed-flow impellers typically have no more than eight stages. Propeller units are typically single-stage, but may be available in two stages, depending on the pump's size and characteristics (Figure 8.19). For most multistage applications, operators should consult manufacturers about the maximum number of stages available for a particular type of pump.

Most manufacturers provide inlet screens in the suction bells of vertical turbine pumps. These "standard" screens have such small openings that the suction can be rapidly and almost completely blocked (Canapathy, 1982), reducing inflow pressures, and playing havoc with pump operations. Operators should omit these screens. If screening is necessary, the utility should install either traveling-water or low-velocity screens at the pump inlet.

## OPERATIONS AND MAINTENANCE.

Pump manufacturers typically will provide literature or guidance on the O&M activities needed to keep a particular pump functioning optimally. The following information is a guideline that operators can use to develop or evaluate a utility's O&M plan.

***Operations.*** The following activities directly affect the performance of liquid pumps.

*Appurtenances.* The type of drive system used is important. Drives can be electric, hydraulic, or engine- or steam turbine-driven. The choice of constant- or variable-speed drive depends on the specific pumping requirements, because pumps themselves can deliver a range of flow rates, depending on their design, speed, and total dynamic head.

*Testing Procedures.* The appropriate factory and field tests for centrifugal pumps depend on the pump's size and importance, design constraints, budget constraints, conditions of service, and installation. Two common factory tests are the hydrostatic test and the performance test. Hydrostatic tests typically are performed at 1.5 times the pump's shutoff head to determine how watertight the pump case and gaskets are. Performance

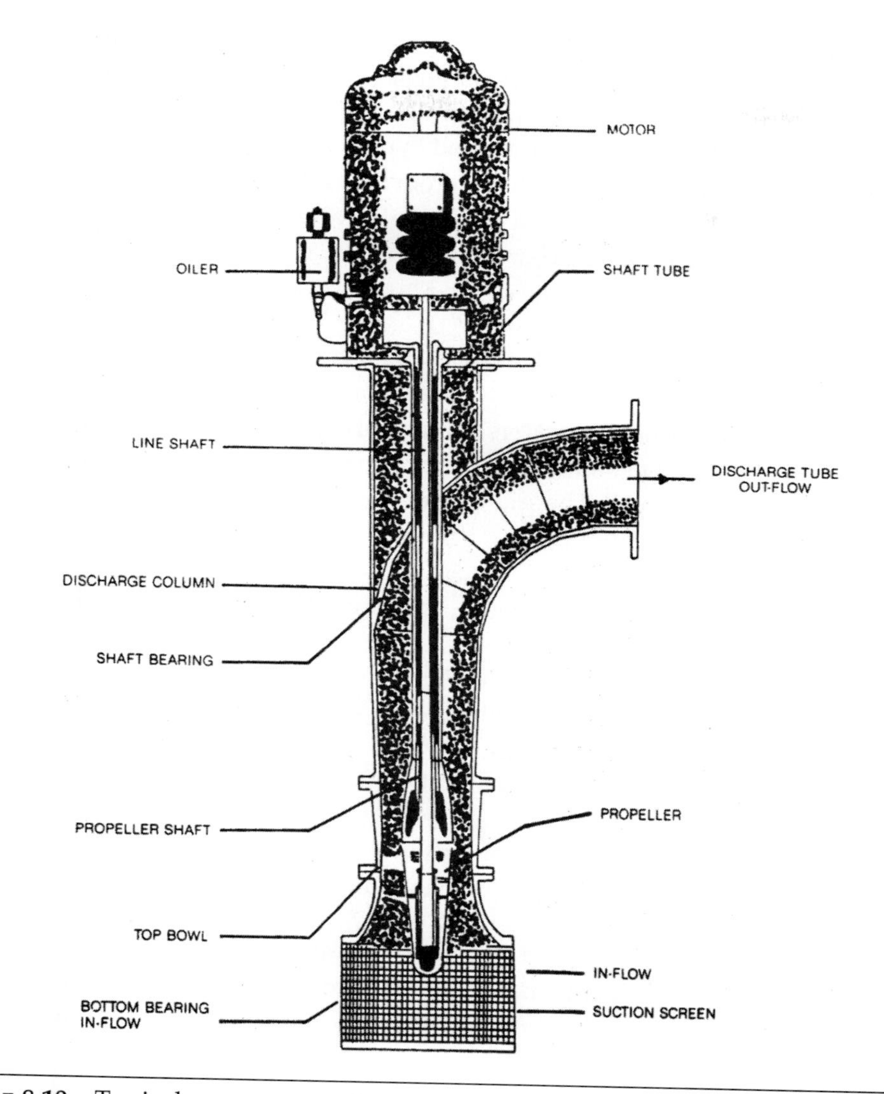


**FIGURE 8.19** Typical wastewater effluent pump with propeller.

tests ensure that the pump meets all the specified design conditions. This test may or may not require witnesses.

Three common field tests are the performance test, vibration analysis, and noise testing. The results of field and factory performance tests will differ because of the equipment configuration and the accuracy of field instruments.

*Startup.* Before starting up a nonclog pump, operators must ensure that wastewater- and electrical-flow patterns are properly established by doing the following:

1. Close the tank or wet-well drain and bypass valves;
2. Open the valve on the suction line from the tank to the wet well;
3. Open the pump's suction and discharge valves;
4. Open the valve on the discharge line; and
5. Set the inlet gate valve to the desired position.

When opening and closing valves, operators should never treat gate valves like throttling valves, because they can trap material and ultimately plug. Once the flow pattern has been established, operators should turn on the pump's seal water (if applicable). They should adjust the seal-water pressure to exceed the discharge pressure by 100 to 140 kPa (15 to 20 psi). The seal-water flow should be adjusted according to the manufacturer's specifications.

*Shutdown.* To shut down a nonclog pump, operators should do the following:

1. Adjust the pump control to "manual";
2. Adjust the hands/off/automatic (H/O/A) switch to "off" (O);
3. Close all suction and discharge valves;
4. Adjust the circuit breaker disconnect at the motor control center to "off" (lock- and tag-out);
5. Shut off the seal water to the pump; and
6. If needed, follow the startup guidelines to put a standby pump into service;

If shutting down the pump for a long time (typically on the order of several months or more—please consult with the manufacturer to confirm the specific period because different types of pumps require different levels of care), operators also should do the following:

1. Drain all tanks and wells, flush pipelines, and thoroughly clean the system;
2. Open the electrical equipment's circuit breakers and lock and tag them out;
3. Rotate the pump shafts several times a month to coat bearings with lubricant, thereby retarding oxidation and corrosion;
4. Run the bubbler-tube air compressor once a month (if applicable); and
5. Apply rust inhibitor as needed.

*Monitoring.* The general O&M monitoring requirements listed above are sufficient.

*Controls.* The general O&M control requirements listed above are sufficient for centrifugal pumps, although operators may want to integrate pumping controls with overall treatment-process-optimization efforts.

***Maintenance.*** The following maintenance activities can affect the longevity of successful pump performance.

*Wear.* All rotating parts in pumping equipment will eventually wear out, but wear can be exacerbated by operating a pump too fast or pumping too many solids. So, operators can profoundly affect the useful life of a pump.

Ideally, centrifugal pumps should operate as close as possible to their BEPs on the pump curve. If the pump runs to the right of the BEP, the high velocities inside the pump could lead to scouring, recirculation, and cavitation—accelerating wear and increasing the radial loads on the drive train. If the pump runs too far to the left of the BEP, separation and recirculation will occur, reducing efficiencies and increasing the chances of clogging, possible overheating (if the unit is pump-fluid cooled), and faster wear.

If abrasion is unavoidable, the centrifugal pump components most susceptible to wear are the suction nozzle, impeller vanes and tip, casing walls next to the impeller tip, and stuffing box seal and shaft. The pump impeller, volute, and seal area should be designed with generous wear allowances and made of abrasion-resistant materials. The pump shaft should include an abrasion-resistant sleeve, and if a mechanical seal is used, it should have abrasion-resistant faces.

Most bearings are designed for useful lives of 30 000 to 50 000 hours, so bearing wear is typically not a problem. Most bearings fail because of poor operating conditions, not because they wear out.

*Standard Maintenance.* Because of the wide variation in construction materials, operating conditions, and duty cycles, pump maintenance programs need to be site-specific, based on the manufacturer's recommendations, and address both preventive and corrective maintenance. Also, operators should be trained to recognize the signs of impending component wear.

In general, daily or weekly inspections and periodic cleanings are good operating procedures and will help operators detect early signs of trouble. Inspections only take a few minutes and could save a lot of money and downtime. Pump inspectors should

- Ensure that noise, vibration, and bearing and stuffing-box temperatures are normal;
- Look for abnormal fluid or lubricant leaks from the gaskets, O-rings, and fittings;

- Make sure that lubricant levels are adequate and the lubricant itself is in good condition;
- Change the lubricant as recommended by the manufacturer (based on pump run-time); and
- See whether shaft seal leaks are within acceptable limits.

Every 6 months, inspectors should

- Check that the stuffing box's gland (packing) moves freely, and clean and oil the gland's studs and nuts;
- Scrutinize the stuffing box for excessive leakage, adjust the gland to reduce leakage (if necessary), and replace the gland if adjusting it does not solve the problem;
- Look for corrosion on all fasteners and foundation bolts, and ensure that all are securely attached; and
- Ensure that couplings or belts are properly aligned, and check for wear.

The timing of complete overhauls depends on the pump's construction materials, operating conditions, run time, and preventive-maintenance schedule. When completely overhauling a pump, operators should:

- Check the impeller, case, wear rings, seals, bearings, shaft, and shaft sleeve for wear and corrosion;
- Replace all gaskets and O-rings;
- If necessary, wash out the bearings' housing with a brush and hot oil or solvent, flush the housing with a light mineral oil to remove all traces of solvent and prevent rust, wipe the bearings with a clean lint-free cloth, and clean them in the same way as the housing.

All pump work must only be done on an out-of-service pump (see the O&M manual for appropriate shutdown procedures). If parts need to be replaced, operators should consult the O&M manual for the proper procedure and double-check product compatibility.

*Predictive Maintenance.* Operators should monitor pump vibrations when the pump is started up and periodically throughout the pump's life. Excessive vibration indicates that something is mechanically or operationally wrong with the pump. Mechanical sources of vibration include improper installation or foundation; pump misalignment; worn bearings; piping strain; loose bolts or parts; and unbalanced or damaged rotating components (impellers, shafts, or sleeves). Hydraulic sources of vibration include cavitation, recirculation, water hammer, air entrained in the pumped fluid, turbulence in the suction well, and pump operation far left or right of the BEP. If ignored, excessive vibration could damage pump components.

*Spare Parts.* The spare parts that should be kept onsite depend on the pump, its application, and its manufacturer. A centrifugal pump, for example, will typically need a complete set of gaskets, O-rings, and packing.

*Troubleshooting.* The recommended troubleshooting activities depend on the pump, its application, and its manufacturer. For example, Table 8.1 is a troubleshooting guide for recessed-impeller centrifugal pumps.

# SOLIDS PUMPING

Maintaining proper solids handling when solids are moved from one process to another could be an operator's most demanding responsibility. One of the most common issues is using solids pumps correctly under various operating conditions.

Wastewater utilities produce a variety of solids with different characteristics (Table 8.2). As treatment processes are modified or upgraded, pumping systems also may need to be modified to accommodate the varying thixotropic nature of solids.

*Thixotropy* is the tendency of certain gels to be semisolid when still and fluid when stirred or shaken [e.g., an ice cream milkshake, which "sets up" in a container and will only flow out when the container is tapped or jarred (U.S. EPA, 1979).] Solids can become thixotropic if the pipelines are not properly flushed before shutdown (the thixotropic effect may occur in a few hours or days, depending on the solids characteristics) and will need more pressure to begin flowing in the pipeline after startup. Basically, when the solids begin to move, their friction is higher, so head losses in the pipeline are greater. As the solids move faster, both the friction and the head losses drop. (This is why a solids pump's total head varies so much, even when the speed is constant.)

The thixotropic effect is partially offset by slippage and seepage—a thin film of liquid on the pipe wall that solids "ride" when they begin flowing. This condition aids solids pumping (U.S. EPA, 1979).

**TYPES OF SOLIDS PUMPS.** There are many types of solids pumps: positive-displacement, centrifugal with recessed impellers, hose, pneumatic ejector, grit, grinder, and air-lift. Effective solids handling depends on properly matching the pump, solids characteristics, and actual system head requirements.

**TABLE 8.1** Troubleshooting guide for recessed-impeller centrifugal pumps.

| Problems | Causes | Solutions |
|---|---|---|
| No liquid delivered | Pump not primed. Speed too low. | Prime pump. Check voltage and frequency. |
| | Air leak in suction or stuffing box. | Repair leak. |
| | Suction or discharge line plugged. | Unplug line. |
| | Wrong direction of rotation. | Correct direction of rotation. |
| | Discharge valve closed | Open discharge valve. |
| Not enough pressure | Speed too low. | Check voltage and frequency. |
| | Air leak in suction or stuffing box. | Repair leak. |
| | Damaged impeller or casing. | Repair or replace. |
| | Wrong direction of rotation. | Correct direction of rotation. |
| Motor runs hot | Liquid heavier and more viscous than rating. | Increase dilution factor. |
| | Packing too tight. | Adjust packing. |
| | Impeller binding or rubbing. | Align impeller properly. |
| | Defects in motor. | Repair or replace motor. |
| | Pump or motor bearing overlubricated. | Lubricate bearing properly. |
| Stuffing box overheats | Packing too tight, not enough leakage of flush liquid. | Adjust packing. |
| | Packing not sufficiently lubricated and cooled. | Adjust packing. |
| | Wrong grade of packing. | Replace packing. |
| | Box not properly packed. | Properly pack box. |
| | Bearings overheated. | Grease properly, check tightness. |
| | Oil level too low or too high. | Adjust oil level to proper level. |
| | Improper or poor grade of oil. | Replace with proper grade of oil. |
| | Dirt or water in bearings. | Clean and regrease bearings. |
| | Misalignment. | Align properly. |
| | Overgreased. | Grease bearings properly. |
| Bearings wear rapidly | Misalignment. | Align properly. |
| | Bent shaft. | Repair or replace. Correct source of vibration. |
| | Lack of lubrication. | Lubricate bearings. |
| | Bearings improperly installed. | Reinstall bearings properly. |

**TABLE 8.1** Troubleshooting guide for recessed-impeller centrifugal pumps (*continued*).

| Problems | Causes | Solutions |
|---|---|---|
| | Moisture in oil. | Replace oil. |
| | Dirt in bearings. | Clean and relubricate. |
| | Overlubrication. | Lubricate properly. |
| Not enough liquid delivered. | Air leaks in suction of stuffing box. | Repair leaks. |
| | Speed too low. | Check voltage and frequency. |
| | Suction or discharge line partially plugged. | Unplug line. |
| | Damaged impeller or casing. | Repair or replace. |
| Pump works for a while then loses suction vibration. | Leaky suction line. | Repair suction line. |
| | Air leaks in suction or at stuffing box. | Repair leaks. |
| | Misalignment of coupling and shafts. | Properly align coupling and shafts. |
| | Worn or loose bearings. | Replace or tighten bearings. |
| | Rotor out of balance. | Balance rotor. |
| | Shaft bent. | Repair or replace shaft. |
| | Impeller damaged or unbalanced. | Repair or balance impeller. |

**POSITIVE-DISPLACEMENT PUMPS.** Most solids pumping at wastewater utilities is done by positive-displacement pumps (Figure 8.7). They typically handle thicker solids (more than 4% solids). Two common types of positive displacement pumps are: controlled-volume and rotating positive-displacement (e.g., progressing cavity and rotary lobe). These pumps can sometimes be used interchangeably, depending on the solids process downstream.

***Controlled-Volume.*** Controlled-volume pumps move a discrete volume of fluid in each cycle. (They are also called *reciprocating pumps* because of how the pump driver moves.) Three types of controlled-volume pumps are discussed here: plunger, piston, and diaphragm.

*Plunger Pumps.* Plunger pumps (Figure 8.20) use a plunger to move fluid through a cylindrical chamber. The plunger moves along an axis to build pressure in the cylinder that actuates check valves on each side of the pump body (Figure 8.21). When the plunger lifts (upstroke), the resulting low pressure automatically closes the discharge

**TABLE 8.2** Sludge types and their characteristics.

| Sludge type | Total solids range, % | Viscosity | |
|---|---|---|---|
| | | Low | High |
| Raw primary sludge | 0.5–5.0 | X | |
| Secondary sludges | | | |
| Waste activated sludge | 0.3–2.5 | X | |
| Trickling filter biological towers | 0.25–1.5 | X | |
| Biological nutrient removal | 0.5–4.0 | X | |
| Chemical sludges | | | |
| Lime sludges | 2.0–10.0 | | X |
| Alum/iron salts sludges | 1.0–5.0 | X | X |
| Thickened sludges | | | |
| Thickened waste activated sludge | 1.5–6.0 | | X |
| Thickened primary sludge | 1.5–8.0 | | X |
| Thickened chemical sludge | 2.0–1.5 | | X |
| Stabilized sludges | | | |
| Anaerobic digested sludge | 1.5–7.0 | X | X |
| Aerobic digested sludge | 0.5–4.0 | X | |
| Lime stabilized sludge | 1.5–10.0 | | |
| Dewatered sludges | | | |
| Belt presses/centrifuge | 10–25 | | X |
| Plate and frame filter presses | 20–40 | | X |

valve and opens the suction valve, allowing fluid to be drawn into the pump chamber. When the plunger lowers (downstroke), the resulting high pressure automatically closes the suction valve and opens the discharge valve, forcing the fluid out of the pump chamber. The volume of the fluid discharged is equal to the area of the plunger multiplied by its stroke length.

The flow rate can be altered by changing the stroke length (within the constraints of the pump body volume) or the number of strokes per minute. To increase capacity further, numerous plungers can be actuated by the same drive mechanism. Wastewater utilities typically use simplex, duplex, triplex, and quadruplex plunger pumps to deliver solids at the design flow rate. Their capacities typically range up to 32 L/s (500 gpm) at 690 to 1030 kPa (100 to 150 psi) of delivered pressure. When pumping dewatered sludge cake, capacities range from 0.6 to 6.3 L/s (10 to 100 gpm) at pressures up to 10 340 kPa (1500 psi).

Ball checks are readily accessible through quick-opening valve covers. Air chambers are installed on the discharge side of the pump (and often on the suction side to reduce the high-impact loading). Mounting the pressure gauges on the chambers helps reduce or eliminate gauge fouling.

**FIGURE 8.20** Triplex plunger pump.

*Piston Pumps.* Piston pumps are similar to plunger pumps, except they displace fluid via a piston rather than a plunger, are hydraulically driven, and have tighter tolerances. They are better suited for high-pressure applications than plunger pumps and are typically used to pump dewatered solids. The piston pump's most important component is a sequencing valve (typically, a poppet valve), which isolates the pump from the material in the pipeline and avoids harmful hammer effects due to pipeline backflow.

*Diaphragm Pumps.* In diaphragm pumps (Figure 8.22), a flexible membrane separates the solids from the driving force [a mechanical device (e.g., a cam shaft or an eccentric) or hydraulic fluid pressure]. The solids flow is controlled by the membrane's reciprocating action and valves or ball checks. The pump's capacities range up to 6 L/s (100 gpm) at discharge pressures up to 690 kPa (100 psi).

Double-Disc pumps™ (Figure 8.23) are based on a free-diaphragm® technology that combines the technologies of plunger and diaphragm pumps; they use two reciprocating connecting rods and discs, which operate in tandem to provide both pumping

**Figure 8.21** Positive displacement sludge pump. (Courtesy ITT Marlow.) (Steel and McGhee, 1979).

and valving. Their discharge capacities and pressures are similar to those of other controlled-volume pumps.

***Rotating.*** These are several types of rotating positive-displacement pumps, including progressing-cavity, rotary-lobe, vane, and gear. Progressing-cavity pumps are most often used to pump solids. Lobe, vane, gear, and flexible-member pumps typically are used to meter fuel or chemical feeds rather than pump solids and other abrasive materials.

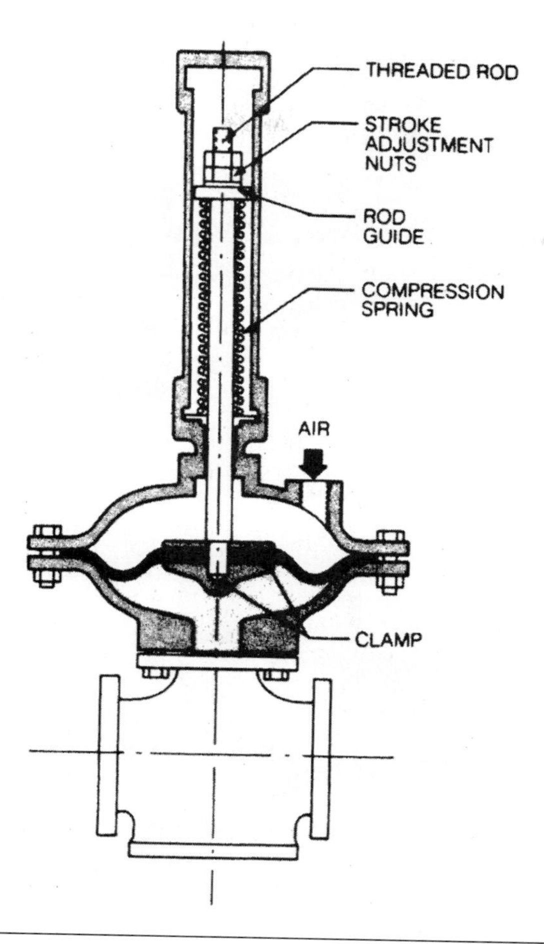


**FIGURE 8.22** Diaphragm sludge pump.

*Progressing-Cavity Pumps.* Self-priming progressing-cavity pumps (also called *progressive-cavity pumps* in industrial applications) deliver a smooth flow of solids at discharge capacities up to 40 L/s (600 gpm) and pressures up to 1700 kPa (250 psi) (Figure 8.24). A continuous screw fitted into a matching stator moves fluid through the pump (the pumping action differs from that of a screw pump because the screw is fully enclosed and does not have an upper bearing. The pump's cast-iron body, chrome-plated tool steel rotor, and nitrite rubber stator (65 to 75 durometer hardness) can handle scum and secondary sludge, but for more abrasive solids containing high amounts of sand, grit, or other erosive materials (e.g., primary sludge), ceramic-coated oversize rotors with 50- to 55-durometer nitrite stators should be used.

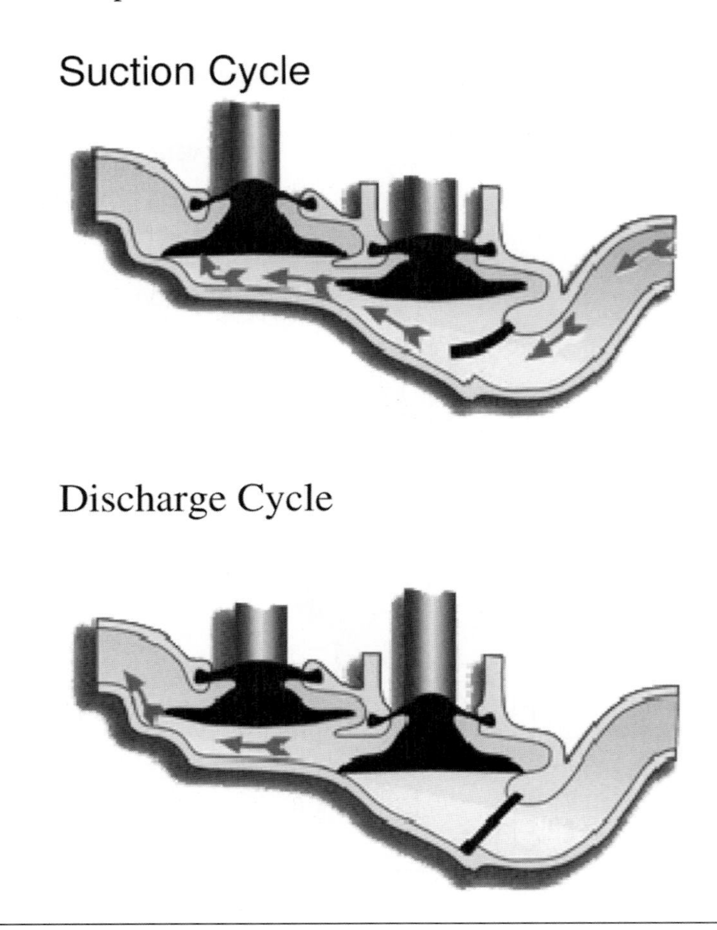


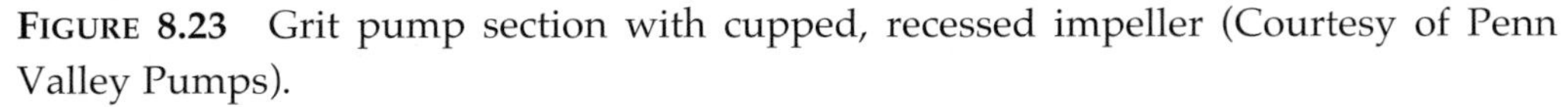

**Figure 8.23** Grit pump section with cupped, recessed impeller (Courtesy of Penn Valley Pumps).

If run dry, the pump will be damaged in a matter of minutes. So, either the pump intake must be kept wet, or operators should use protective devices (e.g., capacitance-type flow detectors, low-pressure sensors or switches, and thermocouples imbedded in the stator).

*Rotary-Lobe Pumps.* Rotary-lobe pumps typically move solids by trapping the material between a rotating lobe and its housing, and then carrying the solids to the discharge port (Figure 8.25). The rotating lobes mesh at the discharge port, eliminating the space and forcing the solids into the pipeline. Timing gears phase the lobes. This phasing, the rotors' shape, and the constant velocity produce a continuous discharge.

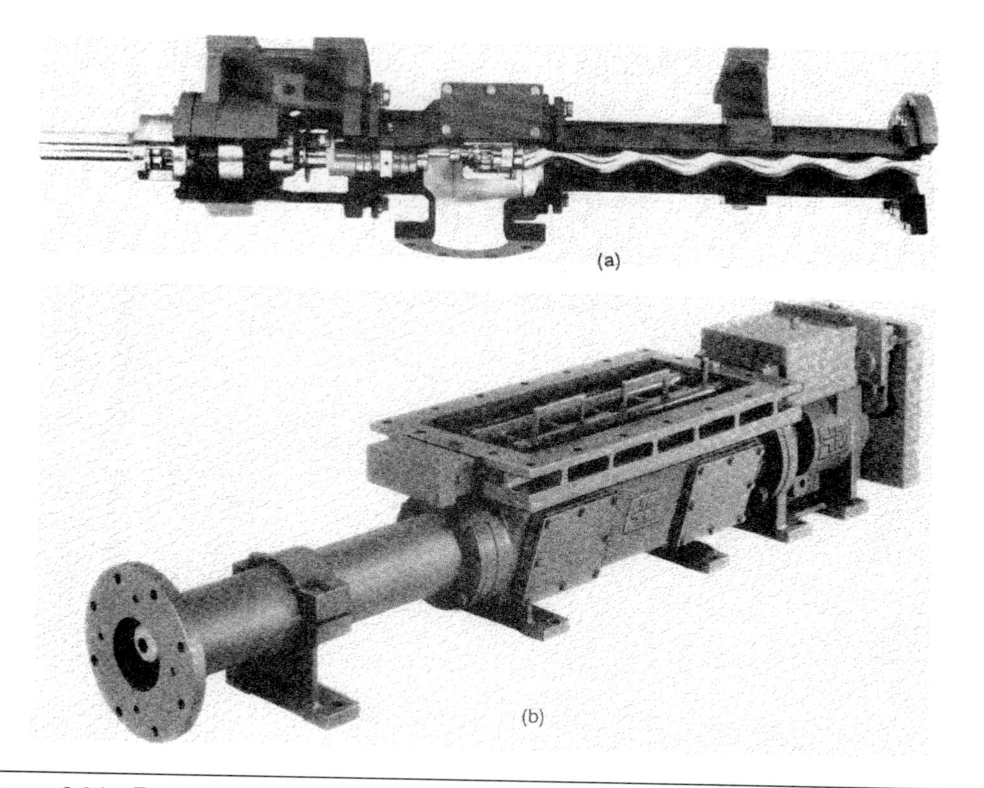


FIGURE 8.24 Progressing cavity pumps: (a) sludge pump and (b) pump with "bridge breaker" for sludge cake.

Other types of rotary-lobe pumps use arc-shaped pistons (rotor wings) in annular cylinders machined in the pump body (Figure 8.25). The pistons are driven in opposite directions by external timing gears. As they move, the rotors expand the cavity on the inlet side, allowing fluid to enter the pump chamber. The rotors then carry the fluid around the cylinder to the outlet side and then contract the cavity, thereby forcing the fluid out of the pump.

***Other Types of Solids Pumps.*** Although the most common type of solids pump is the positive-displacement pump, solids also can be moved via recessed-impeller centrifugal pumps, hose pumps, pneumatic ejector pumps, grit pumps, grinder pumps, and air-lift pumps.

*Recessed-Impeller Centrifugal Pumps.* Two types of centrifugal pumps are typically used to pump solids: the open two-port radial nonclog pump and the vortex flow pump

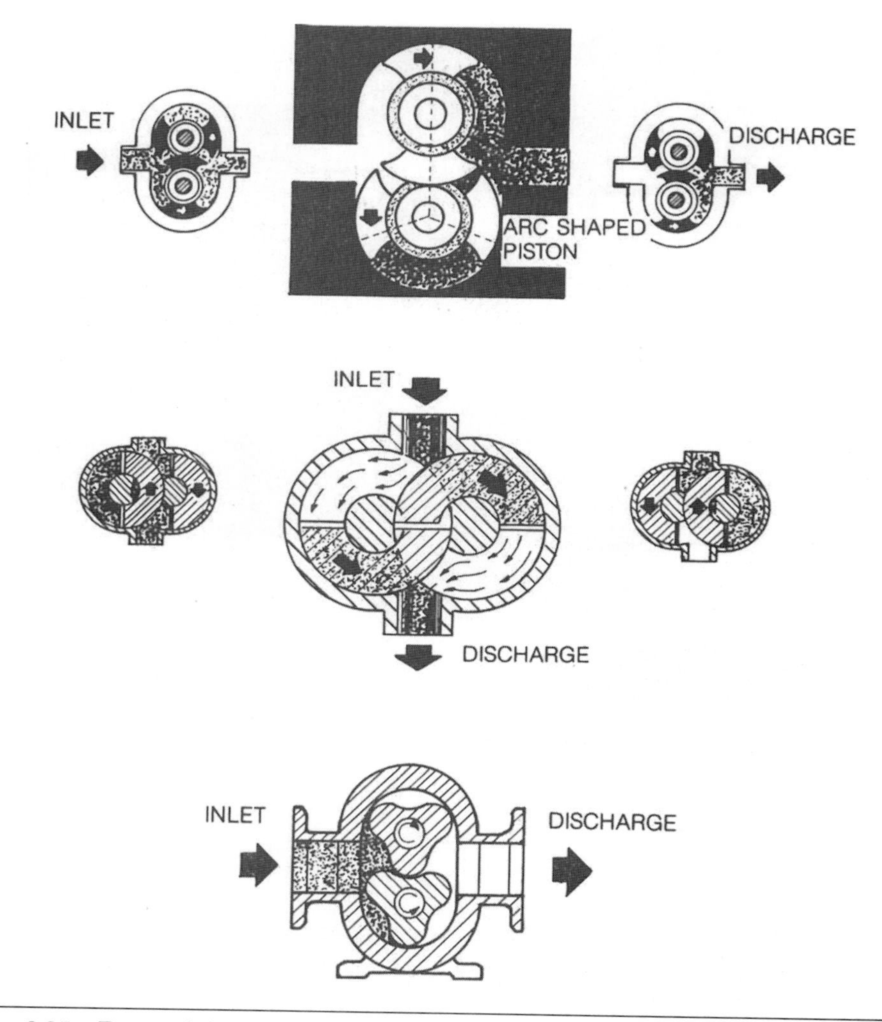


**FIGURE 8.25** Rotary lobe pumps.

with a recessed impeller (Figure 8.26). Both pumps handle grit and liquid readily if the velocity is sufficient to overcome thixotropy. Special linings and impellers are needed to control wear when handling grit slurries or abrasive chemicals (e.g., lime slurry).

*Hose Pumps.* A hose pump operates as a positive-displacement pump in which the hose is completely squeezed between a roller and a track (Figure 8.27). Hose pumps avoid backflow and siphoning without any check valves, and none of their mechanical components touch the fluid being pumped. Servicing is relatively simple, because all

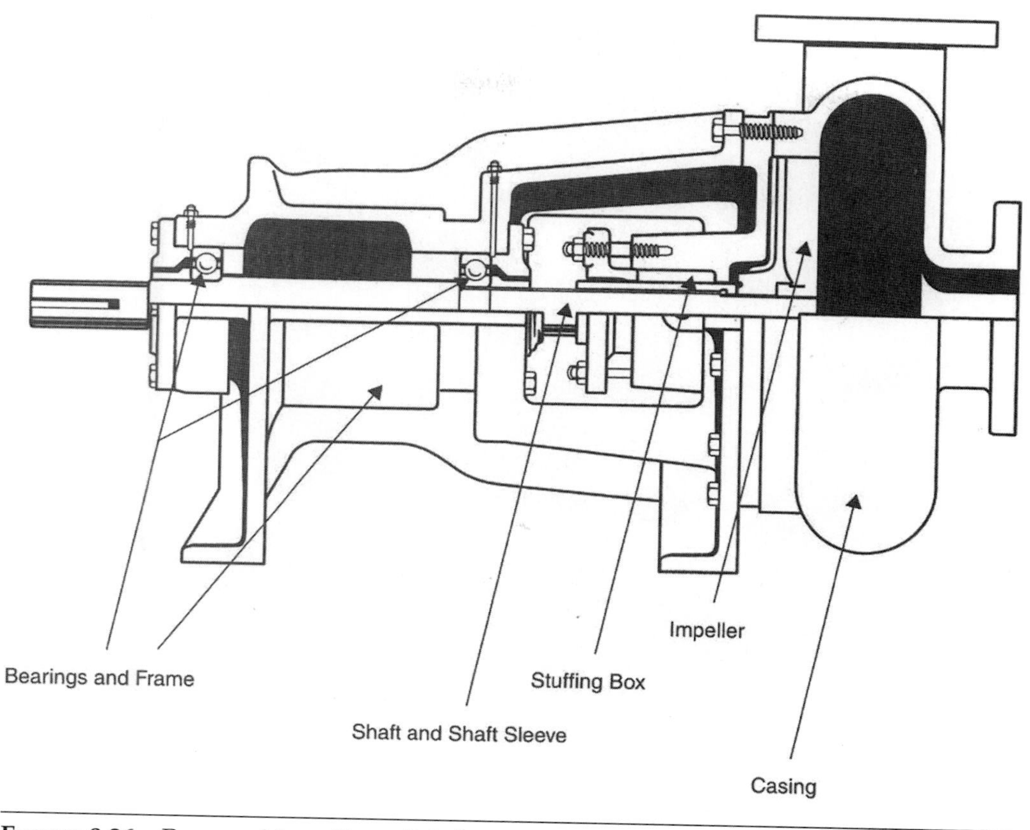


FIGURE 8.26 Recessed impeller solids-handling pump.

moving parts are external to the fluid or piping. However, they can only handle up to 350 gpm. Also, overoccluding (squeezing) the hose stresses it and the mechanical drive components unnecessarily; underoccluding it reduces pump efficiency and allows backflow to occur.

*Pneumatic Ejector Pumps.* A pneumatic ejector uses both air and the fluid to be pumped to pressurize its vessel. Then a discharge valve is opened, and the high pressure "ejects" the fluid–air mixture.

*Grit Pumps.* A "grit" pump is basically a centrifugal pump with a recessed impeller (Figure 8.28). Grit pumping can be more difficult than other solids pumping because the grit's sheer weight can clog pumps and its abrasive nature quickly deteriorates piping. To minimize these problems, the pumps are made of abrasion-resistant alloys and can be run more frequently, so large quantities of grit cannot accumulate.

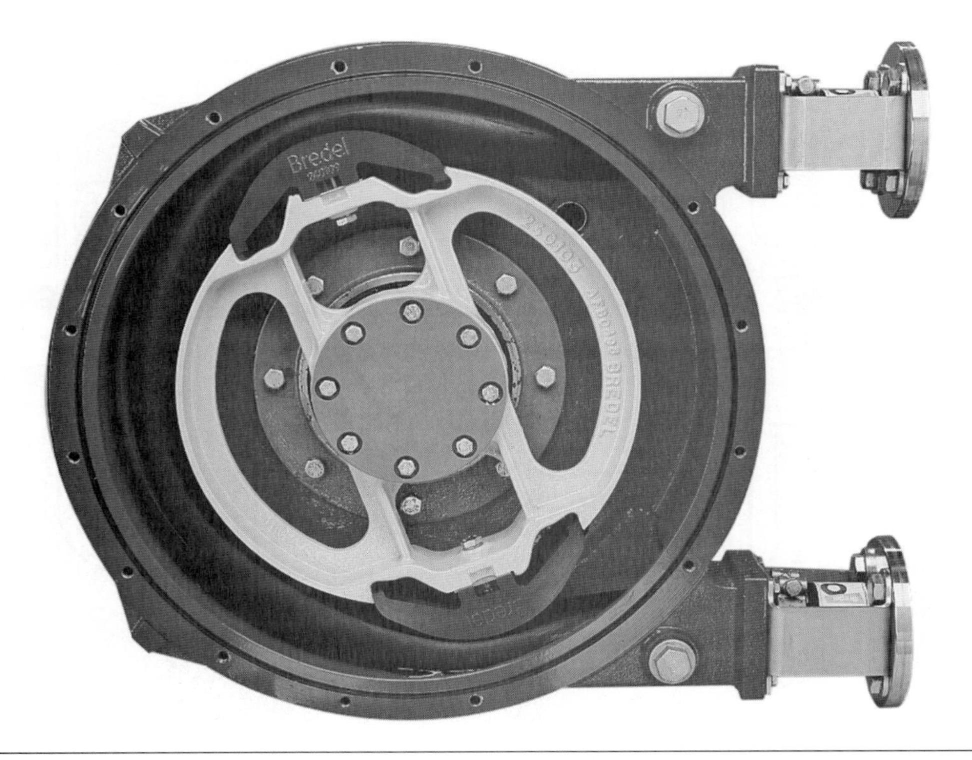


**FIGURE 8.27** Hose pump section (Courtesy of Watson-Marlow Bredel).

*Grinder Pumps.* A mechanical grinder can be fitted across the suction end of either a centrifugal or positive-displacement pump. The grinding mechanism "chews" influent solids into finer solids, minimizing potential clogging or "balls" of material that can clog downstream piping or processes.

*Air-Lift Pumps.* Air-lift pumps were once used to pump return activated sludge from clarifiers; now, they typically lift grit from vortex degritting devices. In these pumps, diffused air is added to the bottom of the air lift, mixing with the solids and making them less dense so the fluid surface rises above the discharge point. The required submergence of the air lift and the small allowable head limits the air lift's capability between the discharge pipe and the tank fluid surface.

The pumping rate is controlled by the airflow rate. Throttling the influent valve will eventually result in plugging and "backblow" into the clarifier or decant tank. A common problem is the tendency to make the flow rate too high (to avoid the spurting, inconsistent flow that is characteristic of the pump's discharge). To avoid this problem,

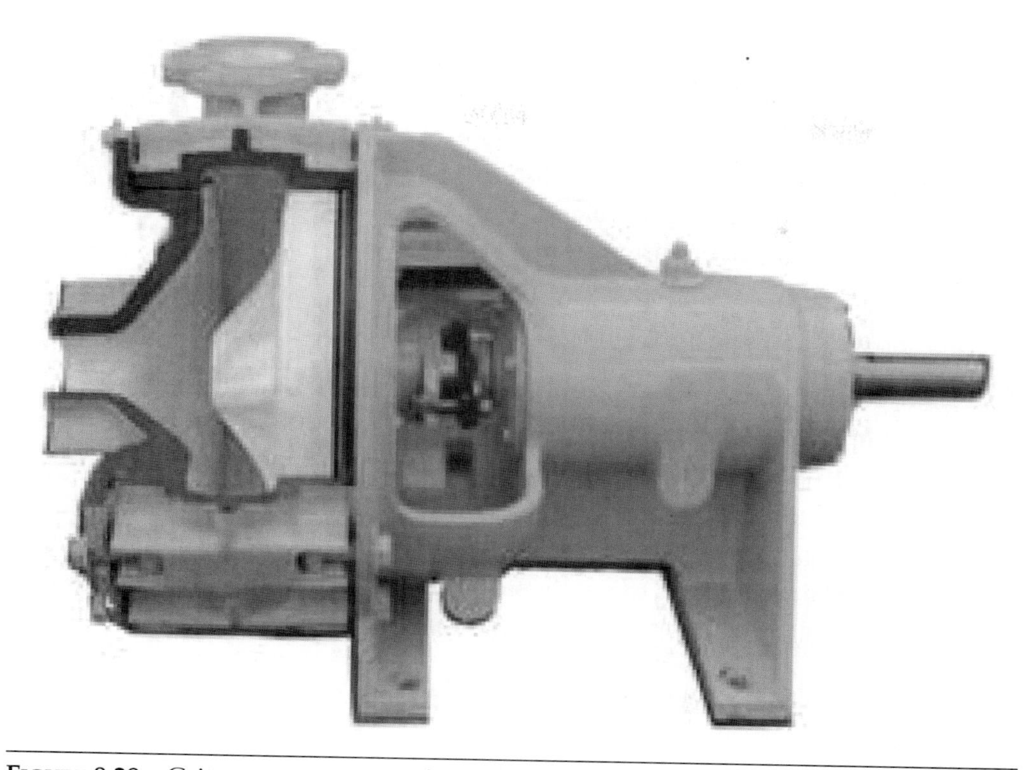

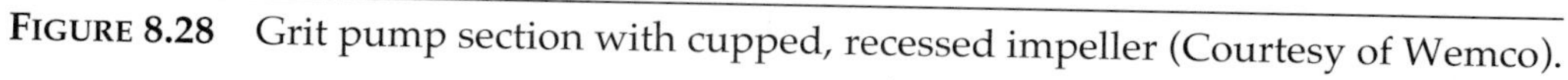

**FIGURE 8.28** Grit pump section with cupped, recessed impeller (Courtesy of Wemco).

operators should periodically measure the pumping rates—if only with a bucket and stopwatch—and calculate the approximate flow rate.

Because fluid is not forced through them mechanically, air-lift pumps are prone to plugging, so they must be rodded periodically to keep them fully operational. If the air lift is plugged or not operating, operators should confirm that the air valve is fully open and then check the pump's air blower to ensure that it is working properly. If the pump still is not working, then operators should rod the air lift as follows:

1. Turn off the air supply;
2. Remove the plug in the T at top of air lift;
3. Rod the lift to break up the blockage;
4. Replace the plug in the T; and
5. Turn on the air supply.

**CATEGORIZATION.** Solids pumps can be classified based on impeller type, suction, location, and orientation.

***Rotor or Impeller.*** Except for centrifugal pumps, solids pumps do not use conventional impellers to transfer energy from the mechanical device to the flow. Instead, progressing-cavity pumps use rotors, rotary-lobe pumps use lobes, positive-displacement pumps use other means of transferring energy. (Impellers are the most distinguishing feature of solids pumps.)

***Suction.*** Solids pumps are typically designed as single-suction units.

***Location.*** Solids pumps are typically installed outside the fluid being pumped. One exception is the air-lift pump, because the air injected into the suction piping must be submerged in the fluid being pumped.

***Orientation.*** Solids pumps—especially progressing-cavity pumps—are typically mounted horizontally, but even progressing-cavity pumps can be mounted vertically. The motors can be mounted on top, at the side, or even axially. (The orientation issue is less significant for solids pumps than for liquids pumps.)

**APPLICATIONS.** ***Controlled-Volume Pumps.*** *Plunger Pumps.* For solids applications, plunger pumps should have a welded-steel base and a hardened cast-iron or steel body and valving. Pistons are often surface-hardened to reduce their wear rate.

Plunger pumps used to be the most popular choice for handling thick, abrasive solids because the pulsating effect of the reciprocating plungers helps reduce line clogging and avoid bridging the sludge hoppers. However, its popularity has declined now that sludge thickening and dewatering processes require a more constant feed flow to mix with polymers and other coagulant aids. Nonetheless, plunger pumps are sometimes used as a backup to help dislodge blockages in the piping systems. They also are used to pump dewatered sludge cake (in which case, they are called *power pumps*).

*Piston Pumps.* Piston pumps typically are used to pump dewatered solids.

*Diaphragm Pumps.* Diaphragm pumps can operate at any rate by adjusting the number of strokes per minute. They have no packing or seals and can run dry without any damage. If compressor capacity is available, air-operated diaphragm pumps can be used. These pumps cost less to install, but need five to six times more power than a pump driven by an electric motor.

Diaphragm pumps can handle the same materials as plunger pumps, without the abrasion problems. They also handle strong or toxic chemicals more safely, because the diaphragm can prevent or substantially minimize leakage that could cause damage or threaten health and safety.

***Rotating Positive-Displacement Pumps.*** *Progressing Cavity Pumps.* A progressing-cavity pump's stator is suitable for most solids applications but may need to be modified to accommodate special solids characteristics or chemical applications. The pump may be mounted horizontally or vertically without affecting its efficiency or operations. The pump delivers at a constant rate that can be changed quickly by changing the pump speed.

Regardless of the type of solids involved, this pump operates simply; operators can control the delivery rate by modulating the pump speed [maximum should be 300 rev/min and 240 kPa (35 psi) per pump stage]. However, they must sharply reduce the speed when pumping abrasive material under high head [maximum should be 200 rev/min and 100 kPa (15 psi) per pump stage].

Progressing-cavity pumps can handle both thickened sludge and even sludge cake with solids concentrations up to 65% (Figure 8.21, Part B). Because they impart minimal shear to the fluid being moved, progressing-cavity pumps also are used to pump polymers, which can be damaged by agitation. With the proper selection of pump materials, the progressing-cavity pump can handle chemical slurries (e.g., lime and ferric chloride).

*Rotary-Lobe Pumps.* Used for many years to pump nonabrasive material, rotary-lobe pumps are now increasingly being used to transport scum and municipal sludge (except screenings, grit, composted solids, and septage) (Cunningham, 1982). During operation, the lobes mesh together while rotating in opposite directions, forming cavities between the rotors and the casing that trap and move the solids (Figure 8.22). Pump speeds range from 200 to 300 rpm (lower speeds are used for more abrasive solids). Although one manufacturer reported achievable total system heads of 515 kPa (75 psi), operators should avoid higher discharge pressures to limit hydraulic slip, which is proportional to the net differential pressure.

**OPERATIONS AND MAINTENANCE.** For any given pump, the pump manufacturer will provide literature or guidance on particular O&M activities that should be followed. The following information is meant as guideline for developing or reviewing an O&M plan.

A well-planned O&M program will prevent unnecessary equipment wear and downtime. Utility staff should begin by determining the actual and projected quantities of solids (e.g., grit, screenings, primary sludge, primary scum, return sludge, waste sludge, and thickened and dewatered solids) to be pumped under various flow conditions. Staff typically can find this information in the utility's O&M manual or engineering design report; otherwise, they can calculate loadings based on the size of the processing units and pumping systems. Then, they should evaluate the solids loadings and seasonal variations at various critical processes. With this information, staff can then determine how many solids pumps should be in service and the timing and length of the pump cycles under various conditions.

If the grit-handling pumps typically run 5 minutes an hour in dry weather, for example, staff may calculate that they need to run 8 minutes an hour after storms to handle the anticipated increase in volume. Likewise, the primary sludge pumps also may need to run longer under wet weather conditions. Also, as solids characteristics change, operators may need to change the polymer doses in the thickeners or respond to changes in alkalinity in the anaerobic digesters. Staff need to plan for such variations so they can respond promptly and appropriately.

***Operations.*** The following operations activities directly affect solids pump performance.

*Appurtenances.* Positive-displacement pumps need a drive system that can operate the pump at the speed needed to perform adequately under all operating conditions. Sometimes, this involves manually and automatically timed starts and stops, as well as varied pump discharge rates. This variable-speed arrangement can be provided via mechanical vari-drives; variable pitch pulleys; direct-current, variable-speed drives; alternating-current, variable-frequency drives; eddy-current magnetic clutches; or hydraulic speed-adjustment systems. Each has various advantages and disadvantages with respect to cost, amount and ease of maintenance required, efficiency, turndown ratio, and accuracy. Because positive-displacement pumps are constant torque machines, operators should ensure that the variable-speed drive's output torque exceeds the pump's torque requirement at all operating points. Although variable-speed drives are often either a necessity or an enhancement to proper plant operation, the challenge is providing the continued maintenance and servicing required.

*Testing Procedures.* Positive-displacement pumps should be field-tested for flow, speed, amperage, voltage, pressure, and no-flow protection devices. These tests should only be performed by an authorized factory representative to avoid system overpressurization.

*Startup.* Before starting up a solids pump, operators should check the following items to ensure that each piece of equipment is installed correctly:

- Pump, driver, coupling, or sheave alignment;
- Water flush connections to the stuffing box; and
- Open valves on both the suction and discharge sides of the pump.

Positive-displacement pumps should not be operated against a closed discharge valve.

Before operating a solids pump for the first time, operators should:

- Ensure that the pump is filled with fluid;
- Confirm that adequate seal water flow and pressure for the stuffing box seal are available (unless mechanical seals are used); and
- Bump the drive to check for rotation.

Once the solids pump is started up, operators should check the following items:

- Inlet and outlet flow rate;
- Noise or vibration;
- Bearing housing temperature;
- Running amperage;
- Pump speed; and
- Pressure.

Dry operations can harm positive-displacement pumps, so they should never be allowed to run dry.

*Shutdown.* Before shutting down positive-displacement pumps, operators should ensure that all isolation or discharge valves are open. Then, follow the startup procedures in reverse order.

*Monitoring.* In addition to general pump maintenance activities, operators should seriously consider high-pressure protection because positive-displacement pumps will always deliver flow if the pump is running. A high-pressure protection device ensures that the pumping equipment will shut down at a preset pressure before a catastrophic failure occurs. If rupture or bursting discs are used, they should be enclosed in a secondary containment structure that can capture pumped material, because the discs break open to relieve pressure. Also, positive-displacement pumps run under tight tolerances or rely on pumped flow to lubricate and cool the pumping elements, so operators should consider a no-flow protection device. This device can be a low-pressure switch, presence–absence protector, flow switch, temperature sensor, or a combination of these.

*Controls.* The primary pumping controls include time clocks, elapsed-time meters, solids-content analyzers, flow-rate meters, pressure gages, pressure switches, flow switches, and sludge blanket indicators. The emphasis should be placed on ensuring that these pumps do not develop too much head or run dry. Operators may use an indicating mode, a controlling mode, or both.

***Maintenance.*** The following maintenance activities affect the longevity of solids pump performance.

*Wear.* Excessive wear is the most common problem for positive-displacement pumps. The wear rate depends on how the pump is operated and where it is used. More abrasive solids and higher rotational speeds, for example, increase pump wear.

*Controlled-Volume Pumps.* In plunger pumps, wear occurs between the piston packing and its cylinder. Excessive leakage around the shaft and plunger packing may exacerbate cylinder wear, which in turn will accelerate the destruction of the packing. If this happens, the entire worn shaft or plunger may need to be reworked or replaced. So, staff should frequently inspect the packing along the cylinder and evaluate the extent of leakage, so they will promptly re-pack or replace it when needed, thereby extending the life of the cylinder.

If the pump runs with a short stroke for long periods, the unit will wear excessively along a short length of the shaft, accelerating packing deterioration and leakage. Ideally, the pump should operate with a long, slow stroke so it moves more material per stroke and cycles less frequently, thereby reducing wear.

Also, the mechanical seals can be replaced with water-lubricated seals that are more suitable for abrasive conditions; however, staff still must evaluate seal leakage and wear to minimize cylinder wear.

*Progressing-Cavity Pumps.* As a progressing-cavity pump's rotors or stators become well worn, the pump capacity diminishes enough to prevent the transfer of solids because of slippage within the pump itself. Excessive wear indicates that the pump's operating speed has been too great. Under abrasive conditions, progressing-cavity pumps should not exceed 200 to 300 rpm.

Depending on wear, the stator and even the rotor may need to be replaced. Because the stator wears faster than the rotor, the rotor can be plated as much as 0.8 to 1.0 mm (0.030 to 0.040 in.) oversize, so worn stators can be reused. This plating, which doubles the effective life of the stators, can be used if wear is excessive.

*Grinders or Grinder Pumps.* Grinders or grinder pumps wear rapidly and will work inadequately if not adjusted, rebuilt, or repaired to offset this wear. Grinders need proper clearance adjustment to maintain adequate grinding and avoid excessive wear. The cutters are designed so they can be easily replaced and their hardened surfaces rebuilt. For pumps with grinders, maintaining proper grinding will forestall wear or damage to the pump itself.

***Standard Maintenance.*** Following are maintenance checklists for various solids pumps.

*Plunger Pumps.* To maintain plunger pumps properly, utility staff should do the following:

- Check primary solids pump motors for heat, noise, or excessive vibration.
- Check the oil level around the plunger pump (add oil if needed).
- Check the oil level in the eccentric oiler.
- Check the alignment of the connecting rod.
- Check the condition of the shear pin (it should not have any bends inside or outside the driving flange or eccentric).
- Check the fluid level in the air chamber.
- Check the gear reducer's operating temperature (and listen for unusual noise).
- Check the oil level inside the plunger (wrist pin lubrication). If the plunger is filled with fluid or sludge, drain and refill with clean oil.
- Grease the bearings on the main drive shaft weekly.
- Check the keys in the driving flange (they should be tight, preventing backlash).
- Flush (through the oiler) the babbitt-lined bearings with about 59 mL (2 oz) of kerosene. Then, refill weekly with the proper lubricant.
- Check the eccentric liner for unusual wear because of lack of lubrication.
- Check the main shaft for rust (if present, remove it and recoat the shaft with a film of grease).
- Check the oil level in the reducer.
- Check valve balls and seats. More frequent inspection may be needed if the slurry is abrasive.
- Check the condition of the coupling between motor and reducer.
- Check the plunger for sidewall wear.
- Replace the oil inside the plunger monthly.
- Check the wrist pin (it should not be excessively loose).
- Check the tightness of the set screws on the drive flange.
- Inspect the motor and reducer seals for leakage. Change the oil in the gear reducer monthly or after every 500 hours of operation.

*Progressing Cavity Pumps.* To maintain progressing cavity pumps properly, utility staff should do the following:

- Check the level and condition of the oil in the gear reducer.
- Confirm that the high-discharge-pressure shutoff switch is functioning properly.

- Check the shaft alignment.
- Check the condition of all painted surfaces.
- Visually inspect mounting fasteners for tightness.
- Clean dirt, dust, or oil from equipment surfaces.
- Check all electrical connections.
- Stop and start equipment, checking for voltage and amp draw and any movement restrictions because of failed bearings, improper lubrication, or other causes.
- Check the drive motor for any unusual heat, noise, or vibration.
- Check mechanical seals and packing for leakage or wear.

*Rotary Lobe Pumps.* To maintain rotary lobe pumps properly, utility staff should do the following:

- Check the preventive maintenance master schedule, lubrication master schedule, and equipment history sheets to schedule lubrication and preventive maintenance tasks.
- Update the equipment history record as work is done.
- Check all equipment fasteners for tightness.
- Check the condition of all visible parts of the pump and its related components.
- Check the pump casing and impellers for wear.
- Check the oil quantity and quality in the gear case.
- Check the controls and starter equipment for overall condition.
- Check and adjust belt tightness regularly.
- Check mechanical seals and packing for leakage or wear.

***Predictive Maintenance.*** Vibration monitoring can be helpful. Because positive-displacement pumps produce a pulsed flow, some vibration is expected and the piping system should be isolated via properly designed foundations, expansion (flexible) joints, and pulsation dampeners. Excessive vibration, however, is typically caused by misalignment, loose foundation bolts, improperly supported piping, NPSH problems, or a mechanical problem.

***Spare Parts.*** The spare parts needed depend on the application, pump type, and manufacturer.

***Troubleshooting.*** The troubleshooting approach depends on the application, pump type, and manufacturer. Tables 8.3 through 8.5 are troubleshooting guides for plunger pumps, progressing cavity pumps, and rotary lobe pumps, respectively, that will help

**TABLE 8.3** Troubleshooting guide for plunger pumps.

| Problems | Causes | Solutions |
|---|---|---|
| Incorrect rotation | Incorrect connections. | Refer to connection diagram and reconnect according to instructions. |
| No sludge flow | Valve ball clogged or not installed. | Open valve chambers and inspect. |
| | Air leak in suction line. | Check line with clear water under pressure. |
| | Valve closed, clogged line. | Inspect valves, backflush sludge lines. |
| Low capacity | Pump stroke set too short. | Set to longer stroke. |
| | Air leak in suction piping. | Check line with clear water under pressure. |
| Excessive power | Pump packing too tight. | Check packing and readjust. |
| Shear pin failure | Excessive discharge pressure. | |
| | Eccentric bolts loose. | Tighten to proper torque. |
| | Flange faces dirty. | Wash face of both flanges and eccentric with kerosene and reassemble. |
| | Pump packing too tight. | Check packing and readjust per instructions. |
| | Eccentric bearing clearance too great. | Remove shims as required. |
| Overheating of gears or bearings | Inadequate lubrication. | Check oil level in gears and eccentric oiler. |
| | Excessive lubrication. | Check gear oil level. If correct, drain and refill with clean oil of recommended grade. Set eccentric oiler at prescribed rate. |
| Noise or vibrations | Suction valve slamming. | Open snifter valve. |
| | Excessive clearance in eccentric bearing. | Remove shims as required. |
| | Air chamber full of water. | Drain. |
| Runs hot | Insufficient lubrication. | Check lubrication level and adjust to recommended levels. |
| | Excessive lubrication. | Check lubricant level and adjust to recommended level. |
| | Wrong lubricant. | Flush out and refill with correct lubricant as recommended. |

**TABLE 8.3** Troubleshooting guide for plunger pumps (*continued*).

| Problems | Causes | Solutions |
|---|---|---|
| Runs noisily | Loose hold-down bolts. | Tighten bolts. |
| | Worn disk. | Disassemble and replace disk. |
| | Level of lubricant in the reducer not properly maintained. | Check lubricant level and adjust to factory-recommended level. |
| | Damaged pins and rollers. | Disassemble and replace ring gear, pins, and rollers. |
| | Overloading of reducer. | Check load on reducer. |
| Output shaft does not turn in | Input shaft broken. | Replace broken shaft. |
| | Key missing or sheared off. | Replace key. |
| | Eccentric bearing broken, lack of lubricant. | Replace eccentric bearing. Flush and refill with recommended lubricant. |
| | Coupling loose or disconnected. | Properly align reducer and coupling. Tighten coupling. |
| Oil leakage | Worn seals caused by dirt or grit entering seal. Breather filter clogged. | Replace seals. Replace or clean filter. |
| | Overfilled reducer. | Check lubricant level and adjust to recommended level. |
| | Vent clogged. | Clean or replace element, being sure to prevent any dirt from falling into the reducer. |
| Motor fails to start | Blown fuses. | Replace fuses with proper type and rating. |
| | Overload trips. | Check and reset overload in starter. |
| | Improper power supply. | Check to see that power supply agrees with motor nameplate and load factor. |
| | Improper line connections. | Check connection with diagram supplied with motor. |
| | Open circuit in winding or control switch indicated by humming sound when switch is closed. | Check for loose wiring connections. Also see that all control contacts are closing. |
| | Mechanical failure. | Check to see if motor and drive turn freely. Check bearings and lubrication. |

TABLE 8.3 Troubleshooting guide for plunger pumps (*continued*).

| Problems | Causes | Solutions |
|---|---|---|
| | Short-circuiting. | Indicated by blown fuses. Rewind motor. |
| | Poor stator coil connection. | Remove end bells; locate with test lamp. |
| | Rotor defective. | Look for broken bars or end rings. |
| | Motor may be overloaded. | Reduce load. |
| Motor stalls | One phase may be open. | Check lines for open phase. |
| | Overloaded motor. | Reduce load. |
| | Low motor voltage. | See that nameplate voltage is maintained. Check connection. |
| | Open circuit. | Fuses shown. Check overload relay, stator, and push-buttons. |
| Motor runs and then shuts down | Power failure. | Check for loose ties, connections to lines, fuses, and controls. Check starter overloads. |
| Motor does not get up to speed. | Voltage too low at motor terminals because of line drop. | Use higher voltage or transformer terminals, former terminals, or reduce load. Check connections. Check conductors for proper size. |
| | Starting load too high. | Check what load motor is supposed to carry at start. |
| | Broken rotor bars or loose rotor. | Look for cracks near the rings. A new rotor may be required as repairs are typically temporary. |
| | Open primary circuit. | Locate fault with testing device and repair. |
| Motor takes too long to accelerate | Excess loading. | Reduce load. |
| | Poor circuit. | Check for high resistance. |
| | Defective squirrel cage rotor. | Replace with new rotor. |
| | Applied voltage too low. | Get power company to increase power tap. |
| Wrong rotation | Wrong sequence of phases. | Reverse connections at motor or at switchboard. |
| Motor overheats while running under load. | Overloaded. | Reduce load. |

TABLE 8.3 Troubleshooting guide for plunger pumps (*continued*).

| Problems | Causes | Solutions |
|---|---|---|
| | Frame or bracket vents may be clogged with dirt and prevent proper ventilation of motor. | Open vent holes and check for a continuous stream of air from the motor. |
| | Motor may have one phase open. | Check to make sure that all leads are well-connected. |
| | Grounded coil. | Locate and repair. |
| | Unbalanced terminal voltage. | Check for faulty leads, connections, and transformers. |
| Motor vibrates after corrections have been made. | Motor misaligned. | Realign. |
| | Weak support. | Strengthen base. |
| | Coupling out of balance. | Balance coupling. |
| | Driven equipment unbalanced. | Rebalance driven equipment. |
| | Defective ball bearing. | Replace bearing. |
| | Bearings not in line. | Line up properly. |
| | Balancing weights shifted. | Rebalance motor. |
| | Polyphase motor running single phase. | Check for open circuit. |
| | Excessive end play. | Adjust bearing or add washer. |
| Unbalanced line current on polyphase motors during normal operation | Unequal terminal volts. | Check leads and connections. |
| | Single-phase operation. | Check for open contacts. |
| Scraping noise | Fan rubbing fan shield. | Remove interference. |
| | Fan striking insulation. | Clear fan. |
| | Loose on bedplate. | Tighten holding bolts. |
| Noisy operation | Air gap not uniform. | Check and correct bracket fits or bearing. |
| | Rotor unbalanced. | Rebalance. |
| Hot bearings, general | Bent or sprung shaft. | Straighten or replace shaft. |
| | Excessive belt pull. | Decrease belt tension. |

TABLE 8.3 Troubleshooting guide for plunger pumps (*continued*).

| Problems | Causes | Solutions |
|---|---|---|
| | Pulleys too far away. | Move pulley closer to motor bearing. |
| | Pulley diameter too small. | Use larger pulleys. |
| | Misalignment. | Realign drive. |
| Hot ball bearings | Insufficient grease. | Maintain proper quantity of grease in bearing. |
| | Deterioration of grease or lubricant contaminated. | Remove old grease, wash bearings thoroughly in kerosene, and replace with new grease. |
| | Excess lubricant. | Reduce quantity of grease; bearings should not be more than 50% filled. |
| | Overloaded bearing. | Check alignment, side and end thrust. |
| | Broken ball or rough races. | Clean housing thoroughly, then replace bearing. |

operators identify pumping problems and determine their solutions. Figure 8.29 is a manufacturer's published troubleshooting guide for rotary lobe pumps.

## MISCELLANEOUS PUMPS OR PUMPING APPLICATIONS

Some pumps and pumping applications (e.g., grinding, chemical metering, and pneumatic ejectors) do not fit neatly into the kinetic or positive-displacement categories.

**GRINDING.** There are basically two types of sludge-grinding pumps: comminuting and pumping, and comminuting and grinding.

***Comminuting and Pumping.*** Typically used to prevent oversized particles from damaging or plugging downstream piping and equipment, a comminuting and pumping device (Figure 8.30) both grinds and pumps liquids and solids. It also may be used to grind scum and screenings. In this device, the solids flow axially to a serrated, wobble-plate rotor, which grinds them and forces them through a matching set of discharge bars in the outlet. The size of the resulting particles depends on the size of the rotor's serrated edges and the spacing of the bars. Because the device functions like a

**TABLE 8.4** Troubleshooting guide for progressing cavity pumps.

| Problems | Causes | Solutions |
|---|---|---|
| Pump does not rotate | Incorrect power supply. | Check motor nameplate data; test voltage, phase, and frequency. |
| | Foreign matter in pump. | Remove foreign matter. |
| | If pump or stator is new, or too much friction. | Fill with liquid and hand turn. If still tight, lubricate stator with glycerine or liquid soap (household commercial product). |
| | Stator swells due to high liquid temperature. | Reduce liquid temperature or use an undersized rotor. |
| | Blockage due to solids in liquid. | Decrease solids-to-liquid ratio. |
| | Liquid settles and hardens after pump shutdown. | Clean and rinse pump after each use. |
| Pump does not discharge | Air in suction pipe. | Tighten connections to stop leaks. |
| | Pump speed too low. | Increase drive speed. |
| | Stator worn excessively. | Replace stator. |
| | Rotor worn excessively. | Replace rotor. |
| | Wrong direction of rotation. | Reverse drive motor polarity. |
| Discharge output low | Incorrect power supply. | Check motor nameplate data; test voltage, phase, and frequency. |
| | Air in suction pipe. | Tighten connections to stop leaks. |
| | Pump speed too low. | Increase drive speed. |
| | Stator worn excessively. | Replace stator. |
| | Rotor worn excessively. | Replace rotor. |
| | Discharge pressure too high. | Open discharge valve; remove obstruction. |
| | Suction pipe leaks. | Tighten pipe connections. |
| | Shaft packing leaks. | Tighten packing glands; replace packing. |
| No discharge | Improper rotation direction. | Reverse pump motor. |
| | Pump not primed. | Relieve internal pressure; carefully prime pump. |
| | Loose belt. | Tighten belt. |

**TABLE 8.5** Troubleshooting guide for rotary lobe pumps.

| Problems | Causes | Solutions |
|---|---|---|
| Discharge under capacity | Sludge vaporizing in line to pump. | Check temperature of sludge; if above 35°C (95°F), correct temperature. |
| | Air entering inlet line. | Check and correct any leaks in inlet valve. |
| | Packing leaking excessively. | Adjust packing 5 to 10 drops/minute. |
| | Sludge too thick, unable to move through lines. | Check thickness of sludge; should not exceed 9%. |
| | Worn rotors. | Check thickness of rotors compared to manufactured specifications; if severely worn, replace rotors. |
| Pump stalls after starting | Sludge too thick. | Check sludge thickness; should not exceed 9%. |
| | Worn or unsynchronized timing gears. | Replace timing gears. |
| | Metal-to-metal contact on rotors. | Replace rotors. |
| Pump overheats | Sludge too thick. | Check sludge thickness; should not exceed 9%. |
| | Sludge temperature too high. | Sludge should not be above 35°C (95°F). |
| | Packing gland overtightened. | Adjust packing gland for 5 to 10 drops/minute; if no adjustment left, replace packing. |
| | Shaft bearing worn or failed. | Inspect bearing; if worn or damaged, replace. |
| | Worn or unsynchronized timing gear. | Replace timing gear. |
| | Gear case oil quantity low or quality poor. | Inspect oil; if contaminated, change; if low, add oil. |
| | Metal-to-metal contact on rotors. | Replace rotors. |

**TABLE 8.5** Troubleshooting guide for rotary lobe pumps (*continued*).

| Problems | Causes | Solutions |
|---|---|---|
| Motor overheats | Sludge too thick. | Check sludge thickness; should not be above 9%. |
| | Packing gland too tight. | Adjust gland for 5 to 10 drops/minute. |
| | Overspeed. | Check motor for speed; adjust as necessary. |
| | Sheaves misaligned. | Properly align sheaves. |
| | Shaft bearing worn or failed. | Inspect bearing; if worn or damaged, replace bearing. |
| | Worn or unsynchronized timing gear. | Replace timing gear. |
| | Gear case oil quantity low or poor quality. | Add oil if needed; replace oil if contaminated. |
| | Metal-to-metal contact on rotors. | Replace rotors. |
| Noise and vibration | Air entering inlet line. | Check inlet lines for leaks; correct leaks. |
| | Sludge too thick. | Check sludge thickness; should not be above 9%. |
| | Sheaves misaligned. | Properly align sheaves. |
| | Loose pump and motor mountings. | Check for loose mounts; correct same. |
| | Shaft bearing worn or failed. | Replace bearing. |
| | Worn or unsynchronized timing gear. | Replace timing gear. |
| | Gear case oil quantity low or poor quality. | Add oil if needed; replace oil if contaminated. |
| | Metal-to-metal contact on rotors. | Replace rotors. |
| Pump element wear excessive | Sludge temperature too high. | Should not exceed 35°C (95°F). |
| | Sludge too thick. | Should not be above 9%. |
| | Shaft bearing worn or failed. | Replace bearing. |
| | Worn or unsynchronized timing gear. | Replace timing gear. |
| | Rotors contacting each other. | Replace rotors. |

TABLE 8.5 Troubleshooting guide for rotary lobe pumps (*continued*).

| Problems | Causes | Solutions |
|---|---|---|
| Overheating | Overload. | Reduce load or replace unit with one having sufficient capacity. |
| | Incorrect oil level. | Fill or drain to specified oil level. |
| | Vent plug not installed or is restricted. | Install vent plug provided or remove and clean in solvent. |
| | Excessive or insufficient bearing clearance. | Adjust tapered roller bearing to provide proper axial clearance. |
| | Incorrect grade of oil. | Drain unit and refill with correct grade of oil. |
| Bearing failure | Overload. | Reduce load or replace unit with one having sufficient capacity. |
| | Overhung load rating exceeded. | Reduce overhung load or replace unit with one having sufficient capacity. |
| | Excessive or insufficient bearing clearance. | Adjust tapered roller bearing to provide proper axial clearance. |
| | Insufficient lubrication. | Check oil level; fill if necessary. |
| Shaft failure | Improper coupling alignment. | Readjust as required. |
| | Overhung load rating exceeded. | Reduce overhung load or replace unit. |
| | High, repetitive shockloads. | Provide coupling with ability to absorb shock or replace unit with one having adequate service factor. |
| Gear wear or failure | Overload. | Reduce load or replace unit with one having sufficient capacity. |
| | High, repetitive shockloads. | Provide shock absorbing coupling or replace unit. |
| | Insufficient lubrication. | Check oil level and fill. |
| | Excessive bearing clearance. | Adjust bearing clearance. |
| | Incorrect grade of oil. | Drain unit and refill with new oil as specified. |
| Oil leaks | Incorrect oil level. | Fill or drain to specified oil level. |
| | Vent plug not installed or is restricted. | Install vent plug provided or remove and clean in solvent. |
| | Worn oil seals. | Check and replace as required. |

TABLE 8.5 Troubleshooting guide for rotary lobe pumps (*continued*).

| Problems | Causes | Solutions |
|---|---|---|
| | Loose pipe plugs, backstop cover, or input bracket. | Check and tighten as required. If problem still exists, remove clean mating surfaces, apply new sealant, and reassemble. |
| Noise | Loose mounting or coupling. | Check and tighten as required. |
| | Worn bearings. | Reduce load or replace with one having sufficient capacity. Reduce overhung load or replace unit with one having sufficient capacity. Adjust tapered roller bearing to provide proper axial clearance. Check oil; fill if necessary. |
| | Worn or damaged gears. | Reduce load or replace unit with one having sufficient capacity. Drain unit and refill with new oil. |
| Belt slip (sidewallks glazed) | Not enough tension. | Replace belts; apply proper tension. |
| Drive squeals | Shock load. | Apply proper tension. |
| | Not enough arc of contact. | Increase center distance. |
| | Heavy starting load. | Increase tension. |
| Belt turned over | Broken cord, caused by prying on sheave. | Replace set of belts correctly. |
| | Overloaded drive. | Redesign drive. |
| | Impulse loads. | Apply proper tension. |
| | Misalignment of sheave and shaft. | Realign drive. |
| | Worn sheave grooves. | Replace sheaves. |
| | Flat idler sheave. | Align idler. Reposition on slack side of drive close to drive sheave. |
| | Excessive belt vibration. | Check drive design. Check equipment for solid mounting. Consider use of banded belts. |
| Mismatched belts | New belts installed with old belts. | Replace belts in matched set only. |
| | Sheave groove worn unevenly; improper groove angle. Gives appearance of mismatched belts. | Replace sheaves. |

**TABLE 8.5** Troubleshooting guide for rotary lobe pumps (*continued*).

| Problems | Causes | Solutions |
|---|---|---|
| | Sheave shafts not parallel. Give appearance of mismatched belts. | Align drive. |
| Belt breaks | Shock loads | Apply proper tension; recheck drive. |
| | Heavy starting loads. | Apply proper tension; recheck drive. Use compensator starting. |
| | Belt pried over sheaves. | Replace set of belts correctly. |
| | Foreign objects in drive. | Provide drive shroud. |
| Belts wear rapidly | Sheave grooves worn. | Replace sheaves. |
| | Sheave diameter too small. | Redesign drive. |
| | Mismatched belts. | Replace with matched belts. |
| | Drive overloaded. | Redesign drive. |
| | Belt slips. | Increase tension. |
| | Sheaves misaligned. | Align sheaves. |
| | Oil or heat condition. | Eliminate oil. Ventilate drive. |

centrifugal pump, the influent should be dilute unthickened solids; however, it can break up large trash particles. The unit typically can handle flows ranging from about 1.5 to 20 L/s (25 to 300 gpm).

***Comminuting and Grinding.*** A comminuting and grinding device (Figure 8.31) is not really a pump because its primary purpose is shredding solids. It only produces enough head to force the solids through the grinder. The device uses a high-energy impact blade to shear and shred solids before they flow through the slotted openings; depending on the application, capacity, and grinding configuration, it can reduce particle sizes to between 6 and 10 mm (0.25 and 0.38 in.). The device typically is used to comminute previously thickened solids, scum, and screenings to protect downstream dewatering devices from clogging. To reduce the pressure on its seals, the device typically is installed on the suction side of a solids pump. It requires intensive and thorough maintenance.

**METERING.** Similar to positive-displacement reciprocating pumps, metering pumps are designed to provide controlled discharges. The discharges can be controlled via the

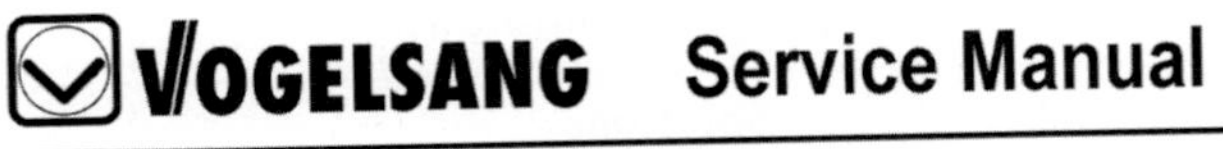


Service Manual
Pump Models:
Series Q Pumps

**Troubleshooting**

| No Discharge | Below Capacity | Irregular Discharge | Lost Prime | Pump Stalls at Startup | Pump Overheats | Motor Overheats | Excessive Power Draw | Noise & Vibration | Pump Element Water | Seizure | Causes | Solution |
|---|---|---|---|---|---|---|---|---|---|---|---|---|
| X | | | | | | | | | | | Incorrect direction of rotation | Reverse Motor |
| X | | | | | | | | | | | Pump not primed | Expel gas from supply line and pumping chamber to introduce liquid. |
| X | X | X | X | | | | | X | | | Insufficient NPSH available | Increase supply line diameter. Increase suction head. Simplify piping and/or reduce length. Reduce pump speed. Decrease liquid temperature. Check effect of increased viscosity on available and permitted power inputs. |
| | X | X | X | | | | | X | | | Product vaporizing in supply line. | |
| X | X | X | X | | | | | X | | | Air entering supply line. | Remake pipe work |
| | X | X | X | | | | | X | | | Gas Supply | Expel gas from supply line and pumping chamber to introduce liquid. |
| X | X | X | X | | | | | X | | | Insufficient head above supply vessel outlet | False product level. Lower outlet position. Increase submergence of supply line. |
| | | X | X | | | | | X | | | Foot valve strainer obstructed or blocked. | Service Fittings. |
| | X | | X | X | X | X | X | X | | | Liquid viscosity above rated figure. | Decrease pump speed. Increase liquid temp. |
| | X | | | | | | | | | | Liquid viscosity below rated figure. | Increase pump speed. Decrease liquid temp. |

FIGURE 8.29 Rotary lobe troubleshooting guide (Courtesy of Vogelsang USA, Inc.).

Service Manual
Pump Models:
Series Q Pumps

| No Discharge | Below Capacity | Irregular Discharge | Lost Prime | Pump Stalls at Startup | Pump Overheats | Motor Overheats | Excessive Power Draw | Noise & Vibration | Pump Element Water | Seizure | Causes | Solution |
|---|---|---|---|---|---|---|---|---|---|---|---|---|
| | | | | | X | | | X | X | X | Liquid temp. above rated figure. | Cool the liquid prior to pumping |
| | | | | X | | X | X | | | | Liquid temp. below rated figure. | Heat the liquid prior to pumping |
| | | | | | | | | X | X | X | Unexpected solids in product. | Clean the system. Add strainer to supply line. |
| | X | X | X | X | X | X | X | X | X | X | Delivery pressure above rated figure. | Check for obstruction. Simplify supply line. |
| | | | | | | | | | | | Seal flushing inadequate | Check that fluid flows freely into seal. Increase flow rate. |
| | X | | | | | X | X | X | | | Pump speed above rated figure. | Decrease pump speed. |
| | | | | | | | | | | | Pump speed below rated figure. | Increase pump speed. |
| | X | | | | X | X | X | X | X | X | Rotor case strained by pipe work. | Align pipes. Add flexible pipes or expansion fittings. |
| | | | | | X | X | X | X | | | Belt Drive Slipping | Tighten belt to specified tension. |
| | | | | | | | | X | | | Flexible coupling misaligned. | Adjust flange alignment. |
| | | | | | X | X | X | X | X | X | Insecure pump drive mountings | Add lock washers and retighten. |
| | | | | X | X | X | X | X | X | X | Shaft bearing wear or failure. | Call Vogelsang or distributor for advice. |
| | | | | | X | X | X | X | | X | Worn unsynchronized timing gears | Call Vogelsang or distributor for advice. |
| | | | | X | X | X | X | X | X | X | Gear case oil level incorrect. | Refer to oil level instructions. |

FIGURE 8.29 *(continued)*.

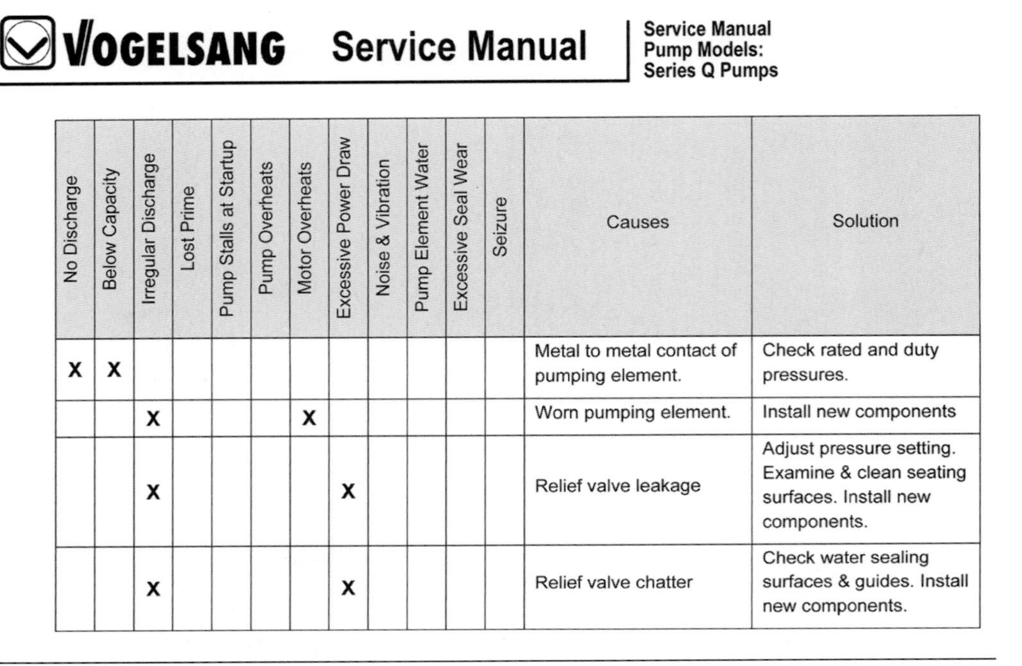


| No Discharge | Below Capacity | Irregular Discharge | Lost Prime | Pump Stalls at Startup | Pump Overheats | Motor Overheats | Excessive Power Draw | Noise & Vibration | Pump Element Water | Excessive Seal Wear | Seizure | Causes | Solution |
|---|---|---|---|---|---|---|---|---|---|---|---|---|---|
| X | X | | | | | | | | | | | Metal to metal contact of pumping element. | Check rated and duty pressures. |
| | | X | | | | X | | | | | | Worn pumping element. | Install new components |
| | | X | | | | | X | | | | | Relief valve leakage | Adjust pressure setting. Examine & clean seating surfaces. Install new components. |
| | | X | | | | | X | | | | | Relief valve chatter | Check water sealing surfaces & guides. Install new components. |

**FIGURE 8.29** (*continued*).

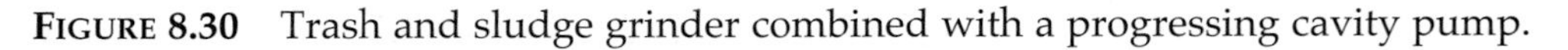

**FIGURE 8.30** Trash and sludge grinder combined with a progressing cavity pump.

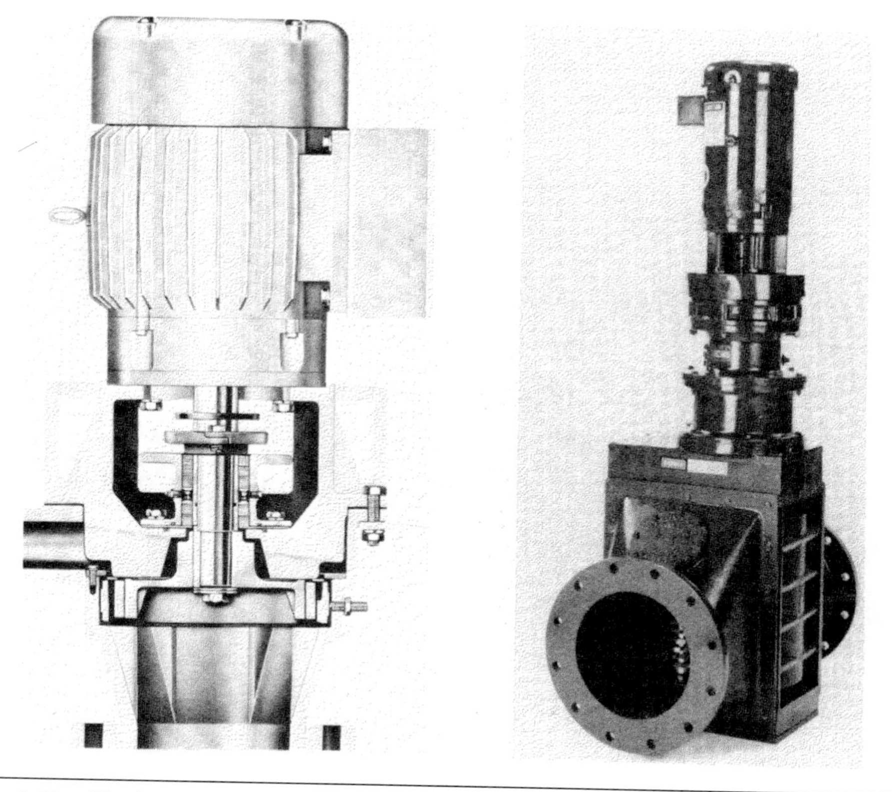

FIGURE 8.31 Sludge grinders.

stroke speed or volume displaced per stroke. Wastewater utilities typically use three types of metering pumps: packed plunger, mechanically actuated diaphragm, and hydraulically actuated diaphragm.

Overall, metering pumps' O&M needs are similar to those of positive-displacement pumps. But they also have suction and discharge check valves, which should be inspected and maintained about once every 6 months to ensure proper discharge metering.

**PNEUMATIC EJECTORS.** Pneumatic ejectors (Figure 8.32) have been used for years in collection systems—primarily in residential areas with low flows—but they required intensive maintenance, so many utilities replaced them with submersible pumping systems. However, the devices are excellent for conveying scum and floatables in wastewater treatment plants. The Anthony Ragnone Treatment Plant in Genesee County, Michigan, has used one to convey scum for 30 years.

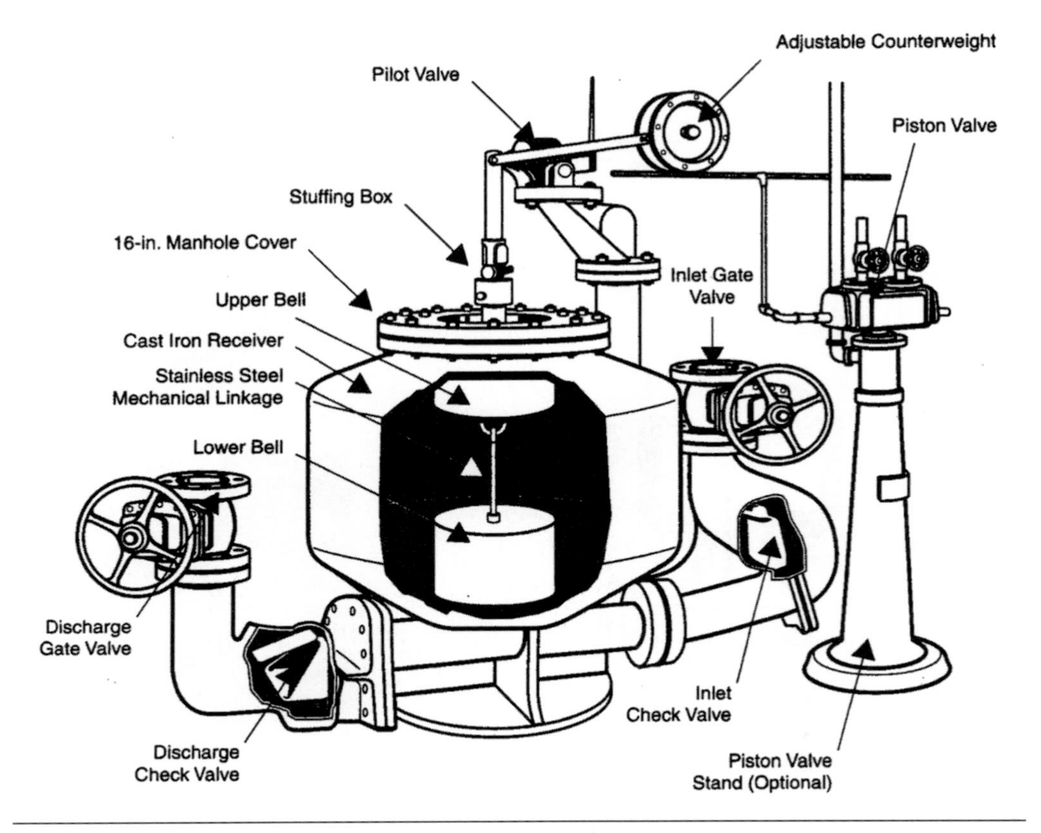


FIGURE 8.32 Pneumatic ejector (in × 25.4 = mm).

***Preliminary Startup.*** Before starting a pneumatic ejector, operators should do the following:

1. Check, clean, and oil all external pivot points around upper bell rods, linkages, and level arms;
2. Confirm that all air, inlet, and discharge valves are still closed;
3. Open the gate valve on the air exhaust line at the piston valves; and
4. Open the inlet valve and half fill the injector pot.

If the counterweight tends to trip when the fluid enters, hold it up to allow the lower bell to fill with fluid. Then, close the inlet gate valve and balance the counterweight against the weight of the upper and lower bells. (This is the approximate location for the counterweight when starting the ejector. Minor counterweight alterations may be necessary once the ejector is in full operation.)

***Startup.*** To start up a pneumatic ejector, operators should do the following:

1. Allow compressors to build pressure to the required amount (the amount will depend on the head needed to convey fluid through the particular pumping system) and leave the controls in the "automatic" position.
2. Open the valve on the air pressure line.
3. Open the valve on the air exhaust line.
4. Open the valve on the discharge line.
5. Open the valve on the inlet line.
6. Close the inlet valve.
7. Bring the ejector up to full operation.

***Shutdown.*** To shut down a pneumatic ejector, operators should do the following:

1. Close the inlet valve.
2. Hold down the counterweight to empty the injector receiver.
3. Close the discharge valve.
4. Close the air exhaust valve.
5. Close the air inlet valve.

Note that the valve in the air pressure line is the first to be opened for start-up and the last to be closed for shutdown.

***Testing Procedures.*** There are several tests for pneumatic ejectors.

*Inlet Check Valve Test.* To test whether the inlet check valve is properly seated, close the discharge valve and allow the injector to fill and the air pressure to be applied as usual. If the ejector receiver empties—indicated by falling bells and rising counterweight—then the inlet check valve is malfunctioning.

*Discharge Check Valve Test.* To test whether the discharge check valve is properly seated, close the inlet valves, pull down the counterweight, and allow the ejector to empty. If it fills again—indicated by rising bells—then the discharge check valve is malfunctioning.

This test only applies when the discharge pipe can hold enough fluid to fill the ejector when it backflows. Otherwise, the ejector must be half or three-quarters full before the inlet valve is closed so the discharge pipe's content will be sufficient to fill the ejector.

*Piston Valve Test.* To test whether the piston valve is functioning properly, remove the heads and note whether the pistons move freely in both directions. (Do this only if the ejector receiver cannot be discharged and exhausted.)

*Slide or Pilot Valve Test.* To test whether the slide or pilot valve is functioning properly, close the inlet valve, move the piston valve's piston toward the exhaust end (closing the main pressure port), and turn on a little air. The 6-mm (0.25-in.) ports in one end of the piston valve should emit a blast of air. Then, move the piston toward the inlet end; the first blast should shut off, and the port in the other end of the piston valve should emit a blast of air. These blasts should alternate as the pilot valve lever is moved back and forth. (A little air may blow past the main piston on the pressure end because, with the heads removed, the piston retains less pressure than it would during actual operations.)

If the discharge pipe or pump is empty, the air may blow straight through the ejector into the discharge pipe. If the piston valves are positioned so the main pressure ports are open, the pressure would be insufficient to reverse the valve and close these ports, no matter which way the lever is moved. If this happens, leave the lever with the bell end down (the proper position if the ejector is empty and allows air to blow through). Quickly shut off the globe valve on the air pressure pipe. This typically reverses the piston valve. If it does not, close the inlet and discharge gate valves and allow air pressure to build up in the ejector. The valve will then reverse, cutting off the air. Open all valves, and repeat this process several times.

*Upper Gear Assembly Cleaning.* The ejector receiver should be blown out manually at least once a week to dispose of any accumulated scum, floatables, and solids. The easiest method is to hold down the counterweight for 2 minutes, clearing out all foreign matter from the ejector receiver.

If the counterweight sticks, remove the cotter pin from the upper bell rod and turn the nut to within one thread of the top. Then, tighten the lower nut against the trunnion. Doing this will postpone the need to completely disassemble the gear for about 6 to 8 months, depending on existing conditions.

When the counterweight sticks again, remove the gear assembly, clean the upper bell and the area around the receiver's throat, and reassemble.

# PUMPING STATIONS

In this manual, a *pumping station* includes all the equipment, appurtenances, and structures needed to move a given fluid from one location to another (excluding the force main after it leaves the pump structure). Proper O&M of pumping systems should address all of these components, not just the pumps themselves.

Pumping stations typically include at least two pumps and a basic wet-well level control system. One pump is considered a "standby" pump, although the controls typically cycle back and forth during normal flows so they receive equal wear.

**TYPES.** Pumping stations can be configured in a wide variety of arrangements, depending on size and application. In this manual, pumping stations are classified based on whether the pumps and appurtenances are installed above, next to, or within the fluid being pumped. The classifications for such pumping-station configurations are: wet pit/dry pit, wet pit only/submersible pumps, and wet pit only/nonsubmersible pumps.

***Wet Pit/Dry Pit.*** In this configuration, two pits (wells) are required: one to hold the fluid, and one to house the pumps and appurtenances. Required for fluids that cannot be primed or conveyed long distances in suction piping, this option is typically used to pump large volumes of raw wastewater, where uninterrupted flow is critical and wastewater solids could clog suction piping. It also is used to pump solids in pipe galleries between digesters or other solids-handling equipment. While construction costs may be higher and a heating, ventilation, and cooling (HVAC) system is necessary when installed below grade, this configuration is best for O&M activities because operators can see and touch the equipment.

***Wet Pit Only/Submersible Pumps.*** In this configuration, one pit (well) holds both the pumps and the fluid being pumped. The pump impeller is submerged or nearly submerged in the fluid; no additional piping is required to convey the fluid to the impeller. This option is common worldwide, and the submersible centrifugal pumps involved can be installed and operated cost-effectively (Figures 8.33 and 8.34).

***Wet Pit Only/Nonsubmersible Pumps.*** In this configuration, one pit (well) holds the fluid. The pumps are installed above the fluid, potentially at or above grade to prevent groundwater infiltration or other flooding. This option is used in areas where the fluid can be "pulled" through suction piping (e.g., treated or finished water) or where shutdowns or failures would not be immediately critical (e.g., a package plant's raw wastewater lift stations, equalization of secondary treated wastewater, and a utility's water systems).

**PIPING AND APPURTENANCE REQUIREMENTS.** The piping and appurtenance requirements for pumping stations typically include:

- Isolation valves for each pump;
- Check valves on the discharge side of each pump;

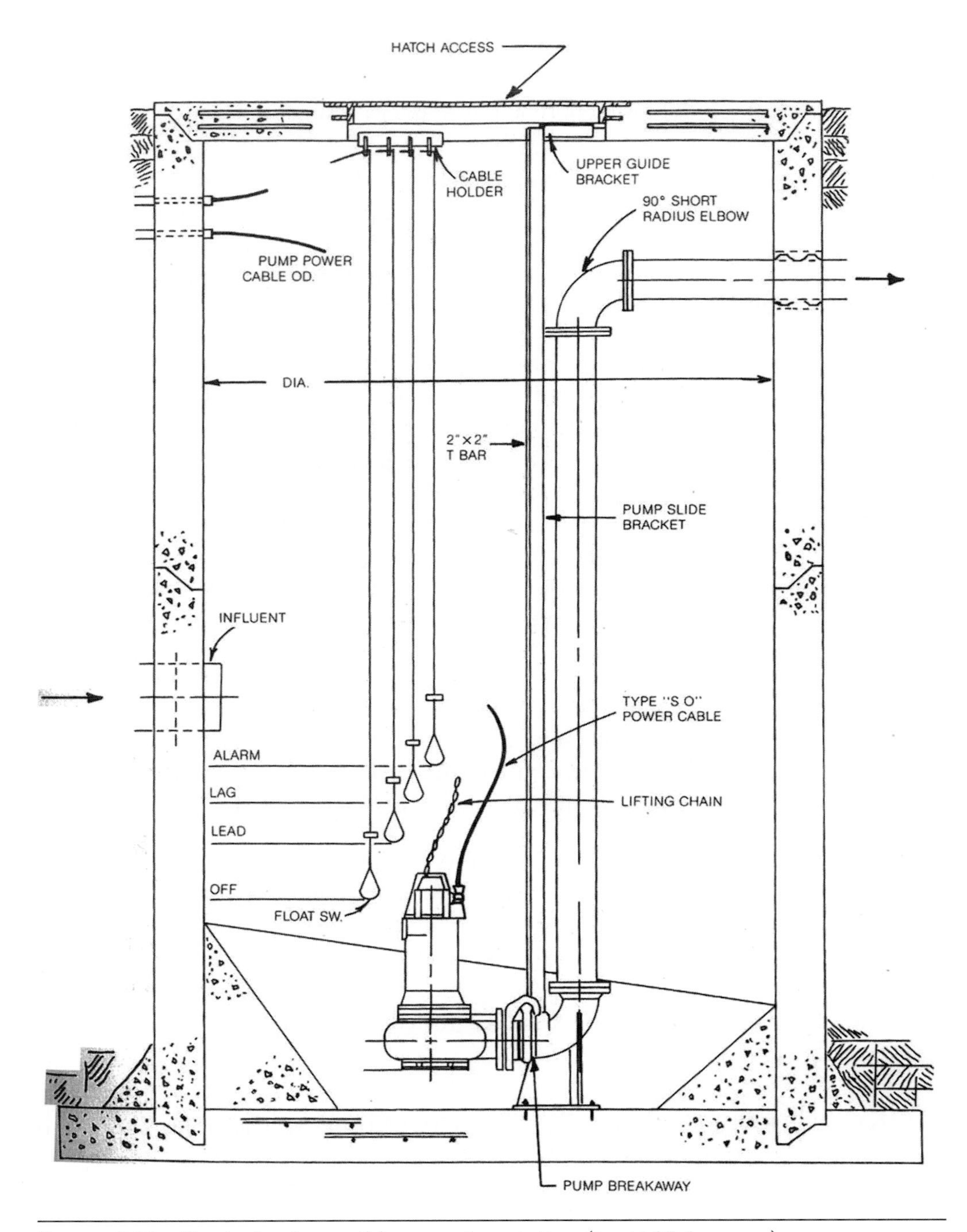


**FIGURE 8.33** Typical submersible pumping station (in. × 25.4 = mm).

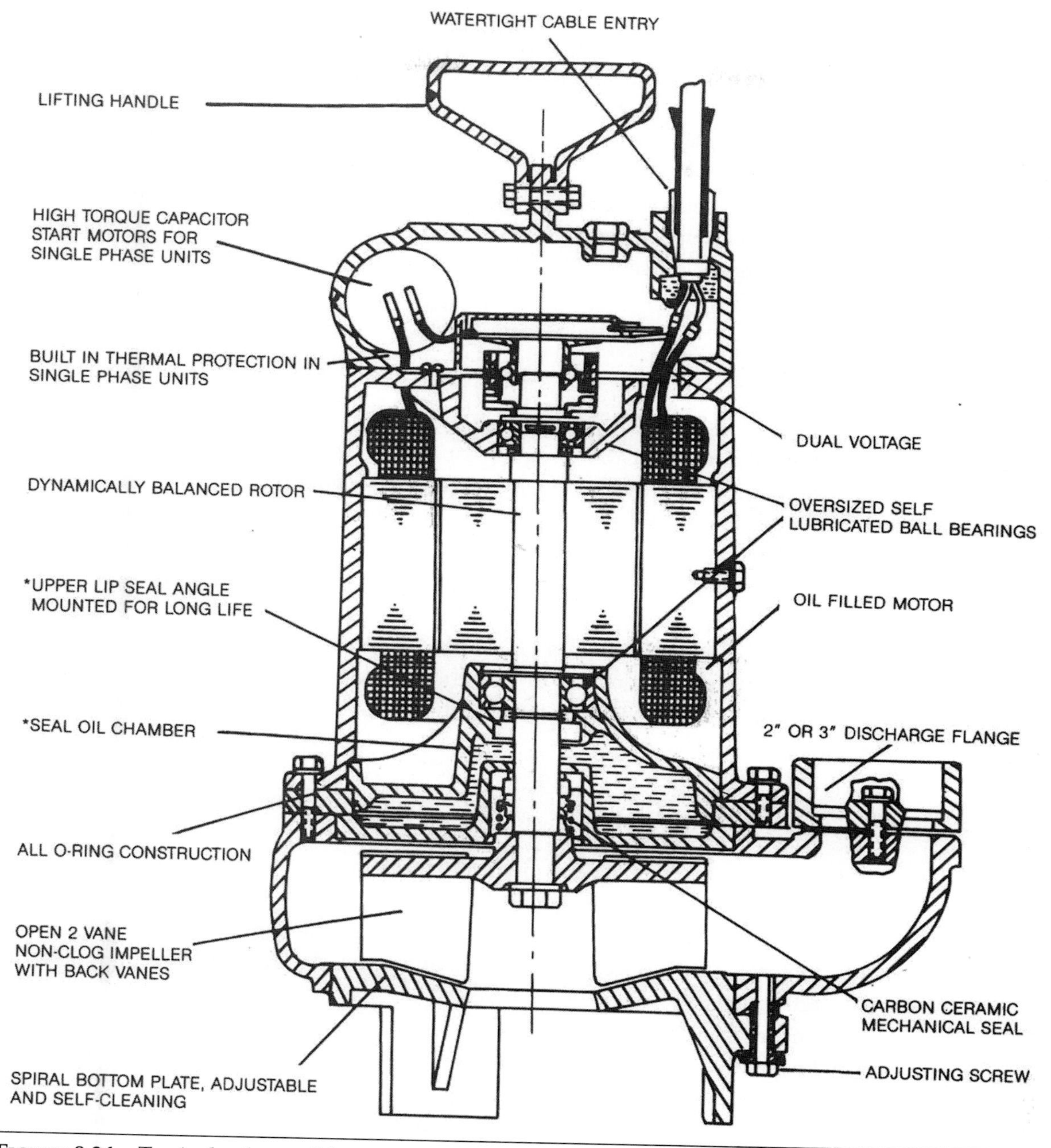


**FIGURE 8.34** Typical submersible pump (in. × 25.4 = mm).

- Expansion joints where the pump connects to suction or discharge piping (to limit external loads on the pumps);
- Flow meter on the force main (to measure the pumping station's total pumped flow);
- Air-release valves (typically installed at high points along a pipeline to relieve high pressures);
- Vacuum-relief valves, which allow air to enter when the pipeline drains and prevent extremely low pressures from developing (used in areas where low pressures can cause the pipeline to collapse);
- Drains;
- Flushing system;
- Monitors; and
- Controls.

As part of a normal O&M program, operators should inspect each appurtenance to ensure that it is operating properly. "Frozen" isolation valves, for example, make pump removal more difficult. Malfunctioning check valves can increase operating costs by either impeding more flow or else leaking flow back to the wet well, thereby increasing pumping cycles.

**PUMPING-STATION MAINTENANCE.** A well-planned maintenance program for pumping systems can reduce or prevent unnecessary equipment wear and downtime. (The following maintenance information applies to both wastewater and solids pumping systems.)

***Basic Maintenance.*** The following is a basic pumping-station maintenance checklist:

- Check the wet-well level periodically (preferably once a day, or more frequently when high flows are expected).
- Record each pump's "run time" hours (as indicated on the elapsed-time meters) at least once in that period and check that the pumps' running hours are equal.
- Ensure that the control-panel switches are in their proper positions.
- Ensure that the valves are in their proper positions.
- Check for unusual pump noises.
- At least once a week, manually pump down the wet well to check for and remove debris that may clog the pumps. [Follow the proper safety precautions (Chapter 5).]

- Inspect the float balls and cables and remove all debris to ensure that they operate properly. Untangle twisted cables that may affect automatic operations.
- If a lift-station pump is removed from service, adjust the lead pump selector switch to the number that corresponds to the pump remaining in operation. (This allows the lead pump levels to govern the operating pump's starts and stops.)
- Periodically inspect the pump-sequencing and alarm control functions as follows:
  1. Adjust each pump's H/O/A selector switch to "off" (O).
  2. Fill the wet well until the "high level" alarm is activated.
  3. Adjust each pump's H/O/A selector switch to "auto" (A). The pumps should start up sequentially. (If they start up simultaneously, insert short time delays between startups in the control algorithm.)
  4. Allow all three pumps to operate automatically. When the fluid level drops to the "low level" setpoint, shut off all the pumps.
  5. Fill the wet well until the lead pump starts, then shut off the fluid. Allow the pump to lower the fluid level until it shuts off.
  6. Again, fill the wet well until the second pump starts, then shut off the fluid. Allow the pump to lower the fluid level until it shuts off.
  7. Repeat this procedure for all pumps.
  8. Double-check that all three pumps' H/O/A selector switches are set to "auto" (A).

***Startup.*** To start up a pumping station, operators should:

1. Turn off the local controls;
2. Turn on the main motor control center breaker;
3. Turn on the local controls for the main, all pumps, and the transformer;
4. Fully open each pump-discharge valve in the valve vault;
5. Fully close the pumpout bypass connection (if any);
6. Adjust the H/O/A selector switch or each pump to "hands" (H) and monitor pump operations for a short period (approximately 10 minutes, but make sure that the wet well does not pump empty);
7. Adjust the H/O/A selector switch to "auto" (A);
8. Repeat steps 6 and 7 for each pump; and
9. Adjust the lead selector switch to the desired position (1/auto/2). (Typically, the position would be "auto".)

To check for proper pump sequencing and level alarm functions, fill the wet well with fluid and monitor the system operations.

***Shutdown.*** To shut down a pump, operators should:

1. Adjust its H/O/A selector switch to "off" (O);
2. Adjust the lead selector switch to the number corresponding to the pump(s) remaining in operation;
3. Place the pump's local disconnect switch to "off" (O);
4. Close its discharge valve; and
5. If repairs are necessary, have maintenance personnel remove the pump from the wet well.

**PIPING AND APPURTENANCE MAINTENANCE.** Properly maintaining pumping-station pipelines, valves, and other appurtenances can minimize pump loads. Excessive head losses on either the suction or discharge side of a pump can increase energy use and the wear rate and, therefore, the O&M costs. Excessive head losses also may lead to process or treatment problems because solids move slower, so the proper solids balance is not maintained. Operators can monitor head losses by routinely checking the pressure gauges on both sides of the pumps.

***Piping.*** When operators notice excessive head losses (indicated by a pressure drop on the suction side of the pump or an increase in pressure on the discharge side), they should determine whether the losses are a result of partial clogging, a restriction somewhere in the line, or materials built up on the pipe wall. To find clogs, operators should start by checking the pressure at various points in the suction and discharge piping, looking for spots with abrupt head loss (e.g., valves or other constrictions). If something is caught in a valve or other appurtenance, backflushing with high-pressure water may remove it. Solids pumping systems typically have flushing systems to help with the O&M of pumps, appurtenances, and pipelines.

If backflushing fails, operators may need to dismantle the piping at the clog site. If this spot clogs repeatedly, it may be prudent to replace the valve or other appurtenance with one less prone to obstructions. (Manufacturers or design engineers typically can recommend a suitable replacement.)

The gradual buildup of scum, grease, or other materials in pipelines can be difficult to dislodge mechanically or chemically. If such buildups are persistent, operators should develop a routine procedure for anticipating and removing them before they disrupt operations. Pipes lined with glass or polytetrafluoroethylene are less prone to buildups, but most plants do not uses such piping. (If the plant does have glass- or polytetrafluoroethylene-lined pipe, operators should be careful not to damage the lining when mechanically removing deposits.)

Scum deposits can be removed from pipes via chemical washing or mechanical scraping. Chemical solutions (e.g., alkaline detergent mixtures) should be pumped into the pipeline and allowed to dissolve scum or grease over hours (or days, if necessary). They typically work best when hot. Afterward, depending on the chemical and downstream process involved, the chemical may have to be drained from the line before the line can be returned to service. (To determine the appropriate handling and disposal requirements, operators should ask suppliers or review the chemical's material safety data sheet.)

Mechanically scraping pipes is a last resort. It typically involves removing the pump from service and draining the piping.

Scum buildup problems typically are addressed via source control (e.g., installing grease traps in the collection system at locations suspected or known to generate grease—restaurants, etc.).

[Note: If the utility's pipes are not color coded, operators should establish a color-code system to simplify operator training and reduce the possibility of discharging solids, wastewater, or other substances into the wrong process unit. They also should stencil directional arrows on the pipes and note what the contents are and where they are going (e.g., "Raw Primary Sludge to Thickener No. 1").]

***Valves.*** Valves should be lubricated regularly (per the manufacturer's instructions), and the valve stems should be rotated regularly to ensure ease of operation. These activities should be part of a regular pump-maintenance program.

A solids pump's flushing system should include a connection on the upstream side of the valve (or gate plunger in diaphragm pumps) so high-pressure water is available to wash out debris caught in the ball checks. Even if flushing is unsuccessful, it will clean solids from the pump before the access ports are opened.

## REFERENCES

Brater, E. F.; King, H. W.; Lindell, J. E.; Wei, C. Y. (1996) *Handbook of Hydraulics*, 7th ed.; McGraw-Hill: New York.

Canapathy, V. (1982) Centrifugal Pump Suction Head. *Plant Eng.*, **89**, 2.

Cunningham, E. R. (Ed.) (1982) Fluid Handling Pumps. *Plant Eng.*, May.

Karassik, I. J. (1982) Centrifugal Pumps and System Hydraulics. *Chem. Eng.*, **84**, 11.

Karassik, I. J.; Krutzsch, W.; Messina, J. (Eds.) (1976) *Pump Handbook*; McGraw-Hill: New York.

Krebs, J. R. (1990) *Wastewater Pumping Systems*; Lewis Publishers, Inc.: Chelsea, Michigan.

Steel, E. W.; McGhee, T. J. (1979) *Water Supply and Sewerage*, 5th ed.; McGraw-Hill: New York.

Street, R. L.; Watters, G. Z.; Vennard, J. K. (1995) *Elementary Fluid Mechanics*, 7th ed.; Wiley & Sons: New York.

U.S. Environmental Protection Agency (1979) *Process Design Manual for Sludge Treatment and Disposal*, EPA-625/1-79-011; U.S. Environmental Protection Agency: Washington, D.C.

# Chapter 9

# Chemical Storage, Handling, and Feeding

# INTRODUCTION

This chapter focuses on the operation and maintenance of equipment used in the handling, storage, and feeding of chemicals for wastewater treatment. Chapter 24, Physical–Chemical Treatment, describes some specific uses of chemicals to improve wastewater treatment. Chemical addition must be evaluated for each specific treatment process while considering its potential effect on downstream treatment units and effluent toxicity. The procedures for choosing chemicals and chemical dosages are discussed under specific treatment processes in other chapters of this manual, in *Design of Municipal Wastewater Treatment Plants* (WEF and ASCE, 1998) and in *Control of Odors and Emissions from Wastewater Treatment Plants* (WEF, 2004).

# CHEMICAL APPLICATION SYSTEMS

All chemicals, whether solid, liquid, or gas, require a feeding system to accurately and repeatedly control the amount applied. Effective use of chemicals depends on accurate dosages and proper mixing. The effectiveness of certain chemicals is more sensitive to dosage rates and mixing than that of others. The design of a chemical feed system must consider the physical and chemical characteristics of each chemical used for feeding, minimum and maximum ambient or room temperatures, minimum and maximum wastewater flows, minimum and maximum anticipated dosages required, and the reliability of the feeding devices. Chemical feed systems typically consist of transferring the chemical from the supplier to the plant storage area, storing the chemical, mixing the chemical with water (occasionally not done), and calibrating the chemical/solution feed rate.

The capacity of the system, potential delivery delays, and chemical use rates are important considerations in both storage and feeding. Storage capacity must take into account the economical advantages of bulk quantity purchases versus the disadvantages of construction cost, spill potential, and chemical degradation with time. Smaller shipments generally result in higher unit chemical costs, increased transportation costs, and greater handling (labor) costs at the treatment plant facility. However, bulk storage and feeding facilities typically require more equipment and/or larger equipment, resulting in higher costs associated with construction and operation of the facility. Storage tanks or bins for solid chemicals must be designed to allow the correct angle of repose of the chemical and to provide its necessary environmental requirements, such as temperature and humidity. Size and slope of feeding lines are important considerations. The selection of materials used for construction of storage tanks, feed equipment, pumps, piping, and valves are also extremely important because many chemicals are corrosive to many materials.

Chemical feeders are sized to meet the minimum and maximum feeding rates required. Manually controlled feeders typically have a common feed turndown range of 10:1 that can be increased to approximately 20:1 with dual-control systems. To provide operational flexibility, the operator should consider future design conditions when selecting the appropriate ranges for chemical feed rates. A common problem is not being able to adjust the pump to low enough feed rates. Chemical feeder control can be manual, automatically proportioned to flow, or dependent on a process parameter (such as pH). A combination of any two of these can also be used. If manual control systems are specified with the possibility of future automation, the feeders selected should be convertible with a minimum of expense. Standby or backup units are generally desirable but may not be necessary for each type of feeder used unless required by the regulating authority. The location and sizing of the chemical feed addition should provide sufficient operational flexibility for potential changes in wastewater quality. Designed flexibility in hoppers, tanks, chemical feeders, and solution lines is the key to maximum benefits at the least cost.

Because dry chemical characteristics vary considerably, the feeder must be selected carefully, particularly in a smaller-sized facility where a single feeder may be used for more than one chemical. Overall, the operator should make provisions to keep all dry chemicals cool and dry. Maintaining a low humidity is particularly important, as hygroscopic (water absorbing) chemicals may become lumpy, viscous, or even rock hard. Other chemicals that absorb water less readily become sticky from moisture on the particulate surfaces, causing increased bridging in hoppers. Typically, only limited quantities of chemical solutions should be made from dry chemicals at any one time because the shelf life of the diluted chemical, especially polymers, may be short. The operator should consult the chemical supplier about the recommended shelf life for each particular chemical and/or chemical solution.

The operator should keep dry chemical handling areas and equipment as dry as possible. Otherwise, moisture will affect the chemicals' density and may result in underfeed. Also, the effectiveness of dry chemicals, particularly polymers, may be reduced. Dust-removal equipment should be used at shoveling locations, bag dump stations, bucket elevators, hoppers, and feeders for neatness, corrosion prevention, and safety reasons. Collected chemical dust may often be used with stored chemicals.

## SAFETY CONSIDERATIONS

Many chemicals used in wastewater treatment plants can be extremely hazardous when not stored, handled, or used properly. Plant operators should always check with state regulatory and local emergency management agencies for reporting and plan development requirements. Federal regulations require that a material safety data sheet

(MSDS) be available to employees. The MSDS provides information on the chemical, including hazards and safety procedures to be used when handling it. The MSDS should be kept in the area where the chemical is used and also on file in a central location at the plant. Operators must use the correct safety equipment and follow the safe handling procedures described in the MSDS. This would include using personal protective equipment such as face shields, eyewear, and protective clothing. Further, the plant should be adequately equipped with emergency eyewashes and showers, dust suppression equipment (for dry chemical systems), and secondary containment structures (for liquid chemical or solution systems).

Table 9.1 presents a summary of the physical properties and the principal safety considerations for handling chemicals typically used in a wastewater treatment plant. Table 9.2 summarizes uses, feed-system types, operating considerations, and other precautions for those same chemicals. For complete instructions, the operator should consult the MSDS for a specific chemical.

# RISK MANAGEMENT

Some gaseous chemicals such as chlorine, sulfur dioxide, or ammonia are hazardous if released to the atmosphere. Compliance with all federal, state, and local regulations is required for the storage of certain chemicals and reporting procedures involved with spills, leaks, and disposal of chemicals. The U.S. Environmental Protection Agency's (U.S. EPA's) Risk Management Program (RMP) rule is specifically concerned with the accidental release of hazardous compounds. The rule was required as part of the Clean Air Act amendments of 1990 (Section 112r) (Prevention, 2002) and requires facilities to identify hazards and manage risks. Both the U.S. EPA and the Occupational Safety and Health Administration (OSHA) have similar regulations for the prevention of hazardous chemical release. The U.S. EPA's RMP generally addresses the protection of public health and the environment, whereas OSHA's Process Safety Management standard is intended for employee protection from chemicals in the workplace (29 CFR 1910.119) (Occupational, 2003).

U.S. EPA's RMP applies to wastewater facilities with processes using chemicals stored above certain threshold quantities. A facility that handles, produces, or stores any of the chemicals indicated in Table 9.3 equal to or above the regulated threshold limits is likely subject to all provisions of the rule. For a complete list of regulated toxic substances and threshold quantities for accidental release prevention, consult 40 CFR 68.130 (Tables 1 through 4) (List, 2003).

If these thresholds are exceeded, the facility operator must document serious accidents (five-year history), analyze worst-case chemical releases, contact and coordinate

TABLE 9.1 Wastewater treatment chemicals—physical properties and safety considerations.

| Chemical name | Synonyms | Physical properties | Safety considerations |
|---|---|---|---|
| Ammonia | Anhydrous ammonia | Boiling point = –33.4 °C (–28.1 °F) highly soluble in water. LEL[a] = 15%, UEL[b] = 28%. Ammonia gas is lighter than air. Colorless gas with pungent, suffocating odor. | Contact with liquid causes severe burns. Extremely irritating to eye and lung tissue. Highly reactive with chlorine, acids. Odor is detectible at 5 ppm, irritating at 25 to 50 ppm. Moderate fire hazard. Ammonia in air and ammonia with chlorine are potential explosion hazards. |
| Chlorine | | Poisonous, reactive, greenish-yellow gas. Boiling point = –34 °C (–29 °F). Slightly soluble in water. Chlorine gas is heavier than air. | Chlorine reacts with moisture to form acids. Very irritating and corrosive to eyes, mucous membranes, and teeth. Acute respiratory distress and asphyxiation can result from exposure. Reacts with many materials to cause fires and explosions. |
| Chlorine dioxide | | Red-yellow or orange gas with a pungent odor. Oxidizer, bleaching agent. Boiling point = 10 °C (50 °F). | Irritates respiratory system. Avoid inhaling. Will damage eyes. Inhalation poison. Violently reacts with organic matter. Explosion hazard. Unstable in sunlight. |
| Defoamers | Antifoam agents | Many commercial products available. Properties vary. | Some antifoam agents may be corrosive or flammable. Avoid contact with eyes and skin. Avoid breathing vapors. |
| Ferric chloride | Iron trichloride, ferric trichloride, ferric perchloride | Available in solid and liquid solution; sp gr of solution, 1% = 1.0084; 45% = 1.487. Very soluble in water. Solutions above 33% will crystalize. | Corrosive liquid. Avoid contact with eyes and skin. Moderately toxic. Inhalation of mist will irritate throat and upper respiratory tract. Releases large amount of heat when anhydrous solid is diluted with water. |
| Ferric sulfate | Iron sulfate | Corrosive in liquid solution. | Avoid contact with eyes and skin. |

**TABLE 9.1** Wastewater treatment chemicals—physical properties and safety considerations (*continued*).

| Chemical name | Synonyms | Physical properties | Safety considerations |
|---|---|---|---|
| Hydrochloric acid | Muriatic acid, hydrogen chloride, chlorohydric acid | Normally available as a solution in concentrations of 31% and 35%; sp gr of solutions are 31% = 1.16 and 35% = 1.18. Highly volatile liquid. Gas vapor is heavier than air. | Highly corrosive liquid and gas. Avoid contact with skin, eyes, and especially respiratory systems. Hydrochloric acid is detectable at 0.1 to 5 ppm, irritating at 5 to 10 ppm. Highly reactive. Liberates potentially explosive hydrogen gas or chlorine gas at high temperatures or on contact with metals. Incompatible with concentrated sulfuric acid. |
| Hydrogen peroxide | Peroxide, hydrogen dioxide | Boiling point of pure liquid = 151 °C (304 °F). Very reactive and potentially unstable. Powerful oxidizer and corrosive. Available commercially in 30, 35, 50, and 70% solutions. Typically stored in concentrations of 50% or less; sp gr of solutions are 30% = 1.112 and 50% = 1.196. | Irritates skin, eyes, and mucous membranes. Avoid breathing mist. Highly reactive. Incompatible with most metals and organic material. Dangerous fire and explosive hazard in higher concentrations. Dust or metal contamination of solutions can result in violent decomposition and potential failure of storage tank. |
| Lime | Calcium hydroxide, hydrated lime, calcium oxide, quicklime | Solid, white powder with bitter taste. Slightly soluble in water. Dry form of quicklime has great affinity for water and liberates large amount of heat when mixed. | Irritating to skin and lungs. Dust problem can result from handling solid. Use dust mask and goggles. |
| Ozone | Triatomic oxygen | Colorless gas with a boiling point = –112 °C (–169 °F). | Oxidizing agent, irritant and toxic. Can irritate eyes, mucous membranes, and respiratory system. ACGIN TWA = 0.1 ppm in air. Odor detectible at 0.01 ppm. Incompatible with oils and other combustible materials. Can intensify fires. |

| | | | |
|---|---|---|---|
| Polymers | Anionic, cationic, non-ionic polymers, | Many available in dry and concentrated forms. | Some polymers are corrosive in water. Some are extremely viscous and slippery. Avoid contact with eyes and skin. Use caution when preparing solutions. |
| Potassium permanganate | Permanganate of potash | Strong oxidant with a characteristic purple color. Bulk density of between 90 and 100 lb/cu ft.[c] | Toxic. Suspected poison. Can react with organics, peroxides, and sulfuric acid. Can form chlorine when in contact with hydrochloric acid. Avoid contact with eyes and skin. Avoid breathing dust. Can react violently and explosively with organic materials. Contact with wood may cause a fire. |
| Sodium bisulfide | Sodium hydrogen sulfide, sodium hydrosulfide, sodium sulfhydrate | White to yellow flakes with hydrogen sulfide odor (rotten eggs). | Irritates eyes, skin, and mucous membranes. Avoid contact, especially with eyes. Contact with acids will generate toxic hydrogen sulfide gas. |
| Sodium hydroxide | Caustic, caustic soda, soda lye | Available in dry solid and solution. Melting point of 50% solution = 11.6 °C (53 °F); sp gr of 50% solution = 1.53. | Very corrosive to body tissues. Avoid inhaling dust or mist. Mixing with water can release large quantities of heat. Incompatible with acids and some metals such as tin, zinc, and especially aluminum. |
| Sodium hypochlorite | Chlorine bleach | Pale-yellow or greenish liquid solution with chlorine odor. Available in 5%, 10%, and 15% solutions. | Strong oxidizer and corrosive to eyes and mucous membrane tissues. Avoid contact with eyes and skin and avoid breathing fumes and mist. Decomposes to chlorine and sodium oxide when heated. Incompatible with acids, ammonia, organics, and some metals. Store out of direct sunlight. |

**TABLE 9.1** Wastewater treatment chemicals—physical properties and safety considerations (*continued*).

| Chemical name | Synonyms | Physical properties | Safety considerations |
|---|---|---|---|
| Sulfur dioxide | Sulfurous anhydride, sulfurous oxide | Colorless gas with strong suffocating odor. Boiling point = –10 °C (14 °F). Heavier than air. | Gas is irritating to mucous membranes and toxic. Inhalation poison. Odor detected at 0.5 pm. OSHA[d] TWA limit of 2 ppm. Avoid breathing. Will form an acid mist with water vapor. |
| Sulfuric acid | Hydrogen sulfate, vitriol, oil of vitriol | Colorless, clear, oily liquid. Available in several concentrations but typically used as 93% solution; sp gr of 93% solution is 1.834. | Highly corrosive. Can burn and char skin when exposed. Especially harmful to eyes. Releases a large amount of heat when diluted with water. Can react with organics, chlorates, permanganates, fuminates, or powdered metals, causing fires or explosions. |

[a]LEL = lower exposure limit.
[b]UEL = upper exposure limit.
[c]lb/cu ft × 16.02 = $kg/m^3$.
[d]OSHA = Occupational Safety and Health Administration.

**Table 9.2** Wastewater treatment chemicals—operating considerations.

| Chemical name | Uses | Feed system type | Operating considerations | Other comments |
|---|---|---|---|---|
| Ammonia | Nutrient addition, disinfection | Gas | Handle cylinders with care. Ventilate top of storage rooms. | Never use copper or brass in ammonia service. Iron or steel is satisfactory. |
| Chlorine | Disinfection, taste and odor control | Gas | Store chlorine containers in a cool, dry, well-ventilated area. Use care when handling chlorine containers. | Use rag dipped in ammonia water to detect leaks. Do not put water directly on a leak. |
| Chlorine dioxide | Disinfection | Gas | Typically generated on site. | |
| Defoamers | Controlling foaming in activated sludge systems or in ponds | Liquid | | |
| Ferric chloride | Sludge conditioning, coagulation, phosphorus removal | Solid or liquid | | Corrosive to most metals, especially aluminum, copper, and carbon steel, and to nylon. |
| Ferric sulfate | Coagulation, phosphorus removal | Solid | | |
| Hydrochloric acid | Neutralization | Liquid | | Iron or steel is not satisfactory in hydrochloric acid service. |
| Hydrogen peroxide | Odor control, supplemental dissolved oxygen, control of bulking | Liquid | Store in cool area. Keep away from combustible materials. | Iron and steel are unsatisfactory in hydrogen peroxide service. Avoid copper. Stainless steel, aluminum, and some plastics perform well. |
| Lime | Coagulation, pH adjustment, sludge conditioning, phosphate removal | Solid | Lime slakers produce large amounts of heat and must be watched carefully for excessive heat buildup. Absorbing water can cause caking or swelling. | |

**Table 9.2** Wastewater treatment chemicals—operating considerations (*continued*)..

| Chemical name | Uses | Feed system type | Operating considerations | Other comments |
|---|---|---|---|---|
| Ozone | Odor control, disinfection | Gas | Typically generated on site. | |
| Polymers | Coagulant, filter aids, sludge conditioning | Solid or liquid | | Iron or steel typically are not satisfactory. |
| Potassium permanganate | Taste and odor control, iron removal | Solid | Do not store in open containers. Avoid mixing with combustible materials. | Contact with wood can cause fire. |
| Sodium bisulfide | Dechlorination, chromium treatment, pH control, bactericide | Solid | Reacts with acids to form hydrogen sulfide gas, which is both flammable and toxic. | |
| Sodium hydroxide | pH control, odor control, cleaning | Liquid | Concentrated caustic freezes at high temperatures. Plugged or frozen caustic lines increase risk of exposure. | Do not use near aluminum. Galvanized piping is not suitable for caustic service. |
| Sodium hypo-chlorite | Disinfection, odor control | Liquid | Can decompose in storage. Keep away from light and heat. | |
| Sulfur dioxide | Dechlorination, pH control, chrome reduction | Gas | Will reliquify in system piping if allowed to cool. | Can be more corrosive than chlorine if wet at elevated temperatures. |
| Sulfuric acid | pH control | Liquid | Store in dry containers. | Iron and steel is permissible in concentrated sulfuric acid service, but dilute sulfuric is very corrosive to steel. |

**TABLE 9.3** Partial list of regulated toxic substances and threshold quantities for accidental release prevention.

| Chemical name | Chemical abstracts service (CAS) no. | Threshold quantity (lb) |
|---|---|---|
| Chlorine | 7782-50-5 | 2500 |
| Ammonium (anhydrous) | 7764-41-7 | 10 000 |
| Ammonia (aqueous with concentration 20% or greater) | 7764-41-7 | 20 000 |
| Sulfur dioxide (anhydrous) | 7446-09-5 | 5000 |
| Propane | 74-98-6 | 10 000 |
| Methane | 74-82-8 | 10 000 |

local emergency response personnel, and submit a RMP to governmental agencies, state emergency response commissions, and local emergency planning committees. A prevention program within the RMP includes the identification of hazards, documented operating procedures, training programs for personnel, maintenance requirements, emergency response programs, and complete accident investigation. When storing more than a specific quantity of gaseous chemicals such as chlorine or sulfur dioxide, requirements for secondary containment and emergency scrubber systems should be considered to prevent or minimize the spread of chemical releases. Emergency response procedures should be developed in cooperation with local emergency officials. Federal regulations require that local emergency planning agencies be notified when specific amounts of certain chemicals are stored (see Chapter 5, Occupational Safety and Health, for further discussion).

# UNLOADING AND STORING CHEMICALS

Special precautions are necessary for the safe unloading and storage of chemicals. Because a complete discussion of these precautions would be impractical, the operator must obtain such information directly from the chemical manufacturer or supplier for each chemical storage and handling system within the treatment plant.

**UNLOADING CHEMICALS.** In general, chemicals are delivered to a wastewater treatment plant in dry, liquid, or gaseous form. Liquid chemicals can be in various combinations of phases and viscosity. For instance, liquid polymers are available in single-phase liquids, multiphase emulsions, Mannich (a special type of highly viscous solution polymer), and gel. Chemicals are also delivered in a variety of containers, including bags, drums, totes, cylinders (gas), ton containers, bulk trucks, and railroad tank cars. Dry chemicals are generally available in either bagged, drummed, mini-bulk, or bulk form. Mini-bulk includes super sacs or bulk bags. When daily requirements are small, bagged chemicals are preferred because their handling and storage are relatively simple, involving either manual labor or mechanical handling. Bagged chemicals, delivered either in loose bags or on pallets in trucks or boxcars, are transferred to storage by hand lifts or forklift. For loose bags, conveyors may be used if there is a long distance between the unloading point and the storage area. For bag shipments on pallets, using a forklift to move the loaded pallets to storage and then to the point of use reduces the manual labor. Drums contain more chemical (typically approximately 181 kg [400 lb] per drum) than bags but are more difficult to handle. Mechanical drum movers and dumpers are available and can be used were chemical is received in drum form.

For intermediate usage rates and where chemicals can be purchased in mini-bulk form, super sacs or bulk bags can be used to reduce the amount of manual labor involved. Mini-bulk containers typically contain 907 kg (1 ton) of chemical, which equals approximately forty 23-kg (50-lb) bags. Feed equipment for mini-bulk containers generally includes support frames to hold the container above the feeder and valves on the bag and/or feeder to control the flow of chemical into the feeder. In addition, forklifts generally are used to move the bags to and from storage and to lift the bag into a support frame located over the feeder. Therefore, forklift access to the feeder through adequately sized aisles and/or roll-up doors is required. Even though the cost of feed equipment and operating space requirements can be somewhat more for mini-bulk containers than for bags or drums, the amount of manual labor required for chemical handling can be reduced.

Bulk shipments of dry or liquid chemicals are delivered in trucks or by rail in boxcars and hopper cars. Bulk unloading facilities typically must be provided at the treatment plant. Rail cars constructed for top unloading will require an air-supply system and flexible connectors to pneumatically displace the chemical from the car. Bottom unloading can be accomplished by using a transfer pump or a pneumatic transfer system. The U.S. Department of Transportation (U.S. DOT) regulations concerning chemical tank car unloading should be observed.

Bulk unloading areas should be arranged to contain the chemical for recovery, neutralization, or disposal in case of a spill. Spill control measures such as spill booms and mats, neutralization chemicals, and waste disposal drums should be readily available near the unloading area. Special training is required for operators responsible for spill response. All unloading areas should be level and allow tank trucks to pull away in a forward direction. Pavement design for heavier truckloads may require the use of concrete for roads and the unloading area. In addition, consideration must be given to appropriate truck turning radius. Before any unloading operation, the operator should verify that the storage tank will hold the contents of the tank car or tank truck and that the tank is well vented. When transferring with air, an air surge will occur at the end of the product transfer, so it is important to have properly sized relief vents. Level indication at the unloading station is highly recommended to help the operator monitor the filling operation and avoid overfilling a bulk tank. Generally, a quick-disconnect should be provided at the unloading station for connection to the supply hose from the chemical tank car or truck. Consulting with the chemical vendor will help the operator determine the type and size of quick-disconnect needed to simplify the chemical transfer.

Properly trained employees must supervise all unloading operations. If the operator must leave the transfer operation, the operation should be shut down. The operator must use protective eyewear, gloves, and clothing during the unloading operations. Emergency showers and eyewashes should be adjacent to the unloading station area and tested at least monthly for proper operation. In addition, smoking should not be allowed in any chemical unloading area, and unloading should be done during daylight hours unless adequate lighting is provided for nighttime unloading operations.

Both U.S. DOT and the U.S. Coast Guard classify chlorine, ammonia, and sulfur dioxide as nonflammable compressed gases. As such, when shipped in the United States by rail, water, or highway, they must be packaged in containers that comply with both U.S. DOT and U.S. Coast Guard regulations regarding loading, handling, and labeling.

Chlorine gas is commonly available in 45- and 68-kg (100- and 150-lb) steel cylinders; in 907-kg (1-ton) steel containers; and (for large quantities) in railroad tank cars, tank trucks, or barges. *The Chlorine Manual* (The Chlorine Institute, Inc., 1986) addresses in detail the appropriate loading and handling of chlorine containers.

Various mechanical devices, such as skids, troughs, and up-ending cradles, simplify handling of 68-kg (150-lb) chlorine cylinders. When unloaded from trucks or platforms, cylinders must not be dropped to ground level. If cylinders must be lifted or lowered and an elevator is not available, specially designed cradles or carrying platforms in combination with a crane or derrick are recommended. Chains, lifting magnets, and rope slings that encircle the cylinders are unsafe and should not be used. For lateral movements, a properly balanced hand truck is useful. Cylinders being moved should always have valve-protection hoods in place. Because these hoods are not designed to hold the weight of cylinders and their contents, the cylinders should never be lifted by their hoods.

Ton containers may be moved by various methods, including rolling, a crane hoist monorail system, or by specially fitted trucks or dollies. When it is necessary to lift them, as from a multiunit railroad car or truck, it is good practice to use a suitable lift clamp or lifting beam in combination with a hoist or crane with at least a 1814-kg (2-ton) capacity.

Receiving and unloading areas and safety precautions applicable to handling single-unit railroad tank cars, tank trucks, or other shipping containers are subject to federal, state, and local regulations.

**STORING CHEMICALS.** The equipment used for storing and handling chemicals varies with the type of chemical used, form of the chemical (liquid or dry), quantity of chemical, and plant size.

Storage tanks and vessels must be labeled to show their contents. Each tank and/or storage area should be clearly labeled indicating both the contents and the National Fire Protection Association (NFPA) Hazard Identification System (National Fire Protection Association, 2001).

The layout and design of a chemical storage area must comply with all local codes and ordinances. Storage areas for chemicals should be clean, temperature controlled, properly ventilated, and protected from corrosive vapors. In some cases, humidity control also may be necessary, especially with dry chemicals. Cylinders and ton containers should be stored in a fire-resistant building away from heat sources, flammable substances, and other compressed gases. Storage and use areas should be equipped with suitable mechanical ventilators for normal occupancy and proper controls and leak-detection equipment to contain and hold extremely hazardous gases. State and local ordinances that have adopted all aspects of the NFPA guidelines will require scrubbers. Additional discussion regarding chlorine and sulfur dioxide emergency gas scrubbers is presented in Chapter 26, Effluent Disinfection. Sulfur dioxide scrubbers are similar to those used for chlorine. At some locations, secondary containment of individual bulk containers can be provided in place of scrubbers. In all cases, however, it is always a good practice to equip gas storage areas with proper ventilation control (i.e., normal ventilation is terminated when a leak is detected) and emergency gas scrubbing equipment (typically sized to neutralize the largest single container). Cylinders and ton containers should not be stored outdoors. Both local and remote alarms should be provided to alert the operator of any potential leak of chemicals such as chlorine, sulfur dioxide, and ammonia in both storage and use areas.

***Bulk Storage.*** Figure 9.1 presents a typical packaged-type bulk-storage tank (or bin) for dry chemicals. Dust collectors should be provided on both manually and pneumatically filled tanks. Airborne chemical dust is not only hazardous to employee health, but also may be an explosion hazard. The construction material for the storage tank

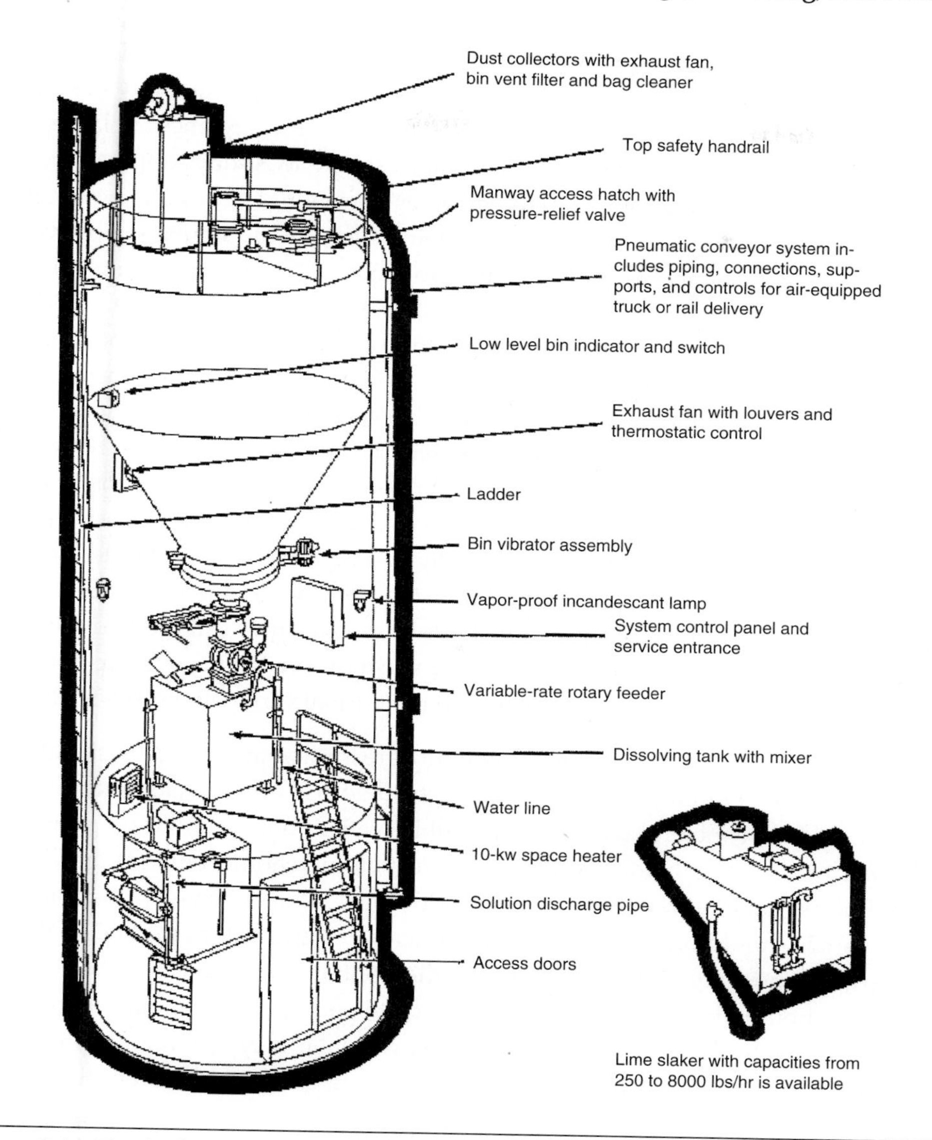


**FIGURE 9.1** Typical packaged-type bulk storage tank and accessories (lb/h × 0.453 6 = kg/h) (National Lime Association, 1995).

and the required slope and baffling at the outlet may vary with the type of chemical stored. Some dry chemicals, such as lime, require airtight bins to reduce the potential for moisture within the storage vessel. Bulk storage tanks for dry chemicals are often equipped with bin vibrators to lessen chemical bridging in the storage vessel.

Bulk-storage tanks for liquid or dry chemicals should be sized according to normal (average) chemical feed usage rates, shipping time required, and quantity of shipment. The total storage capacity should be at least between 1.25 and 1.5 times the largest anticipated supplier shipment and should provide at least a 15- to 30-day supply of the chemical at the design average usage. Gas storage areas also should be sized to accommodate a sufficient number of cylinders and bulk containers to provide similar supply volumes based on average chemical feed usage rates. In many cases, the minimum number of storage days is dictated by regulatory requirements. Storage tanks for most liquid chemicals can be either inside or outside. However, outdoor tanks must be insulated or heated or both if the chemical can crystallize or become sufficiently viscous to impede chemical flow at the minimum ambient temperature for that specific geographic location. In addition, outdoor tanks should be constructed of materials that are resistant to both deterioration from ultraviolet (UV) light and cracking because of wide variations in external temperatures. All liquid storage tanks must have an air vent to avoid excessive pressure or vacuum in the tank. Vents for chemical systems should be carefully located so that air ventilation intakes are not affected. In addition, the vent should not be in an area that can be routinely accessed by personnel. Some liquids, such as hydrochloric acid with a vent scrubber, also should have a pressure- and/or vacuum-relief valve or system to protect the tank from excessive pressure or vacuum if the normal vent path becomes plugged.

***Leak Detection.*** Liquid-storage tanks generally are located at ground level and should include secondary containment for both catastrophic and minor leaks. Secondary containment, such as concrete structures or double-walled tanks, should be provided for spills and/or leaks of chemicals. Containment volume should be a minimum of the largest bulk storage tank volume plus an additional 10%. Whenever possible, pump control panels should be outside the containment area and valves located so that they are operable and accessible from outside the containment area. Pumps and other equipment should always be elevated on concrete pads to avoid damage from minor leakage. Automated leak-detection interlocked with alarm systems can be included in the secondary containment areas for bulk storage tanks. The detection of the chemical fluid allows an operator to be notified of potential problems at storage areas. In addition, regularly scheduled (as frequent as daily) visual inspections should be conducted within containment areas to confirm that spills or leaks have not occurred. The plant should avoid installing chemical storage tanks underground, if possible, and provide secondary containment or double-walled tanks with leak detection systems if underground installation must be used. Without secondary containment, soil and groundwater contamination could result if a leak occurs.

Two relatively common pressurized liquid feed systems are flooded suction systems with an overhead storage tank (Figure 9.2) and suction lift systems using ground

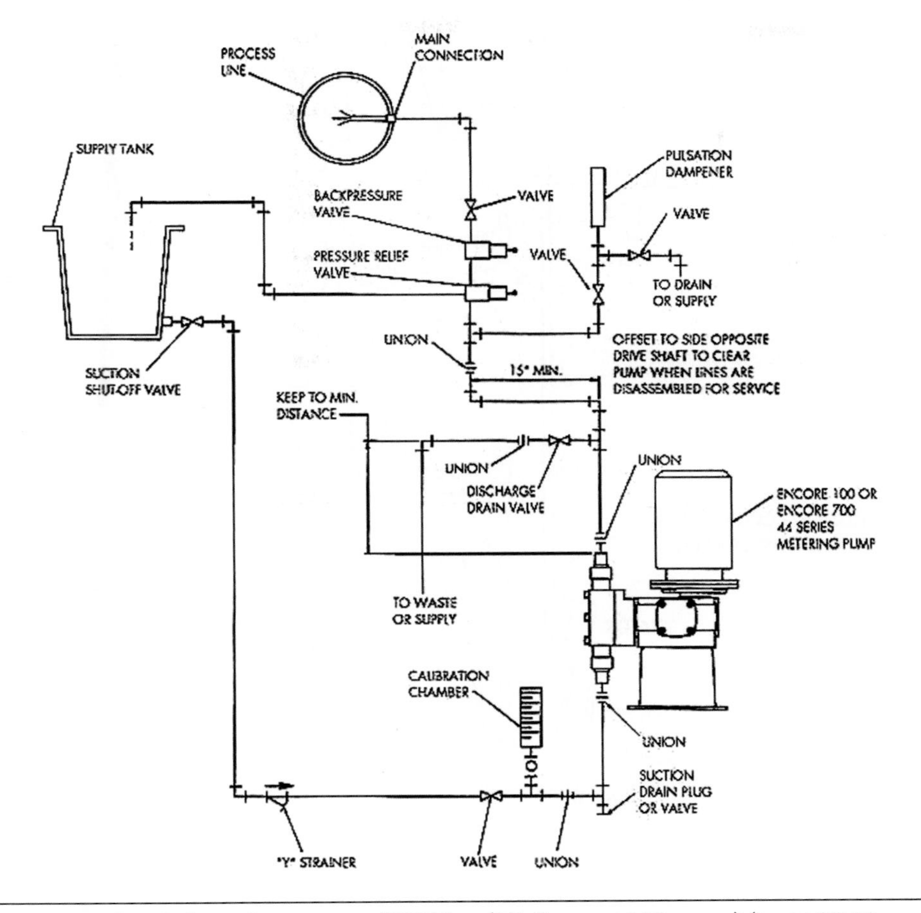


**FIGURE 9.2** Flooded suction system (USFilter/Wallace and Tiernan) (in. × 25.40 = mm).

storage (Figure 9.3). Figure 9.2 shows an overhead storage system used to gravity feed the chemical metering pump. A rotodip-type feeder or rotameter can be used for a gravity feed system also. Figure 9.3 shows a ground storage system with a suction lift transfer pump. The metering pump (diaphragm type) is often used for pressure-feed systems.

***Bag, Drum, and Tote Storage.*** Typically, the determination of what type of storage method is acceptable will depend on plant size, average chemical feed usage rates, chemical storage space available, and the amount of labor that can be allocated to chemical handling. Small plant facilities may only use bags or drums (or possibly totes) to meet an adequate chemical storage. As stated earlier, chemical storage should provide a minimum of 15 to 30 days of chemical storage at average chemical feed us-

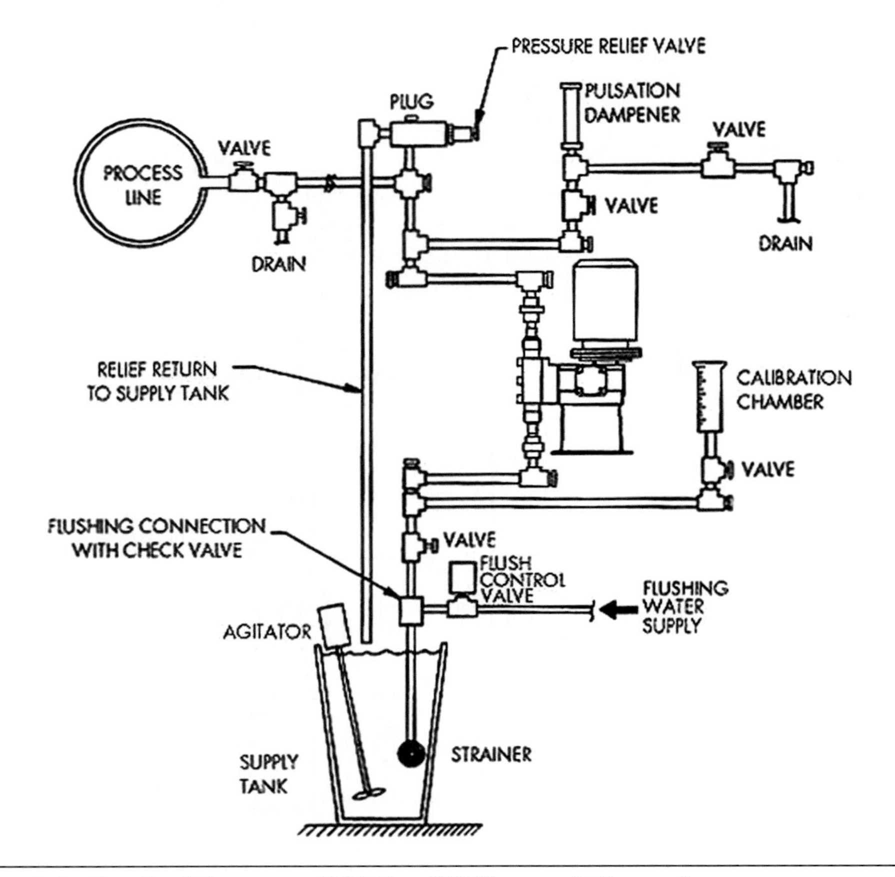


**FIGURE 9.3** Suction lift system (USFilter/Wallace and Tiernan).

age rates. Small plants may have sufficiently low average chemical feed usage rates that the small number of bags, drums, or totes required may not take up much storage space and may not require much labor for handling. However, large plant facilities, with larger average chemical feed usage rates, will require bulk storage tanks to minimize the amount of storage space required to meet minimum storage requirements and the amount of labor required for chemical handling. Bulk storage systems use additional material-handling equipment to provide a more efficient, less labor intensive, and less costly approach to chemical handling as compared with storing and handling a significant number of chemical bags or drums or totes. For an intermediate size of plant, bulk bags or totes can provide a cost-effective approach to chemical handling compared with bulk storage or bag/drum storage. For an intermediate size of plant, the amount of labor required to handle the larger containers is typically less than the labor required to handle the many bags or drums because fewer of the larger containers must be handled. The larger container size also can require less storage space to maintain minimum chemical inventories. However, bulk bags or totes do require some additional chemical handling equipment (compared with handling bags/drums) to handle the large size of container. However, the amount of additional equipment required is typically less than the additional equipment required for bulk storage systems.

Typically, bags, drums, or totes are stored in a dry, temperature-controlled, low-humidity area and should be used in proper rotation: first in, first out. Bags should be stored off the floor level and liquid drum or tote storage areas must be properly contained in case of a spill. Bag- or drum-loaded hoppers are frequently sized for a storage capacity of eight hours at the nominal maximum feed rate, so personnel are not required to fill or charge the hopper more than once per shift.

***Cylinder and Ton Container Storage.*** Whether in storage or in use, cylinders (68-kg [150-lb]) need proper support to prevent accidental tipping. Cylinders should be supported by chaining or anchoring to a fixed wall or support and should be readily accessible and removable. Bulk containers should be stored horizontally, slightly elevated from the floor level, and blocked to prevent rolling. Monorails are often used for movement of bulk containers within the storage area. In addition, beams may be used as a convenient storage rack for supporting both ends of the containers. Bulk containers should not be stacked or racked more than one high. Chlorine cylinders and containers should be protected from impact, and handling should be kept to a minimum. Full and empty cylinders and bulk containers should be stored separately and tagged or identified according to their disposition.

Chaining or anchoring chlorine cylinders in place during use is also important. Chlorine piping, cylinder connections, and other equipment could be damaged during any type of movement if not properly anchored.

Proper, forced, mechanical ventilation in storage and feed areas is important for the safety and health of operating personnel. If the gas is heavier than air, such as chlorine, ventilation should be drawn from the floor. On the other hand, if the gas is lighter than air, such as ammonia, the ventilation should be drawn from near the ceiling. One air change per minute (60 air changes per hour) is considered adequate ventilation when personnel occupy chlorine storage and feed areas, per *Recommended Standards for Wastewater Facilities* (Great Lakes, 1997). The International Fire Code suggests a ventilation rate of approximately $5.08 \times 10^{-3}$ $m^3/m^2 \cdot s$ (1 cfm/sq ft) of building, but this yields fewer air changes per hour (International Fire Code, 1999). Typically, room ventilation is automatically activated (through the use of door switches or other means to detect entry) when the room or area is occupied by plant operators. The point of ventilation discharge should be carefully placed so it does not exhaust near other air inlets or occupied external areas. In addition, where required by law, gas scrubbers may be necessary in the ventilation discharge. Ventilation rates for discharge to an emergency scrubber are typically based on the rupture of the single largest container. For example, the release from a 907-kg (1-ton) container rupture would typically require a ventilation rate of 1.42 $m^3/s$ (3000 cfm).

Leak detection and containment in the storage and feed areas are extremely important. Automatic leak-detection equipment for chlorine gas and sulfur dioxide are effective in detecting relatively small concentrations of the subject chemicals. Interlocks should be provided that shut down the normal ventilation in the event that a leak is detected. Local and remote alarms are typically initiated and the emergency scrubber and fan are started. Leak-repair kits, self-contained breathing apparatus meeting the requirements of the National Institute for Occupational Safety and Health, and all other safety equipment should be available outside the storage/feed area and properly maintained. Only operating personnel that have been thoroughly trained and certified in emergency response should be involved in a chemical-handling emergency.

***Chemical Purity.*** Typically, technical grade chemicals are used in wastewater treatment to minimize the chemical purchase costs. High-purity chemicals are not required and can significantly increase the cost for chemicals. However, care must be exercised to avoid chemicals that contain excessive quantities of hazardous impurities that could result in wastewater treatment plant upsets or violations of the wastewater treatment plant effluent limitations. The effluent limitations that could be exceeded may be either an actual numerical limit listed in the effluent permit for each individual hazardous compound or a limit based on a whole effluent toxicity (WET) test. A very low concentration of many hazardous compounds, such as heavy metals, can be toxic to the biological organisms used in the WET test, resulting in test failure and associated violation of the effluent limitation. For heavy metals such as lead, mercury, copper and zinc, the wastewater treatment plant can have very low effluent limitations and the typical wastewater treatment processes will not be able to adequately remove these compounds to meet the very stringent limits. In addition, many heavy metals can accumulate in the sludge generated by the wastewater treatment processes. Other chemicals such as chlorine and sulfur dioxide can be detrimental to the discharge water quality. Low concentrations of chlorine exhibit effluent toxicity and sulfur dioxide can deplete dissolved oxygen. The actual quantity of allowable contaminate will depend on any effluent limitation and the maximum quantity of chemical used at the wastewater treatment plant.

Once the chemical is received at the treatment plant, care should be exercised to avoid contamination of the chemical. Material safety data sheet documents and supplier instructions should be consulted to determine types of contamination that must be avoided. Depending on the chemical, contamination can result in significant degradation of the effectiveness of the chemical and, in some cases, the formation of extremely hazardous and even life-threatening conditions in the plant. For example, metal contamination (such as rust) in hydrogen peroxide can destroy significant quantities of active chemical and even result in an explosive release of gas.

**TROUBLESHOOTING.** Chemical unloading and storage systems can have operating problems. Table 9.4 presents a brief troubleshooting guide for chemical unloading and storage systems that will help the operator to identify problems and develop possible solutions.

## PUMPING, PIPING, AND HANDLING MATERIALS

Piping and accessories for transporting and feeding various chemicals should be provided only after specific chemicals have been selected for use in the wastewater treatment plant. The operator should use only materials that are compatible with each

TABLE 9.4 Troubleshooting guide for chemical unloading and storage systems.

| Observations | Checks and remedies |
|---|---|
| **Pneumatic conveying (low pressure)** | |
| Material not moving from car or truck to silo. | Check pressure at blower. If high, line may be plugged. Shut off, discharge to system, clear line, and restart. |
| | If pressure is normal or low, material is not entering conveying line. Check discharge gate on truck or rail car. |
| Dust discharging from silos. | Ensure that dust filter on silo is operational. |
| | Check dust filter for broken or torn bags or tubes. |
| | Ensure that silo dust filter capacity is sufficient for blower discharge. |
| **Bucket elevators and screw conveyors** | |
| Material backing up at inlet. | Listen to determine whether system is operating. If so, reduce amount at inlet to unit and continue. |
| | If unit is stopped but motor is running, check for broken shear pins. |
| | If motor is tripped out, check for broken or overloaded conveyor. |

chemical. For example, many chemical-handling systems require special materials for construction and special types of piping; tubing; and transport channels, pumps, valves, and gaskets. Table 9.5 gives a summary of tank, pump, piping, and valve materials compatible with several commonly used chemicals and offers an overview on general material compatibility for these chemicals. The operator should always consult equipment manufacturers and chemical suppliers before selecting materials of construction. Sodium hypochlorite is one of many chemicals that require special handling precautions. As an example of material selection consideration, more detailed information on feeding and handling equipment for this chemical is listed in Table 9.6.

**TABLE 9.5** Materials of construction—chemical handling facilities.

| Chemical | Tanks | Pumps | Pipe | Valves |
|---|---|---|---|---|
| Alum | FRP[a] | Nonmetallic | PVC,[b] CPVC,[c] FRP | Nonmetallic |
| Chlorine | Steel cylinders | N/A | Carbon steel to vaporizer | Carbon steel |
| Ferric chloride | FRP, rubber-lined steel | Nonmetallic or rubber lined | FRP, CPVC, PVC, rubber-lined steel | Rubber-lined, CPVC |
| Ferrous sulfate | FRP | Nonmetallic | PCV, CPVC, FRP | Nonmetallic |
| Hydrogen peroxide | Aluminum alloy 5254, Type 316L stainless steel | Type 316 stainless steel, Teflon | Aluminum, Type 316L stainless steel | Type 316 stainless steel, Teflon |
| Methanol | Carbon steel | Cast steel | FRP, carbon steel | Carbon steel |
| Ozone | N/A | N/A | Type 316 stainless steel | CF-8M |
| Polymers | FRP | Nonmetallic | PVC, CPVC | Nonmetallic |
| Sodium bisulfite | FRP | Nonmetallic | PVC, CPVC, FRP | Nonmetallic or plastic lined |
| Sodium hydroxide | FRP, special construction | Stainless or carbon steel | CPVC, FRP, stainless steel | Stainless steel, nonmetallic |
| Sodium hypochlorite | FRP, special construction | Nonmetallic | FRP, CPVC | Nonmetallic or plastic lined |
| Sulfur dioxide | Carbon steel | N/A | Carbon steel | Carbon steel |
| Concentrated (93%) sulfuric acid | Phenolic lined steel | CN-7M (Alloy 20) | Type 304 stainless, 1.8 m/s maximum | CN-7M for throttling, CF-8M for shut-off |

[a]FRP = fiber-glass-reinforced plastic.
[b]PVC = polyvinyl chloride.
[c]CPVC = chlorinated polyvinyl chloride.

**TABLE 9.6** Materials suitable for use with sodium hypochlorite (The Chlorine Institute, 2000).

| Components | Materials of construction compatible with sodium hypochlorite solutions |
|---|---|
| Rigid pipe | Lined (PP,[a] PVDF,[b] PTFE[c]) steel, CPVC[d]/PVC[e] (Sch 80) and titanium |
| Fittings | Same as rigid pipe |
| Gaskets | Viton™ |
| Valves | PVC, CPVC, PP |
| Pumps (centrifugal) | |
| Body | Nonmetallic (PVC, TFE, Kynar, Tefzel, Halar) |
| Impeller | Same as body |
| Seals | Silicon carbide |
| Storage tanks | Rubber-lined steel, fiber glass, and HDPE[f] |

[a]PP = polypropylene.
[b]PVDF = fluorinated polyvinylidene.
[c]PTFE = polytetrafluoroethylene.
[d]CPVC = chlorinated polyvinyl chloride.
[e]PVC = polyvinyl chloride.
[f]HDPE = high-density polyethylene tanks.

Because of the concern about chemical leaks and spills from system piping, recent trends show that the use of double-walled pipe (with leak-detection sensors) or single-walled pipe placed in a containment trough or trench should be given serious consideration in the design of new systems. A variety of issues involving safety, legal liability, government legislation, insurance risk issues, and so forth, may require the use of double-contained piping systems to reduce soil and groundwater contamination and to protect plant personnel from leakage of hazardous chemicals from conventional piping systems. Providing systems with double containment will avoid serious leaks or spills from the chemical piping. Double-walled pipe-containment systems are designed to contain a leak as it occurs, detect the leak, and alarm the plant operators.

Frequently, welded pipe and fittings are used for hazardous chemical piping whenever possible and recommended to minimize the potential for leaks. Plastic materials generally can be solvent welded, heat welded, or heat fusion bonded. Metallic materials can be heat welded, such as electric arc, or brazed. If flanges must be used, then specially designed secondary containment bags that fit around the flange or shields can be used to contain any chemical spray from leaks.

Chemical piping should include an adequate number of valved drain locations to allow chemical to be drained from piping, valves, and equipment before removal for maintenance. Each valve should include a removable cap to minimize leakage through the valve before and after use.

# CHEMICAL FEEDING SYSTEMS

This section discusses gas, liquid, and solids feeding systems with on-site generation. Table 9.7 lists types of feed systems for specific chemicals. A complete chemical addition system for any given process unit has provisions for receiving and storing the chemical, transferring and metering the chemical, and mixing and injecting the chemical to the process stream.

**GAS FEEDERS.** Chemicals fed as a gas include chlorine, ammonia, and sulfur dioxide. Typically, these chemicals are transported and stored as a liquefied, compressed gas and then metered or fed to the wastewater treatment process as a gas.

***Description of Equipment.*** If chlorine or sulfur dioxide is drawn as a liquid, an evaporator is used to vaporize the chlorine. An example of an evaporator system is shown in Figure 9.4. A chlorine evaporator consists of a chlorine chamber in a hot-water bath tank. Liquid chlorine is fed to the chamber, the heat vaporizes it, and gaseous chlorine is delivered to the point of application (White, 1986). An electric heater, recirculating hot-water bath, or steam can supply heat to the evaporator. Typical operating problems associated with evaporators are frosting or icing because of liquid particles being entrained with the gas leaving the chamber, scaling outside the chamber, and concentration of impurities in the chlorine inside the chamber. Frosting or icing is caused by demand exceeding the capacity of the evaporator and can be alleviated temporarily by reducing chlorine flow. Scaling and impurities will reduce the capacity of the evaporator. Regular shutdowns for cleaning will help prevent unscheduled service interruptions resulting from scaling and impurities.

Gas feeders are classified as either solution feed or direct feed. Solution-feed, vacuum-type feeders are most commonly used in chlorination and in dechlorination with sulfur dioxide (White, 1986). An example of vacuum-type feeders is shown in Figure 9.5. A gas can also be fed in a pressure system (Figure 9.6), but a vacuum system is preferred because of safety issues. Some state regulations, in fact, will allow only remote vacuum systems to eliminate releases from pressurized gas systems.

The main advantage of a gas-feed system is the relative simplicity of the transfer and feed system. Often, the vapor pressure of the material itself can be used to transfer the material, greatly reducing the equipment requirements.

The chemical can be injected directly to the process as a gas; however, there are several potential problems associated with direct feed. Some wet gases are extremely corrosive, leading to maintenance and reliability problems with the mixer. Also, direct-feed systems could allow water to back up into the gas lines at low gas feeds, damaging the equipment.

**TABLE 9.7** Chemical-specific feeding recommendations (*Wastewater Treatment*, 1997).

| Common name/ formula use | Best feeding form | Chemical-to-water ratio for continuous dissolving[a] | Types of feeders | Accessory equipment required | Suitable handling materials for solutions[b] |
|---|---|---|---|---|---|
| Alum: $Al_2(SO_4)_3 \cdot XH_2O$<br>Liquid<br>1 gal 36°Be = 5.38 lb of dry alum: 60 °F<br>Coagulation at pH 5.5 to 8.0<br>Sludge conditioner<br>Precipitate $PO_4$ | Full strength under controlled temperature or dilute to avoid crystallization<br>Minimize surface evaporation; causes flow problems<br>Keep dry alum below 50% to avoid crystallization | Dilute to between 3 and 15% according to application conditions, mixing, etc. | *Solution*<br>Rotodip<br>Plunger pump<br>Diaphragm pump<br>1700 pump<br>Loss in weight | Tank gauges or scales<br>Transfer pumps<br>Storage tank<br>Temperature control<br>Eductors or dissolvers for dilution | Lead or rubber-lined tanks, Duriron, FRP,[c] Saran, PVC-1, vinyl, Hypalon, epoxy, 16 SS, Carpenter 20 SS, Tyril |
| Aluminum sulfate: $Al_2(SO_4)_3 \cdot 14H_2O$ (alum, filter alum)<br>Coagulation at pH 5.5 to 8.0<br>Dosage between 0.5 and 9 gpg<br>Precipitate $PO_4$ | Ground, granular, or rice<br>Powder is dusty, arches, and is floodable[d] | 0.5 lb/gal<br>Dissolver detention time 5 min for ground (10 min for granules) | *Gravimetric*<br>Belt loss in weight<br>*Volumetric*<br>Helix<br>Universal<br>*Solution*<br>Plunger pump<br>Diaphragm pump<br>1700 pump | Dissolver<br>Mechanical mixer<br>Scales for volumetric feeders<br>Dust collectors | Lead, rubber, FRP, PVC-1, 316 SS, Carpenter 20 SS, vinyl, Hypalon, epoxy, Ni-resistant glass, ceramic, polyethylene, Tyril, Uscolite |
| Ammonia anhydrous: $NH_3$ (ammonia)<br>Monel,<br>Chlorine-ammonia treatment<br>Anaerobic digestion<br>Nutrient | Dry gas or as aqueous solution:<br>see "Ammonia, aqua" | — | Gas feeder | Scales | Steel, Ni-resistant,<br>316 SS, Penton, Neoprene |

TABLE 9.7 Chemical-specific feeding recommendations (*Wastewater Treatment*, 1997) (*continued*).

| Common name/ formula use | Best feeding form | Chemical-to-water ratio for continuous dissolving[a] | Types of feeders | Accessory equipment required | Suitable handling materials for solutions[b] |
|---|---|---|---|---|---|
| Ammonia, aqua: $NH_4OH$ (ammonium hydroxide, ammonia water, ammonium hydrate)<br>Chlorine-ammonia treatment<br>pH control<br>Nutrient | Full strength | — | *Solution*<br>Loss in weight<br>Diaphragm pump<br>Plunger pump<br>Bal. diaphragm pump | Scales<br>Drum handling equipment or storage tanks<br>Transfer pumps | Iron, steel, rubber, Hypalon, 316 SS, Tyril (room temperature to 28%) |
| Calcium hydroxide: $Ca(OH)_2$ (hydrated lime, slaked lime)<br>Coagulation, softening<br>pH adjustment<br>Waste neutralization<br>Sludge conditioning<br>Precipitate $PO_4$ | Finer particle sizes more efficient, but more difficult to handle and feed | Dry feed: 0.5 lb/gal maximum<br>Slurry: 0.93 lb/gal (i.e., a 10% slurry) (light to a 20% concentration maximum) (heavy to a 25% concentration maximum) | *Gravimetric*<br>Loss in weight<br>Belt<br>*Volumetric*<br>Helix<br>Universal<br>*Slurry*<br>Rotodip<br>Diaphragm<br>Plunger pump[e] | Hopper agitators<br>Non-flood rotor under large hoppers<br>Dust collectors | Rubber hose, iron, steel, concrete, Hypalon, Penton, PVC-1<br>No lead |
| Calcium hypochlorite: $Ca(OCl)_2 \cdot 4H_2O$ (H.T.H., Perchloron, Pittchlor)<br>Disinfection<br>Slime control<br>Deodorization | Up to 3% solution maximum (practical) | 0.125 lb/gal makes 1% solution of available $Cl_2$ | *Liquid*<br>Diaphragm pump<br>Bal. diaphragm pump<br>Rotodip | Dissolving tanks in pairs with drains to draw off sediment<br>Injection nozzle<br>Foot valve | Ceramic, glass, rubber-lined tanks, PVC-1, Penton, Tyril (room temperature), Hypalon, vinyl, Usolite (room temperature), Saran, Hastelloy C (good)<br>No tin |

| | | | | | |
|---|---|---|---|---|---|
| Calcium oxide: CaO (quicklime, burnt lime, chemical lime, unslaked lime)<br>Coagulation<br>Softening<br>pH adjustment<br>Waste neutralization<br>Sludge conditioning<br>Precipitate $PO_4$ | 0.25 to 0.75 in. pebble lime<br>Pellets<br>Ground lime arches and is floodable<br>Pulverized will arch and is floodable<br>Soft burned, porous best for slaking | 2.1 lb/gal (range from 1.4 to 3.3 lb/gal according to slaker, etc.)<br>Dilute after slaking to 0.93 lb/gal (10%) maximum slurry | *Gravimetric*<br>Belt<br>Loss in weight<br>*Volumetric*<br>Universal<br>Helix | Hopper agitator and non-flood rotor for ground and pulverized lime<br>Recording thermometer<br>Water proportioner<br>Lime slaker<br>High-temperature safety cut-out and alarm | Rubber, iron, steel, concrete, Hypalon, Penton, PVC-1 |
| Carbon, activated: C (Nuchar, Norit, Darco, Carbodur)<br>Decolorizing, taste and odor removal<br>Dosage between 5 and 80 ppm | Powder: with bulk density of 12 lb/cu ft<br>Slurry: 1 lb/gal | According to its bulkiness and wetability, a 10 to 15% solution would be the maximum concentration | *Gravimetric*<br>Loss in weight<br>*Volumetric*<br>Helix<br>Rotolock<br>*Slurry*<br>Rotodip<br>Diaphragm pumps | Washdown-type wetting tank<br>Vortex mixer<br>Hopper agitators<br>Non-flood rotors<br>Dust collectors<br>Large storage capacity for liquid feed<br>Tank agitators<br>Transfer pumps | 316 SS, rubber, bronze, Monel, Hastelloy C, FRP, Saran, Hypalon |
| Chlorine: $Cl_2$ (chlorine gas, liquid chlorine)<br>Disinfection<br>Slime control<br>Taste and odor control<br>Waste treatment<br>Activation of silica[f] | Gas: vaporized from liquid | 1 lb to 45 to 50 gal or more | Gas chlorinator | Vaporizers for high capacities<br>Scales<br>Gas masks<br>Residual analyzer | *Anhydrous liquid or gas*<br>Steel, copper, black iron<br>*Wet gas*<br>Penton, Viton, Hastelloy C, PVC-1 (good), silver, Tantalum<br>*Chlorinated $H_2O$*<br>Saran, stoneware, Carpenter 20 |

**Table 9.7** Chemical-specific feeding recommendations (*Wastewater Treatment*, 1997) (*continued*).

| Common name/ formula use | Best feeding form | Chemical-to-water ratio for continuous dissolving[a] | Types of feeders | Accessory equipment required | Suitable handling materials for solutions[b] |
|---|---|---|---|---|---|
| | | | | | SS, Hastelloy C, PVC-1, Viton, Uscolite, Penton |
| Chlorine dioxide: $ClO_2$<br>Disinfection<br>Taste and odor control (especially phenol)<br>Waste treatment 0.5 to 5 lb $NaClO_2$ per mil. gal $H_2O$ dosage | Solution from generator<br>Mix discharge from chlorinizer and $NaClO_2$ solution or add acid to mixture of $NaClO_2$ and NaOCl. Use equal concentrations: 2% maximum | Chlorine water must contain 500 ppm or more of $Cl_2$ and have a pH of 3.5 or lower<br>Water use depends on method of preparation | *Solution*<br>Diaphragm pump | Dissolving tanks or crocks<br>Gas mask | *For solutions with 3% $ClO_2$*<br>Ceramic, glass, Hypalon, PVC-1, Saran, vinyl, Penton, Teflon |
| Ferric chloride:<br>$FeCl_3$ - anhydrous<br>$FeCl_3 - 6H_2O$ = crystal<br>$FeCl_3$ - solution<br>(Ferrichlor, chloride, or iron)<br>Coagulation pH 4 to 11<br>Dosage: 0.3 to 3 gpg (sludge conditioning 1.5 to 4.5% $FeCl_3$)<br>Precipitate $PO_4$ | Solution or any dilution up to 45% $FeCl_3$ content (anhydrous form has a high heat of solution) | *Anhydrous to form:*<br>45%: 5.59 lb/gal<br>40%: 4.75 lb/gal<br>35%: 3.96 lb/gal<br>30%: 3.24 lb/gal<br>20%: 1.98 lb/gal<br>10%: 0.91 lb/gal<br>(Multiply $FeCl_3$, by 1.666 to obtain $FeCl_3 \cdot 6H_2O$ at 20 °C) | *Solution*<br>Diaphragm pump<br>Rotodip<br>Bal. diaphragm pump | Storage tanks for liquid<br>Dissolving tanks for lumps or granules | Rubber, glass, ceramics, Hypalon, Saran, PVC-1, Penton, FRP, vinyl, epoxy, Hastelloy C (good to fair), Uscolite, Tyril (Rm) |
| Ferric sulfate:<br>$Fe_2(SO_4)_3 \cdot 3H_2O$<br>(Ferrifloc)<br>$Fe_2(SO_4)_3 \cdot 2H_2O$<br>(Ferriclear)<br>(iron sulfate) | Granules | 2 lb/gal (range)<br>1.4 to 2.4 lb/gal for 20-minute detention (warm water permits shorter detention) | *Gravimetric*<br>Loss in weight<br>*Volumetric*<br>Helix<br>Universal<br>*Solution* | Dissolver with motor-driven mixer and water control<br>Vapor remover solution tank | 316 SS, rubber, glass, ceramics, Hypalon, Saran, PVC-1, vinyl, Carpenter |

| | | | | | |
|---|---|---|---|---|---|
| Coagulation pH 4 to 6 and 8.8 to 9.2<br>Dosage: 0.3 to 3 gpg<br>Precipitate $PO_4$ | | Water insolubles can be high | Diaphragm pump<br>Bal. diaphragm pump<br>Plunger pump<br>Rotodip | | 20 SS, Penton, FRP, epoxy, Tyril |
| Ferrous sulfate: $FeSO_4 \cdot 7H_2O$<br>(Copperas, iron sulfate, sugar sulfate, green vitriol)<br>Coagulation at pH 8.8 to 9.2<br>Chrome reduction in waste treatment<br>Wastewater odor control<br>Precipitate $PO_4$ | Granules | 0.5 lb/gal (dissolver detention time 5 min minimum) | *Gravimetric*<br>Loss in weight<br>*Volumetric*<br>Helix<br>Universal<br>*Solution*<br>Diaphragm pump<br>Plunger pump<br>Bal. diaphragm pump | Dissolvers<br>Scales | Rubber, FRP, PVC-1, vinyl, Penton, epoxy, Hypalon, Uscolite, ceramic, Carpenter 20 SS, Tyril |
| Hydrogen peroxide: $H_2O_2$<br>Odor control | Full strength or any dilution | — | Diaphragm pump<br>Plunger pump | Storage tank, water metering and filtration device for dilution | Aluminum, Hastelloy C, titanium, Viton, Kel-F, PTFE, chlorinated polyvinylchloride (CPVC) |
| Methanol: $CH_3 \cdot OH$<br>Wood alcohol<br>denitrification | Full strength or any dilution | — | Gear pump<br>Diaphragm pump | Storage tanks | 304 SS, 316 SS, brass, bronze, Carpenter 20 SS, cast iron, Hastelloy C, buna N, EPDM, Hypalon, natural rubber, PTFE, PVDF, NORYL, Delrin, CVPC |

**TABLE 9.7** Chemical-specific feeding recommendations (*Wastewater Treatment*, 1997) (*continued*).

| Common name/ formula use | Best feeding form | Chemical-to-water ratio for continuous dissolving[a] | Types of feeders | Accessory equipment required | Suitable handling materials for solutions[b] |
|---|---|---|---|---|---|
| Ozone: $O_3$<br>Taste and odor control<br>Disinfection<br>Waste treatment<br>Odor: 1 to 5 ppm<br>Disinfection: 0.5 to 1 ppm | As generated<br>Approximately 1% ozone in air | Gas diffused in water under treatment | Ozonator | Air-drying equipment<br>Diffusers | Glass, 316 SS, ceramics, aluminum, Teflon |
| Phosphoric acid, ortho: $H_3PO_4$<br>Boiler water softening<br>Alkalinity reduction<br>Cleaning boilers<br>Nutrient feeding | 50 to 75% concentration (85% is syrupy; 100% is crystalline) | — | *Liquid*<br>Diaphragm pump<br>Bal. diaphragm pump<br>Plunger pump | Rubber gloves | 316 SS (no F)<br>Penton, rubber, FRP, PVC-1, Hypalon, Viton, Carpenter 20 SS, Hastelloy C |
| Polymers, dry<br>High-molecular-weight synthetic polymers | Powdered, flattish granules | Maximum concentration 1%<br>Feed even stream to vigorous vortex (mixing too fast will retard colloidal growth)<br>1 to 2 hours detention | *Gravimetric*<br>Loss in weight<br>*Volumetric*<br>Helix<br>*Solution (Colloidal)*<br>Diaphragm pump<br>Plunger pump<br>Bal. diaphragm pump | Special dispersing procedure<br>Mixer: may hang up; vibrate if needed | Steel, rubber, Hypalon, Tyril<br>Noncorrosive, but no zinc<br>Same as for $H_2O$ of similar pH or according to its pH |

| | | | | | |
|---|---|---|---|---|---|
| Polymers, liquid and emulsions[g]<br>High-molecular-weight synthetic polymers<br>Separan NP10 potable grade, Magnifloc 990; Purifloc N17 Ave.<br>Dosage: 0.1 to 1 ppm | *Makedown to:*<br>Liquid<br>0.5 to 5%<br>*Emulsions:*<br>0.05 to 0.2% | Varies with charge type | Diaphragm pump<br>Plunger pump<br>Bal. diaphragm pump | Mixing and aqueous tanks may be required | Same as dry products |
| Potassium permanganate: $KMnO_4$<br>Cairox<br>Taste odor control<br>0.5 to 4.0 ppm<br>Removes Fe and Mn at a 1-to-1 ratio | Crystals plus anticaking additive | 1.0% concentration (2.0% maximum) | *Gravimetric*<br>Loss in weight<br>*Volumetric*<br>Helix<br>*Solution*<br>Diaphragm pump<br>Plunger pump<br>Bal. diaphragm pump | Dissolving tank<br>Mixer<br>Mechanical | Steel, iron (neutral and alkaline), 316 SS, PVC-1, FRP, Hypalon, Penton, Lucite, rubber (alkaline) |
| Sodium aluminate: $Na_2Al_2O_4$, anhydrous (soda alum)<br>Ratio $Na_2O/Al_2O_3$<br>1/1 or 1.15/1 (high purity)<br>Also $Na_2Al_2O_4{\cdot}3H_2O$ hydrated form<br>Coagulation<br>Boiler $H_2O$ treatment | Granular or solution as received<br>Standard grade produces sludge on dissolving | Dry 0.5 lb/gal<br>Solution dilute as desired | *Gravimetric*<br>Loss in weight<br>*Volumetric*<br>Helix<br>Universal<br>*Solution*<br>Rotodip<br>Diaphragm pump<br>Plunger pump | Hopper agitators for dry form | Iron, steel, rubber, 316 SS, Penton, concrete, Hypalon |
| Sodium bicarbonate: $NaHCO_3$ (baking soda)<br>Activation of silica<br>pH adjustment | Granules or powder plus TCP (0.4%) | 0.3 lb/gal | *Gravimetric*<br>Loss in weight<br>Belt<br>*Volumetric*<br>Helix<br>Universal<br>*Solution*<br>Rotodip<br>Diaphragm pump<br>Plunger pump | Hopper agitators and non-flood rotor for powder, if large storage hopper | Iron and steel (dilute solutions: caution), rubber, Saran, SS, Hypalon, Tyril |

**TABLE 9.7** Chemical-specific feeding recommendations (*Wastewater Treatment*, 1997) (*continued*).

| Common name/ formula use | Best feeding form | Chemical-to-water ratio for continuous dissolving[a] | Types of feeders | Accessory equipment required | Suitable handling materials for solutions[b] |
|---|---|---|---|---|---|
| Sodium bisulfite, anhydrous: $Na_2S_2O_5$ ($NaHSO_3$) (sodium pyrosulfite, sodium meta-bisulfite)<br>Dechlorination: about 1.4 ppm for each ppm $Cl_2$<br>Reducing agent in waste treatment (as Cr) | Crystals (do not let set)<br>Storage difficult | 0.5 lb/gal | *Gravimetric*<br>Loss in weight<br>*Volumetric*<br>Helix<br>Universal<br>*Solution*<br>Rotodip<br>Diaphragm pump<br>Plunger pump<br>Bal. diaphragm pump | Hopper agitators for powdered grades<br>Vent dissolver to outside | Glass, Carpenter 20 SS, PVC-1, Penton, Uscolite, 316 SS, FRP, Tyril, Hypalon |
| Sodium carbonate: $Na_2CO_3$<br>(soda ash: 58% $Na_2O$)<br>Water softening<br>pH adjustment | Dense | Dry feed<br>0.25 lb/gal for 10-minute detention time, 0.5 lb/gal for 20-minute<br>Solution feed<br>1.0 lb/gal<br>Warm $H_2O$ and/or efficient mixing can reduce retention time if material has not sat around too long and formed lumps—to 5 min | *Gravimetric*<br>Loss in weight<br>*Volumetric*<br>Helix<br>*Solution*<br>Diaphragm pump<br>Bal. diaphragm pump<br>Rotodip<br>Plunger pump | Rotolock for light forms to prevent flooding<br>Large dissolvers<br>Bin agitators for medium or light grades and very light grades | Iron, steel, rubber, Hypalon, Tyril |

| | | | | | |
|---|---|---|---|---|---|
| Sodium chlorite: $NaClO_2$ (technical sodium chlorite)<br>Disinfection, taste, and odor control<br>Industrial waste treatment (with $Cl_2$ produces $ClO_2$) | Solution as received | Batch solutions 0.12 to 2 lb/gal | *Solution*<br>Diaphragm<br>Rotodip | Chlorine feeder and chlorine dioxide generator | Penton, glass, Saran, PVC-1, vinyl, Tygon, FRP, Hastelloy C (fair), Hypalon, Tyril |
| Sodium hydroxide: NaOH (caustic soda, soda lye)<br>pH adjustment, neutralization | Solution feed | NaOH has a high heat of solution | *Solution*<br>Plunger pump<br>Diaphragm pump<br>Bal. diaphragm pump<br>Rotodip | Goggles<br>Rubber gloves<br>Aprons | Cast iron, steel<br>For no contamination, use Penton, rubber, PVC-1, 316 SS, Hypalon |
| Sodium hypochlorite: NaOCl (Javelle water, bleach liquor, chlorine bleach)<br>Disinfection, slime control<br>Bleaching | Solution up to 16% Available $Cl_2$ concentration | 1.0 gal of 12.5% (available $Cl_2$) solution to 12.5 gal of water gives a 1% available $Cl_2$ solution | *Solution*<br>Diaphragm pump<br>Rotodip<br>Bal. diaphragm pump | Solution tanks<br>Foot valves<br>Water meters<br>Injection nozzles | Rubber, glass, Tyril, Saran, PVC-1, vinyl, Hastelloy C, Hypalon |
| Sulfur dioxide: $SO_2$<br>Dechlorination in disinfection<br>Filter bed cleaning<br>Approximately 1 ppm $SO_2$ for each ppm $Cl_2$ (dechlorination)<br>Water treatment $Cr^{+6}$ reduction | Gas | — | *Gas*<br>Rotameter<br>$SO_2$ feeder | Gas mask | Wet gas: Glass, Carpenter 20 SS, PVC-1, Penton, ceramics, 316 (G), Viton, Hypalon |

**TABLE 9.7** Chemical-specific feeding recommendations. (*Wastewater Treatment*, 1997) (*continued*).

| Common name/ formula use | Best feeding form | Chemical-to-water ratio for continuous dissolving[a] | Types of feeders | Accessory equipment required | Suitable handling materials for solutions[b] |
|---|---|---|---|---|---|
| Sulfuric acid: $H_2SO_4$ (oil of Vitriol, Vitriol)<br>pH adjustment<br>Activation of silica<br>Neutralization of alkaline wastes | Solution at desired dilution<br>$H_2SO_4$ has a high heat of solution | Dilute to any desired concentration: NEVER add water to acid but rather always add acid to water | *Liquid*<br>Plunger pump<br>Diaphragm pump<br>Bal. diaphragm pump<br>Rotodip | Goggles<br>Rubber gloves<br>Aprons<br>Dilution tanks | *Concentration >85 %:*<br>Steel, iron, Penton, PVC-1 (good), Viton<br>*40 to 85%:*<br>Carpenter 20 SS, PVC-1, Penton, Viton<br>*2 to 40%:*<br>Carpenter 20 SS, FRP, glass, PVC-1, Viton |

[a]To convert g/100 mL to lb/gal, multiply figure (for g/100 mL) by 0.083. Recommended strengths of solutions for feeding purposes are given in pounds of chemical per gallon of water (lb/gal) and are based on plant practice for the commercial product.
The following table shows the number of pounds of chemical to add to 1 gallon of water to obtain various percent solutions:

| % Solution | lb/gal | % Solution | lb/gal | % Solution | lb/gal |
|---|---|---|---|---|---|
| 0.1 | 0.008 | 2.0 | 0.170 | 10.0 | 0.927 |
| 0.2 | 0.017 | 3.0 | 0.258 | 15.0 | 1.473 |
| 0.5 | 0.042 | 5.0 | 0.440 | 20.0 | 2.200 |
| 1.0 | 0.084 | 6.0 | 0.533 | 25.0 | 2.760 |
| | | | | 30.0 | 3.560 |

[b]Iron and steel can be used with chemicals in the dry state unless the chemical is deliquescent or very hygroscopic, or in a dampish form and is corrosive to some degree.

[c]FRP, in every case, refers to the chemically resistant grade (bisphenol A+) of fiber-glass-reinforced plastic.

[d]*Floodable* as used in this table with dry powder means that, under some conditions, the material entrains air and becomes "fluidized" so that it will flow through small openings, like water.

[e]When feeding rates exceed 100 lb/h, economic factors may dictate use of calcium oxide (quicklime).

[f]For small doses of chlorine, use calcium hypochlorite or sodium hypochlorite.

[g]Informtion about many other coagulant aids (or flocculant aids) is available from Nalco, Calgon, Drew, Betz, North American Mogul, American Cyanamid, Dow, etc.

Note: gal $\times$ 3.785 $\times 10^{-3}$ = $m^3$; in. $\times$ 25.40 = mm; lb/cu ft $\times$ 16.02 = kg/$m^3$; lb/gal $\times$ 0.119 8 = kg/L; and ppm = mg/L.

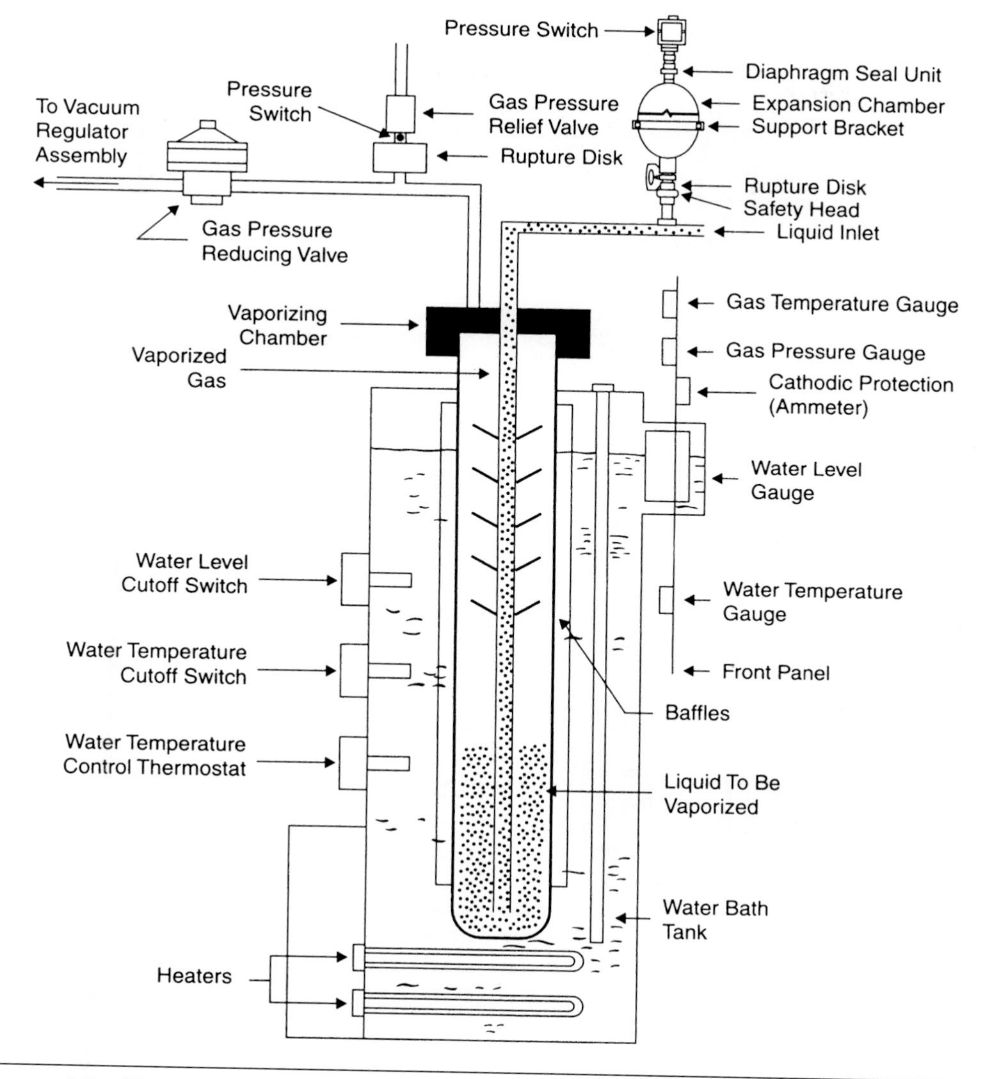


**FIGURE 9.4** Typical gas evaporator.

Most typical gas-feed systems mix the gas with water in an eductor (injector) to form a solution before injecting it to the process. The advantage of a solution-feed system is that it can limit corrosion problems to a small piece of equipment, such as an ejector, that can be coated economically or be made of corrosion-resistant materials.

The mixing of the solution with the process stream is a critical aspect of solution feeding. Without adequate mixing, the solution can lose much of its effectiveness, caus-

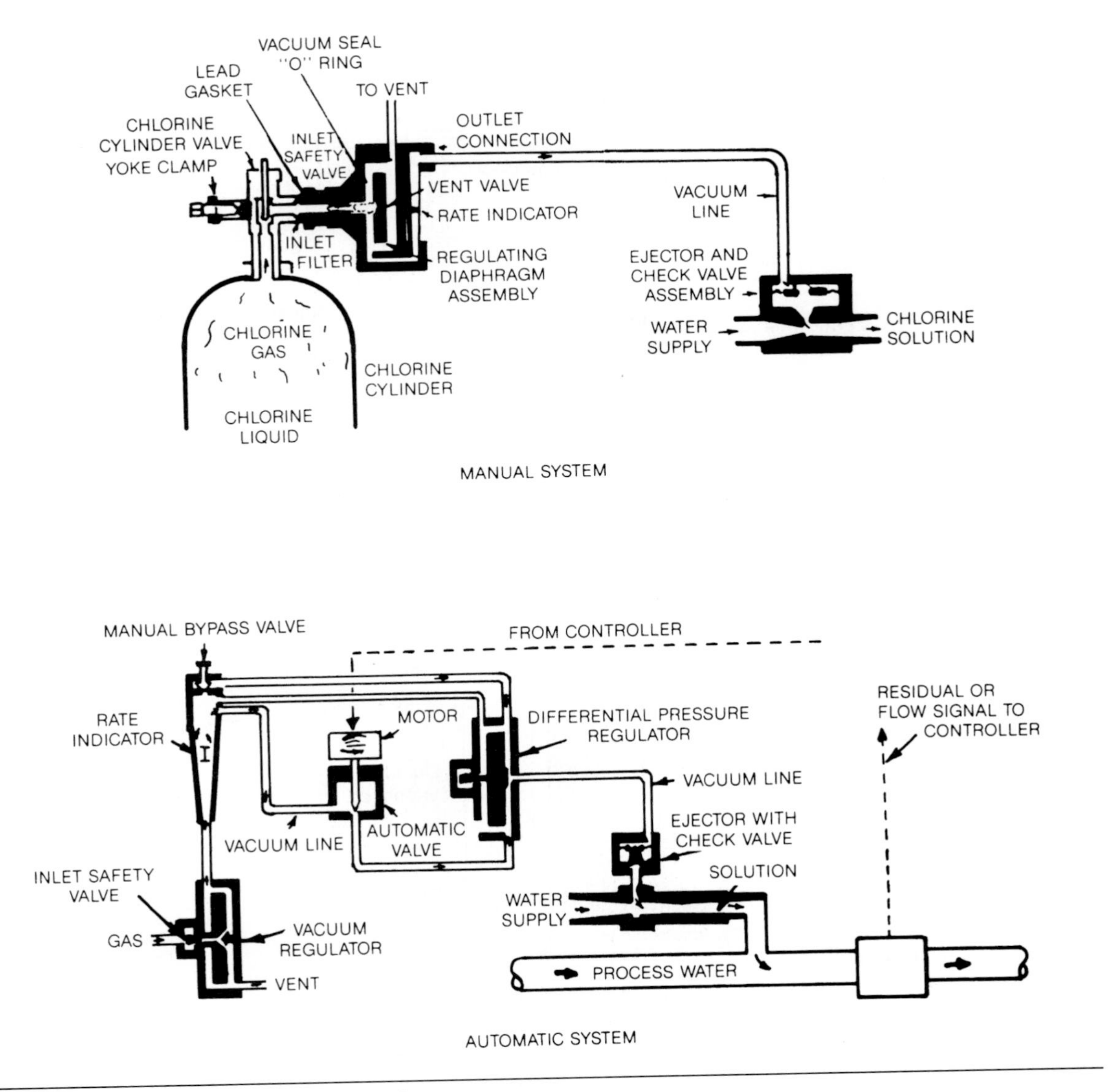


**FIGURE 9.5** Typical gas vacuum-feed systems.

ing overfeeding of the chemical and even process upsets in extreme situations. Careful selection of the injection point is required so that adequate mixing in the plant can be achieved. Under some highly turbulent conditions, adequate mixing within a pipe can require a pipe length as short as 10 pipe diameters. In-line static mixers also can be used to enhance mixing; however, consideration should be given to the mixing appli-

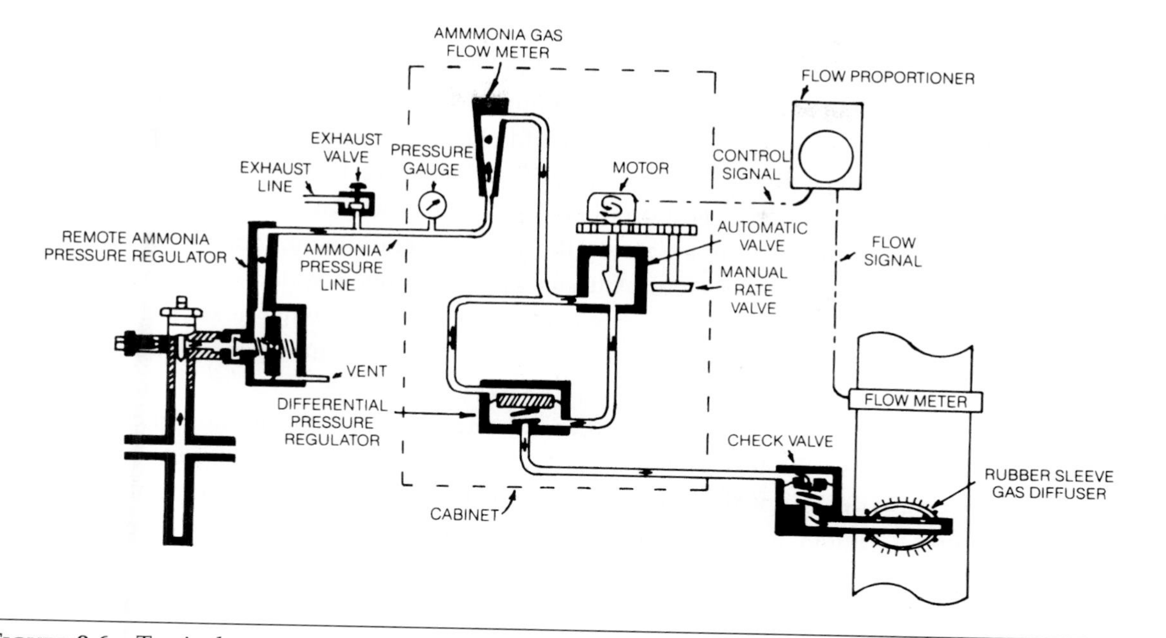


FIGURE 9.6 Typical ammonia gas pressure-feed system.

cation because certain process streams could foul the mixer and render it ineffective or nonoperational. In addition, the effect on any process control systems should be evaluated before using only a static mixer. Static mixers can provide good axial mixing across the diameter of the pipe but generally provide little, if any, radial or longitudinal mixing. Frequently, chemical feed systems use diaphragm-type chemical metering pumps that provide a pulsing flow and not a continuous flow. Therefore, if the control parameter, such as pH, is measured after the static mixer, the measured pH may show significant fluctuations with each flow pulse delivered by the pump. These control parameter fluctuations may prevent proper operation of the control system unless a mixed tank is used to smooth out the control parameter fluctuations. For injection to larger-diameter pipes, a pipeline diffuser can be used to inject the solution evenly across the diameter. Solution injected at a tee or saddle may tend to stay along the wall of the pipe and not enter the bulk fluid at the center of the pipe.

Pumps and valves can aid mixing. A hydraulic jump is sometimes used to mix chemical solutions with the process but requires careful analysis of hydraulic conditions under all possible flow regimes that may occur. Static mixers are used to improve mixing efficiencies. Mechanical devices such as vertical mixers, in-line mixers, and side-entry mixers can also be installed.

***Operational Considerations.*** Operational problems with sulfonators, chlorinators, and ammoniators can generally be traced to problems with the operation of the injector or clogging of the feed equipment by impurities. When problems occur in feeding with vacuum-type feeders, the operator should check the injector water supply to ensure that it conforms with the flowrate and pressure required by the equipment. If the injector is operating properly and producing sufficient vacuum, the operator should trace back through the vacuum system to the metering equipment, looking for an obstruction in the gas supply system. Automatically controlled systems can have a control malfunction that will drive the rate control device closed when it should be open. In that case, the unit should be switched to manual control until the automatic controls can be repaired.

Another operational consideration concerns verifying the amount of chemical fed each day. Gas cylinders should be mounted on scales and their loss in weight noted each day to determine the amount of chemical feed and check the chemical flow meter reading. Cylinder weighing also allows the operator to anticipate an empty cylinder and change it promptly without interrupting treatment. Equipment scales are available to handle multiple cylinders and automatic switch-over from an empty cylinder to a full cylinder.

**LIQUID FEEDERS.** Sodium hypochlorite, some polymers, phosphoric acid, and ferric chloride are typically examples of chemicals fed as liquids. Caustic soda (sodium hydroxide) and hydrogen peroxide are also fed as liquids but are not commonly used in municipal wastewater treatment plants. Like the gas systems, liquid-feed systems have some means of receiving and storing the chemical, transporting and measuring it, and mixing and injecting it to the process. Liquid systems generally require pumps for conveyance.

The use of sodium hypochlorite (NaOCl) solution has become more prevalent in many plants today as a replacement for chlorine gas ($Cl_2$) as a disinfectant. There are certain advantages in the use of hypochlorite, such as safety issues, but it is important to consider some basic factors before converting to a liquid hypochlorite feed system. For example, to replace a 907-kg (1-ton) container of chlorine, at least 7.6 $m^3$ (2000 gal) of hypochlorite solution may be required. Hypochlorite solutions decompose over time and lose available chlorine for disinfection, so storage considerations are extremely important. In addition, there are several chemical differences. Hypochlorite solutions exhibit high pH (greater than 12) and add alkalinity to the water; however, chlorine creates an acidic solution when added to water and decreases the alkalinity.

Because chlorine gas is highly toxic, sodium hypochlorite provides an alternative even though the solution is corrosive and is a severe skin and eye irritant. One important safety consideration when using sodium hypochlorite is proper venting. Certain metals can cause the sodium hypochlorite solution to decompose to oxygen and salt. Pressure buildup in tanks, piping sections, and valves can be a concern if not properly vented.

***Description of Equipment.*** A typical solution-feed system consists of a bulk storage tank, transfer pump, day tank (sometimes used for dilution), and liquid feeder. Some liquid chemicals can be fed directly without dilution, and these may make the day tank unnecessary, unless required by a regulatory agency. Nonetheless, dilution water can be added to prevent plugging, reduce delivery time, and help mix the chemical with the wastewater. However, sometimes, the dilution water can have adverse chemical effects. For instance, dilution water that has not been softened can potentially cause calcium carbonate scale to build up on the piping. Special consideration should be given to the final water chemistry of the solution before adding dilution water. Figure 9.7 shows a typical solution-feed schematic.

Liquid feeders are typically metering pumps. Metering pumps are generally of the positive-displacement type using either plungers or diaphragms. An example of a diaphragm pump is presented in Figure 9.8. For details on pump maintenance and operation, the operator should consult the pump manufacturer's literature or the *Metering Pump Handbook* (McCabe et al., 1984). Positive-displacement pumps can be set to feed over a wide range (10:1) by adjusting the pump stroke length. In addition, the feed range can sometimes be extended by adjusting the pump speed as well. Sometimes control valves and rotameters may be sufficient; in other cases, the rotating dipper wheel-type feeder may be satisfactory. For uses such as lime slurry feeding, centrifugal pumps with open impellers or double-diaphragm pumps can be used. The type of liquid feeder used depends on the viscosity, corrosivity, solubility, suction, discharge head, and internal pressure-relief requirements.

The chemical addition rate can be set manually by adjusting a valve or the stroke/speed on a metering pump. The operators should obtain or develop a set of calibration curves showing the percent of full stroke versus the pump discharge. Alternatively, those adjustments can be made automatically with instrumentation. Automatic-feed systems can be designed to control the feed flow based on a process variable such as influent flow, residual chlorine concentration, or pH. For instance, a control scheme could pace the pump speed based on wastewater flow and use another measured parameter to "trim" the pump stroke. Whatever the control scheme used, to ensure control of the chemical addition the operator needs to carefully check the process treatment units, maintain storage and day tank inventories, and understand the equipment.

Pressure relief should be provided for positive-displacement metering pumps to prevent line failures if all discharge valves or pump isolation valves are closed. Evi-

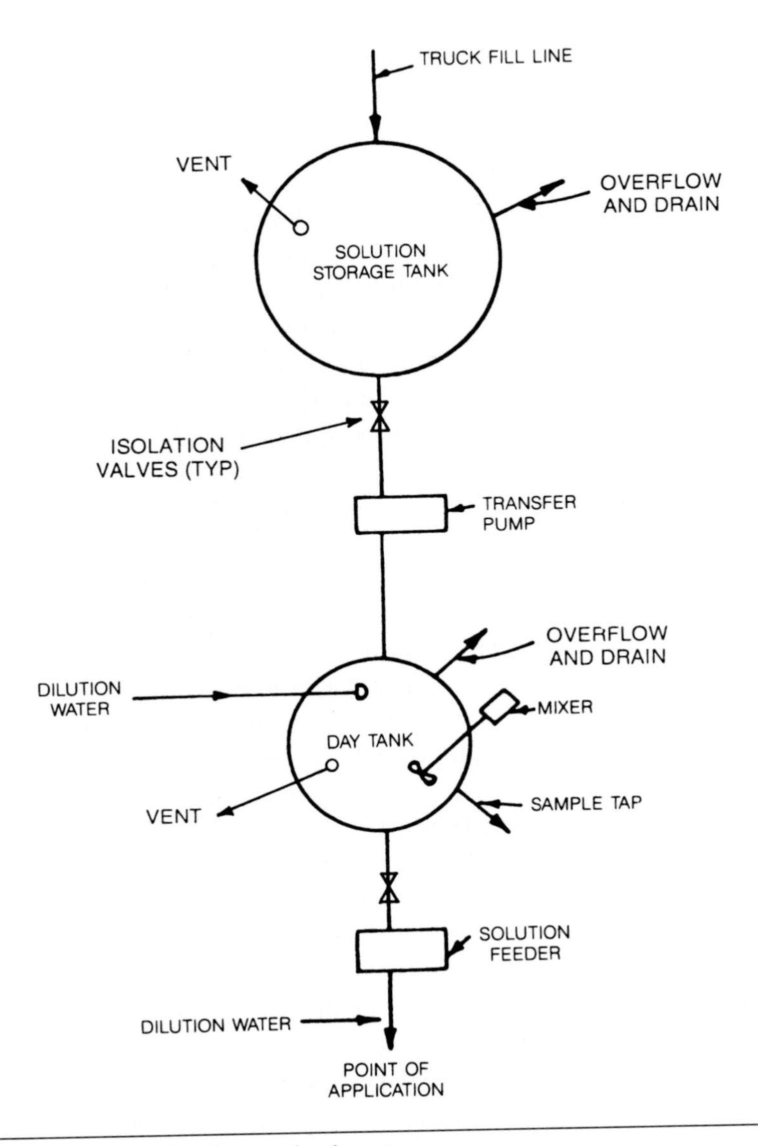


**FIGURE 9.7** Typical solution-feed system.

dence of discharges from expansion-relief valves should be reported at once and the cause investigated and corrected. Backpressure valves are also required with positive-displacement pumps to ensure proper pressure differential between the suction and discharge valves and to provide sufficient backpressure to seat the pump discharge check valve. Most metering pumps should have a 34- to 69-kPa (5- to 10-psi) differen-

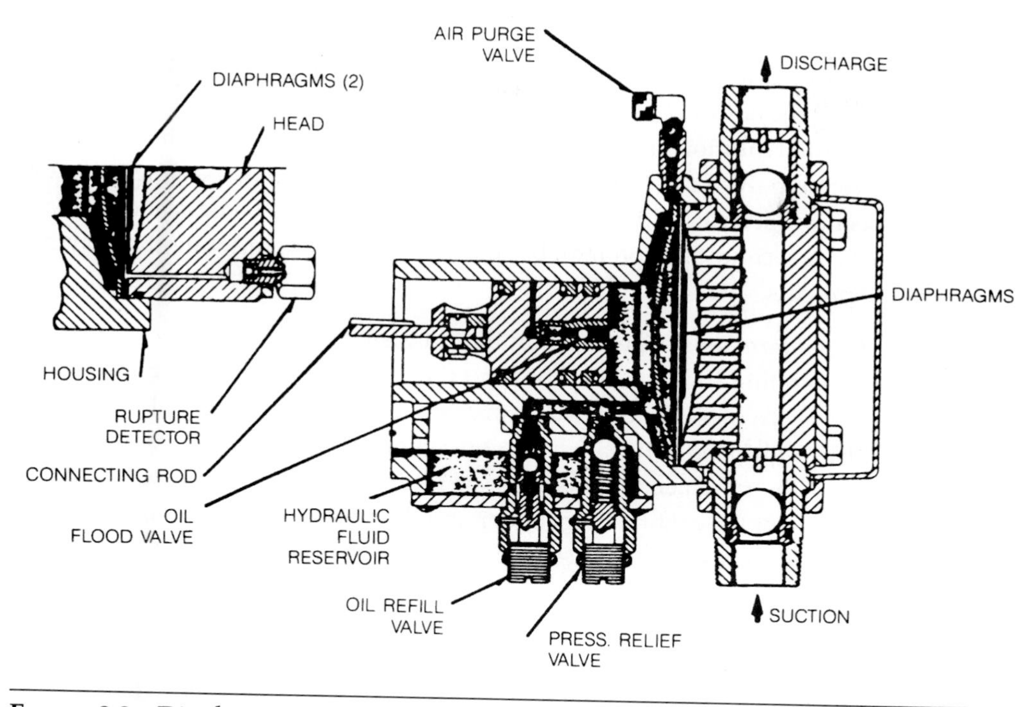


**FIGURE 9.8** Diaphragm pump.

tial across the valves. If this differential is not available, the installation of a backpressure valve can develop additional head pressure close to the pump discharge connection (McCabe et al., 1984). A good liquid-feed system also includes valving so that the lines, pumps, and meters can be isolated from the process, cleaned and/or drained, and prepared for maintenance.

In addition, the operator should make provisions for other accessories as part of a chemical metering pump system using positive-displacement pumps. The installation of a pressure gauge in the discharge line is used for monitoring pressure at the pump and for assisting with setting relief-valve pressures or charging pulsation dampeners (McCabe et al., 1984). Pulsation dampeners are typically located in the discharge piping as close to the pump discharge connection as possible. Dampeners are also sometimes placed in the suction side of the pump. The location of the discharge pulsation dampener is critical to absorb the maximum fluid accelerations created by a positive-displacement pump. The pulsation dampener should be isolated with a valve to allow for maintenance. Most plastic pipe chemical-feed systems should include provisions for pulsation dampeners.

Polymers added to aid settling often require special attention for adequate mixing. For optimum performance, the manufacturer's instructions for preparing the dilute solution and locating the addition point should be followed explicitly. These instructions may include an initial dilution of the polymer with water followed by an hour or more of mixing to age the polymer before feeding the solution. Feeding a polymer solution that has not been properly aged will reduce the chemical's effectiveness and increase processing costs. A second dilution may be required just before injecting the polymer solution to the process.

***Operational Considerations.*** Chemical and liquid feeders, such as metering pumps, can have many operating problems that prevent the feeder from delivering the correct amount of chemical. To monitor feeder delivery, all liquid feeders, especially metering pumps, should be equipped with a calibration cylinder. The cylinder can be either empty or full, and the time needed to empty or fill the cylinder should be noted. By knowing the volume of the calibration cylinder and the time for filling or emptying the cylinder, the operator can calculate the exact pump output. This determination serves as a check on the pump capacity.

Many metering pump systems handle chemicals that coat or build a layer of residue or slurries that can settle out solids during operation. Strainers are helpful in removing large particulate, but the operator must keep these cleaned. Periodic flushing to remove residues and deposits is often required. Piping and valve arrangements should allow the system to be isolated so that a clear liquid, such as water, can be used to pressurize the system for flushing the residue or solid buildup. Such flushing systems can be operated manually using hand-operated valving or can be automatically operated using solenoid valves with a timer control system. Systems where the metering pumps and piping are periodically shut down will require flushing connections to remove solids. In addition, an allowance for tees and wye cleanouts should be included for the piping system where longer horizontal piping runs cannot be adequately flushed.

A metering pump will lose capacity and become erratic when the suction or discharge valves become worn or when poor hydraulic conditions exist. These conditions will be indicated by the cylinder test described above. Also, debris in the chemicals being fed may obstruct or block the check valves, thus impeding their operation and decreasing the pump's performance.

Some chemicals such as hydrogen peroxide and sodium hypochlorite need metering pumps specially designed to handle offgassing. Offgassing is gas produced from the chemical during storage and feeding. Other types of offgas-relief systems can be incorporated to the feed pump piping also.

**DRY CHEMICAL FEEDERS.** Lime, alum, and activated carbon are typical of the kinds of chemicals used with a dry chemical-feed system. These systems are complex because of their many storage and handling requirements. The simplest method of feeding dry or solid chemicals is by hand. Solid chemicals may be preweighed and added or poured by the bagful into a dissolving tank. This method generally applies only to small plants where dry chemical-feed equipment is used.

***Description of Equipment.*** A dry installation (Figure 9.9) consists of a feeder, a dissolver tank, and a storage bin or hopper. Dry feeders are of either the volumetric or gravimetric type. Volumetric feeders are generally used only where low initial cost and low feed rates are desired, and less accuracy is acceptable. These feeders deliver a constant, preset volume of chemical and do not respond to changes in material density. The volumetric feeder is initially calibrated by trial and error and then readjusted periodically if density of the material changes.

Most volumetric feeders are included in the positive-displacement category. All designs of this type use some form of moving cavity of a specific or variable size. A belt, screw, or auger can provide the cavity. The chemical falls into the cavity and is almost fully enclosed and separated from the hopper's feed. The rate at which the cavity moves and empties and the cavity size govern the amount of chemical fed (National Lime Association, 1995). Some types of volumetric feeders can be subject to flooding. This can be especially important for those feeding from bins or large hoppers. Flooding occurs when chemical is forced through the feeder (such as by chemical weight or momentum) in an uncontrolled fashion such as through the center of a helical screw auger. Rotary valves or other devices may be required upstream of these feeders to prevent a flooding condition.

When extreme accuracy and reliability are required, the gravimetric or weigh feeder is recommended. Most of the feeding principle differences lie in the system of scales, levers, and balances necessary to maintain a continuous flow of material at a predetermined weight versus time rate. Although the material may change in form, size, or density, the gravimetric feeder automatically compensates for the difference.

Gravimetric feeders take many forms but generally are classified into three groups: the pivoted-belt group, the rigid-belt group, and the loss-in-weight group. Gravimetric feeders are not "official" scales. The feeder feeds first and then checks and adjusts the feed through weighing (National Lime Association, 1995). For feeding critical processes, weigh feeders should be checked periodically as noted in the discussion of operational considerations.

Slide gates, knife gate valves, or other devices are typically installed between the bin or hopper and the feeder. These devices allow the feeder to be isolated from the bin

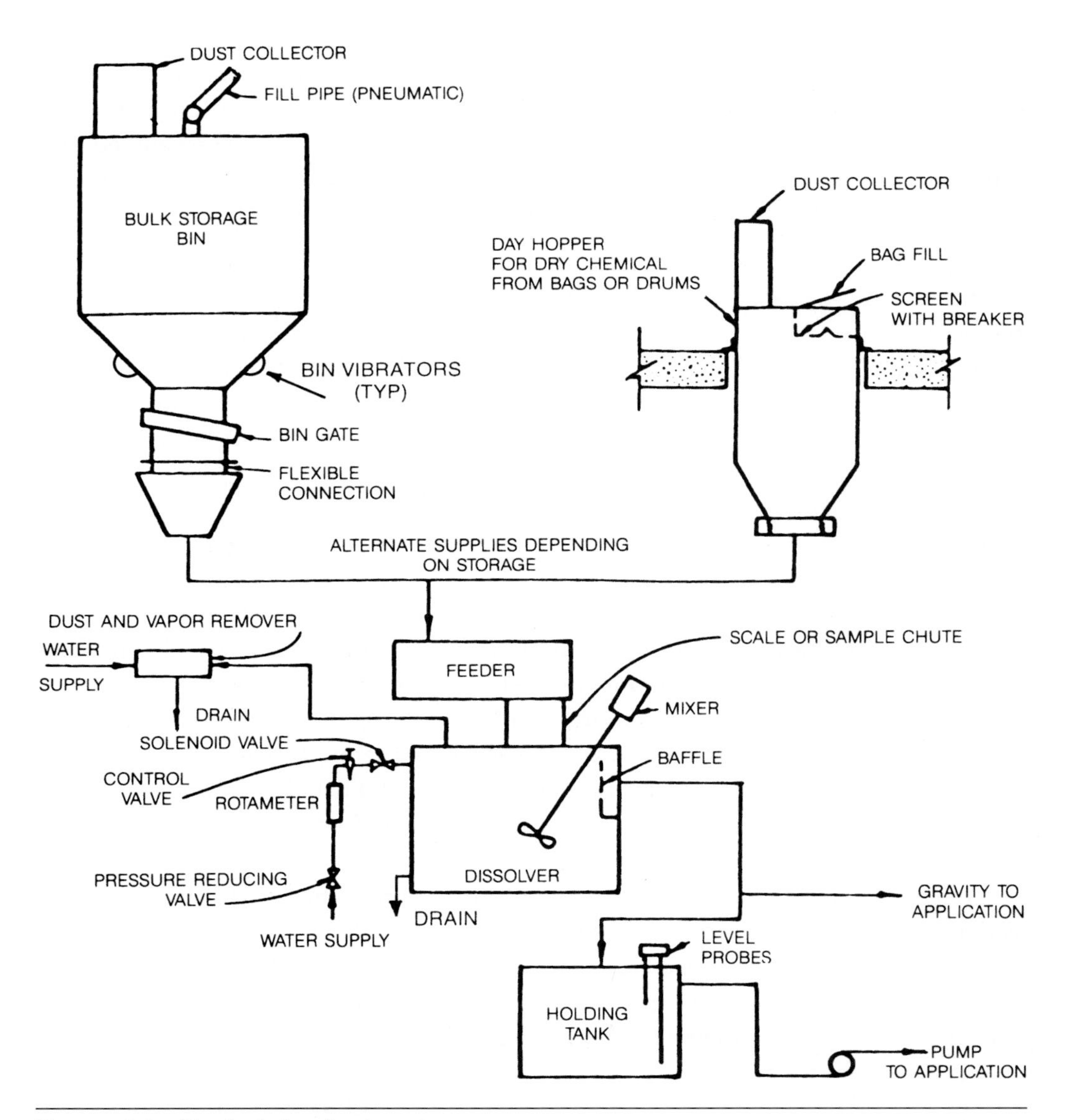


**FIGURE 9.9** Typical dry-feed system.

or hopper so that the feeder can be maintained or replaced without emptying the bin or hopper. A flexible coupling also is frequently installed below the gate or valve to isolate vibrations produced by the feeder.

Dissolvers are a key component of dry-feed systems because any metered chemical must be wetted and mixed with water to provide a chemical solution free of lumps and undissolved particles. Most feeders, regardless of type, discharge their material to a small dissolving tank equipped with a nozzle system or mechanical agitator, depending on the solubility of the chemical being fed. The surface of each particle needs to be completely wet before it enters the feed tank to ensure thorough dispersal and avoid clumping, settling, or floating. When feeding some chemicals, such as polymers, into dissolvers, care must be taken to keep moisture inside the dissolver from backing up into the feeder. Moisture or moist air in the feeder can collect on the surface of the dry chemical in the feeder and result in clumps of chemical that clog the feeder. Methods of isolation can be used between the feeder and dissolver or heaters can be installed in the feeder to dry any moisture that gets into the feeder.

A dry chemical feeder can, by simple adjustment and change of speed, vary its output tenfold. The dissolver must be designed to closely match its application. A dissolver suitable for handling a feed rate of 4.5 kg/h (10 lb/h) may not be suitable at a rate of 45 kg/h (100 lb/h). Typically, the dissolver must provide a minimum detention time to allow the chemical to properly dissolve and must operate with a chemical concentration less than a maximum chemical concentration. The maximum chemical concentration must be below the chemical solubility in the water used to dissolve the chemical. Therefore, even though the chemical feeder may be made to feed a much higher feed rate, the dissolver may not be large enough to provide the minimum detention time or maximum chemical concentration required for proper dissolving. However, long detention times in dissolvers also should be avoided to minimize the delay between changing the feeder feed rate and a change in dosage of chemical in the wastewater. The longer the detention time, the longer the delay and the higher the potential for either adding too much or too little chemical for a period of time.

Most dry feeders are of the belt, grooved-disk, screw, or oscillating-plate type. The feeding device (belt, screw, disk, etc.) is typically driven by an electric motor. Many belt feeders, particularly the gravimetric type, also contain a material flow-control device such as a movable gate or rotary inlet for metering or controlling flow of the chemical to the feed belt. An example of a screw-type volumetric feeder is shown in Figure 9.10.

***Operational Considerations.*** Dry chemical-feeder output should be checked periodically by taking a "catch". That is, the output from the feeder is caught in a pan or similar device for a known period (such as one minute) and then the pan is weighed. The chemical output per day can then be calculated based on the precise measure of the amount added during the short interval.

A feed curve for the particular feeder should be developed, determining relationships between the capacity setting on the machines and the actual kilograms (pounds) fed per hour. The "catch" can then verify the expected capacity from the feed curve.

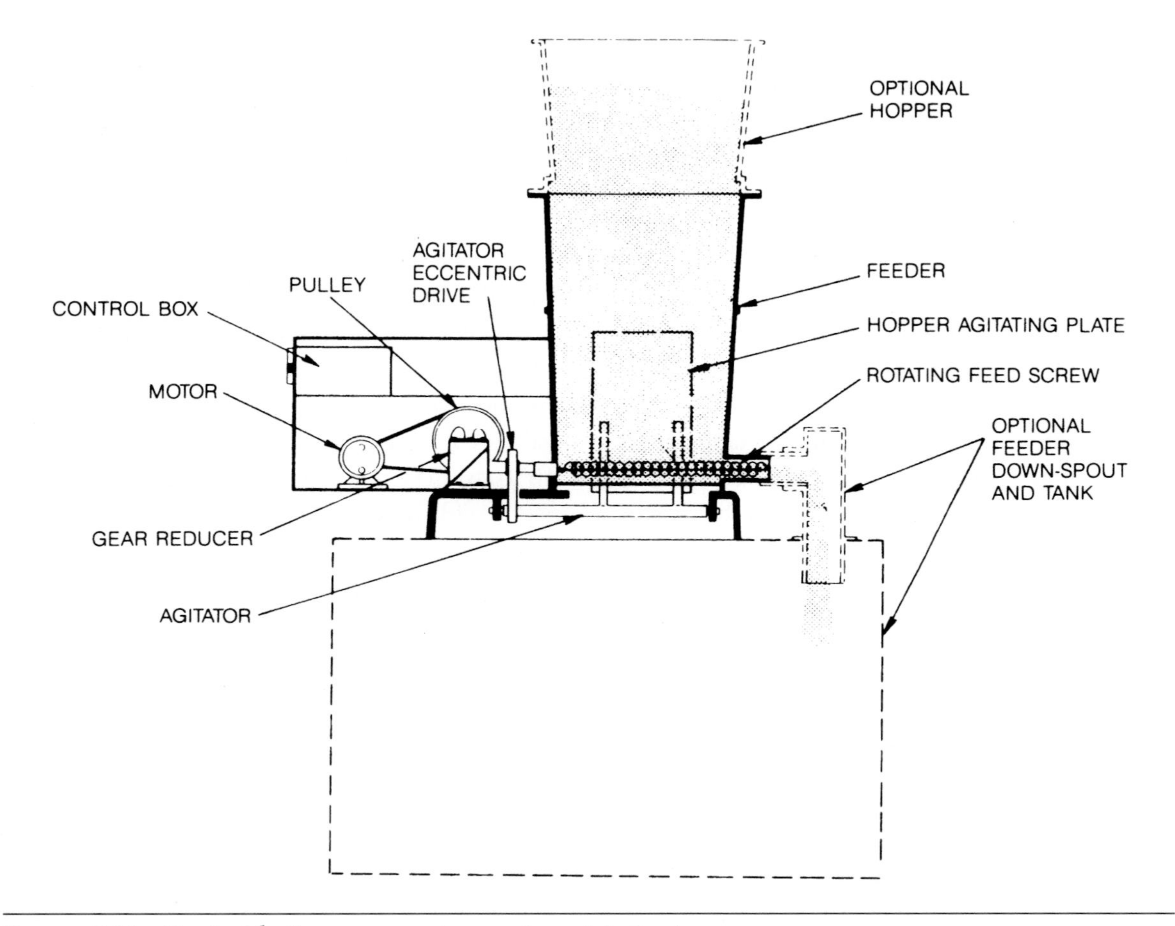


**FIGURE 9.10** Typical helix-or screw-type volumetric feeder.

**ON-SITE GENERATION.** Generation of chemical at the treatment plant site has limited application but is available for some chemicals such as sodium hypochlorite. On-site generation can have a number of advantages over chemical delivery and storage. On-site generation can eliminate the dependence on commercial chemical suppliers and concerns associated with transportation, handling, and storage of chemical. In addition, on-site generation can reduce or eliminate degradation that may occur during chemical storage.

Sodium hypochlorite is a chemical that can be generated at the treatment plant site and can be generated on demand through the electrolysis of a brine solution. Figure 9.11 shows a flow diagram of one system that is available to generate hypochlorite.

Standard skid-mounted generation units are available, as shown in Figure 9.12, that can be installed in the treatment plant. Local regulatory issues can limit the use of chlorine gas, requiring the use of sodium hypochorite for disinfection. Because sodium hypochlorite can experience significant degradation during storage, on-site generation can potentially reduce the total quantity of hypochlorite used and potentially the total chemical cost when compared to off-site purchase. However, a site-specific cost analy-

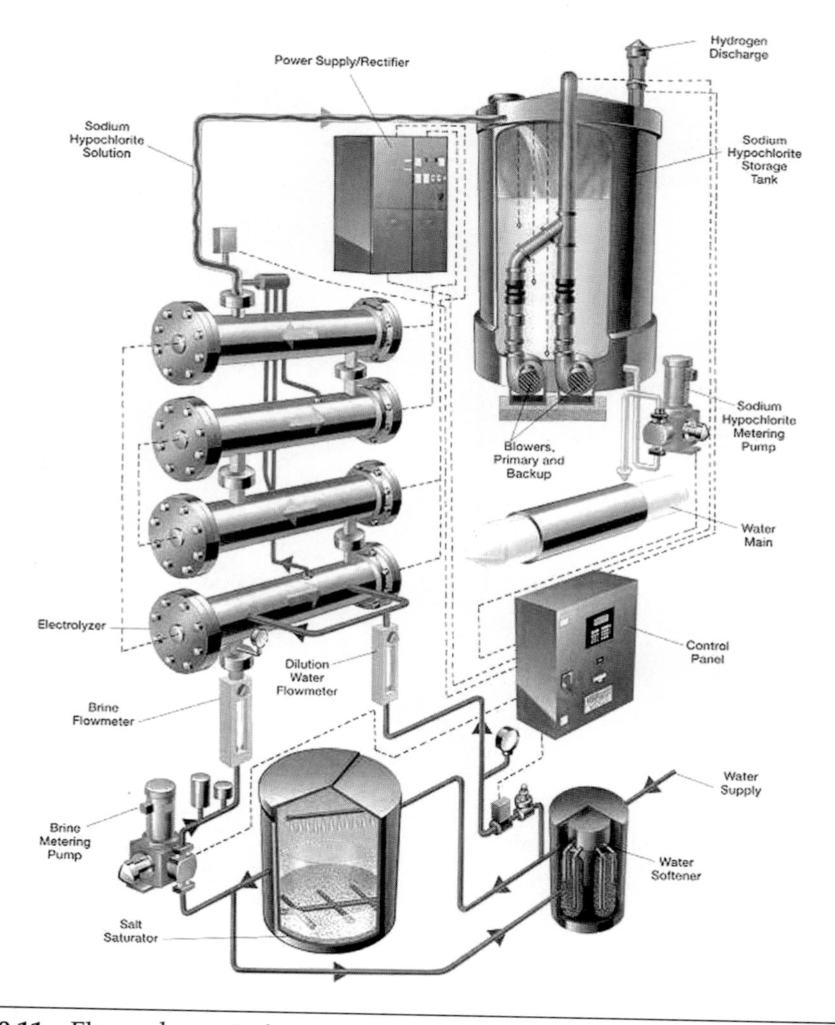


FIGURE 9.11 Flow schematic for on-site hypochlorite generation.

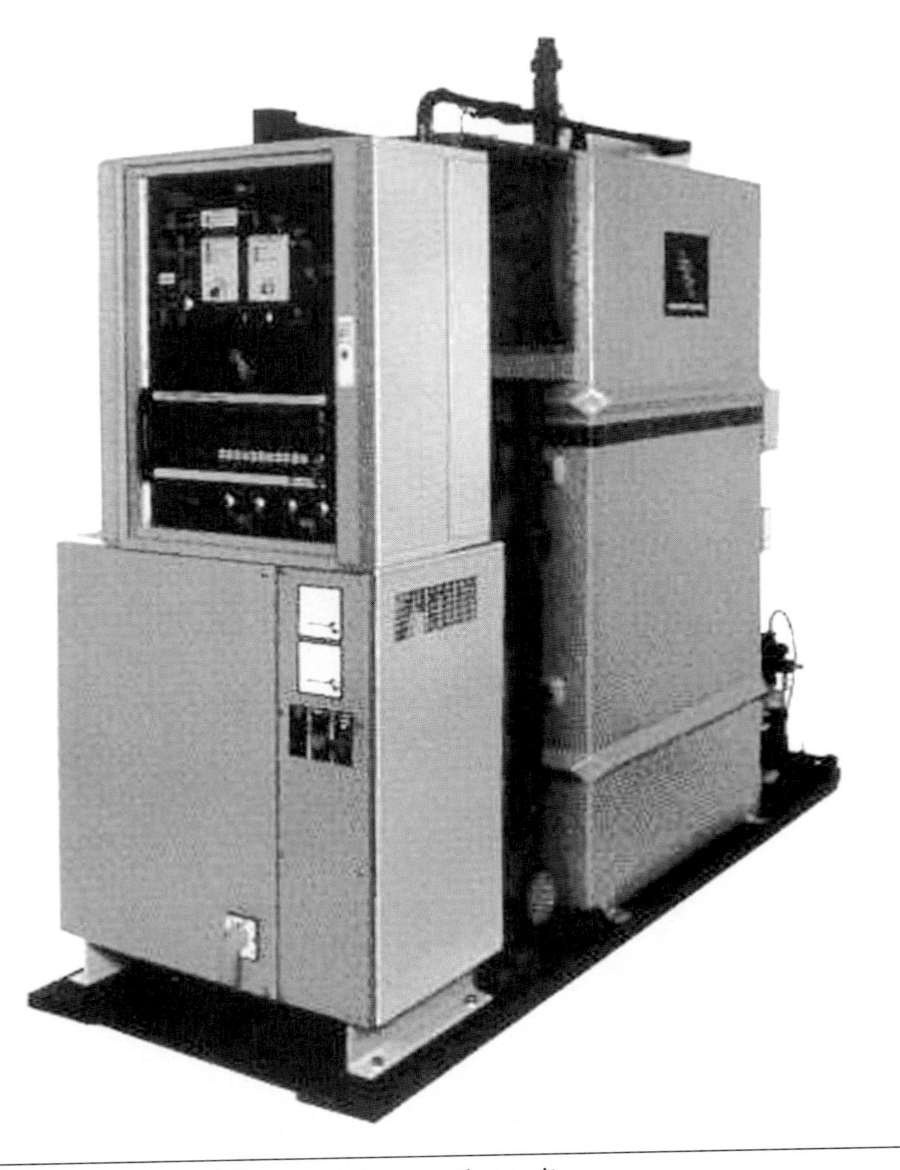

FIGURE 9.12 Standard skid-mounted generation units.

sis should be performed to confirm the cost effectiveness of on-site generation and determine the amount of savings, if any, available to pay off the additional equipment cost.

# HELPFUL HINTS

- Verify chemical delivered for proper type, amount, and concentration (if applicable).
- Always supervise the unloading. Never leave the chemical delivery truck unattended.
- Coordinate type of pipe connection and size with both chemical supplier and transportation company (this is important because many chemical suppliers broker the transportation). Verify that compressors for dry chemical unloading are available from the transportation company.
- When unloading dry bulk chemical (such as lime, alum, and soda ash), check interlocks, before unloading chemical, with dust collection systems to ensure proper operation. Also check dust filters periodically.
- Have waste buckets or some other method for collecting liquid waste remaining in transfer hoses after chemical transfer to prevent as much chemical as possible from getting on the ground.
- Double-check all bulk transfer connections for tightness.
- Monitor the level measurement equipment in the bulk storage tanks carefully during chemical transfer to avoid overfilling the tanks.
- Periodically check that tank vents are clear and unobstructed during liquid unloading operations.
- Coordinate with the chemical supplier to provide hydraulic lift gates for offloading of pallets or bags. Many plants may not have proper equipment to offload directly from the truck bed.
- Obtain proper weight tickets from the transportation company to verify the actual weight of chemical received to pay for only the quantity of chemical received. Check the accuracy of the weight tickets by comparison with the volume of chemical placed into the bulk storage tanks by noting the chemical level in the bulk storage tanks immediately before and after the chemical delivery.
- Verify with the transportation company that the trucks are equipped with the proper pressure-relief systems.
- Check bags delivered for tears, dampness, and other deformations and/or damage. Check drums received for leaks, dents, and other deformations and/or damage.
- Provide positive secondary containment for all chemicals.
- Check piping and vents for obstructions in liquid storage tanks.
- Periodically check for liquid leakage at storage (and day) tanks and associated containment area(s).

- Keep spill cleanup supplies, neutralizers, etc., fully stocked and accessible in the event of a minor leak.
- Periodically clean and calibrate level measurement and indication instrumentation in liquid and dry storage tanks.
- Periodically check alarm interlocks for low and high levels in tanks.
- Equip tanks with drain connections and isolation valves. Use of lockable drain valves is recommended.
- Equip tanks for OSHA-approved access (ladders, platforms, etc.) to monitor and maintain level instruments, vents, dust collectors, etc.
- Humidity control is critical for dry bag storage areas. Purchasing bagged chemicals with interior plastic liners is recommended.
- Exercise care when "cracking" drum bungs in the event of excess pressure.
- Periodically check and exercise chlorine (and sulfur dioxide) emergency gas scrubbing equipment, including but not limited to fans, controls, and chemical neutralizers.
- Use eductors for dry chemical makeup.
- Confirm adequate makeup water pressure and capacity.
- Confirm mixing requirements with the chemical supplier (i.e., slow speed mixing for polymer is generally required; however, the mixer must be able to generate sufficient torque to handle the higher viscosity).
- Use the appropriate respiratory protection equipment, as required, for chemical makeup to provide protection from dusting, acid flumes, etc.
- Equip mix tanks with drain connections and isolation valves.
- Inspect and clean mixing equipment and level instrumentation for buildups. This is very important for chemicals such as lime.
- Install calibration columns to verify feed rate accuracy from metering pump equipment.
- Provide isolation valves and drain valves, such as at the pump suction, for ease of maintenance and equipment removal.
- Inspect the condition of foot valve operators and suction piping. Excessive wear, corrosion, etc., may cause chemical metering pump systems to lose prime and not function properly.
- Minimize the length of suction and discharge piping when feeding lime slurry. Calcium carbonate buildup is a significant issue and a significant maintenance problem.
- Provide flush connections for cleaning and maintaining water slurry feed piping systems for chemical such as lime.
- Locate the chemical-feed system as close as possible to the feed application point.

- As a good operating practice, consider developing a spill prevention and countermeasures control plan (SPCC) for all chemicals even though federal regulations currently require a SPCC for petroleum and petroleum-derivatives only.
- Verify chemical compatibility with the materials of construction for pumps, piping, fittings, valves, tubing, gaskets, seals, etc. Consult the chemical supplier for guidance associated with proper material selection.
- Label, in accordance with federal and state regulations, all chemical storage tanks for hazard identification by plant personnel and emergency response personnel (fire department, hazardous material team, etc.).

# CHEMICAL DOSAGE CALCULATIONS

Presented below are several examples of chemical dosage calculations, each based on the following equation:

$$\text{Dosage, ppm} = \frac{(\text{kg chemical/d [lb/d]})}{(\text{mil. kg wastewater treated/d [lb/d]})} \tag{9.1}$$

The alternate form of the above equation is

$$\text{kg chemical/d (lb/d)} = (\text{dosage, ppm}) \times (\text{mil. kg wastewater treated/d [lb/d]}) \tag{9.2}$$

This calculation method applies to any chemical measured in kilograms per day (pounds per day).

**Example 9.1.** Calculate the chlorine dosage in ppm.

**Given:**

Average daily wastewater flow treated = 1890 $m^3/d$ (0.5 mgd)

Average daily chlorine use = 6.80 kg/d (15 lb/d)

**Solution:**

**In metric units**

$$\begin{aligned}\text{Dosage, ppm} &= \frac{(\text{kg chemical/d})}{(\text{mil. kg wastewater treated/d})}\\ &= \frac{(6.80\ \text{kg/d})}{(1890\ \text{m}^3/\text{d} \times 0.001\ \text{mil. kg/m}^3)}\\ &= \frac{(6.80\ \text{kg/d})}{(1.89\ \text{mil. kg/d})} = 3.6\ \text{ppm}\end{aligned}$$

**In U.S. customary units**

$$\text{Dosage, ppm} = \frac{(\text{lb chemical/d})}{(\text{mil. lb wastewater treated/d})}$$
$$= \frac{(15\ \text{lb/d})}{(0.5\ \text{mgd} \times 8.34\ \text{lb/gal})}$$
$$= \frac{(15\ \text{lb/d})}{(4.17\ \text{mil. lb/d})} = 3.6\ \text{ppm}$$

**Example 9.2.** Calculate the dosage of ferric chloride used as a coagulant for primary clarification.

**Given:**

Average daily flow rate = 21 200 $m^3$/d (5.6 mgd)
Ferric chloride use = 151.4 L of 40% by weight solution/d (40 gal of 40% by weight solution/d)
1 L (gal) of 40% solution = 1.33 kg (11.1 lb)

**Solution:**

**In metric units**

$$\text{Dosage, ppm} = \frac{(\text{kg chemical/d})}{(\text{mil. kg wastewater treated/d})}$$
$$= \frac{151.4\ \text{L/d} \times 1.33\ \text{kg/L} \times (40\%/100)}{(21200\ \text{m}^3/\text{d} \times 0.001\ \text{mil. kg/m}^3)}$$
$$= \frac{(80.5\ \text{kg/d})}{(21.2\ \text{mil. kg/d})} = 3.8\ \text{ppm}$$

**In U.S. customary units**

$$\text{Dosage, ppm} = \frac{(\text{lb chemical/d})}{(\text{mil. lb wastewater treated/d})}$$
$$= \frac{40\ \text{gal} \times 11.1\ \text{lb/gal} \times (40\%/100)}{(5.6\ \text{mgd} \times 8.34\ \text{lb/gal})}$$
$$= \frac{177.6\ \text{lb/d}}{46.7\ \text{mil. lb/d}} = 3.8\ \text{ppm}$$

**Example 9.3.** If the 3.8-ppm dosage were increased to 6 ppm with the same wastewater flow, how many liters per day (gallons per day) of ferric chloride would be needed?

**Solution:**

**In metric units**

$$\text{Liters of chemical needed} = \frac{6\text{ ppm}}{3.8\text{ ppm}} \times 151.4\text{ L/d} = 239\text{ L/d}$$

**In U.S. customary units**

$$\text{Gallons of chemical needed} = \frac{6.0\text{ ppm} \times 40\text{ gal}}{3.8\text{ ppm}} = 63.2\text{ gal}$$

**Example 9.4.** Calculate the concentration percentage (strength) of a diluted polymer batch.

**Given:**

Polymer used = 68 L/d (18 gal)
kg polymer/L (lb polymer/gal) = 1.25 (10.4)
Water used = 7570 L/d (2000 gal)

**Solution:**

**In metric units**

$$\begin{aligned}
\text{Batch concentration, }\% &= \frac{\text{kg polymer}}{\text{kg polymer} + \text{kg water}} \times 100 \\
&= \frac{68\text{ L/d} \times 1.25\text{ kg/L}}{(68\text{ L/d} \times 1.25\text{ kg/L}) + (7570\text{ L/d} \times 1\text{ kg/L})} \times 100 \\
&= \frac{85\text{ kg/d}}{85\text{ kg/d} + 7570\text{ kg/d}} \times 100 \\
&= \frac{85\text{ kg/d}}{7655\text{ kg/d}} \times 100 \\
&= 1.1\%
\end{aligned}$$

**In U.S. customary units**

$$\text{Batch concentration, } \% = \frac{\text{lb polymer} \times 100}{\text{lb polymer} + \text{lb water}}$$
$$= \frac{18\ \text{gal} \times 10.4\ \text{lb/gal} \times 100}{(18\ \text{gal} \times 10.4\ \text{lb/gal}) + (2000\ \text{gal} \times 8.34\ \text{lb/ gal})}$$
$$= \frac{187.2\ \text{lb} \times 100}{187.2\ \text{lb} + 16\ 680\ \text{lb}}$$
$$= 1.1\%$$

# REFERENCES


The Chlorine Institute, Inc. (1986) *The Chlorine Manual*, 5th ed.; The Chlorine Institute: Washington, D.C.

The Chlorine Institute, Inc. (2000) *Sodium Hypochlorite Manual*, 2nd ed; The Chlorine Institute: Washington, D.C.

Great Lakes—Upper Mississippi River Board of State and Provincial Public Health and Environmental Managers (1997) *Recommended Standards for Wastewater Facilities*; Health Research, Inc.: Albany, New York.

International Fire Code (1999). International Code Council, Inc.: Falls Church, Virginia; December.

List of Substances (2003) *Code of Federal Regulations*, Section 68.130, Title 40.

McCabe, R. E.; Lanckton, P. G.; Dwyer, W. V. (1984) *Metering Pump Handbook*, 2nd ed.; Industrial Press, Inc.: New York.

National Fire Protection Association (2001) *Standard System for the Identification of the Hazards of Materials for Emergency Response*, Standard No. 704; National Fire Protection Association: Quincy, Massachusetts.

National Lime Association (1995) *Lime Handling, Application and Storage*, 7th ed., Bull. 213; National Lime Association: Arlington, Virginia.

Occupational Safety and Health Standards (2003) *Code of Federal Regulations*, Section 1910.119, Title 29.

Prevention of Accidental Releases (2002) *Code of Federal Regulations*, Section 112r, Title I.

*Wastewater Treatment System Design Augmenting Handbook* (1997). MIL-HDBK-1005/16; Department of Defense: Washington, D.C.

Water Environment Federation; American Society of Civil Engineers (1998) *Design of Municipal Wastewater Treatment Plants*, Manual of Practice No. 8; Water Environment Federation: Alexandria, Virginia.

Water Environment Federation (2004) *Control of Odors and Emissions from Wastewater Treatment Plants*, Manual of Practice No. 25; Water Environment Federation: Alexandria, Virginia.

White, G. C. (1986) *Handbook of Chlorination*, 2nd ed.; Van Nostrand Reinhold: New York.

# Chapter 10

# Electrical Distribution Systems

# INTRODUCTION

The electrical distribution system has three major functions. First, the system transfers power from the transmission system to the distribution system. Second, the system reduces the voltage to a value suitable for connection to local loads. Lastly, the electrical distribution system protects the entire network by isolating electrical faults.

This chapter first provides basic terminology and concepts, followed by a discussion of typical electrical distribution systems and typical distribution-system components. Basic maintenance and troubleshooting are also covered.

Relay coordination and harmonics are covered in the next two sections. Staffing and training and high-voltage safety are also discussed. Related topics, such as utility metering/billing, energy-cost reduction, energy audits, and power-factor correction, are also presented in this chapter. Cogeneration, which is being used at many wastewater facilities, is also covered in this chapter.

# BASIC TERMINOLOGY AND CONCEPTS

**DIRECT CURRENT.** The two basic forms of electric current are *direct current* and *alternating current*. For direct current, the driving force (voltage) across any electrical load remains nearly constant and electricity, measured in amperes, flows in one direction. Batteries and direct-current generators provide direct current. Although direct current is typically not a major power source in wastewater treatment plants (WWTPs), it may nonetheless be used for charging batteries, certain instrumentation, breaker tripping power, or, in some cases, operating direct-current machinery. In some instances, direct-current power provides excitation current for synchronous motors or generators. One common application in wastewater treatment plants (WWTPs) is the use of direct-current power to operate chemical metering pumps.

**ALTERNATING CURRENT.** *Alternating current*, the most common commercial source of electricity, flows first in one direction and then in the opposite direction. Alternating-current voltage follows the patterns illustrated in Figure 10.1 (Jackson, 1989). The frequency of alternation in the United States is practically universal at 60 cycles/sec or 60 Hz. The alternating-current voltage across an electrical load increases to a maximum value. It then passes through zero to the same maximum value in the opposite direction. In theory, current-flow changes follow a sine wave (shaped by the sine of an angle as the angle increases from 0 to 360 degrees). In practice, the wave form may deviate slightly from a true sine wave.

**PHASED POWER.** Almost without exception, utilities deliver power as three-phase alternating current. *Three-phase power* involves three hot conductors, with voltage in each conductor 120 degrees out of phase with the other two. Although two-phase and other phase systems are possible, they are rarely used industrially. Single-phase current flows between one phase and another and between any phase and the ground. Where single-phase current is selected (e.g., lighting), circuits are designed to balance the loads (i.e., equal amperage among phases).

**POWER FACTOR.** *True electrical power*, typically expressed in watts (W) or kilowatts (kW), delivered to a circuit represents the actual power (as indicated by a watt meter) being consumed (Figure 10.2). Apparent power always exceeds the actual power in a circuit containing inductance or capacitance. *Apparent power*, typically

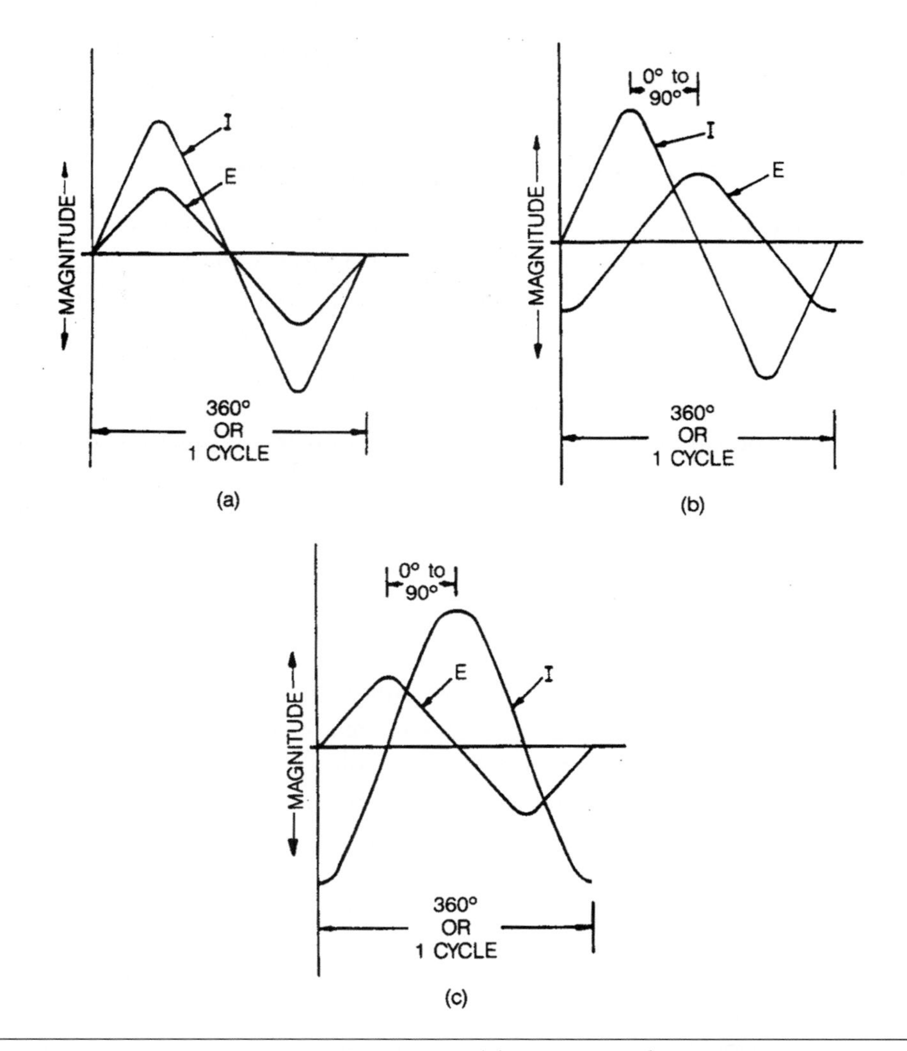


**FIGURE 10.1** Alternating current patterns: (a) circuit with coinciding voltage and current waveforms, (b) current waveform leading voltage waveform, and (c) current waveform lagging voltage.

expressed in kilovolt-amperes (kVA), is the product of the effective voltage on the load (as measured with a voltmeter) and the effective current in the circuit (as measured with an ammeter). The ratio of the true power to apparent power, called the "*power factor*" (cosine of the phase angle), can represent a measure of energy efficiency. An ideal power factor would be 1.0 (100%), but the power factor for typical plant systems ranges from 0.8 to 0.85 (80 to 85%) lagging. The power factor (PF) is calculated as follows:

$$\text{PF} = \frac{\text{kW}}{\text{kVA}} = \text{cosine } \varnothing \tag{10.1}$$

For example, if a boring mill operated at 100 kW and the apparent power was 125 kVA, then

$$\frac{(100 \text{ kW})}{(125 \text{ kVA})} = (\text{PF}) = 0.80$$

It is important to note that the power factor in a nonlinear environment does not follow these formulas or tables without filters or chokes installed on the harmonic generators.

**TRANSFORMERS.** There are two types of transformers used to decrease or increase voltage: step-down or step-up. A *transformer* consists of a set of windings around a core (coil), with a primary side that receives power and a secondary side that releases power.

A three-phase transformer containing three sets of coils may be connected in either a delta form or a Y form. In a delta form, each phase connects to another phase

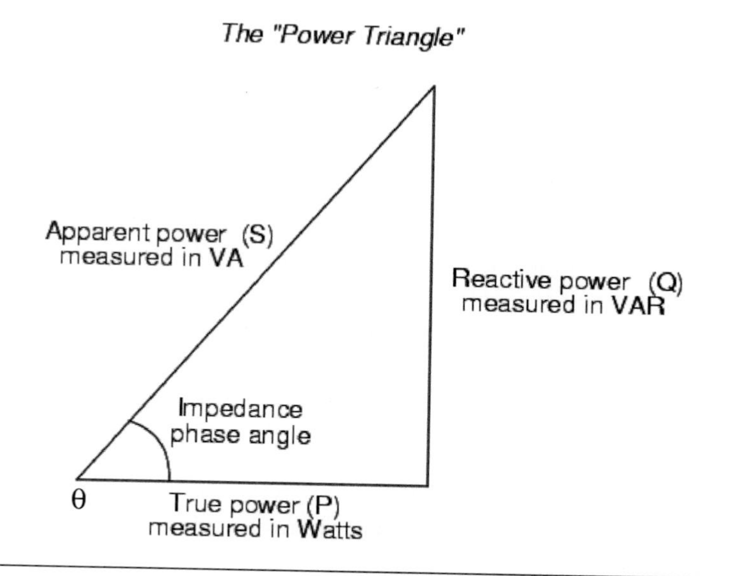


FIGURE 10.2 An example of the "Power Triangle" (Kuphaldt, 2007).

through a winding. In a Y form, each phase connects through a winding to a common point that is grounded as a safety feature. The primary and secondary connections are not necessarily the same. For example, the most common industrial step-down transformer includes a delta connection on the primary side and a Y connection on the secondary side.

Because power represents voltage times amperage, the higher the voltage delivered to a plant for a given power supply, the lower the amperage. In turn, the amperage squared multiplied by the resistance equals the power loss in wiring. As a result, the power company distributes power at the highest practical voltage to allow use of lower amperage and smaller wire.

High voltages in a plant increase the hazard of working with electricity and require more expensive insulated equipment. Therefore, a step-down transformer near the plant boundary will reduce the delivery voltage to a level usable in the plant. The most common form of power used in small plants comes from the Y winding on a transformer secondary, yielding 480 V between phases and 277 V to the ground. Higher voltages may be used within a plant to ease power distribution or to operate large motors. Typically, more than 373-kW (500-hp) systems or distribution systems as large as 4160 V (5 kV) can serve large plants.

**RELAY COORDINATION.** A *coordination study* is the process of determining the optimum characteristics, ratings, and settings of the power system's protective devices. The optimum settings are focused on providing systematic isolation of the faulted section of the system, leaving the remaining system in operation.

**HARMONICS.** *Harmonics* represent one of the component frequencies of a wave or alternating current, which is an integral multiple of the fundamental frequency. Based on a 60-cycle system, a third harmonics would have a frequency of 180 cycles. A harmonic analysis evaluates the steady-state effects of nonsinusoidal voltages and currents on the power system and its components. Some of the sources of these waveshape disturbances are direct-current rectifiers, adjustable-speed drives (ASDs), arc furnaces, welding machines, static power converters of all kinds, and transformer saturation.

**BASIC ELECTRICAL FORMULAS.** The relationships among current (I), voltage (E), and resistance (R) in a direct-current circuit are given as follows:

$$E = I \times R \tag{10.2}$$

For example, calculate the voltage if the amps are 2.0 and the resistance is 5 ohms.

Volts = 2.0 amps × 5 ohms
Volts = 10.0

Power consumed by a load in a direct-current circuit is expressed as follows:

$$P = I^2 \times R \tag{10.3}$$

For example, calculate the power if the amps are 2.0 and the resistance is 5 ohms.

Power = $2^2 \times 5$
Power = 20 watts

In an alternating-current circuit (single-phase, assuming the circuit only contains resistance), power is expressed as follows:

$$P = E \times I \tag{10.4}$$

For example, what is the power for a single-phase circuit if the volts are 10 and the amps are 2.0?

Power = 10 volts × 2.0 amps
Power = 20 watts

With a balanced three-phase system, the expression for power is

$$P = E \times I \times \text{Power Factor} \times 1.732 \tag{10.5}$$

For example, what is the power if the voltage is 240 across a three-phase motor, the current draw in any leg is 5 amps, and the power factor is 0.8?

Power = 240 volts × 5 amps × 0.8 × 1.732
Power = 1663 watts

# TYPICAL ELECTRICAL DISTRIBUTION SYSTEMS

**INTRODUCTION.** The input for a distribution substation is typically at least two transmission or subtransmission lines to provide redundancy. Input voltage may be 115 kV, for example, or whatever is common in the area. The output represents a number of feeders. Distribution voltages are typically medium voltage (between

2.4 and 33 kV) depending on the size of the area served and the practices of the local utility. System safety and reliability are paramount. For this reason, there is substantial redundancy in most distribution systems.

**TYPICAL LAYOUT.** The electrical-distribution system for any plant can be shown on the single-line distribution diagrams for that particular plant. These diagrams show power sources for all of the units drawing power and all feeders consuming power. Figure 10.3 shows a diagram of a hypothetical plant distribution system in block form, and Figure 10.4 shows the same system in electrical engineering format. Table 10.1 lists symbols used on the diagram, as well as other commonly used symbols.

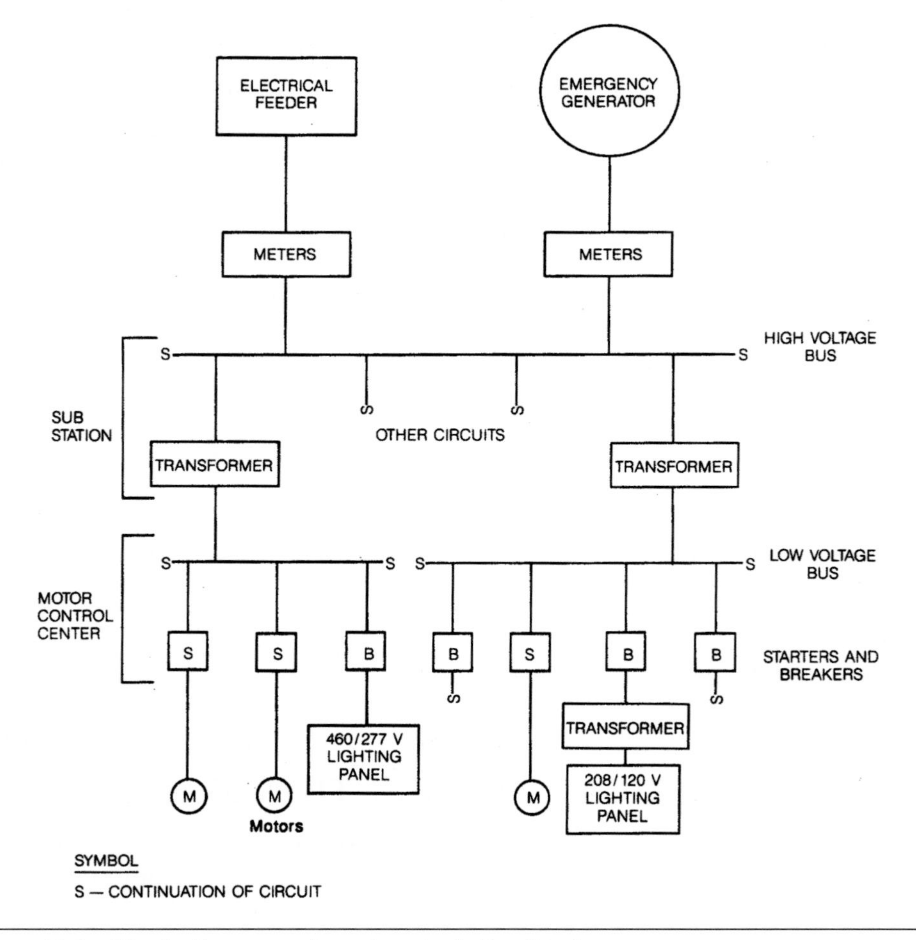


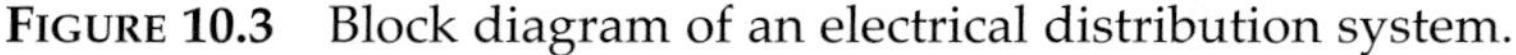

FIGURE 10.3 Block diagram of an electrical distribution system.

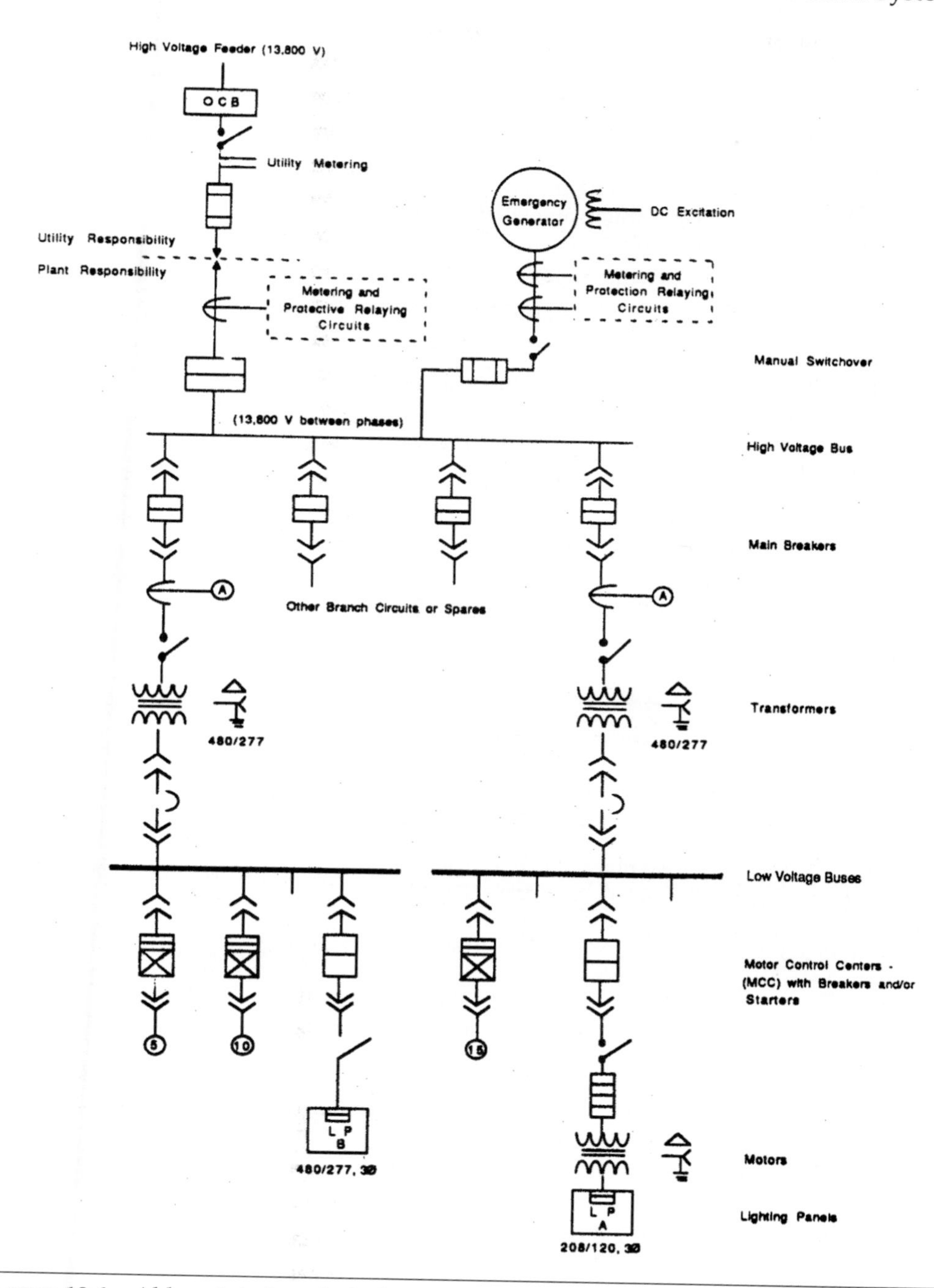


FIGURE 10.4 Abbreviated typical electrical distribution system (metering and protective equipment omitted).

**Table 10.1** Typical electrical symbols.

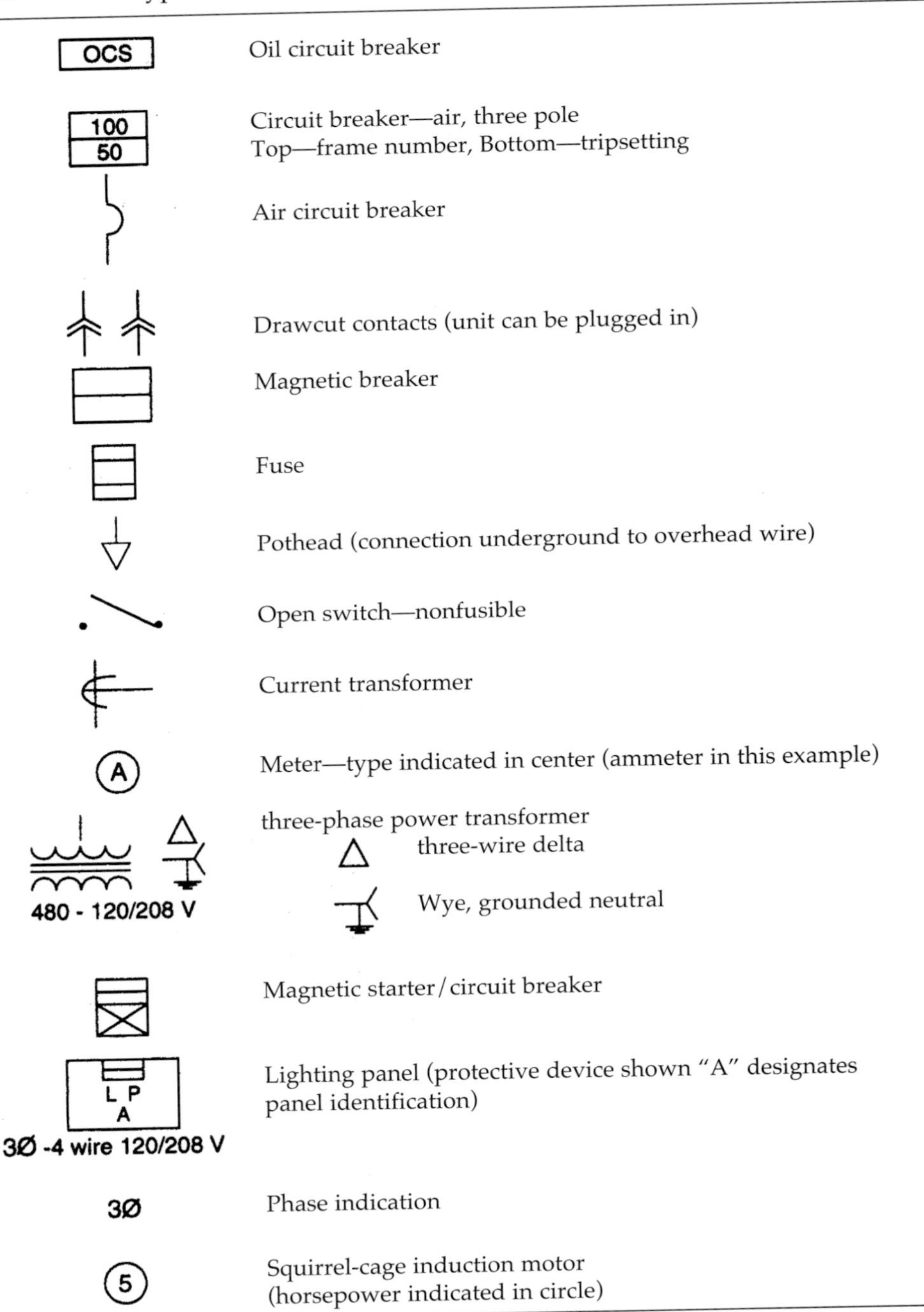

|  |  |
|---|---|
| OCS | Oil circuit breaker |
| 100 / 50 | Circuit breaker—air, three pole<br>Top—frame number, Bottom—tripsetting |
| | Air circuit breaker |
| | Drawcut contacts (unit can be plugged in) |
| | Magnetic breaker |
| | Fuse |
| | Pothead (connection underground to overhead wire) |
| | Open switch—nonfusible |
| | Current transformer |
| A | Meter—type indicated in center (ammeter in this example) |
| 480 - 120/208 V | three-phase power transformer<br>△ three-wire delta<br>Wye, grounded neutral |
| | Magnetic starter/circuit breaker |
| L P A<br>3Ø -4 wire 120/208 V | Lighting panel (protective device shown “A” designates panel identification) |
| 3Ø | Phase indication |
| 5 | Squirrel-cage induction motor (horsepower indicated in circle) |

In the example shown, the plant provides a transformer to change the voltage of its delivered power to a lower voltage for plant distribution. In many cases, however, the utility may provide the transformer, thereby supplying the plant with the normal distribution voltage for plant use.

In the diagram shown, the utility's high-voltage feeder and the plant's transformer deliver power to the plant. The utility provides its own metering and certain protective devices. The utility maintains the equipment it supplies, while the plant maintains all other equipment. For the case shown, the high-voltage power comes to a high-voltage bus. The plant's emergency generator also feeds the high-voltage bus in this example. Additionally, there can be other feeders operating either simultaneously, in parallel, or with a manual or automatic switchover.

In the largest stations, all the incoming lines have a disconnect switch and a circuit breaker. In some cases, the lines will not have both (i.e., only a switch or a circuit breaker is needed). Typically, these will also have a current transformer to measure the current coming in or going out on a given line.

Once past the switching components, the lines of a given voltage all tie in to a common bus. This is a number of thick metal bus bars (almost always three bars) because three-phase current is almost universal.

The most sophisticated substations have a double bus, in which the entire bus system is duplicated. Most substations, however, will not have a double bus because it is typically used for ultrahigh reliability in a substation whose failure could bring down the entire system.

Once buses are established for the various voltage levels, transformers may be connected between the voltage levels. These will again have a circuit breaker, much like transmission lines, in case a transformer has a fault (commonly referred to as a "short circuit").

Additionally, a substation always has the control circuitry and protective devices needed to command the various breakers to open in case some component fails.

A typical electrical-distribution system in smaller plants is 480/277 V (Y system). This system, with the neutral connection (a common connection in Y), economically supplies lighting loads at the 277-V level.

## COMPONENTS OF A DISTRIBUTION SYSTEM

**FEEDERS.** Electricity is transmitted over large areas using ultrahigh voltage and low currents to limit voltage drop. This typically occurs through overhead lines that are protected with lightning arresters. Once delivered to substations, the voltage is stepped down to the delivered voltage, which is typically 480 V-alternating current (V-ac). Once in the plant, engineers design a distribution system to power all the equipment in the plant. In an effort to create a high-reliability design, engineers will protect the transmission media by adding redundancy, breakers, lightning arresters, buried cable, emergency generators, and uninterruptible power systems (UPSs).

**AUTOMATIC TRANSFER SWITCH.** The automatic transfer switch (ATS) ensures system reliability by providing a continuous supply of power. An ATS can automatically sense the best source of power from two feeds and then direct appropriate opening and closing of motor-driven disconnect switches (circuit breakers) to connect that source to the load.

The primary purpose of the ATS system is to automatically sense loss or decrease of voltage, determine that an acceptable alternate source of voltage exists, and switch the load to that alternate source. A voltage of 120 V-ac obtained from potential transformers in preferred and alternate sources is continuously sampled to determine which source should feed the load. Should a fault current be detected in either the preferred or alternate source at the time of closure, the ATS will prevent the potentially damaging action.

**SWITCHING FUNCTION.** An important function performed by a substation is *switching*, which refers to the connection and disconnection of transmission lines or other components to and from the system. Switching events may be "planned" or "unplanned". A transmission line or other component may need to be de-energized for maintenance or for new construction (e.g., adding or removing a transmission line or a transformer).

To maintain reliability of supply, no company ever brings down its whole system for maintenance. All work that needs to be performed—from routine testing to adding entirely new substations—must be done while keeping the whole system running.

More importantly, perhaps, a fault may develop in a transmission line or any other component. Examples of this include a line that is hit by lightning and develops an arc, or a tower that is blown down by high winds. The job of substations is to isolate the faulted portion of the system. There are two main reasons for this. First, a fault tends to cause equipment damage and, second, it tends to destabilize the entire system. For example, a transmission line left in a faulted condition will eventually burn down. A transformer left in a faulted condition will eventually blow up. While these events are happening, the power drain makes the system more unstable. Disconnecting and isolating the faulted component quickly tends to minimize both problems.

**SERVICE TRANSFORMERS.** Service transformers "step down" the voltage from ultrahigh voltage to usable voltage. Many plants' equipment run on 480 V-ac and then have a local transformer to step the voltage down to 120 V-ac to power lighting panels. The primary side of the step-down transformer is the high-voltage side, and the secondary side is the customer load side. Typically, there is a main breaker that protects the transformer from the utility grid and can isolate the plant for load testing of generators. A power company typically will have a complex series of system monitors and breakers to prevent any one system from disrupting other clients on the distribution network. The secondary side will typically have disconnect switches and breakers to cut the power on demand or when a short is detected on the load side.

Service transformers typically step down the voltage from 26 400, 13 200, or 4160. These transformers are typically owned and maintained by the utility company. In some cases, the customer may elect to own and maintain the transformer. Typically, customer ownership is chosen when an attractive high-tension rate is offered. The customer must weigh the reduced energy costs against ownership and maintenance expenses.

**TIE BREAKER.** A tie breaker is used to connect two discrete sections of a unit substation or motor control center, which operate independently. Tie breakers are typically open. Usually, a Kirk Key (Kirk Key Interlock Company, Massillon, Ohio) interlock system is used to allow only two of the existing three breakers (two mains and a tie) to be operated at any one time.

**PROTECTIVE RELAYS.** Protective relays monitor and disconnect loads from the power source if protective-device parameters are exceeded. The main types of protective relays are overcurrent, feeder, voltage frequency, and motor and programmable logic controllers (PLCs).

**STANDBY OR EMERGENCY POWER SUPPLY.** Standby or emergency power can be supplied either through an "in-house" generator or a separate feeder from the utility. The standby or emergency generator shown in Figure 10.3 has a manual switchover system, an arrangement obviating the need for complex equipment that increases the installation cost and could malfunction. If the plant cannot withstand even a short period without power or if the plant is unattended during certain periods, then the switchover device should be automatic.

To avoid the cost of owning and maintaining an emergency generator, a WWTP may receive power from a separate utility-owned line, preferably a power line not fed from the same feeder or substation. Although switchover can be automatic, it is typically manual, again to protect the utility. Depending on a plant's agreement with the utility, either the plant or the utility may control switchovers.

**SWITCHGEAR.** The term "switchgear", which is typically used in association with the electric power system, or grid, refers to the combination of electrical disconnects and/or circuit breakers used to isolate electrical equipment. Switchgear is used both to de-energize equipment to allow work to be done and to clear faults downstream. Switchgear is located anywhere that isolation and protection may be required (e.g., generators, motors, transformers, and substations). Although switchgear can be installed in any area protected from the weather, an enclosed, indoor location is preferable. While some outdoor installations are insulated by air, this requires a large amount of space.

**SUBSTATIONS.** The substation includes a main circuit breaker, the high-voltage bus, and several feeder breakers (called *circuit breakers*) supplying power to motor control centers (MCCs) and other electrical facilities. In some cases, the substation also includes a transformer.

Figure 10.5 shows a main step-down transformer in a large [more than 380 000-m$^3$/d (100-mgd)] WWTP. It is a delta/Y-type with 67 000 V on the primary side and 13 800-V

FIGURE 10.5 High-voltage transformer.

phase-to-phase on the secondary. High-voltage power comes in through insulated terminals on the top, and low-voltage power goes out through cables in the top duct to the substation bus. The transformer shown is cooled by oil, which circulates through an exterior cooling section. At another substation in the same plant (Figure 10.6), there are transformers that reduce voltage from 13 800 to 4160 V.

Figure 10.7 shows a 13 800-V breaker in the substation connected to the main transformer in Figure 10.5. The breaker is in the panel on the right (door open), and various meters and protective devices are shown to the left.

**MOTOR CONTROL CENTER.** The motor control center includes a series of steel cabinets that contain motor starters and breakers for other services. Typically, these devices are installed under cover because general-purpose enclosures are cheaper and easier to maintain. Power is distributed from the MCCs to motors and lighting panels. Typically, a transformer feeds one or more MCCs that are located near the units served. The MCCs, purchased as packaged units, include buses into which starters can be plugged. The MCCs can also accommodate special devices, such as PLCs, adjustable-frequency drives, and reduced-voltage starters.

FIGURE 10.6 Intermediate-voltage transformer.

**FIGURE 10.7** High-voltage circuit breaker.

The *starter*, a device that connects power to an electric motor, allows the motor to start. Motor starters contain coils, contacts, overload heaters, and springs. All these components can be replaced and/or repaired. Motor starters operate when the coil energizes. The starter controls the opening and closing of contacts that allow the flow of electrical power to the motor. Starters with built-in overload relays help protect the electric motor when the load becomes too great and threatens to overload and possibly damage the motor. In an overload situation, this type of protection can shut down the motor, thereby preventing it from overheating. Figure 10.8 shows an MCC in a WWTP. This motor control center has various cubicles that house starters with different capacity ratings.

**MOTORS.** While electric motors are only one component of an electrical drive, they are one of the most important. To obtain maximum efficiency and reliability from electric motors, considerable care must be given to their application, control, and protection. Industrial motor standards are published by the Institute of Electrical and Electronic Engineers (Piscataway, New Jersey), the American National Standards

FIGURE 10.8 Motor control center.

Institute (Washington, D.C.), the National Electrical Manufacturers Association (NEMA) (Rossyln, Virginia), and Underwriters Laboratories Inc. (Northbrook, Illinois). These standards are continually improved and updated. Motors are rated by horsepower and duty cycle; it is important to know the load that will be driven and the torque requirements when selecting an electric motor. Motors can be either direct current or alternating current. Direct-current motors are suitable in load applications supplying large torque, while alternating-current motors, either synchronous or induction, are used more extensively commercially. The squirrel-cage-induction alternating-current motor is the most common, with a characteristic speed–torque relationship determined by its construction.

**ADJUSTABLE-SPEED DRIVES.** The following five types of ASDs are available:

- *Variable-frequency drives* are electronic devices that control the speed of the motor by controlling the frequency of the voltage at the motor. These devices are used in a wide range of applications and can provide constant-torque and variable-torque operation. They are most efficient where the flow rates, speed, torque, etc. are not constant (e.g., wastewater pumping, aeration blower operation, or sludge handling). These units also affect power quality by producing harmonics in the system; such conditions can be corrected by line filters designed for the application. While ASDs save energy, they are complex electronic devices that require trained repair experts.

- *Direct-current ASDs* are electronic devices that control direct-current motors by changing the voltage applied to the motor. Direct-current ASDs are a traditional ASD device and are used almost exclusively for constant load (e.g., cranes, elevators, and hoists) and close-speed control.
- *Eddy-current drives* are electrical devices that use an electro-magnetic coil on one side of coupling to induce a magnetic field across a gap, creating an adjustable coupling. Eddy-current drives have been in wide use for more than 50 years in variable-torque, rough-duty, and high-starting torque applications. They are found in material-handling conveyors; heating, ventilating, and air conditioning (HVAC) pumps; and fans.
- *Hydraulic drives* are devices operating much like an automotive hydraulic transmission. Typical applications are found in constant-torque situations, difficult environments, and rough-duty applications. Hydraulic drives are found driving large pumps and conveyors.
- Mechanical devices can be used to control speed. Mechanical speed-control products include gearing, mechanical transmissions, and belt drives with variable-pitch pulleys.

**BRANCH POWER PANEL.** Branch power panels are supplied with power from a bucket of a MCC, which is transformed to low voltage (120 V-ac single-phase power) to supply circuit-breaker protection for building small power (e.g., lighting, receptacles, small fans, and motors).

**LIGHTING PANELS.** There are two basic types of lighting panels. Power comes to one unit as 480/277 V three-phase; the lighting operates at 277 V. This panel is used mostly for yard lighting, as well as lighting the larger process buildings. The other type of lighting panel serves a building, laboratory, or other facility with a number of convenient outlets for operating small equipment and power tools. A transformer precedes such a lighting panel to reduce the voltage to 120/208 V three-phase or 120/240 V single-phase.

**LIGHTING CONTROLS.** Lights can be turned on and off with simple manual switches, photoelectric cells, time clocks, motion detectors, and timers. Photocells, frequently used for outside lighting, require cleaning or replacement every 2 or 3 years. This task may be cumbersome because the photocells are typically in areas with difficult access and often require special equipment and cleaning. Accordingly, astrological time clocks are becoming popular. They run on an annual basis with the "power-on" time for night use increasing as winter approaches. The obvious disadvantage of such units is the need for clock resetting after a power outage.

Motion detectors conserve energy because the lights turn off when the space is vacant. Nonetheless, motion detectors are seldom used industrially. Simple timers that can be installed in a switch box operate more reliably than motion detectors. The timers can be set for the area's expected period of use. If a plant has a process computer, it can be used to turn lights on and off according to preprogrammed instructions. In all cases, override switches should be provided to accommodate access during emergency operation procedures.

**LIGHTING.** ***Lighting Intensity.*** Adequate lighting is needed for safety, efficient operation, and security at night. Individual efficiency has been shown to drop drastically when lighting levels are reduced at workstations.

The measure of lighting intensity on a surface is known as a lux (lx), or foot-candle (ft-c). In general, office and laboratory areas require lighting ranging from 345 to 1075 lx (32 to 100 ft-c). Reading and close work require higher ratings up to 1075 lx. Shops require 538 to 1615 lx (50 to 150 ft-c), with localized spotlighting requiring higher values. Outside areas, such as parking lots, require from 11 to 215 lx (1 to 20 ft-c), depending on the expected levels of activity. Because all lighting diminishes with dirt buildup on the fixtures and filament deterioration, measurements should follow a few months of usage to determine the adequacy of the actual lighting level. Recommended lighting levels for various work functions can be obtained from any manufacturer of lighting equipment.

***Incandescent Lights.*** The types of lights most commonly used in WWTPs are incandescent, fluorescent, and high-intensity discharge (HID). Incandescent light, the oldest form of lighting, uses the most energy per lux and has the shortest life. Incandescent bulbs operating at a lower voltage than rated last longer, but their efficiency is lower. Because incandescent fixtures are more compact and less expensive, incandescent lights are often used for spotlighting and lighting closets and other areas that require lighting for only short periods. Additionally, because of their lower cost, incandescent lights with a heavy enclosure are often used in areas where explosive vapors may be present.

***Fluorescent Lights.*** Fluorescent lighting is typically provided via long tubes. These tubes are more expensive than the simple electric light bulb, and fluorescent fixtures typically cost more than incandescent fixtures. Nevertheless, fluorescent lighting is typically more economical than incandescent lighting because it uses power more efficiently; in addition, a tube lasts as much as eight times longer than an incandescent bulb.

Fluorescent lighting provides good color, similar to that of incandescent lighting. Its uses include illuminating WWTP offices and laboratory areas. Some fluorescent lights require a short time for the arc to be struck; however, most industrial lights have instant-start capability.

***High-Intensity Discharge Lights.*** Three common types of HID units are mercury vapor, high-pressure sodium vapor, and metal halide. All of these types require a few minutes before reaching full intensity because they incorporate a circuit to heat and vaporize the metal or compound used in the lamp. Mercury vapor lamps have a long life and are cheaper than other HID units. Metal halide lamps are more efficient than mercury vapor lamps, but they cost more and typically are unavailable in small sizes. Although metal halide or mercury vapor lights may be needed in rare cases where color rendition is required, high-pressure sodium vapor units are typically best for HID lighting applications because of their low cost, long life, and efficient power use.

Although low-pressure sodium vapor units, which are more efficient than the other types discussed, still exist, they are cumbersome and have a much shorter life than the high-pressure sodium type. Most HID lights are used in areas that must remain illuminated for a long period of time (e.g., yard lighting required for the entire night and lighting in work or process areas where it is impractical to turn the lights on and off frequently). In such areas as pump rooms and switch-gear rooms, where lights can be turned off periodically to save energy, the fluorescent light is preferred because of the time required by HID lights to reach full intensity. In the past, HID lights, because of their brilliance, were not recommended where the luminaire would be close to a person's eyes. However, recently developed luminaires allow use of HID lights in areas with low ceilings.

***Emergency Lighting.*** Emergency lighting is required for illuminating critical control areas and for allowing egress from an area if the normal lights go out. An emergency generator that starts automatically with a power failure is wired separately to turn on emergency lights in critical areas. Instead of an emergency generator, battery packs are often used for evacuation. Sometimes the lights are attached to the battery unit. Battery packs are trickle charged using an alternating-current source and a built-in rectifier. They are connected to the same alternating-current circuit that supplies the normal lighting and thereby sense the loss of alternating current power.

**CONTROL CIRCUITRY.** Plant supervisors should understand the control diagrams for the equipment installed in their plant; therefore, they should request complete legends for the control-circuit diagrams.

Control circuitry for starters, shown on ladder diagrams, provides for control power as single-phase, with one side as ground and the other side as potential. Control diagrams for two hypothetical circuits are shown in Figure 10.9.

In the case of Circuit A, pressure on the start button allows current to flow through the stop circuit and closed switches [limit switch (LS) and time switch (TS)] to the main coil in the starter (M) and then through the closed overload contacts—one for each phase. As the main coil becomes energized, it pulls the three-phase contactor for the motor or other devices into the closed position, allowing the unit to start. Simultaneously, contacts labeled "MA" are opened or closed, depending on their original state. Contact MA, shown below the start button, allows the main coil to stay energized, even after the release of the start button. When the stop button is pushed, or the limit switch, the time switch, or an overload opens, the coil is de-energized and the motor stops. For example, the limit switch, external to the unit, may open on a high-liquid level. The time switch will open after a preprogrammed period elapses.

The circuit (Figure 10.9) illustrates a different type of starting switch, such as a level switch. With the switch in the off position, nothing runs. In the hand position, the motor runs. In the automatic position, the motor only runs if the remote limit switch is closed. The automatic line contains an interlock from Circuit A. This means the motor or other device for Circuit A must be energized before Circuit B can operate. For example, a vacuum filter can be interlocked to operate only when the conveyor that removes the dewatered solids is operating.

Both diagrams include overload contacts, which are actuated by overload heaters that sense motor current. Because the overload contacts depend on heat buildup to operate, they will open if a high current flows briefly or if a current higher than the rated maximum motor current continues for a longer period. Typically, the current rating of the overload contacts can be changed by replacement with a different size, or, in some cases, they can be manually adjusted.

Operators must learn why an overload occurred and correct the abnormal condition before restarting a tripped unit. Restarting a tripped unit two or more times within a short period may cause the motor or another electrical component to overheat and fail.

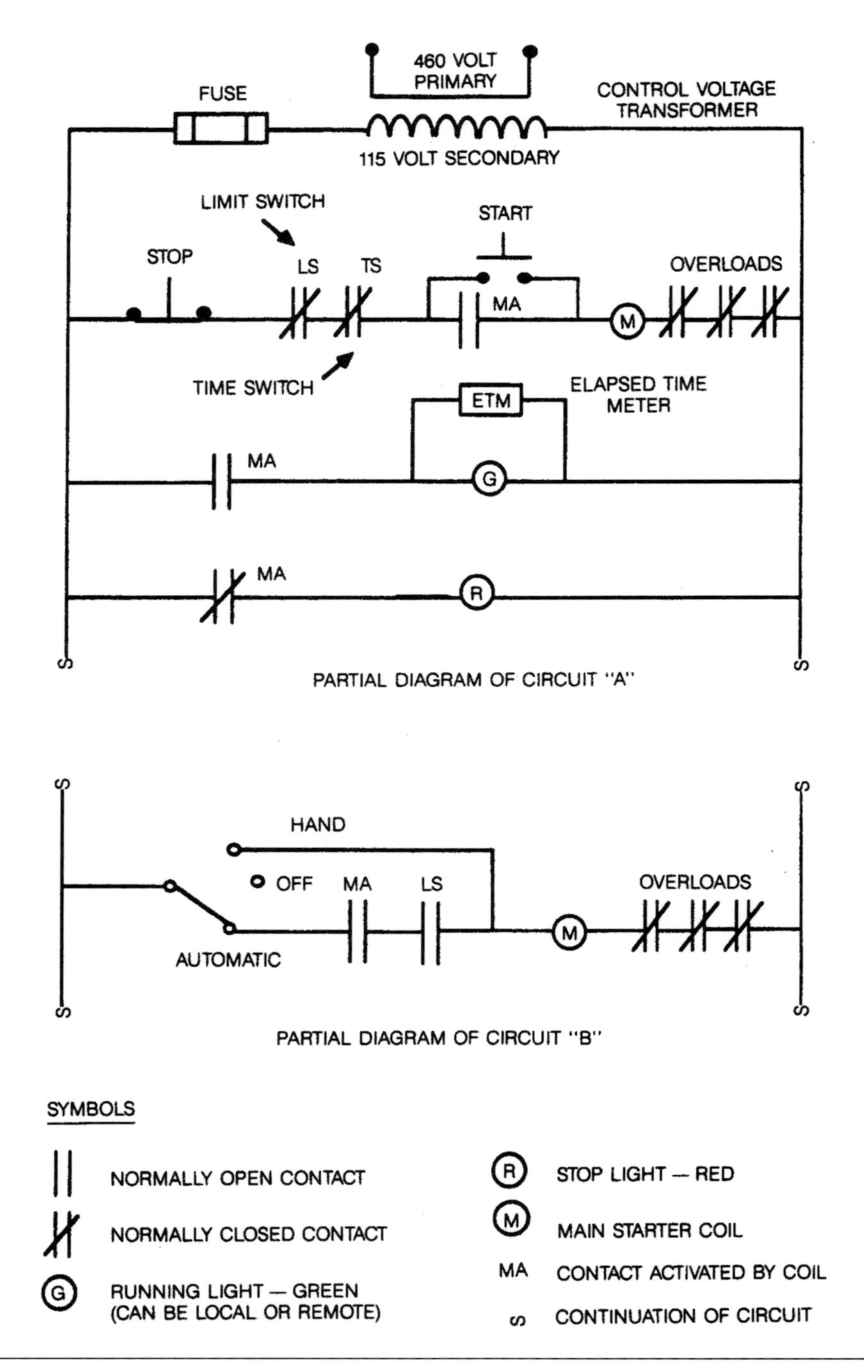


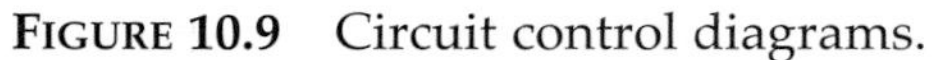

**FIGURE 10.9** Circuit control diagrams.

The best means of motor protection is to apply the type of start switch used in Circuit A. However, a disadvantage of Circuit A is that it deactivates on power failure; the motor must then be restarted manually when power returns. Conversely, the motor in Circuit B, with its hand-off-automatic switch, will start when power returns and the other switches are closed. A disadvantage of Circuit B is that the motor is not adequately protected from overload. If an overload occurs, the heater will restore contact after it cools. Then, the motor will attempt to start, and the heater will kick it out again. After three or more start–stop cycles, the motor may burn up. To prevent such multiple starts, additional protection can be installed.

If a motor coil develops an open circuit, the motor will still run, but in a single-phase mode; thus, it will eventually burn out. To protect the motor from an open circuit, a phase-failure relay will trip the starter. Various other types of timers and interlocks can be incorporated into starter circuits. Good maintenance practice includes having spare contacts for a starter to allow the addition of other functions in the future.

**PROGRAMMABLE LOGIC CONTROLLERS.** Programmable logic controllers are solid-state devices (a member of the computer family) that incorporate most of the control features of hard-wired relays. However, PLCs offer the advantage of being readily changed. They are capable of storing instructions to directly control such functions as sequencing, timing, counting, arithmetic, data manipulation, and communication for process control. A programmable controller has two basic sections: a central processing unit and an input–output interface system to field devices. Incoming signals from such devices as limit switches, analog sensors, selector switches, etc. are wired to the input interfaces; devices to be controlled (e.g., motor starters, solenoid valves, pilot lights, etc.) are connected to the output interfaces. The central processing unit accepts input data, executes the stored control program, and updates the output devices.

Programmable logic controllers can monitor electrical distribution systems and make critical decisions in milliseconds to maintain the health of the system. They can reallocate loads, start and stop emergency generators, and alert operators to potential problems.

**UNINTERRUPTIBLE POWER SYSTEMS.** Critical equipment should be protected from power losses with UPSs. They ensure continuity of power. Alternating-current power is supplied to a battery/charger system. The output of the battery is then connected to an inverter, which converts direct-current power to alternating-current power at 60 cycles. In the event of an alternating-current power failure, the battery maintains power to the system via the inverter. The input voltage to the UPS depends on the load served. Typically, the input voltage is 120-V single-phase or 480-V three-phase. The device is powered from a branch power circuit. Uninterruptible power systems are used to power computers, paging systems, and alarm systems.

**INSTRUMENTATION AND CONTROL POWER.** Critical instruments and PLCs can be powered by UPS systems. Most of the remote instrumentation in a plant is powered without UPS. In these cases, power comes from a branch power circuit.

Typical instrumentation and control power is either 120 V-ac or 24 V-direct current (V-dc) power. In some cases, the instrument is loop-powered, which means that the power to run the instrument comes from the 24 V-dc power supply within the PLC input/output card.

**CAPACITORS.** The concept of a capacitor as a kilovar generator is helpful in understanding its use for power factor improvement. A capacitor may be considered a kilovar generator because it supplies the magnetizing requirements (kilovars) of induction motors.

This action may be explained in terms of the energy stored in capacitors and induction devices. As the voltage in alternating-current circuits varies sinusoidaly, it alternately passes through zero voltage points. As the voltage passes through zero voltage and starts toward maximum voltage, the capacitor stores energy in its electrostatic field, and the induction device gives up energy from the electromagnetic field. As the voltage passes through a maximum point and starts to decrease, the capacitor gives up energy and the induction device stores energy. Thus, when a capacitor and an induction device are installed in the same circuit, there will be an interchange of magnetizing current between them. The leading current taken by the capacitor neutralizes the lagging current taken by the induction device. Because the capacitor relieves the supply line of supplying magnetizing current to the induction device, the capacitor may be considered to be a kilovar generator because it actually supplies the magnetizing requirements of the induction device.

Power-factor correction capacitors are installed in either of two locations: (1) at a MCC to correct for the total load on the MCC, which is a compromise because all of the connected loads are not energized at the same time; and (2) directly, at all motors that are 19 kW (25 hp) and larger. The latter method then corrects the power factor of the load to which it is connected when energized.

**CONDUIT AND WIRING CONSIDERATIONS.** For safety and mechanical protection, most wiring in WWTPs is enclosed in rigid or flexible conduit. Open-tray wiring is typically restricted to large cables or to shops where wiring changes are frequent and the atmosphere is not detrimental to the conductor installation.

Wiring can be either copper or aluminum. With aluminum wiring, copper-to-aluminum junctions require careful construction. When copper prices drop, the incentive to use aluminum is reduced.

Insulation ratings are typically higher than the voltages being used. For example, 460-V systems typically have 600-V insulation. Special insulation is available to accommodate high or low temperatures or corrosive conditions.

*The National Electrical Code Handbook* (2005) specifies the minimum allowable wire and conduit sizes, number of wires of each size in a conduit of a given size, wire junctions, and many other aspects of wiring. Local jurisdictions may impose more rigorous standards; many local power companies also have their own rules for distribution-system wiring. Any changes or additions to electrical equipment must conform to the latest edition of the National Electrical Code or other applicable standards. Electrical installations typically require inspection to ensure that they comply with local and federal rules.

Stress cones prevent insulation failure at the termination of a shielded cable. This is caused by the high concentration of flux and the high potential gradient that would otherwise exist between the shield termination and the cable conductor. Precautions must be taken to ensure that the termination is free of dirt and foreign matter, and insulating compounds and tapes must be applied in a prescribed manner. A typical use for stress cones would be in pad-mounted transformers, switchgear, high-voltage motor installations, or in most applications where shielded cable is being terminated. Typically, stress cones are not provided by the motor manufacturer, but are made by the motor installer.

**GROUNDING.** The main reason why grounding is used in electrical-distribution networks is safety. Indeed, when all metallic parts in electrical equipment are grounded, there are no dangerous voltages present in the equipment case if the insulation inside the equipment fails. If a live wire touches the grounded case, then the circuit is effectively shorted, and the fuse or circuit breaker will isolate the circuit. When the fuse is blown, the dangerous voltages are safely dissipated.

Safety is the primary function of grounding, and grounding systems are designed to provide the necessary safety functions. Although grounding also has other functions in some applications, safety should not be compromised in any case. Grounding is often used to provide common ground reference potential for all equipment, but the existing building grounding systems might not provide good enough ground potential for all equipment, which could lead to potential ground difference and ground-loop problems that are common in computer networks and audio/video systems.

**PROTECTIVE DEVICES.** Overcurrent relays are the most typically used devices for short-circuit (direct-connection, phase-to-phase, or phase-to-ground) protection of industrial power systems. Overcurrent relays have adjustable current settings. Time-delay relays permit momentary overcurrent without opening the breaker. A phase failure relay will interrupt power to equipment with either current loss in a phase or severe imbalance among the power-line phases.

If lightning causes overvoltage in a circuit, a lightning arrestor limits the overvoltage by providing a conducting path to the ground of low impedance. Many other devices for protection against current disruption are also available.

Circuit-opening devices provide a means of disconnecting a faulty circuit or equipment from the distribution system. Devices include circuit breakers, fuses, interrupter switches, load-break switches, disconnect switches, and contactors. Distribution protection typically consists of either equipment or circuit protection.

Protection devices that are not limited to the specific circuit with the fault protect the rest of the system from the faulty element. For example, if a motor overheats because of overloading, it can be protected by a *contactor*—a magnetically operated relay switch that responds to either high current or high temperature. If the current to be interrupted exceeds the interrupting capacity of the contactors, a circuit breaker or fuse should be installed in the circuit.

## MAINTENANCE AND TROUBLESHOOTING

**GENERAL.** The vast majority of electrical maintenance should be predictive or preventive. This section focuses exclusively on these activities. There are four cardinal rules to follow in any maintenance program: (1) keep it clean, (2) keep it dry, (3) keep it tight, and (4) keep it frictionless.

In general, the most probable cause of electrical equipment failure is foreign contamination (e.g., dust particles, lint, or powdered chemicals). Dirt buildup on moving parts will cause slow operation, arcing, and subsequent burning. Moreover, coils can short-circuit. Dirt will always impede airflow and result in elevated operating temperatures.

Electrical equipment always operates best in a dry atmosphere, where corrosion is eliminated. Moisture-related grounds and short circuits are also eliminated. Liquids other than water (e.g., oil) can cause insulation failures or increase dirt and dust buildup.

Most electrical equipment operates at a high speed or is subjected to vibration. The term "tightness" is not a contest of human strength; rather, it must always be a calculated or specified value and be obtained with a torque wrench. When aluminum conductors are used, each mechanical junction should be made with a specific torque. Insufficient torque will result in localized heat at the joint and, ultimately, failure. Excessive torque will result in deformation.

Any piece of equipment or machinery is designed to operate with minimum friction. Dirt, corrosion, or excessive torque will often cause excessive friction.

Of the four cardinal rules, none is essentially electrical in nature. The failure of a bearing in a motor can lead to an ultimate motor winding failure that is electrical, but the root cause of the failure could have been mechanical.

The goal of any electrical preventive maintenance program is to minimize electrical outages and ensure continuity of operation. The bulk of the program centers on proactive tasks. A successful program includes not only these proactive tasks with appropriate documentation, but also a consistent commitment to obtain reliable equipment that has been properly installed. This commitment should be evident in any design construction upgrade or replacement work. It is extremely difficult to maintain equipment that has not been soundly engineered or that has become outmoded, obsolete, or overloaded during plant growth. Because a program is implemented by personnel, not computers, qualified field personnel are required for a successful program.

**PREVENTIVE AND PREDICTIVE MAINTENANCE SPECIFICS.** A good preventive maintenance program includes checking MCCs annually. The time interval between inspections can vary depending on area cleanliness, atmosphere, operating temperatures, vibration, and overall operating conditions. Maintenance involves cleaning the equipment and checking for obvious defects (e.g., burned relays, damaged insulation, fire, ant infestation, unsealed conduits entering boxes, and loose leads). The maintenance crew should inspect relays for loosened terminals and lock screws; examine the coils, resistors, wiring, toggles, and holding devices; and look for dirty or burned contacts, dirty or worn bearings, and foreign material in magnetic air gaps. Power should be turned off during the work. However, if power shutdown is impossible, a vacuum cleaner with a plastic tip can be used. Other specific maintenance instructions are contained in the *Switch Gear and Control Handbook* (Smeaton, 1998).

All mechanisms at unit substations should be checked using the manufacturer's recommended schedule and procedure. Breakers should be drawn out and checked for signs of arcing. Additionally, all appropriate components should be cleaned, and adjustments should be made to the mechanisms as required.

Every 2 to 3 years, the overload devices should also be checked. Although equipment can be purchased if the plant performs this work, contracting the work typically is more practical. Many major equipment manufacturers can provide this service.

For oil-filled transformers, the oil level and temperature should be checked in accordance with manufacturer's recommendations. At least every 3 years, the oil should be sampled and replaced or re-refined. The oil should also be checked for moisture and alkaline content. Electrical service companies can perform the sampling, even while the transformer remains in service. Transformer cooling fans should be checked periodically (follow the operations and maintenance manual guidelines). Transformer top bushings should be checked for contamination at intervals recommended by the manufacturer. Some transformers have a nitrogen blanket that eliminates moisture buildup in the oil. The nitrogen gas bottle should also be checked at intervals recommended by the manufacturer.

Standby generators should be exercised with load monthly. Testing should allow the generator to reach its operating temperature. The inspection should include a check of the starter batteries or compressed air starter.

Critical motors should be tested for integrity of insulation; additionally, every 5 to 6 years they should be sent out for testing of the windings, bearing inspection, cleaning, re-varnishing, and baking. Table 10.2 provides a preventive maintenance checklist for capacitors.

**TABLE 10.2** Capacitor preventive-maintenance annual checklist.

| Component | Condition | Result |
|---|---|---|
| Case | Physical damage<br>Overheating<br>Discoloration<br>Leaks, puncture | Replace. Check with U.S. Environmental Protection Agency authority for polychlorinated biphenyl-filled apparatus. |
| Structure | Ventilation | Area kept clean |
| Site | Safety | Isolation, fenced enclosure |
| | Rust peeling | Refurbish, paint |
| | Identification | Engraved nameplates |
| Fuses | Continuity | Self-indicating fuses, check with ohmmeter |
| Bleeder-resistor | | Check with manufacturer for resistor values |
| Nominal operating conditions | Operating voltage and currents | 7-day profile with recording meters ($\pm$10% of rated values)<br>7-day temperature profile (50 to 90 °C if indoors)<br>Calculate kilovolt-amperes and compare to nameplate values |
| Insulation | Breakage, cracks | Replace damaged housing<br>Megger from bushings to case, 1000 megohms minimum |

Cables carrying 600 V or more should be tested for insulation integrity at intervals recommended by the manufacturer. The voltage should be slowly raised while reading the microamps. The plot (microamps versus voltage) should be a straight line. A rapid change of slope is indicative of failure. If the cable has a minor failure, going immediately to full load can blow the cable. Therefore, it is important to inspect cables for signs of corona (a white powdery residue). This residue can perpetuate and cause cable insulation failure; so, all residue should be cleaned and neutralized. Most electrical service contractors will use direct current to check the cable.

A frequently overlooked aspect of preventive maintenance is checking electrical grounds. Electrical grounds of machinery, equipment, and buildings should be checked every 2 to 3 years and corrected as necessary.

Emergency lights should be checked at least quarterly. Battery-powered lights should preferably be self-diagnostic (i.e., they should have devices to indicate when a fault exists). At least once each quarter, emergency lights should be turned on for at least 30 minutes. If this procedure is not followed, the battery will deteriorate.

Lighting fixtures should be cleaned annually or more often if recommended by the manufacturer. For fluorescent fixtures, both the diffuser (outer cover) and reflector should be cleaned. When one tube burns out in a fluorescent fixture, all of the tubes should be replaced because of the expected short remaining lives of the others. Where there are large groups of fixtures in a given area, the most economical approach is to relamp all fixtures after a period of time somewhat shorter than their rated life, or, if a light meter is available, when brightness diminishes to a predetermined level.

The following is a supplemental list of preventive and predictive maintenance activities for various electrical components (refer to manufacturer recommendations for task frequency):

*Liquid-filled transformers*

- Verify oil type and level
- Check top bushings for contamination
- If applicable, inspect nitrogen supply system
- Cycle fans to ensure proper operation
- Verify grounding
- Conduct Doble power-factor test
- Perform Doble excitation test
- Conduct insulation resistance test
- Perform dielectric fluid quality test
- Analyze dissolved gas

*Dry transformers (more than 1000 kVA)*

- Doble power-factor test (medium-voltage units)
- Doble tip-up test (medium-voltage units)
- Doble excitation test (medium-voltage units)
- Insulation resistance test
- Turn-to-turn ratio test
- Core, coil, and vent cleaning and vacuuming

*Oil-filled circuit breakers*

- Doble power-factor test
- Insulation resistance test
- Breaker cleaning and lubrication
- Mechanical and electrical function check
- Oil integrity test

*Medium-voltage circuit breakers*

- Doble power-factor test
- Insulation resistance test
- Breaker cleaning and lubrication
- Mechanical and electrical function check
- Contact resistance test

*Medium-voltage cables*

- Direct-current high-potential test
- Doble power-factor test
- Doble tip-up test
- Connection inspection and tightening
- Terminal inspection

*Medium-voltage, metal-enclosed switches*

- Insulation resistance test
- Contact resistance test
- Connection cleaning, inspection, and tightening
- Proper operation of space heater
- Mechanical operation test

*Medium-voltage starters*

- Insulation resistance test
- Contact resistance test
- Connection cleaning, inspection, and tightening
- Proper operation of space heater
- Mechanical/electrical operation tests

*Medium-voltage motors*

- Polarization index test
- Winding resistance test
- Insulation resistance test

*Arresters*

- Doble power-factor test
- Porcelain/polymer surface cleaning
- Connection tightening

*Substation batteries and chargers*

- Specific gravity test
- Battery load test
- Connection cleaning and inspection
- Battery-charger operational test

*Capacitor banks*

- Connection cleaning and inspection
- Porcelain surface inspection and cleaning
- Fuse and fuse-holder cleaning and inspection
- Operational checks

*Ground resistance*

- Three-point fall-of-potential test
- Ground connection inspection

*Protective relays*

- Relay cleaning and inspection
- Connection tightening
- Contact cleaning and inspection
- Injection test and calibration

*Indicating meters—voltage/current*

- Meter cleaning and inspection
- Connection tightening
- Injection test and calibration

*Watt-hour meters*

- Meter cleaning and inspection
- Connection tightening
- Injection test
- Calibration

*Low-voltage power circuit breakers*

- Primary / secondary current injection test
- Contact resistance test
- Insulation resistance test
- Contact assembly cleaning and inspection
- Mechanical component cleaning, inspection, and lubrication

*Low-voltage, molded-case, bolted-in circuit breakers*

- Visual inspection
- Cleaning, inspection, and exercise

*Low-voltage switchgear*

- Insulation resistance test
- Cleaning and visual inspection
- Bolted connection inspection (check for signs of overheating)
- Verification of operational functions
- Indicating-lamp check

*Low-voltage switches*

- Contact resistance test
- Insulation resistance test
- Connection cleaning, inspection, and tightening
- Mechanical operation test

*Low-voltage MCCs*

- Insulation resistance test
- Cleaning and visual inspection
- Bolted connection inspection (check for signs of overheating)
- Verification of operational functions
- Indicating-lamp check

*Aerial switches*

- Porcelain surface cleaning and inspection
- Connection cleaning and inspection
- Contact surface cleaning and inspection
- Mechanical operation tests
- Inspection via binoculars

*Aerial buswork*

- Porcelain surface cleaning and inspection
- Bus connection cleaning and inspection
- Fuse holder cleaning and inspection

*Power distribution system*

- Thermographic infrared inspection while system is energized and under full load

**TROUBLESHOOTING.** Troubleshooting requires a good quality volt-ohmmeter and the usual mechanic's tools. With the volt-ohmmeter, a mechanic can determine whether wiring is grounded and whether insulation is in good condition. On submersible pumps, for example, phase-to-ground can be checked periodically. A resistance reading that drops below previous readings may indicate the beginning of leakage into the windings, requiring overhaul of the pump seal.

If a motor starter trips a motor, leads can be disconnected and the windings checked, reading each phase-to-ground. If the volt-ohmmeter shows a low ohm reading, the phase is grounded. Even if checking the motor phase-to-ground reveals no short, the motor may still have failed phase-to-phase. This possibility cannot be checked routinely, unless the resistance of each winding is known. If the phase-to-ground check does not indicate a defective motor, then further checking must begin with the starter. The staff electrician disconnects motor wires from the starter and again checks phase-to-ground on the motor leads. Bad wiring may be indicated; however, if that is not the case, the motor leads should be disconnected and the starter energized. If the starter holds, then a phase-to-phase breakdown may exist, possibly requiring motor rewind or replacement. Table 10.3 (WPCF, 1984) provides a comprehensive troubleshooting guide for electric motors.

A review of the manufacturer's operations and maintenance manuals for the plant's equipment may disclose other electrical maintenance tasks that can be accomplished by a capable general maintenance person. Unless the facility has a well-trained electrician, complex electrical troubleshooting, particularly for high-voltage equipment, should be contracted. The plant supervisor should arrange for one or more electrical contractors to provide 24-hour service.

## RELAY COORDINATION

A coordination study is the process of determining the optimum characteristics, ratings, and settings of the power system's protective devices. The optimum settings are focused on providing systematic interruptions to the selected power system segments during fault conditions.

Coordination means that downstream devices (breakers/fuses) should activate before upstream devices. This minimizes the portion of the system affected by a fault or other disturbance. At the substation level, feeder breakers should trip before the main. Likewise, downstream panel breakers should trip before the substation feeder supplying the panel.

A protective device coordination study is performed to select proper protective devices (e.g., relays, circuit breakers, and fuses) and to calculate protective relays and circuit-breaker trip unit settings.

A protective device coordination study is performed either during the design phase of a new system to verify that protective devices will operate correctly in an existing system, or when the protective devices are not operating correctly. It is important to note that utility companies can change to a higher level of fault current, which would necessitate changing relay settings.

TABLE 10.3 Troubleshooting guide for electric motors.

| Symptoms | Cause | Result* | Remedy |
|---|---|---|---|
| 1. Motor does not start. (Switch is on and not defective.) | a. Incorrectly connected. | a. Burnout | a. Connect correctly per diagram on motor. |
| | b. Incorrect power supply | b. Burnout | b. Use only with correctly rated power supply. |
| | c. Fuse out, loose or open connection. | c. Burnout | c. Correct open circuit condition. |
| | d. Rotating parts of motor may be jammed mechanically. | d. Burnout | d. Check and correct:<br>1. Bent shaft<br>2. Broken housing<br>3. Damaged bearing<br>4. Foreign material in motor. |
| | e. Driven machine may be jammed. | e. Burnout | e. Correct jammed condition. |
| | f. No power supply. | f. None | f. Check for voltage at motor and work back to power supply. |
| | g. Internal circuitry open. | g. Burnout | g. Correct open circuit condition. |
| 2. Motor starts but does not come up to speed. | a. Same as 1-a, b, c above. | a. Burnout | a. Same as 1-a, b, c above. |
| | b. Overload. | b. Burnout | b. Reduce load to bring current to rated limit. Use proper fuses and overload protection. |
| | c. One or more phases out on a three-phase motor. | c. Burnout | c. Look for open circuits. |
| 3. Motor noisy (electrically). | a. Same as 1-a, b, c above. | a. Burnout | a. Same as 1-a, b, c above. |

**Table 10.3** Troubleshooting guide for electric motors (*continued*).

| Symptoms | Cause | Result* | Remedy |
|---|---|---|---|
| 4. Motor runs hot (exceeds rating) | a. Same as 1-a, b, c above. | a. Burnout | a. Same as 1-a, b, c above. |
| | b. Overload | b. Burnout | b. Reduce load. |
| | c. Impaired ventilation. | c. Burnout | c. Remove obstruction. |
| | d. Frequent start or stop. | d. Burnout | d. 1. Reduce number of starts or reversals.<br>2. Secure proper motor for this duty. |
| | e. Misalignment between rotor and stator laminations. | e. Burnout | e. Realign. |
| 5. Noisy (mechanically) | a. Misalignment of coupling or sprocket. | a. Bearing failure, broken shaft, stator burnout due to motor drag. | a. Correct misalignment. |
| | b. Mechanical imbalance of rotating parts. | b. Same as 5-a. | b. Find imbalanced part, then balance. |
| | c. Lack of or improper lubricant. | c. Bearing failure. | c. Use correct lubricant, replace parts as necessary. |
| | d. Foreign material in lubricant. | d. Same as 5-c. | d. Clean out and replace bearings. |
| | e. Overload. | e. Same as 5-c. | e. Remove overload condition. Replace damaged parts. |
| | f. Shock loading. | f. Same as 5-c. | f. Correct causes and replace damaged parts. |
| | g. Mounting acts as amplifier of normal noise. | g. Annoying. | g. Isolate motor from base. |
| | h. Rotor dragging due to worn bearings, shaft, or bracket. | h. Burnout | h. Replace bearings, shaft, or bracket as needed. |
| 6. Bearing failure | a. Same as 5-a, b, c, d, e | a. Burnout, damaged shaft, damaged housing. | a. Replace bearings and follow 5-a, b, c, d, e. |
| | b. Entry of water or foreign material into bearing housing. | b. Same as 6-a. | b. Replace bearings and seals, and shield against entry of foreign material (water, dust, etc.). Use proper motor. |

*Many of these conditions should trip protective devices rather than burnout motors.

**Table 10.3** Troubleshooting guide for electric motors (*continued*).

| Symptom | Caused by | Appearance |
|---|---|---|
| 1. Shorted motor winding | a. Moisture, chemicals, foreign material in motor, damaged winding. | a. Black or burned coil with remainder of winding good. |
| 2. All windings completely burned | a. Overload.<br>b. Stalled.<br>c. Impaired ventilation.<br>d. Frequent reversal or starting.<br>e. Incorrect power. | a. Burned equally all around winding<br>b. Burned equally all around winding<br>c. Burned equally all around winding<br>d. Burned equally all around winding<br>e. Burned equally all around winding |
| 3. Single-phase condition. | a. Open circuit in one line. The most common causes are loose connection, one fuse out, loose contact in switch. | a. If 1800-rpm motor—four equally burned groups at 90° intervals.<br>b. If 1200-rpm motor—six equally burned groups at 60° intervals.<br>c. If 3600-rpm motor—two equally burned groups at 180°.<br>NOTE: If Y connected, each burned group will consist of two adjacent phase groups. If delta connected, each burned group will consist of one phase group. |
| 4. Other | a. Improper connection.<br>b. Ground | a. Irregularly burned groups or spot burns. |

Many burnouts occur shortly after motor is started up. This does not necessarily indicate that the motor was defective, but typically is due to one or more of the above mentioned causes. The most common of these are improper connections, open circuits in one line, incorrect power supply, or overload.

## HARMONICS

A harmonic analysis evaluates the steady-state effects of nonsinusoidal voltages and currents on the power system and its components. These harmonic currents create heat, which, over time, will raise the temperature of the neutral conductor, causing nuisance tripping of circuit breakers, overvoltage problems, blinking of incandescent lights, computer malfunctions, etc.

Among the electrical devices that appear to cause harmonics are personal computers, dimmers, laser printers, electronic ballasts, stereos, radios, televisions, fax machines, and any other equipment powered by switched-mode power supply equipment. This is not to say that harmonics will cause all these problems, only that it is possible.

These problems can be prevented somewhat by using a dedicated circuit for electronic equipment. Also, on a branch circuit, an isolated ground wire should be used for sensitive electronic and computer equipment. A more expensive alternative is to rectify and filter the mains, thereby effectively removing all low-frequency harmonics, including the fundamental. Oversized neutrals represent another possible means to prevent overheating of this wire. In power-distribution systems, electricians are typically interested in measuring the current; thus, a "true-root mean square" current measuring clamp-on meter is typically used.

## STAFFING AND TRAINING

Electrical staff duties include all aspects of maintenance (preventive, predictive, corrective, and emergency). It is rare that a facility or district will have plant personnel who possess the ability and knowledge to service, renew, and overhaul each piece of equipment presently in use or projected to be used in the plant. Supplementing the staff's skill base with outside specialists and maintenance contractors is a normal practice.

In a small facility, some electrical work may be done by the operator. In many small facilities, the operator also serves as the mechanic, the electrician, and the instrument technician. In larger facilities, specialization typically prevails, and there are a variety of titles that are used in the electrical trades. However, whether the facility is small or large, it is essential that only qualified, well-trained personnel work on instrument or electrical equipment.

At least one journeyman electrician is typically employed in medium-to-large facilities. Facilities of this size range may employ various levels of electricians (e.g., an electrician's helper, an electrician, a high-voltage electrician, an instrument technician helper, and an instrument technician).

An electrician's helper works under the supervision of an experienced electrician. Often, the helper is in an apprentice program. An electrician is certified by the state and/or the municipality. A first-level electrician would work on lower voltages (120 to 600 V). The 600-V threshold is not rigid; it depends on the municipality or state. A high-voltage electrician works on systems in excess of 600 V.

Like an electrician's helper, an instrument-technician helper works under the supervision of an experienced instrument technician. Often, the helper is in an apprentice program. The instrument technician has typically completed an apprenticeship program and provides on-the-job training for assigned helpers.

Sometimes, job classifications are established for the major divisions of electrical specialties. These categories are solid-state controls and drives; lighting; and switchgear (e.g., high-, medium-, and low-voltage motors and motor control circuits).

Normally, electricians handle 120 V and above, and instrument technicians handle below 120 V. Instrument technicians routinely work on the following systems: phone, intercom, alarm, security, and computers.

Most electricians learn their trade through apprenticeship programs. These programs combine on-the-job training with related classroom instruction. Apprenticeship programs may be sponsored by joint training committees made up of local unions of the International Brotherhood of Electrical Workers (Washington, D.C.) and local chapters of the National Electrical Contractors Association (Bethesda, Maryland); company management committees of individual electrical contracting companies; or local chapters of the Associated Builders and Contractors (Arlington, Virginia) and the Independent Electrical Contractors Association (Alexandria, Virginia). Because of the comprehensive training received, those who complete apprenticeship programs qualify to do both maintenance and construction work.

Apprenticeship programs typically last 4 years, and include at least 144 hours of classroom instruction and 2000 hours of on-the-job training each year. In the classroom, apprentices learn electrical theory as well as installation and maintenance of electrical systems. On the job, apprentices work under the supervision of experienced electricians. To complete the apprenticeship and become electricians, apprentices must demonstrate mastery of the electrician's work.

Most localities require electricians to be licensed. Although licensing requirements vary from area to area, electricians typically must pass an examination that tests their knowledge of electrical theory, the National Electrical Code, and local electric and building codes. Training, however, does not end once an employee attains journeyman status. Mandated by the certifying state, training typically consists of a 15-hour electrical code update course and 6 hours of electrical electives every 3 years. This continuing education effort is typically supplemented with other internal courses. Continued training of 40 hours per year for electrical trades is not uncommon. Examples of this include vendor training for newly installed equipment and new safety procedures.

## HIGH-VOLTAGE SAFETY

If not handled properly, electricity—particularly high-voltage electricity—can be extremely dangerous. The following basic suggestions should be followed by everyone working with or near high-voltage circuits:

- **Consider the result of each act.** There is absolutely no reason to take chances that will endanger your life or the lives of others. Always consider what you are going to do and how it might affect you and others around you.
- **Keep away from live circuits.** Do not change parts or make adjustments inside machinery or equipment when high voltage is energized. Always de-energize the system.
- **Do not service high-voltage electrical equipment alone.** Another person should be present when high-voltage equipment is being serviced. This person should be capable of rendering first aid in the event of an emergency.
- **Do not tamper with interlocks.** Do not depend on interlocks for protection; always shut down the equipment or de-energize the electrical system. Never remove, short-circuit, or tamper with interlocks except to repair the switch.
- **Do not ground yourself.** Make sure you are not grounded when adjusting equipment or using measuring equipment. When servicing energized equipment, use only one hand and keep the other hand behind you.
- **Do not energize equipment if there is any evidence of water leakage.** Repair the leak and wipe up the water before energizing.
- **If work is required on energized circuits, the following practical safety rules should apply:**
  - Only authorized and experienced personnel should work on energized electrical systems. If you don't know about electricity, you could be playing a deadly game.
  - Ample lighting is an absolute necessity when working around high-voltage electrical current that is energized.
  - The employee doing the work should be insulated from the ground with some suitable nonconducting material (e.g., dry wood or a rubber mat of approved construction).
  - The employee doing the work should, if at all possible, use only one hand in accomplishing the necessary repairs.
  - Identify all circuit breakers to indicate which equipment or branch outlets they control so the system or equipment can be de-energized immediately in case of an emergency.

- A person qualified in first aid for electric shock should be near the work area during the entire repair period.
- **Wear all recommended personal protective equipment, including flash protective attire and equipment.**

## UTILITY METERING AND BILLING

Next to labor, electricity is typically one of the most costly items in the operation of a WWTP. There are many ways to lower electrical costs without compromising service or water quality. Understanding how the local electric utility computes your bill is the first step in controlling energy costs.

**BILLING FORMAT.** Utility rate structures are proposed by utilities and accepted or modified by a public service commission in the state where the utility is located. Plant supervisors should become familiar with schedules for their facilities and be sure that the plant is charged for electricity at the most economical rate schedule available. Schedule information will also help the supervisor avoid operations that unnecessarily increase costs. To determine whether the rate structure being applied is the most economical available, plant supervisors should discuss alternatives with the utility concerned or employ a rate consultant.

Most utilities submit bills to their customers on a monthly basis. As meter reading follows a Monday through Friday schedule, a bill might not cover the normal calendar month because the number of weekends included in the monthly billing period will vary. The average bill typically contains two basic charges: an energy charge and a demand charge. A flat fee may also be part of the bill, although it typically represents a minor portion of the total bill and is the same for each customer in a specific classification.

**ENERGY CHARGES.** The energy charge is based on the quantity of electricity supplied and is typically a rate of a given number of dollars per kilowatt hours ($/kWh) of power use.

Most public service commissions also allow a utility to apply a fuel adjustment charge (typically $0.01 to $0.05 in the United States), which can be either a credit or an additional charge reflecting changes in the cost of fuel. This charge varies from month to month. Use of the fuel adjustment charge permits the utility to continually recover changing fuel costs without seeking rate adjustments from the public service commission.

**DEMAND CHARGES.** Demand, measured in kilowatts, is the maximum rate at which electricity is used. Demand represents the maximum number of kilowatts drawn, typically measured in 15- to 30-minute intervals within a monthly billing cycle. Summation of customer demands indicates the total generation and distribution capacity that the electrical utility must maintain. Thus, the demand charge, based on the plant's measured demand, compensates the utility for its incremental investment in the additional generation and distribution capacity necessary to meet its theoretical maximum load.

The utility company's demand meter measures the WWTP's demand. Most utilities record the demand for each 15 to 30 minutes electronically on tapes and transcribe the tapes at their offices. Although the utility does not routinely furnish the demand information to its customers, it will provide printouts, if requested, for a monthly or annual fee. This information also may be obtained from the utility account representative.

The customer is charged for the highest demand recorded during the billing period. Some utilities apply a "ratchet" to demand. This means that the demand charge or a certain percentage of the demand in any billing period will be applied as the minimum demand for calculating demand charges during the following year (11 months maximum or summer months high) if actual demand is lower during any of those months.

**POWER FACTOR CHARGES.** Because a poor power factor requires utilities to install larger generating equipment and increases their line losses, utilities frequently apply a penalty for a poor power factor and sometimes allow a credit for a high power factor. The penalty or credit typically is a certain percentage adjustment for each one-tenth of a point that the power factor drops or rises above the established figure. Power factor adjustment can be applied either to the energy and demand or to the demand only.

**OTHER RATE FEATURES.** Frequently, utilities set a summer use rate that differs from their winter rate. Utilities with a heavy air-conditioning load, for example, may have their peak electrical demand loads during the summer. They will, therefore, establish a higher summer rate. Conversely, utilities that experience high heating loads and peak demands during the winter will charge more then.

Another frequently applied adjustment, time-of-day rates, imposes higher energy and demand charges to that period of the day when the utility experiences the highest average demand. The utility may divide the day into two to five periods. To charge based on time-of-day rates, the utility must install metering to record the consumption during each period. To reduce time-of-day charges, the operator may, if practical, postpone some operations (e.g., pumping or solids handling) until the peak periods have passed.

As mentioned, rate structures are based on a customer's type of load and the delivered voltage. Typically, higher delivered voltages will result in a lower rate. In this case, the customer must furnish transformers to reduce the voltage in the plant.

Utilities may offer special services (e.g., yard lighting and spare feeders). Special services and charges, therefore, must either be standardized and approved by the public service commission or be included in special contractual arrangements between the utility and the customer.

## ENERGY COST REDUCTION

**GENERAL.** Elements in reducing energy costs include rates and riders, demand control, and efficiency. First, know your rate structure and take full advantage of any cost savings without sacrificing your process. Second, learn about all applicable rate structures and riders offered by your utility. Some utilities may allow for more cost-effective operation.

For example, higher delivered voltages will typically result in a lower rate. This lower rate can be obtained by furnishing transformers to reduce the voltage in the plant.

Additionally, utilities sometimes use another system to reduce their total demand. The alternate system entails granting a credit or a preferred rate to customers who agree to reduce electrical consumption immediately upon the utility's request. Typically, the utility allows the customer a reasonable period after notification to make the necessary adjustments. In a WWTP, few if any processes can be stopped to reduce demand. If the plant's emergency generator has the appropriate air pollution permits, it can be used to lower the power consumption from the utility.

If a facility's rate structure has a demand charge, the use of demand control can reduce energy costs. Minor changes in operational procedures can reduce demand charges. For example, if spare equipment is to enter service, the equipment being shut down should be turned off first to avoid running both motors at the same time. Demand meters should be installed and trends studied. An operator should also review the feasibility of modifying an operation to minimize demand. Where practical, a demand control system should be installed that monitors demands and either alerts the operator to forecasted demand exceedances or adjusts certain predefined equipment to ensure that demand limits are not exceeded.

**MOTORS.** Replacement of existing motors with high-efficiency motors has significant potential for energy savings; the average savings on these energy-efficient motors ranges from 3 to 5%. The payback period also depends on the service conditions and running time of the motor. Incremental savings are greatest on smaller motors or motors that have to be rewound. Overall savings on larger motors [more than 75 kW (100 hp)] are smaller on a percentage basis because of the high replacement cost; payback periods on the replacement of standard motors vary from 2 to 10 years. High-efficiency motors have more copper for lower loss in the windings, better design to reduce no-load loss, and better bearings and aerodynamics to reduce friction and windage loss.

The Energy Policy Act of 1992, a federal statute, requires that certain types of new motors sold in the United States as of October 24, 1997, must meet or exceed specified efficiency ratings. Table 10.4 provides a comparison between standard- and premium-efficiency motors. The efficiency of the motor is highest when they are operated close to 100% of their load.

**TRANSFORMERS.** Energy-efficient transformers should be used, especially in the case of dry transformers. However, efficiency is seldom specified when buying transformers. Values of 95% or higher are typical, and differences between high- and low-efficiency units are only 1 to 2%, with a significant first-cost premium for more efficient units. Complete life-cycle costs must be carefully examined, along with the economics of high-efficiency, dry-type transformers.

Transformers can be expected to operate 20 to 30 years or more and, as such, buying a unit based only on its initial cost is uneconomical. Transformer life-cycle cost (also called "total owning cost") takes into account not only the initial transformer

TABLE 10.4 A comparison between standard- and premium-efficiency motors.

| Hp | **Standard-efficiency motors** Average efficiency at 100% load | **EPA energy-efficient motors** Minimum nominal efficiency at 100% load | **NEMA premium motors** Minimum nominal efficiency at 100% load |
|---|---|---|---|
| **5** | 83.3 | 87.5 | 89.5 |
| **10** | 85.7 | 89.5 | 91.7 |
| **20** | 88.5 | 91.0 | 93.0 |
| **25** | 89.3 | 92.4 | 93.6 |
| **50** | 91.3 | 93.0 | 94.5 |
| **100** | 92.3 | 94.5 | 95.4 |
| **200** | 93.5 | 95.0 | 96.2 |

EPA = Energy Policy Act of 1992.

cost, but also the cost to operate and maintain the transformer over its life. This requires that the total owning cost (TOC) be calculated over the lifespan of the transformer, as follows:

$$\text{TOC} = \text{initial cost of transformer} + \text{cost of the no-load losses} + \text{cost of the load losses} \tag{10.6}$$

Lastly, transformers should be sized to match the load as closely as possible to reduce the no-load component of the total loss.

**ENERGY EFFICIENT LIGHTING.** Energy can be saved by using lights only when needed, selecting the correct type of lighting, and properly maintaining the lighting system. During plant or office operation, certain areas are unoccupied for varying periods of time. If the lamps are turned off as people leave the area during these various periods, there will be lower energy consumption than if the lamps were left on. The length of time the lamps are off and the type of light source should affect the decision as to whether these lamps should be left burning or turned off. More importantly, however, is the type of lamp involved. Whether savings can be achieved by turning lamps off can only be determined by evaluating the energy savings against the increased costs of lamp replacement. Any evaluation must consider the inconvenience and cost of turning lamps on and off, possible costs for installing switching devices, and the waiting time required for restarting lamps.

Incandescent lamps should be turned off when not needed. Energy will be conserved and electrical costs will be reduced without shortening the life of the lamp. High-intensity lamps typically are not turned off during short unoccupied periods because a 3- to 15-minute warm-up time is required during starting. They should be switched off only when the subsequent startup can be pre-planned so as not to interfere with work assignments. Although each HID lamp has a long life, frequent starting (less than 5 burning hours per start) will reduce lamp life by as much as 30%.

Fluorescent lamps are often turned off to reduce electrical energy consumption and will most often produce savings in energy costs that more than offset the increased costs of lamps and lamp-replacement labor. The life of these lamps depends on the number of hours the lamps operate per each start. "Life" is rated by the industry on a 3-hour burning cycle. The more hours that a lamp is on per start, the more its life is increased. The average life is the point at which 50% of the lamps survive. Table 10.5 depicts the average life of several types of fluorescent lamps per burning hours.

Whenever a new facility will be constructed or an existing facility revamped, an opportunity exists to select the most efficient lighting scheme. First, match the function

TABLE 10.5 Relative life of fluorescent lamps at various burning hours (burning hours per start).

| Lamp | 3 | 6 | 10 | 12 | 18 | Cont |
|---|---|---|---|---|---|---|
| 40 white rapid start | 18 000 | 22 000 | 25 000 | 26 000 | 28 000 | 34 000 |
| High output | 12 000 | 14 000 | 17 000 | 18 000 | 20 000 | 22 500 |
| Very high output | 9000 | 11 300 | 13 500 | 14 400 | 16 200 | 22 500 |
| Slimline | 12 000 | 14 000 | 17 000 | 18 000 | 20 000 | 22 500 |

of the room to the required lighting needs and ensure that the new room finishes (walls and ceilings) are reflective to maximize available light. Second, where feasible, take advantage of natural light. Next, select the most cost-effective lighting system (first and second cost) for the defined requirements. Include programmable timers, override switches, and motion detectors in the design control systems to further attain cost-effective operations.

There are a variety of lighting systems available. Some general information relative to energy efficiency for incandescent lamps, fluorescent lamps, and HID lights is provided herein.

In terms of incandescent lamps, higher wattage general-service lamps are more efficient than lower wattage lamps. Using fewer high-wattage lamps in a given area will save energy. For example, one 100-W general-service lamp produces 1750 lumens, or two 60-W general-service lamps produce 860 lumens each or 1720 lumens total. For the same wattage, general-service lamps are more efficient than extended-service lamps, although the extended-service lamp's life is longer than the general-service lamp (100-W general service lamp/750- to 100-hour life/17.5 lumens/watt 100-W extended-service lamp/2500-hour lamp life/14.8 lumens/watt).

To obtain equal lighting, approximately 18% more lamps and energy are required when using extended-service lamps. Because of their low efficiency, extended-service lamps are recommended only where maintenance labor costs are higher or lamps are difficult to replace.

Fluorescent lamps have three to five times the efficiency of incandescent lamps and compare favorably to most HID sources. Fluorescent-lamp efficiencies vary depending on lamp length, lamp loading, and lamp phosphor.

Three common types of HID units are mercury vapor, high-pressure sodium vapor, and metal halide. Mercury-vapor lamps have a long life and are cheaper than other HID units. Although metal halide lamps are more efficient than mercury-vapor lamps, they cost more and typically are unavailable in small sizes. High-pressure sodium vapor units are typically best for HID lighting applications because of their low cost, long life, and efficient power use.

A good lighting system that is well-maintained will provide the visual conditions for maximum performance to maintain lighting levels for which the system was originally designed. The major lighting system maintenance activities are cleaning and relamping.

Proper cleaning schedules will maximize light for the same level of energy expended. Dirt can account for light losses from 30 to 50%. Therefore, fixtures and lamps should be washed regularly using the proper cleaning solution. Cleaning frequency will depend on the amount and type of dirt in the air and whether the fixture is the ventilated or nonventilated type. Wall and ceiling cleaning and repainting are also techniques to maximize light output for the same power expenditure.

## ENERGY AUDITS

Plants can do much to improve energy usage via better energy management, better instrumentation and control systems, and energy-efficient equipment. In addition to facility-initiated measures, incentive and special rate programs may be available from the local utility.

An energy audit typically starts with creating an energy audit team consisting of plant personnel, an electrical utility representative, and possibly an outside process or energy expert. After reviewing the plant energy bills and collecting plant energy and operating data, the team should conduct a detailed field investigation of plant processes and equipment to find out the amount of energy used and the "when and how" of energy demand. The field investigation includes lighting, HVAC, pumping, and unit processes.

The team should then review outside institutional programs. Many electrical utilities and government entities have rebates, grants, or loan programs that should be considered when prioritizing payback on projects.

The next team task is to use the collected information to develop energy-conservation measures and implementation strategies. The team must look at capital and operating costs, cost-to-benefit ratios, energy savings, process requirements and complexity, and effluent quality. The most essential parts of such a program are the energy audit follow-up activities to ensure that the measures or projects are implemented and the savings achieved. Utilities, government, and associations are a good source of sample audit templates.

# POWER FACTOR CORRECTION

When an electric load has a power factor lower than unity, the apparent power delivered to the load is greater than the real power that the load consumes. Only the real power is capable of doing work, but the apparent power determines the amount of current that flows into the load for a given load voltage.

Energy losses in transmission lines increase with increasing current. Therefore, power companies require that customers, especially those with large loads, maintain the power factors of their respective loads within specified limits or be subject to additional charges. Engineers are often interested in the power factor of a load as one of the factors that affects power-transmission efficiency.

For a linear circuit operating from a sinusoidal voltage, the current must be a sinusoid at the same frequency. When the current is exactly in phase with the voltage, the power factor is 1. This corresponds to a purely resistive load. A toaster is an example of an approximately linear device; an electric motor is not. For a nonlinear circuit (e.g., a switchmode power supply), the current is not necessarily sinusoidal. In that case, there is current at harmonics of the voltage frequency.

Power-factor correction returns the power factor of an electric alternating current power transmission system to very near unity by switching in or out banks of capacitors or inductors that cancel the inductive or capacitive effects of the load. For example, the inductive effect of motor loads may be offset by locally connected capacitors. It is also possible to effect power-factor correction with an unloaded synchronous motor connected across the supply. The power factor of the motor is varied by adjusting the field excitation and can be made to behave like a capacitor when overexcited.

# COGENERATION

If a plant needs constant heating or mechanical energy, cogeneration of electricity should be evaluated. Indeed, large potential savings are available for facilities capable of producing onsite power and heat. Such systems require a protective relaying system that meets all the utility company requirements to protect the system. Cogeneration in wastewater plants typically involves the burning of digester gas to generate electricity and produce heat for processes or buildings. Cogeneration systems are typically installed in larger [more than 132-L/s (3-mgd)] plants, and require detailed investigation.

In most cogeneration and trigeneration power and energy systems, the exhaust gas from electric-generation equipment is ducted to a heat exchanger to recover the thermal energy in the gas. These exchangers are air-to-water heat exchangers, in which the exhaust gas flows over some form of tube and fin heat-exchange surface and the heat from the exhaust gas is transferred to make hot water or steam. The hot water or

**TABLE 10.6** Cogeneration and trigeneration power and energy technology.

| Technology | Operating history | Nitrogen oxide emissions | Size (kWh) | Gas cleaning requirements |
|---|---|---|---|---|
| Internal combustion engines | Common | Common in last few years | 250–2500 | Moderate |
| Microturbines | Recent applications | Low | 30–250 | Extensive |
| Gas turbines | Common | Low | >3000 | Extensive |
| Fuel cells | New | None | 200–1000 | Extreme |
| Stirling engines | New | Very low | 55 | None |

steam is then used to provide hot water or steam heating and/or to operate thermally activated equipment (e.g., an absorption chiller for cooling or a desiccant dehumidifier for dehumidification). Current technologies used include internal combustion engines, microturbines, gas turbines, fuel cells, and stirling engines. Table 10.6 provides further details about these technologies.

## REFERENCES

Jackson, H. W. (Ed.) (1989) *Introduction to Electric Circuits,* 7th ed.; Prentice-Hall: Englewood Cliffs, New Jersey.

Kuphaldt, T. R. (2007) *Lessons In Electric Circuits;* Design Science Library; http://www.ibiblio.org/obp/electricCircuits/ (accessed Jun 2007); Vol. II, Chapter 11, Power Factor.

Smeaton, R. W. (Ed.) (1998) *Switch Gear and Control Handbook;* Institute of Electrical and Electronic Engineers: Piscataway, New Jersey.

*The National Electrical Code Handbook* (2005) National Fire Protection Association: Quincy, Massachusetts.

Water Pollution Control Federation (1984) *Prime Movers: Engines, Motors, Turbines, Pumps, Blowers, and Generators,* Manual of Practice No. OM-5; Water Pollution Control Federation: Washington, D.C.

## SUGGESTED READINGS

Controls Link, Inc., homepage; relay information. http://www.controlslink.com/whatwedo/arcflash.php (accessed May 2007).

Information on harmonics. http://members.tripod.com/~masterslic/harmonics.html (accessed May 2007).

Nilsson, J. W.; Riedel, S. A. (2002) *Introductory Circuits for Electrical and Computer Engineering;* Prentice Hall: New York.

U.S. Department of Labor, Bureau of Labor Statistics, *Occupational Outlook Handbook,* information for electricians. http://www.bls.gov/oco/ocos206.htm#training (accessed May 2007).

Watson, S. K. (2006) Cogeneration Technologies, Trends for Wastewater Treatment Facilities. *Waterworld,* **22** (6), 14–15.

# Chapter 11

# Utilities

## INTRODUCTION

In-plant utilities perform a vital support role in the operation of wastewater treatment facilities. Some utilities assist with equipment and process operations, and others support the safety and welfare of plant personnel. Utilities such as water supplies, compressed air, communications systems, heating, ventilating, and air conditioning systems, fuel supply systems, and roadways continuously serve as an integral part of the plant's daily functions. Others, such as fire protection, storm drainage, and flood protection, serve on an occasional, seasonal, or emergency basis. Loss of any of these services can impair operations, cause facility failure, or impose risks and discomfort on plant staff.

## MAINTENANCE

Routine and regular maintenance of utilities helps ensure long-term and trouble-free operation. Good maintenance practices include the items and frequencies listed in the tables in this chapter. More frequent maintenance may be necessary if indicated by operating experience or if specified by the equipment manufacturer.

The operator should observe appropriate safety precautions and practices when operating and maintaining rotating or electrical equipment (see Chapter 5). Adequate space, lighting, and ventilation are necessary for proper and safe inspection and efficient operation of equipment. The operator should use effective ear protection for extended exposure to equipment with high noise levels. More complete and detailed descriptions of proper operating and maintenance procedures are available in the manufacturer's instruction book furnished with the equipment. Instruction manuals should be in the hands of those responsible for performing the maintenance.

## WATER SYSTEMS

Two types of water—potable and nonpotable—are used in wastewater treatment operations. Potable water (water suitable for drinking) is used for showers, hot-water heaters, water coolers, water closets, urinals, lavatories, washing machines, kitchen sinks, safety showers, mop sinks, fire hydrants, and laboratory sinks with vacuum breakers. Hot water for any of the above units is not drawn from boilers used for supplying hot water to a sludge heat exchanger, sludge heating coils, or other devices where sludge or other contaminants may be present. Nonpotable water (water not ensured to be suitable for drinking) may be used for flushing or filling idle tanks, fire protection, toilet flushing in some states, or for other purposes with limited human contact.

The plant's files should include the water source and name of the supplier. Locations of pipelines, hydrants, booster pumps, and shutoff valves should be noted. Each item should be tagged for emergency use. The different water supplies should be clearly labeled. If special wrenches or tools are required to open or close these valves, the tools should be identified and mounted in a nearby location.

Drinking water supplies must be separate from all other water sources. Interconnections between potable and nonpotable water sources are strictly prohibited, and cross-connections must be prevented. Hose outlets using drinking water service need antisiphon devices to prevent accidental back siphoning or contamination of the drinking water supply. The operator should avoid attaching hoses directly to fire hydrants when flushing tanks if the fire hydrants are supplied with potable water.

Most states require measures to prevent potable water contamination, such as an air-gap tank or backflow preventer for water from a public water supply as it enters the plant site. In addition, protective measures must be taken to supply potable water for service water for wastewater equipment cleaning or any other purpose involving water contact with a contaminant. Air-gap tanks and backflow preventers need regular service and inspection to ensure proper operation.

Sources of nonpotable water include treated wastewater effluent and potable water after it has passed through a break tank with an air-gap inlet or an approved backflow preventer. Nonpotable water is used for seals, scrubbers, backflushing, tank flushing, cooling towers, boilers, digester heating, chlorine solution, irrigation, chemical and polymer dilution, and various other in-plant demands. Although the equipment and distribution systems that provide nonpotable water vary, they always include pumps, piping, valves, and controls. Such systems may also include accessories such as strainers, receiving tanks, flow meters, hydrants, water treatment facilities, and hydropneumatic tanks.

Every hose bib, faucet, hydrant, and sill cock in the treatment plant requires clearly and permanently posted nonpotable water signs to indicate that the water is not safe to drink. Personnel need instructions to prevent accidental consumption of nonpotable water.

Seal water, used for cooling, flushing, and sealing the stuffing boxes of rotating equipment, typically comes from bladder tanks, hydropneumatic systems, or direct pressure systems. Bladder tanks consist of a receiving vessel fitted with a flexible internal membrane (bladder) that separates the nonpotable water supply from the air cushion above. Hydropneumatic tanks, connected to a nonpotable water supply, are pressurized with compressed air. Controls allow for regulation of the compressed air and water that enter the hydropneumatic tank so the necessary capacity and pressure can be maintained. Direct-pressure systems include a pump to draw water from a break tank that feeds the seal water system. A pressure relief valve and bypass line recirculate unused water to the break tank. This method is less energy efficient than the systems discussed above.

Monitoring and recording water usage regularly on a plant-wide basis for each of the several process systems and for other major water-using equipment will signal water-use changes that may require corrective action.

In general, all buried piping will have some type of cathodic protection to prevent corrosion. Inspection and maintenance frequencies for the cathodic protection system and of water system components are shown in Table 11.1. Where possible, the water system should be laid out in a loop with frequent isolation valves so repairs can be conducted without disrupting large sections of the plant.

## DRAINAGE SYSTEMS

Two types of drains are often used in wastewater treatment facilities: plant drains and storm drains. It is important to keep these two systems separate to prevent contamination of surface waters and to reduce the likelihood of being required to have a National Pollutant Discharge Elimination System (NPDES) permit for the storm drainage discharge. Plant drains for dewatering tanks and equipment convey the fluids elsewhere in the plant for appropriate treatment. Operational considerations include the rate and duration of dewatering and their effects on plant processes. Portable pumps may be required to completely dewater tankage. Plant drainage systems should include properly located cleanouts or manholes that are checked and cleaned regularly to avoid debris accumulations that restrict the capacity of the drains. For the same reasons, drains need regular flushing.

**Table 11.1** Water system inspection and maintenance.

| Inspection and maintenance tasks | Suggested frequency[a] |
|---|---|
| Backflow prevention device | M |
| Calibrate meter pump(s) | A |
| Inspect hydropneumatic tank and equipment | W |
| Clean strainers | W/R |
| Operate hydrants | A |
| Operate valves | A |
| Test cathodic protection system | S |
| Monitor water usage | D |

[a]D = Daily; W = weekly; M = monthly; S = semiannually; A = annually; and R = as required.

Where groundwater-pumping systems exist to prevent flotation of structures and tankage, the operator should check the water level at each of the groundwater-monitoring stations. If any water level is above the bottom of the tank or structure, or if the system lacks a monitoring station, the operator should pump the groundwater sump or well dry before dewatering the tank. Failure to follow these procedures can cause structural failures and broken connecting services. If the tankage has hydrostatic relief devices, these units should be checked each time the tank is empty to ensure proper seating and free operation of the units.

Sanitary drainage from building showers, lavatories, floor drains, toilets, and other plumbing fixtures is collected and discharged at the plant headworks for treatment. Floor drains should be flushed weekly to ensure unrestricted drainage through the system and to fill traps to prevent sewer gases from entering the buildings.

Sump pump systems, composed of a pit, pump, and liquid-level controls, help remove collected water and wastes from structures. These systems may have either submersible or vertical centrifugal pumps driven by externally mounted motors. Duplex pump systems are often provided in critical areas. In the sump pits, high water level alarms are necessary to alert personnel of sump pump failure. Routine operation and maintenance includes weekly to monthly flushing of the sumps and checking of the floats and controls.

Stormwater drainage systems convey surface runoff away from building tanks, walks, and roadways to an off-site drainage course or storm sewer. Collection inlets, area drains, culverts, storm sewers, swales, ditches, roof drains, gutters, and downspouts should be kept free of debris. Stormwater ponding should be avoided unless the pond is designed to control stormwater release to the conveyance systems.

## FUEL SYSTEMS

Several fuels provide the energy needed for plant operation, vehicles, and emergency power. These include fuel oil, natural gas, liquid propane gas, digester gas, gasoline, and diesel. Most of these fuels, except natural gas and perhaps digester gas, require a suitably sized storage vessel that may be installed either above or beneath the ground. Underground storage tank systems normally include product distribution piping and associated equipment for filling, vertical, and level indication.

Underground storage tanks for petroleum products must comply with strict federal and state regulations. These regulations, administered by the county health officer, are intended to prevent leaking tanks that can cause groundwater contamination, endanger lives, increase fuel usage, and require expensive cleanup procedures. Since

1946, all underground tank owners have been required to obtain permits to install and operate tanks with hazardous petrochemical products. Notification and permitting also apply to existing tanks that have been taken out of service since January 1, 1974, but have not been removed from the ground.

As of May 8, 1986, all underground storage systems must meet stringent secondary containment requirements, must have proper corrosion protection to prevent releases, and must be continuously monitored for leakage failure of tank or pipelines for the operational life of the tank system. Fuel delivery systems must conform with state and U.S. Department of Transportation rules and regulations, local agency rules for air emissions, and supplier requirements.

Natural gas is generally supplied by pressurized pipelines. Any natural gas leak is readily detected by its characteristic odor. If the odor is detected, the operator should determine the source, and if severe, evacuate the area immediately, notify the local utility and the fire department to isolate the supply to the area, and turn off electrical power to avoid any source of ignition.

Methane generated by the decomposition of solids in the absence of oxygen can be used to fire boilers, heaters, incinerators, or engines. Excess gas can be stored for sale, burned to power equipment or heating, or burned by special gas flaring systems.

Both natural gas and methane (digester) gas are highly toxic and can produce explosive conditions in contained areas; areas where digester gas may collect require combustible gas analyzers and alarms. If gas is detected, the operator should evacuate the area immediately, turn off power at a remote switch or breaker, and ventilate the area. Only personnel equipped with a self-contained breathing apparatus, consistent with the confined-space entry protocol (see Chapter 5), are allowed to enter the area until gas analyzers indicate that safe conditions have been restored. In no case should any other personnel be allowed in an explosive atmosphere.

Safety controls on fuel systems include pressure regulators, flame traps, automatic shutoff valves, pressure-vacuum release valves, and antisiphon valves. Semiannual inspection and maintenance of each item is necessary to ensure its proper operation.

# COMPRESSED AIR

Compressors are machines designed to deliver air or gases at a pressure higher than their normal pressure. Under controlled conditions, air is compressed to provide a power source for operating pneumatic equipment, pumps, and tools. Compressors furnish instrument air supplies, air jets for cleaning equipment, purge air, and air for wastewater ejectors, as well as pressurizing gases for transport and distribution.

**TYPES OF COMPRESSORS.** Compressor types are typically classified as displacement or dynamic. The displacement type can be subdivided into reciprocating positive displacement or rotary positive displacement types (Figure 11.1).

Displacement compressors use pistons, vanes, or other pumping mechanisms to draw air or gases into the unit, increase pressure by reducing the volume, and discharge the air into a receiver. An air receiver stores air under pressure and dampens the pressure pulsations in the air system. By storing air, the receiver reduces the frequency of loading and unloading, or starting and stopping the compressor. Since receivers collect oil and water from the compressed air, they need regular draining.

Portable compressors are self-contained, with drive and air receivers mounted on the unit. Vacuum pumps withdraw air from a tank or pipeline to reduce its air pressure below atmospheric and discharge it to the atmosphere. In principle, vacuum pumps reverse the operation of a compressor.

Accessories and protective devices in compressor systems prevent system and equipment damage, control flow and pressure, remove oil contamination, dry the air (for instrumentation and other uses), reduce noise, separate moisture, and serve other

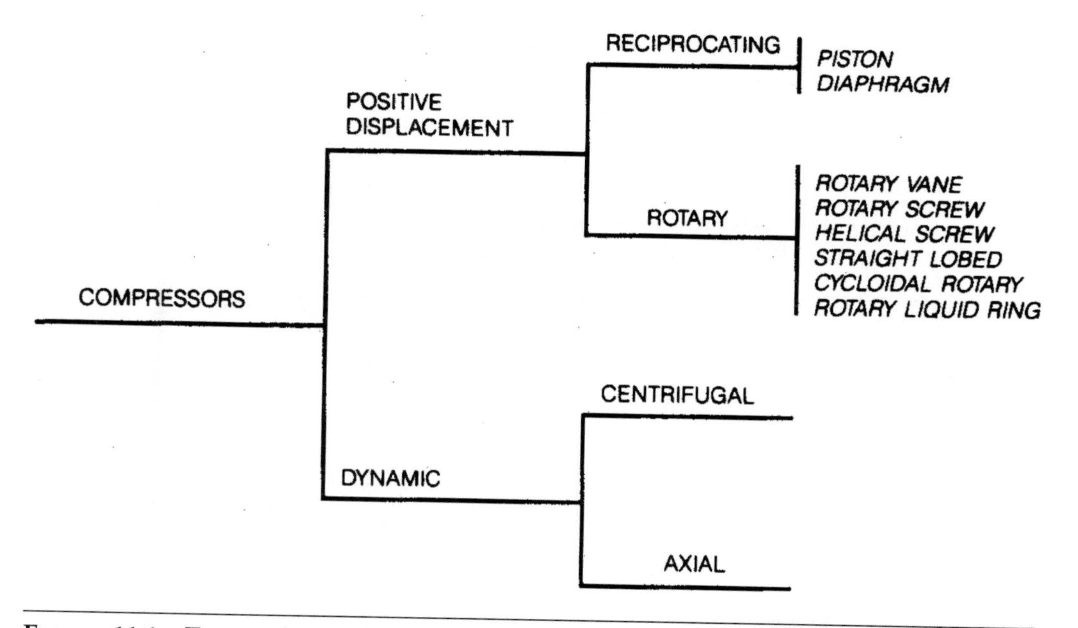


**FIGURE 11.1** Types of air compressors.

functions. A simple compressor system may include an intake filter, intake silencer, compressor, discharge silencer, intercooler or aftercooler, moisture-oil separator, air receiver, safety relief valve, pressure relief valve or pressure regulator, and pressure switches for compressor on-off control (Figure 11.2). Control systems may include elements for monitoring and sequencing in addition to protective devices for the equipment.

## COMPRESSOR MAINTENANCE.

Regular compressor maintenance helps ensure long-term and trouble-free operation. Good maintenance practices include the items and frequencies listed in Table 11.2. More frequent maintenance may be necessary if indicated by operating experience or specified by the equipment manufacturer.

The operator should observe appropriate safety precautions and practices when operating and maintaining rotating or electrical equipment (see Chapter 5). Adequate space, lighting, and ventilation are necessary for proper and safe inspection and efficient operation of the units. Operators should use effective ear protection for extended exposure to equipment with high noise levels. More complete and detailed descriptions of proper operation and maintenance (O&M) procedures are available in the manufacturer's instruction book furnished with the equipment. Instruction manuals should be in the hands of those responsible for the inspection and continuing service of the units.

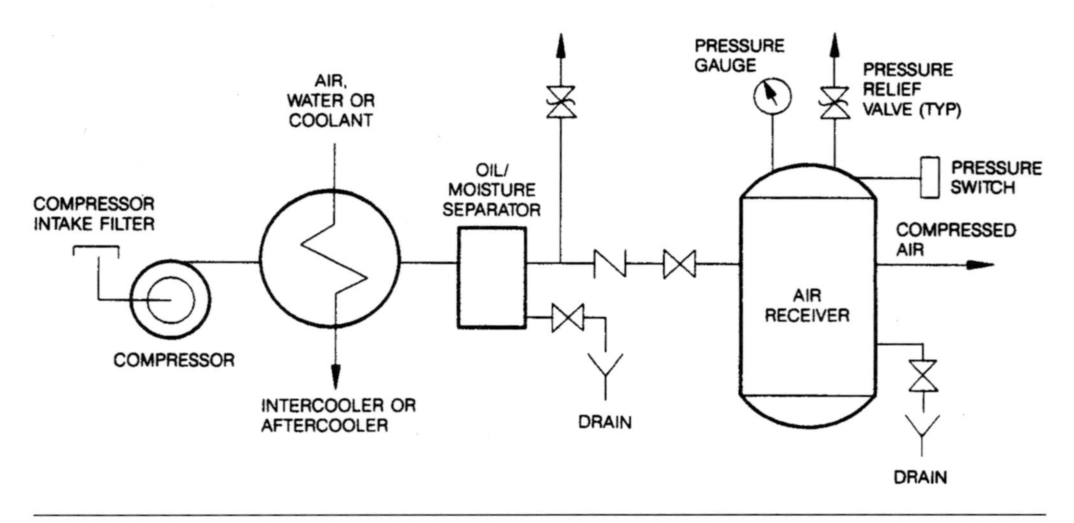


**FIGURE 11.2** Typical compressed air system.

TABLE 11.2 Compressor inspection and maintenance chart.

| Inspection and maintenance tasks | Suggested frequency[a] |
|---|---|
| Check for unusual noise and vibration | S |
| Tighten bolts, belt drives, and chains (see Chapter 12) | R |
| Check, clean, or replace dirty air filters | Q |
| Drain air receivers, intercoolers, separators | Q |
| Check crankcase and drive oil and grease levels; do not overlubricate | M |
| Operate safety valves and regulators manually | S |
| Clear dust, dirt, and debris from compressor and drive external surfaces | S |
| Check all condensate traps | S |
| Check time to pressurize the system | S |
| Check pressure switch and pressure settings | S |
| Inspect piping, valves, unloaders, belt alignment, and electrical connections | S |
| Log temperature, pressure, cooling water, temperature and flow, lubrication use, pressure, vibration, running time, and service repairs | M |
| Check electric motor operation (see Chapter 12) | R |

[a]M = monthly; Q = quarterly; S = semiannually; and R = as required.

**AIR DRYERS.** Air dryers, often included with compressed air systems to remove moisture (dry air) for services such as pneumatic instrument control, usually use one of three common drying methods:

- Refrigeration,
- Regenerative desiccant (heat or heatless), and
- Deliquescent.

Refrigeration dryers cool the compressed air to condense the water vapor and drain off the resulting liquid.

Regenerative desiccant (water-absorbing substances) dryers use a moisture-adsorption medium and a dual-tower arrangement with automatic valving and control circuits. As the active tower removes moisture from the air, the other tower undergoes a purge cycle to remove the moisture from the adsorption media. After completion of the purge cycle, the towers switch, with the previously active tower entering a purge sequence. Heatless types of these dryers require 10 to 15% of the air to purge the adsorption media in the inactive tower. Heater types require 1% of the air for purging the adsorption media in the inactive tower, together with activation of heating elements.

Deliquescent dryers are large vessels containing patented pellet desiccants. Compressed air passes over and through the desiccant where the water vapor is adsorbed.

These units require periodic replacement of the desiccant and removal of accumulated water. As a signal for replacement, some desiccants are designed to change color when they become exhausted.

A typical air-dryer arrangement is presented in Figure 11.3. Normally, each air-dryer system requires a prefilter. Most units have monitors that indicate failure of the units to supply dry air. Regular inspection and maintenance, such as replacement of exhausted desiccants or checking the air temperature on refrigeration dryers, are essential for proper performance of the dryers.

## FIRE PROTECTION SYSTEMS

For any fire, no matter how small, the operator should call the local fire department immediately. Fire extinguishers, fire hoses, sprinkler systems, and fire hydrants should be available to put out fires. Special portable fire extinguishers must be used for electrical and chemical fires. The first five minutes constitute the most critical period for controlling a fire; if appropriate, plant staff should try to extinguish or control small fires until firefighters arrive. It may not be appropriate or advisable for plant staff to try to extinguish small fires in small facilities with limited staff on shift. After their arrival, only firefighters should remain in the area. In case of a major fire, the area must be evacuated immediately.

The National Fire Protection Association (NFPA) classifies fires as Class A, B, C, or D. Class A fires (ordinary combustibles) burn wood, paper, textiles, or trash. Class B

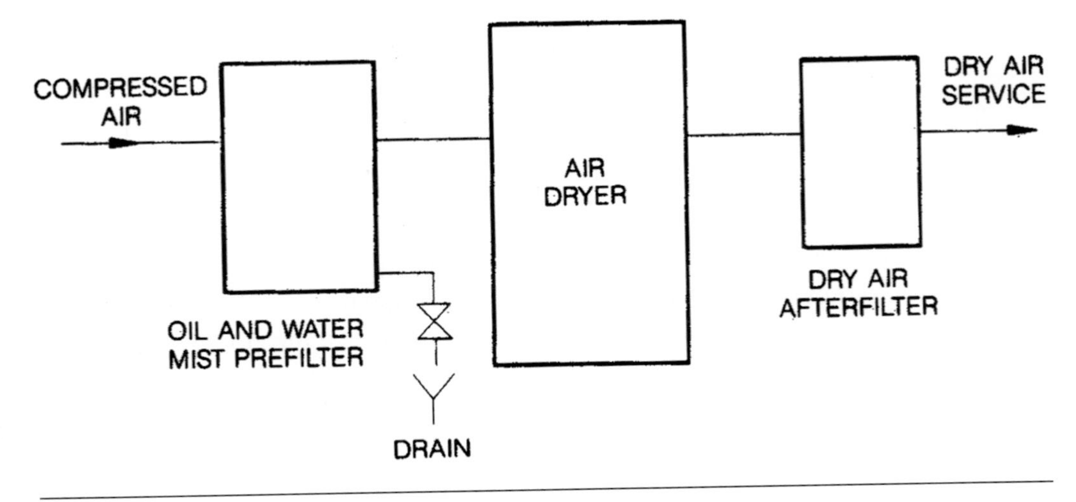


FIGURE 11.3 Air-dryer system.

fires (flammable liquids) burn flammable liquids such as gasoline, paint, solvents, grease, and oils. Class C fires (electrical equipment) are those in or near electrical equipment. Class D fires (combustible metal) burn combustible metals such as sodium, titanium, uranium, zirconium, lithium, magnesium, and sodium-potassium alloys.

Water from hose connections or Class A fire extinguishers is appropriate for Class A fires. Fire extinguishers with special solutions or dry chemicals are required to extinguish the other classes of fires (Figure 11.4). Portable fire extinguishers of the appropriate class must be placed in strategic areas to extinguish the type of fire that may occur in each area. The proper type of fire extinguisher for Class A, B, C, or D fires must be present, fully charged, and in good working condition.

Fire protection emphasizes yearly inspection of fire extinguishers for corrosion, damage, loss of charge, repair, or placement of the units as necessary. Spare units should be on hand. Tags stating the most recent maintenance and recharge data must be attached to each unit.

Plant staff must be informed that fighting fires is serious business. They must be instructed in proper use of fire extinguishers and warned that using fire extinguishers in poorly ventilated areas without proper safety equipment can cause asphyxiation. The operator must ensure that smoke detection sensors, thermal sensors (where applicable), and appropriate alarms are strategically placed to alert plant staff of danger and the need for corrective actions. Signs prohibiting smoking and open flames must be clearly posted in areas where combustible materials are stored or where flammable gases may accumulate.

Fire hydrants, located throughout a plant for fire department use, must remain unobstructed and easily identifiable. Other special fire protection equipment at the plant site may include sprinkler systems, deluge systems, fire pumps, standpipes, hose cabinets, and hose reels. Maintenance requirements for each system need regular review to ensure that alarms function and equipment is ready for operation. Automatic sprinklers and fire-hose systems must remain fully pressurized at all times.

The local fire department should inspect the plant yearly to review fire protection equipment and evaluate fire prevention procedures. The operator should consult the local fire department to resolve any questions related to fire control, prevention, or protection. Fire department telephone numbers should be prominently posted at each plant telephone.

## HEATING, VENTILATING, AND AIR CONDITIONING

The air within the plant's buildings is handled by heating, ventilating, and air-conditioning (HVAC) systems ranging from simple to complex. Air-handling systems can

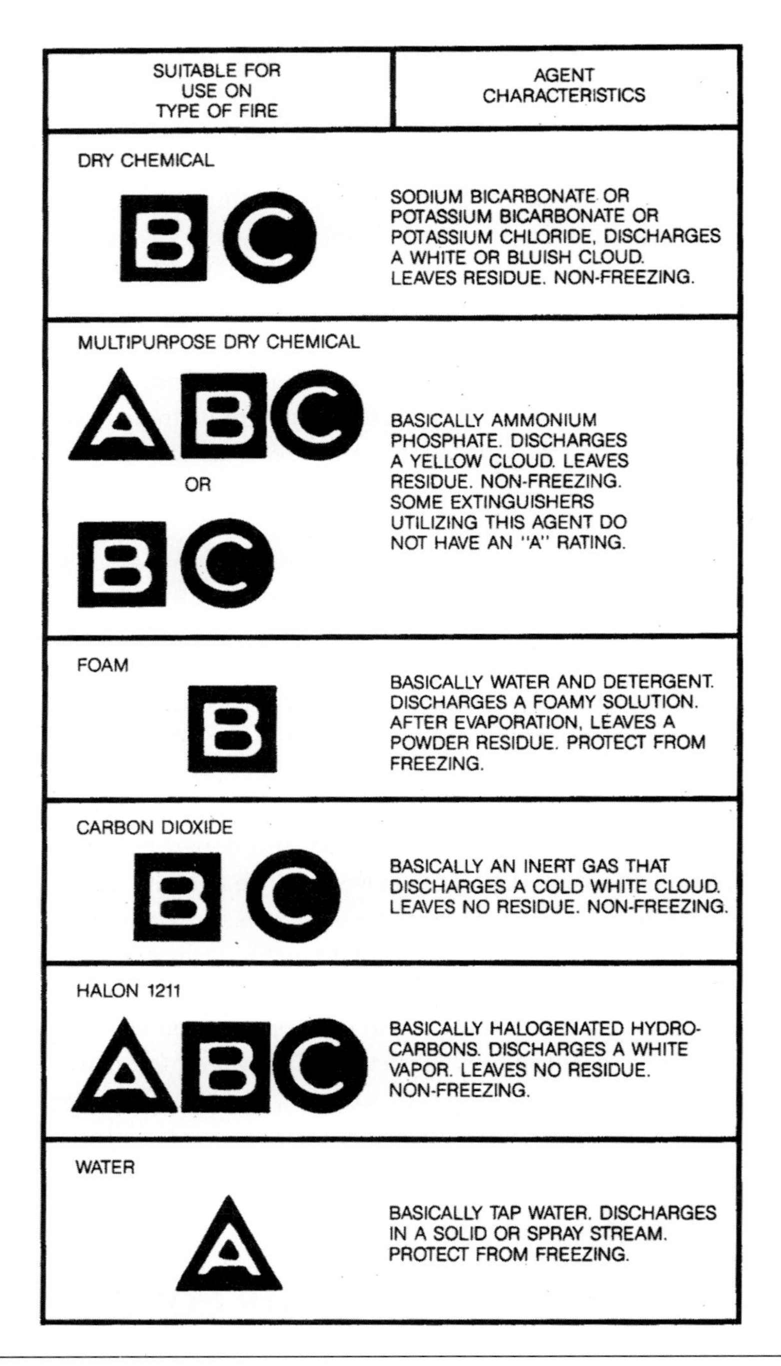

| SUITABLE FOR USE ON TYPE OF FIRE | AGENT CHARACTERISTICS |
| --- | --- |
| DRY CHEMICAL<br>B C | SODIUM BICARBONATE OR POTASSIUM BICARBONATE OR POTASSIUM CHLORIDE, DISCHARGES A WHITE OR BLUISH CLOUD. LEAVES RESIDUE. NON-FREEZING. |
| MULTIPURPOSE DRY CHEMICAL<br>A B C<br>OR<br>B C | BASICALLY AMMONIUM PHOSPHATE. DISCHARGES A YELLOW CLOUD. LEAVES RESIDUE. NON-FREEZING. SOME EXTINGUISHERS UTILIZING THIS AGENT DO NOT HAVE AN "A" RATING. |
| FOAM<br>B | BASICALLY WATER AND DETERGENT. DISCHARGES A FOAMY SOLUTION. AFTER EVAPORATION, LEAVES A POWDER RESIDUE. PROTECT FROM FREEZING. |
| CARBON DIOXIDE<br>B C | BASICALLY AN INERT GAS THAT DISCHARGES A COLD WHITE CLOUD. LEAVES NO RESIDUE. NON-FREEZING. |
| HALON 1211<br>A B C | BASICALLY HALOGENATED HYDRO-CARBONS. DISCHARGES A WHITE VAPOR. LEAVES NO RESIDUE. NON-FREEZING. |
| WATER<br>A | BASICALLY TAP WATER. DISCHARGES IN A SOLID OR SPRAY STREAM. PROTECT FROM FREEZING. |

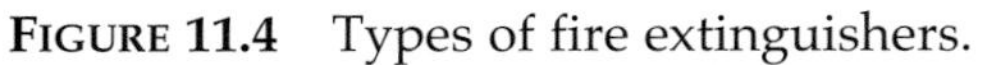

**FIGURE 11.4** Types of fire extinguishers.

include only ventilating units that supply unheated or uncooled outside air; heating and ventilating units that supply heated air as needed; and HVAC units that provide heated or mechanically cooled air, as necessary.

Ventilation air supports life, prevents accumulation of explosive gases, reduces heat buildup, and helps maintain safe working conditions. During hot weather, air movement helps cool the body by evaporating perspiration. In buildings, ventilation air eventually reduces temperatures and humidity as hot, indoor air is exchanged for cooler, outside air. Heat supplied through heating coils, ducts, tubes, or elements comes from sources such as boilers, hot-air furnaces, electric elements, heat pumps, or heat-recovery devices. Mechanical air-conditioning systems are designed to provide comfort for employees and visitors or controlled environments for systems, products, or equipment. Air-conditioning systems cool, dehumidify, filter, and circulate room air. Dehumidifiers are either mechanically refrigerated appliances or devices containing chemical absorbents that remove moisture from the air. The reduction of air moisture by air conditioning or dehumidification helps to enhance human comfort and prevent rust, rot, mold, and mildew.

**VENTILATING.** Fans, sometimes with renewable filters, provide ventilation air. Ventilation may be provided by supply fans, exhaust fans, or both. Operation of ventilation equipment may be continuous or intermittent. Intermittent systems are often interconnected with lighting systems so that they activate automatically when personnel enter the area and switch on the lights. Likewise, the fans deactivate when the lights are shut off. Two other fan-control systems include a timer and a thermostat. The timer is often used where a remote station is not visited frequently or is operated seasonally.

Many areas must be ventilated independently to avoid the problem of exhausting obnoxious or hazardous gases from one space into another. For example, pumping station wet wells that contain screens, mechanical devices, and other equipment needing maintenance or inspection require a ventilation system separate from that ventilating the dry wells. Also, laboratory facilities and fume hoods must have separate ventilation systems to prevent distribution of odors or obnoxious gases to other areas within the building.

Ventilation rates for each structure are based on its use and occupancy (WEF, 1992). In all areas, adequate ventilation must be provided to ensure safe working conditions for plant personnel. Personnel must not enter screen rooms or wet wells unless the mechanical ventilation equipment is functioning properly.

Some conditions require high rates of ventilation. For wet wells, forced air must enter the space at a rate of at least 12 air changes/hour and, if ventilation is intermittent, the rate must be at least 30 air changes/hour (WEF, 1994). One air change equals the total volume of air in the ventilated space. Dry wells require at least 6 air changes/hour or, if ventilation is intermittent, at least 30 air changes/hour.

If ventilation systems fail in confined spaces where a hazardous atmosphere may develop, the operator should exercise extreme caution. A hazardous atmosphere may develop wherever hazardous or explosive gases such as hydrogen sulfide, methane, gasoline, oils, solvents, or other hydrocarbons are present or may enter; where toxic gases such as hydrogen sulfide or methane may accumulate; or where oxygen may be depleted.

The operator should not attempt to enter these areas without testing the atmosphere for toxic or combustible gases and oxygen deficiency. If it is necessary to enter a space where asphyxiation or exposure to explosive or toxic gases might occur, operators should follow the protocol for entry to confined spaces as described in Chapter 5. Following the safety rules can save the life of the operator and those of fellow workers.

**HEATING.** In general, there are three types of heating for buildings or process equipment: hot water, steam, and hot air. In addition, infrared systems are used to heat small areas and some special processes. The most common energy sources for heating are coal, oil, gas, and electricity. Steam or hot-water boilers require special handling to ensure proper operation. Burner controls for boilers are equipped with flame safeguard systems that prevent explosive conditions. Only trained staff or knowledgeable equipment mechanics should service and maintain these units. Steam and hot-water systems are normally used when there is a high heating load. Electrical and infrared heating are used for lower heating loads and special situations.

Some large process equipment units and drives produce excess heat, which, if properly reclaimed, can be used to supplement heating of building spaces during cold weather. To conserve energy, heat-recovery equipment should be considered where heating and ventilation systems are required to use more than 50% outdoor air; that is, the rate of supply is one-half the rate of exhaust. Heat-recovery systems reclaim heat in exhaust air streams and use the reclaimed heat to preheat outside ventilation air. Similarly, heat-recovery systems serving air-conditioned spaces use the cooler exhaust air to precool ventilation air. Types of heat-recovery systems include air-to-water coil, heat pipe, and rotary heat-exchanger wheel.

An air-to-water coil system transfers heat from the exhaust air to the outside air as it enters and passes through the system. The system consists of a water-filled coil in the exhaust fan system connected by piping to a similar coil in the supply fan system. A circulation pump, valving, and temperature controls are required to operate the system.

A heat-pipe system is an air-to-air heat-transfer device consisting of a bank of refrigerant-filled tubes that transfer heat from one end of the tube bank to the other. Warm air heats the refrigerant, which vaporizes and moves to the opposite end of the tubes located in the outside airstream. Heat transfers to the incoming outside air from the refrigerant vapor as it condenses. The condensate then flows back to the exhaust side of the tube. Rotary heat-exchanger wheel systems consist of a desiccant-impregnated wheel that rotates slowly between the incoming, outside airstream and the exhaust air. The wheel transfers heat from the warm exhaust air to the cooler outside air being drawn into the system.

**AIR CONDITIONING.** Air conditioning is based on a simple law of gases—when a gas is compressed, it heats; conversely, when a gas expands, it cools. A condenser-compressor unit circulates refrigerant through piping, valves, condensing coil, and cooling coil (evaporator). Within the cooling coil, a liquid refrigerant absorbs heat as it evaporates, thereby chilling the surrounding air. A blower draws warm air through a filter and across the cooling coil and then forces the cooled air into the room. Air conditioning serves to dehumidify and cool the air.

Dehumidifiers draw moisture-laden air over refrigerated coils to condense the water vapor and then drain off the condensate. A water-absorption medium, periodically replaced or regenerated, can also remove moisture from air. Fans draw humid air through the dehumidifier and return dry air to the room.

**AIR MONITORING.** Sensors are used to detect smoke or heat, temperature, combustible hydrocarbons (methane or gasoline), hydrogen sulfide, carbon monoxide, sulfur dioxide, ammonia, ozone, and chlorine. Detection equipment may be interconnected with the heating and ventilating equipment. Safe plant operation requires maintenance of all sensors to keep them in perfect operating condition.

To conserve energy, HVAC systems can be equipped with automatic setback systems. In all cases, the thermostats should be set as high as practical in summer and as low as reasonable in winter. Many plants maintain temperatures in areas not continually occupied as low as 10 to 13 °C (50 to 55 °F) during the winter and as high as 38 °C (100 °F) during the summer.

Computer rooms require special control of temperature, humidity, and air purification. Other areas with electrical or electronic equipment may also require dehumidification or air conditioning.

Exhaust from some ventilated spaces may require odor treatment before release to the atmosphere. Several odor treatment systems can be used, ranging from scrubbing towers with chemicals to fume incinerators. Each type of system requires special attention to ensure effective operation (see Chapter 13).

**HEATING, VENTILATING, AND AIR CONDITIONING MAINTENANCE.** Table 11.3 lists several of the important maintenance tasks required for HVAC equipment. More detailed and specific O&M information is available from the manufacturer's instruction guides. Maintenance frequencies should at least equal those indicated, but should be higher if specified by the equipment manufacturer or suggested by operating experience. The operator should observe all safety precautions and practices recommended by the manufacturer when operating and maintaining rotating, electrical, or fuel-system equipment.

## COMMUNICATION SYSTEMS

Reliable communication systems are vital during normal plant operations and during emergencies. The telephone offers the principal mode of communication between wastewater treatment personnel and others outside the plant. Local telephone service companies provide the communication links, and plants can purchase or lease their own telephones. Sophisticated telephone systems can restrict certain telephone units to local calls while allowing unrestricted access from other units. Cellular telephones can also be a vital communication function for facilities.

Communication systems within a plant generally include several telephones, including a few with paging speaker systems. These systems may also have such items as portable beepers, pagers, or two-way radio communicators. Special telephones, required for hazardous areas, may be owned and maintained by the plant or may be leased from others. Battery recharge systems are required to keep the portable units operative. The radio communication system serves as an effective backup system when normal telephones are out of service.

## LIGHTNING PROTECTION

Many parts of the country experience frequent lightning strikes. Large arrestors are applied to electrical systems to protect the apparatus from the effect of overvoltages. Lightning rods are used to protect buildings and other property.

## ROADWAYS

Plant roadways provide cars, trucks, and other vehicles with access to the plant's buildings, structures, and equipment. The roadways must have stable, all-weather surfaces on base courses strong enough to bear truckloads and withstand frost action, if necessary. Roadway surfaces may be asphalt pavement, concrete, macadam, crushed

TABLE 11.3 Heating, ventilating, and air conditioning inspection and maintenance.

| Inspection and maintenance tasks | Suggested frequency[a] |
|---|---|
| Fuel oil pumps | |
| Check pump seals; adjust or replace as required (see Chapter 12) | R |
| Check clearances within pump for free turning without appreciable end play | A |
| Remove exterior dirt and foreign materials from pump fittings | A |
| Inspect parts for wear; replace worn parts | A |
| Check strainers and water separators; clean regularly | M |
| Heating coils | |
| Tighten loose nuts, bolts, and screws | R |
| Check crank arm pivots and damper rods for wear; replace if necessary | S |
| Flush strainers, dirt pockets, and drip legs | M |
| Purge air from coil through vent | S |
| Unit heaters | |
| Clean casing, fan blades, fan guard, and diffuser | A |
| Tighten fan guard, motor frame, and fan bolts | R |
| Check fan clearances; maintain free rotation | A |
| Air-handling units and fans | |
| Check filters; clean or replace as necessary | S |
| Check for unusual noise and excessive vibration; perform during all other inspection and maintenance work | R |
| Inspect fan wheels, shafts, housing scroll, and side liners for corrosion or abrasive wear; replace as necessary | A |
| Tighten bolts and connections | R |
| Clean housing, wheels, louvers, and inlet and outlet ductwork | A |
| Check condition and tension of belts; replace cracked, frayed, or worn belts (see Chapter 12) | R |
| Check alignment of system components | A |
| Clean dampers and check linkage for freedom of movement, corrosion, and abrasion | A |

[a]D = daily; W = weekly; M = monthly; S = semi-annually; A = annually; and R = as required.

stone, compacted gravel, soil cement, or other suitable surfaces. Roadways are crowned and sloped to shed water and may include curbs and gutters to control and direct runoff. Grades, vertical clearances, and curve radii must accommodate delivery trucks and maintenance equipment.

Any potholes, cracks, settlements, depressions, raveling, warping, scaling, or other damage needs prompt repair. Maintenance may take the form of patching, sealing, dragging, or resurfacing. Asphalt pavements and concrete repair areas must be clean and dry before patching, sealing, or resurfacing to ensure a good, tight bond. Maintenance includes keeping roadways clear of debris and litter and trimming or removing trees, shrubs, and other obstructions to ensure a clear line of sight at intersections. A seal coat should be applied to asphalt pavement approximately every 5 years.

Signs are necessary to indicate speed limits, no-parking areas, fire zones, delivery zones, crosswalks, visitor parking, and other traffic-control directions. Paved roads need stripes and other pavement markings for traffic safety and control. Guardrails, bollards, or other traffic barriers need periodic inspection and repair.

In northern climates, maintaining safe year-round access to the plant facilities requires snow removal and deicing of roadways. The operator should avoid using salt on concrete roads or walks because chlorides in the salt eventually deteriorate the concrete. Snow fencing along some roads may be necessary to prevent snowdrifts. Good maintenance includes checking the plows, snowblowers, and deicer inventory before each winter season.

## FLOOD PROTECTION

Most wastewater facilities are located at low elevations within watersheds and many exist within floodplains. Without adequate protection, floods may cause loss of life, injuries, and heavy damages, including loss of equipment, severed transportation lines, and cut communication links. Appropriate flood-protection facilities should keep treatment plants fully operational and accessible during a 25-year flood (a flood having one chance in 25 of being equalled or exceeded during any year). Plant facilities should be protected against a 100-year flood (a flood having one chance in 100 of being equalled or exceeded during any year).

All buildings and structures susceptible to flood damage need protection by either a levee or some other form of floodproofing. Floodproofing, temporary or permanent, involves keeping water out as well as reducing the effects of water entry. Sandbags to build temporary dams are one form of temporary floodproofing.

Protection from floods may best be achieved by permanent dikes, levees, or flood walls. Each system provides a barrier to prevent floodwaters from entering a protected area. Drains or conduits through the barriers must have gates or valves to prevent backflow to the dry side, and interior drainage must be pumped over the barrier during floods when the valves on the drainage conduits are closed to prevent backflow.

An alternative method of floodproofing includes closures, panels, and seals on doors, windows, and other openings to prevent water from entering buildings. Floodproofing with closures and seals, if applied to unsound structures, might result in more damage than might occur without floodproofing. Proper design of the closure must ensure sufficient structural capacity to withstand imposed hydrostatic forces. In addition, the building or structure must be evaluated for stability to resist overturning, sliding, or flotation.

A less desirable method of reducing flood damage is to intentionally flood the structure to balance the internal and external hydrostatic pressures. With this method, a flood-protected drainage system is needed to ensure positive drainage of the spaces after the flood.

Main power service to the wastewater treatment facilities should be located above the high-water mark for the plant protection design flood (typically the 100-year flood). If floods prevent direct access to buildings or structures except by boat, the power company should provide a remote main-power disconnect in an accessible area.

Sump pumps used to drain areas within the structures need to include auxiliary, independent alarms to warn personnel of high water. The sump pumps need to be connected with flood-protected, standby electrical generating equipment.

Plant management responsibilities include development of standard operating procedures for carrying out flood-protection measures. These procedures should specify tasks, sequences, and personnel assignments. Yearly inspections of flood preparations and drills to execute the flood-protection plan are essential parts of a good maintenance program. Summaries of flood-protection procedures should be prominently posted.

Plant management needs advance information from flood forecasting and warning sources as early as possible to carry out the flood-protection plan (WPCF, 1986). Established lines of communication and control are necessary for executing the plan, accounting for vacations, holidays, weekends, and sick leaves. Systems and equipment necessary for the flood-protection plan must be kept in a state of readiness at all times. Inadequate planning or equipment may result in a disaster when a flood does occur.

## SOLID WASTE DISPOSAL

A wastewater treatment facility generates solid waste other than treatment sludges and residues. Paper, cans, cardboard, bottles, rags, kitchen waste, yard waste, bags, pails, and pallets represent a partial list of plant solid wastes requiring disposal or recycling. Good housekeeping is the first step in solid waste management.

Solid waste is normally removed by a service company or plant personnel who collect and haul the solid waste to landfills or recycling centers. In general, open fires for on-site solid waste disposal are prohibited. Solid waste must be stored on-site in appropriate containers to reduce the risk of fires and vermin infestations.

Plant staff should review the types of waste generated and encourage recycling or the use of returnable bulk containers. Some communities have programs to collect aluminum, glass, paper, and yard waste. Participation in these programs conserves resources and promotes community goodwill. With some investigation, the operator may find that some chemicals, solutions, or other items can be obtained in returnable containers. This will reduce the volume of solid waste handled. In addition, many wastewater treatment facilities generate waste oil and hazardous waste that require special handling; regulations vary from state to state. The plant safety officer should review the applicable state laws and develop a hazardous waste management plan that complies with the state's requirements.

## REFERENCES

Water Environment Federation (1992) *Design of Municipal Wastewater Treatment Plants*. Manual of Practice No. 8, Alexandria, Va.; ASCE Manuals and Reports on Engineering Practice No. 76, New York, N.Y.

Water Environment Federation (1994) *Safety and Health in Wastewater Systems*. Manual of Practice No. 1, Alexandria, Va.

Water Pollution Control Federation (1986) *Plant Manager's Handbook*. Manual of Practice No. SM-4, Washington, D.C.

# Chapter 12

# Maintenance

# INTRODUCTION

The purpose of this chapter is to provide basic information on current best practices for maintenance organizations. Libraries could be filled with the wealth of information published over the last decade on efficient maintenance practices. This chapter attempts to reduce the oceans of information into a concise summary, with numerous references and suggested readings at the end of the chapter for those interested in further reading.

# BACKGROUND

The face of maintenance, the way world-class industries approach the basic repair function, has changed prolifically during the past 30 years. From the 1960s, through most of the 1980s, progressive companies practiced time-based preventive maintenance techniques under the premise that a correlation exists between equipment age and failure rate and that the probability of failure can be determined statistically. The first studies aimed at challenging this traditional approach came from the airline industry. In the early 1960s, airline maintenance practices of inspections and complete overhauls accounted for 30% of all operating costs. Forthcoming studies revealed that nearly 90% of all component failures were random, making scheduled replacements ineffective for prolonging life. By 1965, traditional overhauls requiring complete disassembly and remanufacture were replaced with "conditional overhauls" that focused on correcting only the immediate cause of failure. In 1968, the introduction of the Boeing 747 was accompanied by the first attempt at reliability-centered maintenance. By 1972, the practice of scheduled overhaul was discontinued entirely. From this need to make maintenance decisions independent of equipment age, condition-monitoring technologies emerged.

Further evidence that machine component failures have no correlation to age is shown in Figure 12.1. This graph shows how the life distribution of 30 identical 6309 deep-groove ball bearings installed on bearing test machines and run to failure results in large variations and supports abandonment of time-based maintenance strategies for increasing bearing life.

# MAINTENANCE BEST PRACTICES

By definition, "best practices" are those methods that represent the best way to accomplish work and lead to superior performance. Maintenance strategies fall into one of four categories

- Run-to-failure,
- Preventive (time-based),

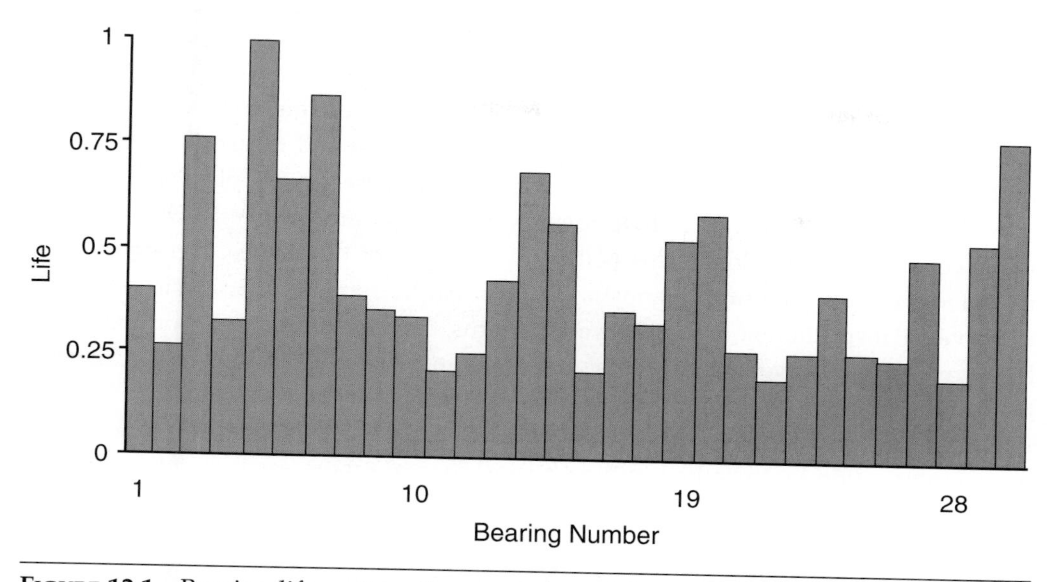


FIGURE 12.1 Bearing life scatter diagram (adapted from NASA, 2000).

- Condition-based, and
- Proactive.

Best-in-class maintenance organizations will use a mix of all of these strategies. Equipment criticality, process redundancy, and life-cycle cost analysis guides proper selection. As equipment history is gathered and reviewed, appropriate strategies evolve accordingly. As organizations move from run-to-failure towards condition-based and proactive maintenance, maintenance costs are typically reduced to one-third of original levels. The key element of a proactive program is root-cause failure analysis.

## FAILURE ANALYSIS

Root-cause failure analysis (RCFA) is essential when trying to eliminate problems and failures from repeatedly occurring. Unfortunately, maintenance departments are often judged by how quickly a unit is returned to service and end up reinstalling conditions that lead to failure. Rarely are machine breakdowns attributable to a single cause. With in-depth analysis, multiple significant contributors can be identified. The roots of failure are

- Physical roots,
- Human roots, or
- Management system (latent) roots.

Preserving failure data is essential in an RCFA. For example, before replacing a failed pump, some of the items to record include operating data before the incident (pressure, temperature, flowrate, level, etc.), coupling condition, shaft alignment as found at failure, shaft runout, and anchor bolt condition. Sampling oil and grease from the failed unit and having lubrication analysis conducted is also useful. The physical examination of a failed component can help identify the failure mechanism, such as fatigue, wear, corrosion, or overload. This is the physical root of the failure. If needed, most manufacturers offer failure diagnostic services. An in-depth gear analysis would include visual inspection noting tooth wear patterns, the measurement of tooth pitch and lead error, quantitative chemical analysis of the material, and testing of the material tensile strength and hardness.

The physical reasons why a component failed, as well as the human errors of omission or action that resulted in the physical root, are more easily identified than a latent root. For example, a failed bearing (physical root) is found to have a damaged cage from inadequate lubrication. The mechanic who delayed greasing (human root) was reassigned to emergency repairs that preempted planned maintenance on the bearing (latent root). A management philosophy that does not allow "system" weaknesses to exist or go unchecked can eliminate huge groups of problems.

## MACHINE INSTALLATION FUNDAMENTALS

The first step in any condition-monitoring program starts before any real-time operating data are collected. Pumps need to be sized for operation at or near the best efficiency point, dynamically balanced, mounted on stable foundations, coupled using precision shaft alignment tolerances, checked for pipe strain, and receive proper lubrication if they are expected to achieve run times of 5 years or more. Ideally, condition monitoring is not needed because machine assembly and installation practices are of such high quality. Of course, this is rarely the case. Bearings are not pressed on shafts properly, housing clearances are too large, and bracket sag goes unaccounted for during alignment. These are but a few of the problems that a good condition-monitoring program will detect.

**SHAFT ALIGNMENT.** Essential and fundamentally sound precision alignment of coupled shafts will increase the reliability of rotating equipment. Recall that bearing life is exponentially (a cubic function) related to force, making small misalignments a much larger problem than one would think. A practical benchmark for shaft alignment tolerances can be found in *Shaft Alignment Handbook* (Piotrowski, 1995; page 145).

Remember that shaft alignments are typically performed with the machine at rest, in a nonoperating state. While running under load, a machine will grow thermally and deflect according to foundation strength, pipe strain, and shaft speed. For the vast majority of equipment in a treatment plant, offline alignment is acceptable. For process critical equipment, or shaft speeds exceeding 1800 rpm, monitoring offline to running condition position changes may be worthwhile. Typically accomplished using an optical jig transit or proximity probes, recorded shaft positions at running speed can be compared to offline starting points and adjustments can be made as needed. Possible applications would be an aeration blower or a biogas compressor. Lockheed Martin (1997a) provides a detailed discussion of "offline to running condition" alignment techniques.

Finally, note that any alignment tolerances provided by a coupling manufacturer are absolutely not transferable to the driven shafts. Couplings do not correct, absorb, dampen, or otherwise fix in any way, shaft misalignment. Allowable coupling misalignment (provided by the coupling manufacturer) has nothing to do with the survivability of the pump system. The system includes not only the coupling, but also shafts, seals, and bearings. The misalignment tolerances shown on page 145 of *Shaft Alignment Handbook* (Piotrowski, 1995) are guidelines for a complete system and are more useful than component specifications. Note that the allowable misalignment is expressed in units of mils per inch, with the denominator representing the distance between the coupling flex points or points of power transmission.

**BALANCING.** Proper balancing of components and assembled rotating equipment is essential for long life. An unbalanced impeller creates a force on machine bearings according to the following:

$$F = 1.77 \times W \times R \times \left(\frac{\text{rpm}}{1000}\right)^2 \tag{12.1}$$

Where

$F$ = force (lbf),
$W$ = unbalance weight (oz),
$R$ = radius of unbalance weight (in.), and
rpm = rotating speed (revolutions per minute).
Note: lbf $\times$ 4.448 = N.

As the shaft speed goes up, the resulting force increases as a square function, making balancing important for high-speed equipment. For example, a centrifuge operating at 2600 rpm with an 8-oz imbalance at a 15-in. radius results in the following force:

$$F = 1.77 \times 8 \times 15 \times \left(\frac{2600}{1000}\right)^2 = 1436 \text{ lbf} \tag{12.2}$$

Note: lbf × 4.448 = N.

The additional load produced by this force reduces expected bearing life. Formulas for calculating bearing life are as follows:

$$L_{10} = \frac{1000\,000}{(\text{rpm} \times 60)} \times \left(\frac{C}{P}\right)^a \tag{12.3}$$

Where

$L_{10}$ = minimum life (hour),
rpm = rotating speed (revolutions per minute),
$C$ = published basic dynamic load rating (lbf),
$P$ = radial load on bearing (lbf), and
$a$ = 3 (ball bearings), 3.33 (roller bearings).

The unbalance force ($F$) is added to ($P$) in eq 12.3, and the exponential relationship (a cubic function) magnifies small balance problems into large ones. Detailed discussions of the above parameters as well as procedures for calculating radial loading ($P$) can be found in most bearing manufacturer catalogs. For an example, assume a spherical roller bearing is rotating at 500 rpm under a radial load ($P$) of 33 000 lbf. The catalog value of the basic dynamic load rating ($C$) is 254 000 lbf and the bearing life is calculated as follows:

$$L_{10} = \frac{1000\,000}{(500 \times 60)} \times \left(\frac{254\,000}{33\,000}\right)^{3.33} = 29\,807 \text{ h} \tag{12.4}$$

Often, unbalance is reported in displacement units of mils (1 mil = 0.001 in.). This is a measure of the vibration response to the unbalance force. Representing the actual unbalance quantity, in units of ounce-inches, is done by relating the two measurements. Testing the residual unbalance by placing a known weight at a known radius and measuring the displacement allows the unbalance to be expressed either way. Shop balancing is done at speeds usually ranging from 250 to 500 rpm, as unbalance is the same at all speeds. The resulting force caused by the unbalance increases with speed, making proper balancing critical at higher operating speeds.

Commonly used balance tolerances are presented below. Which one to use is application-dependent.

***Centrifugal Force <10% Static Journal Load.*** This specification states that the force resulting from unbalance must be less than 10% of the weight supported at each bearing. Typically, it is assumed that the total rotor weight is supported equally by two bearings and that the weight supported at each bearing is the total rotor weight divided by two.

***MIL-STD-167-1 (U.S. Navy).*** This tolerance came from the military (1974), needing operating machinery quiet enough to avoid sonar detection. For rotors operating at more than 1000 rpm, the permissible unbalance is

$$U_{per} = \frac{4W}{N} \tag{12.5}$$

Where

$U_{per}$ = permissible unbalance (oz-in.),
$W$ = total rotor weight (lbm), and
$N$ = rotating speed (rpm).
Note: oz × 28.349 = g; in. × 25.4 = mm.

This is an empirical formula, requiring data to be entered in the units shown for a correct solution. For a 1000-lbm rotor operating at 3600 rpm, the allowable residual unbalance is

$$U_{per} = \frac{4 \times 1000}{3600} = 1.11\,\text{oz-in./plane} \tag{12.6}$$

Note: oz × 28.349 = g; in. × 25.4 = mm.

***American Petroleum Institute.*** The American Petroleum Institute standard is, in effect, one-half of the MIL-STD. The formulas are identical except for the method of calculating the rotor weight ($W$). Here the total rotor weight is assumed to be supported by two journals; hence one-half the total rotor weight is used in the calculation.

***ISO 1940-1 (International Organization for Standardization).*** This standard refers to balance quality of rotating rigid bodies and classifies rotor types by assigning them a balance quality grade, referred to as a "$G$" number. Grades are separated by a factor of 2.5, as shown in the abbreviated listings of Table 12.1. American National Standards Institute (ANSI) has also adopted this standard as ANSI S2.19-1999. Permissible unbalance is calculated from the following:

$$U_{per} = \frac{G \times 6.015 \times \frac{W}{2}}{N} \tag{12.7}$$

Where

$U_{per}$ = permissible unbalance (oz-in.),
$G$ = balance quality grade number,
$W$ = total rotor weight (lb), and
$N$ = rotating speed (rpm).

Note: oz $\times$ 28.349 = g; in. $\times$ 25.4 = mm.

**TABLE 12.1** International Organization for Standardization balance quality grades for rigid rotors.

| Rotor classification (balance quality) | Rotor description (general machinery types) |
|---|---|
| $G$ 40 | Car wheels, wheel rims, drive shafts. |
| $G$ 16 | Automotive drive shafts, parts of crushing and agricultural machinery. |
| $G$ 6.3 | Parts or process plant machines. Marine main turbine gears. Centrifuge drums, fans, and assembled aircraft gas turbine rotors. Paper and print machine rolls. Flywheels, pump impellers, general machinery, and machine tool parts. Standard electric motor armatures. |
| $G$ 2.5 | Gas and steam turbines, blowers, rigid turbine-generator rotors, turbo-compressors, machine tool drives, computer memory drums, and discs. Medium and large electric motor armatures with special requirements. Armatures of fractional horsepower motors and turbine-driven pumps. |
| $G$ 1.0<br>Precision balancing | Jet engine and charger rotors and tape recorder and phonograph drives. Grinding machine drives and small electrical armatures with specific requirements. |
| $G$ 0.4 | Spindles, discs, and armatures of precision grinders. Gyroscopes. |

This also is an empirical formula, requiring data to be entered in the units shown for a correct solution. The standard suggests a balance quality grade of 6.3 for pump impellers. For a 1000-lb rotor operating at 3600 rpm, the allowable residual unbalance is

$$U_{per} = \frac{6.3 \times 6.015 \times \left(\frac{1000}{2}\right)}{3600} = 5.26 \text{ oz - in. / plane} \qquad 12.8$$

Note: oz × 28.349 = g; in. × 25.4 = mm.

**PERIPHERALS AND MISCELLANEOUS PRINCIPLES.** Rotor balancing and proper shaft alignment are important for long machine life. Many other factors play a significant role. Properly designed machine foundations will support applied loads without settlement or crushing, maintain true alignment conditions, and absorb and dampen unwanted vibrations. Useful guidelines for a good foundation include

1. The concrete foundation mass should be 5 to 10 times the mass of supported equipment (the density of concrete is roughly 2400 kg/m$^3$ [150 lb/cu ft]).
2. Imaginary lines drawn downward 30 deg from the shaft centerline should pass through the bottom of the foundation (not the sides).
3. The foundation should be 76 mm (3 in.) wider than the baseplate, 152-mm (6 in.) for pumps greater than 373 kW (500 hp).
4. Use concrete with compressive strength between 20 685 and 27 580 kPa (3000 and 4000 psi).
5. Let newly poured foundations cure 6 to 8 days; this allows concrete compressive strengths to attain 70 to 80% of their final value.

Piping strain in excessive amounts can make a precision balance and shaft alignment job pointless. Pump flanges are fluid connections and never should be used as pipe anchor points. To check for piping strain, place dial indicators on both shafts of the piped system and loosen the foot bolts. No more than 2 mils (0.002 in.) movement should result. Table 12.2 shows practical benchmarks to help aid in evaluating a piping stress measurement.

Additionally, piping configuration can influence long-term reliability. Centrifugal pumps, in particular, are susceptible to turbulent flows and high fluid velocities at the suction point. Generally, suction piping should be at least 1.5 times the pump's inlet

**TABLE 12.2** Piping strain guidelines.

| Total indicator runout (in.*) | Significance |
|---|---|
| <0.002 | Ideal condition |
| 0.002–0.005 | Acceptable |
| >0.005 | Provide needed piping supports |
| 0.020 | Refabricate piping |

*in. × 25.4 = mm.

diameter. Use eccentric reducers with the straight side on top to transition to the pump inlet, and try to provide a straight run of six times the pump inlet diameter.

Shaft runout, or total indicator runout, should always be measured before installing any rotating piece of equipment. Both face and radial runout is measured using dial indicators and measures the eccentricity and perpendicularity of a shaft with respect to the centerline of rotation. Guidelines for acceptable shaft runout are linked to rotating speed, with higher revolutions per minute needing more precision. Table 12.3 shows practical guidelines for shaft runout.

Soft foot conditions exist when the mounting feet of a machine and the mating baseplate do not make full surface contact. Many problems can be traced to soft foot conditions as they can offset the centerline of shaft rotation, warp machine casings, and reduce internal component clearances. Soft foot is measured by placing dial indicators at the machine feet, loosening the hold-down bolts, and checking for movement. More than 2 to 3 mils of movement indicates a soft foot condition that should be corrected. Be sure to use shims that provide full footprint support across the face, and do not use more than four to five shims in a stack. Either of these conditions can make shaft alignment more difficult.

Often overlooked, proper bolt fastening secures the machine in place, absorbs impact forces, and prevents mechanical looseness from developing. A torque wrench is needed for good installation practices. Proper settings are dependent on the size and

**TABLE 12.3** Shaft runout guidelines.

| Rotating speed (rpm) | Max total indicator runout (in.*) |
|---|---|
| 0–1800 | 0.005 |
| 1800–3600 | 0.002 |
| 3600 and up | <0.002 |

*in. × 25.4 = mm.

grade of the bolt. To approximate the wrench torque needed for proper tensioning, use the following:

$$T = 10^{b + m \log d} \tag{12.9}$$

Where

$T$ = torque (ft-lb),
$d$ = bolt diameter (in.), and
$b, m$ = exponents from Table 12.4.

Note: ft × 0.3048 = m; lb × 0.4536 = kg.

Remember that these are guidelines only, for use with fasteners "as received" from the mill. In all cases, manufacturer specifications on torque settings should be followed. Also, most fasteners will retain a small quantity of oil from the manufacturing process. Applying special lubricants can reduce the amount of friction in the fastener assembly, allowing the specified torque to produce a far greater tension than desired.

## CONDITION-MONITORING TECHNOLOGIES

For the vast majority of rotating equipment, integrated condition-monitoring technologies including vibration, thermography, ultrasonics, tribology, and electrical condition monitoring provide the most cost-effective strategy. All of these techniques have the common objective of showing early signs of deterioration, before failure of function. Condition monitoring will not prevent machine deterioration. It will stop unneeded repairs and allow repair before failure of function.

**TABLE 12.4** Fastening guidelines[a] (adapted from Oberg et al., 1996).

| SAE grade number | Tensile strength (psi[b]) | Bolt diameter (in.[c]) | $b$[d] | $m$[d] |
|---|---|---|---|---|
| 2 | 74 000 | 0.25–3 | 2.533 | 2.94 |
| 5 | 120 000 | 0.25–3 | 2.759 | 2.965 |
| 7 | 133 000 | 0.25–3 | 2.948 | 3.095 |
| 8 | 150 000 | 0.25–3 | 2.983 | 3.095 |

[a]For standard unplated industrial fasteners. For cadmium-plated nuts and bolts, multiply by 0.8.
[b]psi × 6.895 = kPa.
[c]in. × 25.4 = mm.
[d]See eq 12.9 for explanation of exponents.

Note that preventive maintenance (calendar or time-based) is still a useful part of an overall effective maintenance program. Applications in which abrasive or corrosive wear occurs are still good candidates for time-based activities.

## VIBRATION ANALYSIS

Before transistorized electronics ushered in the digital age, mechanics used to see if a coin would stand on end when placed on a machine housing. Although only a pass or fail test, it was, nonetheless, an attempt at measuring vibration levels. In the 1930s, T.C. Rathbone, an industry insurance underwriter, measured overall machinery vibration levels using an oscilloscope and calculated frequency components by hand. Vibration analyzers today are highly sophisticated and, of all available nonintrusive testing methods, provide the most information.

Vibration monitoring quantifies machine movement in response to a force. The captured time waveform can be plotted as amplitude versus time, or data can be transformed using a fast Fourier transform (FFT) and expressed as amplitude versus frequency. Any random vibration signal can be represented by a series (a Fourier series) of individual sine and cosine functions that can be summed to yield an overall vibration level. The amplitude of this vibration signal defines the severity of the problem. Plotting the amplitude versus the frequency (the Fourier spectrum) allows for identification of discrete frequencies contributing most to the overall vibration signal, commonly referred to as a "signature analysis" or a "frequency spectrum". Machine looseness, misalignment, imbalance, and soft foot conditions are all fairly easily identified in the frequency spectrum generated by an analyzer.

The instrument used for measuring vibration is a transducer, which converts a sensed mechanical motion to an electrical signal. Transducers can measure displacement (mils), velocity (in./s), and acceleration (*G*s or 32.2 in./$s^2$). All three measurements are describing the same thing, the amplitude of the motion. Once one value is found, the others can be mathematically calculated from the following relationships:

$$v = 2 \times \pi \times f \times d \tag{12.10}$$

$$a = 2 \times \pi \times f \times v \tag{12.11}$$

Where

$d$ = peak displacement (mm [in.]),
$f$ = cycles per second (Hz),
$v$ = peak velocity (in./s), and
$a$ = peak acceleration (in./$s^2$).

Recall that a Hertz is defined as a cycle per second, making 60 rpm = 1 Hz. A machine with a shaft speed of 1800 rpm (30 Hz) that measures displacement amplitude of 10 mils (0.01 in.) has velocity and acceleration of

$$v = 2 \times \pi \times 30 \times 0.01 = 1.88 \text{ in./s} \tag{12.12}$$

$$a = 2 \times \pi \times 30 \times 1.88 = 354 \text{ in./s}^2 \tag{12.13}$$

Note: in. × 25.4 = mm.

Presently, the accelerometer is the preferred transducer for machine monitoring because the frequency response is good over a broad range of operating speeds. They measure acceleration directly and are equipped with electronic integrators to give velocity and displacement. Displacement signals are emphasized, or more pronounced, at lower frequencies (<20 Hz), and acceleration emphasizes high frequency (>1000 Hz) peaks. The best indicator of overall machinery condition is velocity.

Over the past few years, the vibration industry has made great strides in detecting impending bearing failures at their onset, long before loss of function. Early surface imperfections generate short-duration, high-impact spikes of energy. Digital technology using very high sampling rates can capture the impacts. Time-domain waveforms are particularly useful in this area. By examining amplitude, symmetry, and particularly pattern recognition, time waveforms are an indicator of the true amplitude of bearing impacts (the FFT analysis will bias the amplitude low). This technique also has good application in low-speed equipment, gears, looseness, and orbit analysis of sleeve bearings. Figure 12.2 shows the time waveform, as well as the corresponding spectral plot, from a failing bearing on a primary sludge pump. Note the characteristic pattern of a large spike flowed by a "ringing down" in the time waveform and the nonsynchronous energy with a noisy floor in the spectral plot.

## LUBRICATION AND WEAR PARTICLE ANALYSIS

Lubrication and wear particle analysis provides information on the condition of the lubricant itself, as well as the wear condition of the friction surfaces the lubricant is protecting. Analysis of the lubricant quantifies the chemical contamination, its molecular condition, dissolved elements, and state of additives. Common testing includes viscosity measurement, acid number, and base number. Typically, oil analysis can indicate that a problem exists but may not be able to identify the specific problem.

The amount, makeup, shape, and size of the wear particles present in the lubricant provide information about the internal machine condition and the severity of compo-

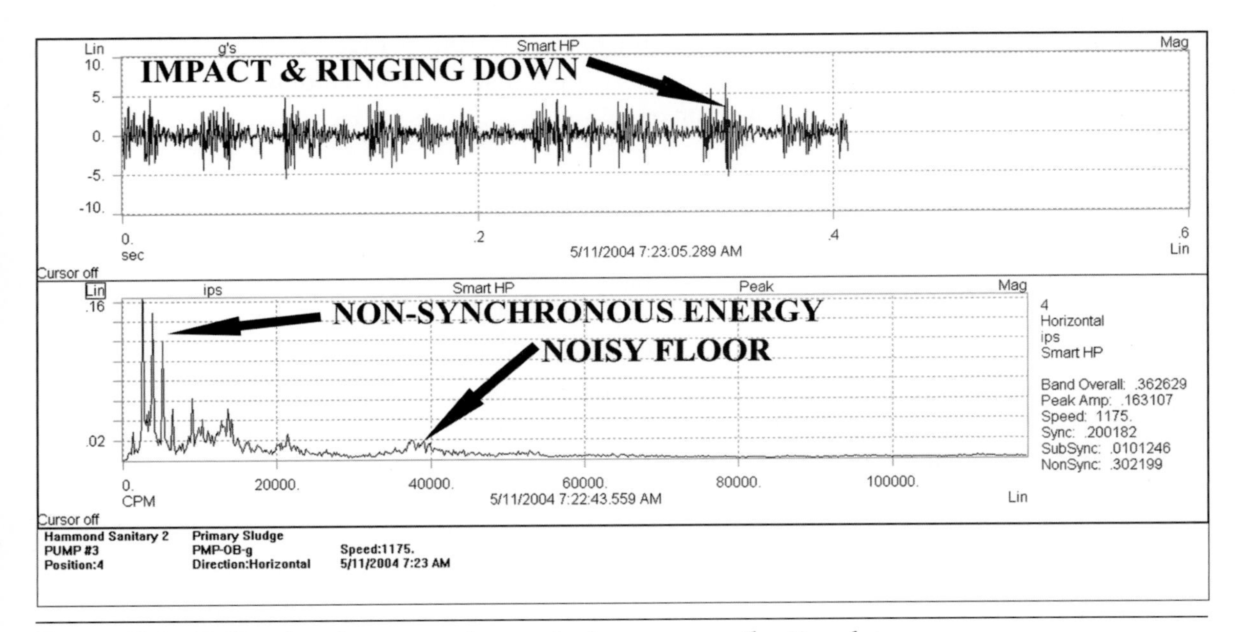


**FIGURE 12.2** Failing bearing on a primary sludge pump—vibration data.

nent wear. There are several different analysis techniques. Changes in the ratio of small to large particles, the total weight of the particles, or the concentration of ferrous particles indicate that the wear process has begun. In most cases, samples drawn from active, low-pressure lines ahead of filters will yield an accurate diagnosis of the specific source, severity, and location of the problem. The importance of sampling location cannot be emphasized enough. The objective is to collect oil samples that represent what is passing through the machines bearings. Large reservoirs and downstream of filters are poor choices.

Every attempt should be made to follow lubrication specifications provided by the equipment manufacturer, particularly recommended volumes and change intervals. In some cases, too much grease will cause overheating and premature failure. Note that a sealed bearing, as supplied from the factory, will typically have the bearing cavity filled to 25 to 35% with grease. In other applications, the manufacturer requires three or four tubes of grease per bearing, but does not explain that the purpose of this volume of grease is to purge the bearing. Results of analysis on used oil or grease samples should be used to adjust the frequency of changes.

# THERMOGRAPHIC ANALYSIS

Commonly identified with electrical equipment monitoring, thermograpy is also a useful tool for monitoring plant machinery. Thermography measures infrared radiation energy emissions (surface temperatures) to detect anomalies. Infrared cameras have resolution to within 0.1 °C and digitally store captured images. Both the absolute and relative temperatures can be obtained on virtually all types of electrical equipment, including switchgear, connections, distribution lines, transformers motors, generators, and buswork. Mechanically, infrared thermography applications are numerous and include all equipment with friction points such as bearings couplings, piston rings, brake drums, and heat exchangers. Like all of the other condition-monitoring technologies, data trending of equipment images will mark changes and help establish baselines. Table 12.5 provides broad guidelines for absolute temperature limits of some common mechanical components.

**TABLE 12.5** Mechanical component temperature limits.

| Component | Temperature (°C) |
|---|---|
| **Bearings, roller element type** | |
| Races/rolling elements | 125 |
| Retainers (plastic) | 120 |
| Cages/retainers/shields (metal) | 300 |
| **Bearings, plain type** | |
| Tin/lead-based Babbitt | 149 |
| Cadmium/tin-bronze | 260 |
| **Seals (lip type)** | |
| Nitrile rubber | 100 |
| Acrylic lip | 130 |
| Silicone/fluoric | 180 |
| PTFE* | 220 |
| Felt | 100 |
| Aluminum (laboratory) | 300 |
| **Mechanical seal materials** | |
| Glass-filled Teflon | 177 |
| Tungsten carbide | 232 |
| Stainless steel | 316 |
| Carbon | 275 |
| V-belts | 60 |

*PTFE = polytetrafluoroethylene.

Motor ratings are established by their maximum allowable operating temperature and are a function of the insulation system. Motor life is reduced by 50% for every 10 °C rise above their rating. Temperatures on the outside of a motor will typically run 20 °C cooler than the inside. Table 12.6 provides useful information for conducting thermal analysis of motors.

## ULTRASONIC ANALYSIS

Ultrasonic sound waves have frequencies above 20 kHz, the threshold at which human hearing stops. Every machine emits a unique sonic signature, which can be monitored for change by an ultrasonic detector. Detectors are typically hand-held devices that resemble a pistol and weigh approximately 0.9 kg (2 lb). Through a process called heterodyning, the ultrasonic signal is modified and processed into the audible range. Headphones allow a technician to hear the processed signal, and a meter quantifies the amplitude in decibels (dB).

A useful application of ultrasonic technology is determining when to grease bearings and how much to apply. This can be a challenging issue for technicians, as trying to determine when it was last done, how much was added, and the type of lubricant used is not always readily available, or trustworthy if it is. Once a baseline is established through historical readings, readings on similar equipment, or observations while lubricating, a change of 8 dB over normal lubricated conditions warrants attention. Deliver lubricant to the bearing until the signal returns to normal levels. Using headphones, lubricant is supplied until the sounds drop off and then begin to rise again. At this exact moment, lubricant delivery stops.

In 1972, NASA completed experiments using ultrasonics for early bearing fault detection. The technical brief (NASA, 1972), describes the experiments conducted using ball bearings and measuring the response in the 24 to 50 kHz range. In general, amplitude response increases more than 12 dB above baseline are an early indicator of

TABLE 12.6 Maximum motor temperatures (°C) at 40 °C ambient.

| Class | Internal | External |
|---|---|---|
| A | 105 | 85 |
| B | 130 | 110 |
| F | 155 | 135 |
| H | 180 | 160 |

deterioration. When tuned into the lower end of the ultrasound range, troubleshooting slow-speed bearings, typically fewer than 600 rpm, is possible.

Additionally, sonic signatures enable technicians to identify and locate compressed air leaks, steam trap leaks, and tank leaks. Compressed gas or fluid leaks through a small opening create turbulence on the downstream side, which can be captured by a scanning ultrasound device.

## ELECTRICAL SURGE TESTING

The surge test is the cornerstone of electrical condition monitoring, providing information on the health of electrical windings. This specialized test locates weaknesses in the turn-to-turn, coil-to-coil, and phase-to-phase insulation of a motor's windings. Many electrical failures start with weakness in the turn insulation (noted at over 80% in some of the literature), and the surge test is the only test capable of identifying this deterioration. The test instrument applies energy pulses to the windings and monitors the stability of the waveform for signs of weakness. Surge comparison produces no numbers to be trended—it is a pass or fail test, requiring careful repetition to determine the location and severity of an observed fault.

Many field electricians will perform a megohm test (commonly referred to as "meggering") on a motor to check winding insulation health. This is not the same thing as a surge test. Surge testing checks for faults in the coils themselves, meggering monitors the insulation quality. It is possible for a motor to continue operating even though it is failing a surge test, but the identified problem should be addressed as appropriate.

## MOTOR CURRENT SIGNATURE ANALYSIS

A nonintrusive test, motor current signature analysis, is useful in detecting mechanical and electrical problems in rotating equipment. The basis for the technology is that an electric motor driving a mechanical load acts as a transducer, varying current draw as the load changes. Analysis of these current variations may be trended over time, ultimately providing early warning of machine deterioration or process alteration.

## PROCESS CONTROL MONITORING

Bear in mind that condition monitoring will not assess the operating efficiency of a pump-motor system. Often costing more than maintenance activities, inefficient pumping should be of interest to all maintenance personnel. Process parameter monitoring is a valuable aide to condition monitoring. Parameters of interest are temperatures, line pressures, and flowrates. A useful equation for determining pump efficiency follows:

$$hp = \frac{gpm \times TDH}{3960 \times efficiency} \tag{12.14}$$

Note: hp × 0.75 = kW.

Where

gpm = gallons per minute,
TDH = total dynamic head (ft), and
efficiency = pump efficiency (%, expressed as a decimal).

Pressure gauges installed on the pump suction and discharge pipes will give the total dynamic head, and a flow meter on the discharge provides the information needed to measure pump efficiency. Using equation 12.14 in the units given, the efficiency of a pump discharging 25 000 gpm, against 17 ft-head, powered by a 150-hp motor, is 72%. The calculation follows:

$$Efficiency = \frac{25\,000 \times 17}{3960 \times 150} = 0.72$$

## PREVENTIVE MAINTENANCE

Note that preventive maintenance (calendar or time-based) is still a useful part of an overall effective maintenance program. Applications in which abrasive or corrosive wear occurs are still good candidates for time-based activities. It is important to continually review a preventive maintenance program, ensuring that tasks are valid and provide measurable benefits. Equipment criticality and historical reliability should be used to eliminate unnecessary tasks and identify where to focus preventive tasks. Condition monitoring cannot replace preventive maintenance tasks. All equipment needs to be lubricated, cleaned, adjusted, and painted and have minor component replacements. The interval at which these tasks are completed can, and should, be adjusted with analysis of condition-monitoring data. Without exception, compliance with an established preventive maintenance schedule should be aggressively pursued.

## RELIABILITY-CENTERED MAINTENANCE

As new monitoring programs are initiated, or existing ones are expanded, trying to decide which technology to use, the success at indicating incipient failures, and the frequency of use can all be somewhat overwhelming. The reliability-centered maintenance (RCM) process provides a structured approach for evaluating these questions. Pioneered by Stanley Nowlan and Warren Heap while working for United Airlines, their document "Reliability Centered Maintenance", published in 1978, remains one of the most solid RCM discussions available. The authors proposed examining the consequences of failure as a starting point for maintenance decision-making and using condition-monitoring data to modify decisions as operating information develops. A basic discussion of RCM decision criteria is presented below. More detailed discussion of RCM methods can be found in the references and suggested readings provided at the end of the chapter.

Reliability-centered maintenance strategy focuses on three rules for evaluating maintenance decisions:

1. Identifying a failure mode,
2. Applicability, and
3. Effectiveness.

**RULE 1—IDENTIFYING A FAILURE MODE.** The intent of all condition-monitoring technologies is to provide quantifiable objective evidence of potential failure before functional failure. With this in mind, any condition-monitoring technology used must be aimed at detecting a specific failure. The idea here is to identify what failure you are trying to prevent.

**RULE 2—APPLICABILITY.** To be considered applicable, the technology should monitor a parameter that correlates to the failure mode identified. Parameter measurements must be accurate enough to provide a reliable trigger for repair work. Also, the technology must detect problems early enough for corrective action to be made—before functional failure.

**RULE 3—EFFECTIVENESS.** The effectiveness rule evaluates the consequences of the failure. Critical failures affecting human safety or the environment cannot be tolerated and warrant immediate response. All other failures should use cost-effective technologies, where the investment is lower than the cost to repair the failures. A useful tool for selecting which condition-monitoring technology to use is an RCM decision tree, as shown in Figure 12.3.

When applied carefully and properly, condition-monitoring technology will reduce maintenance costs. When used as a tool and integrated to a reliability-centered program, costs will be one-third of a run-to-failure approach. Table 12.7 provides cost benchmarks.

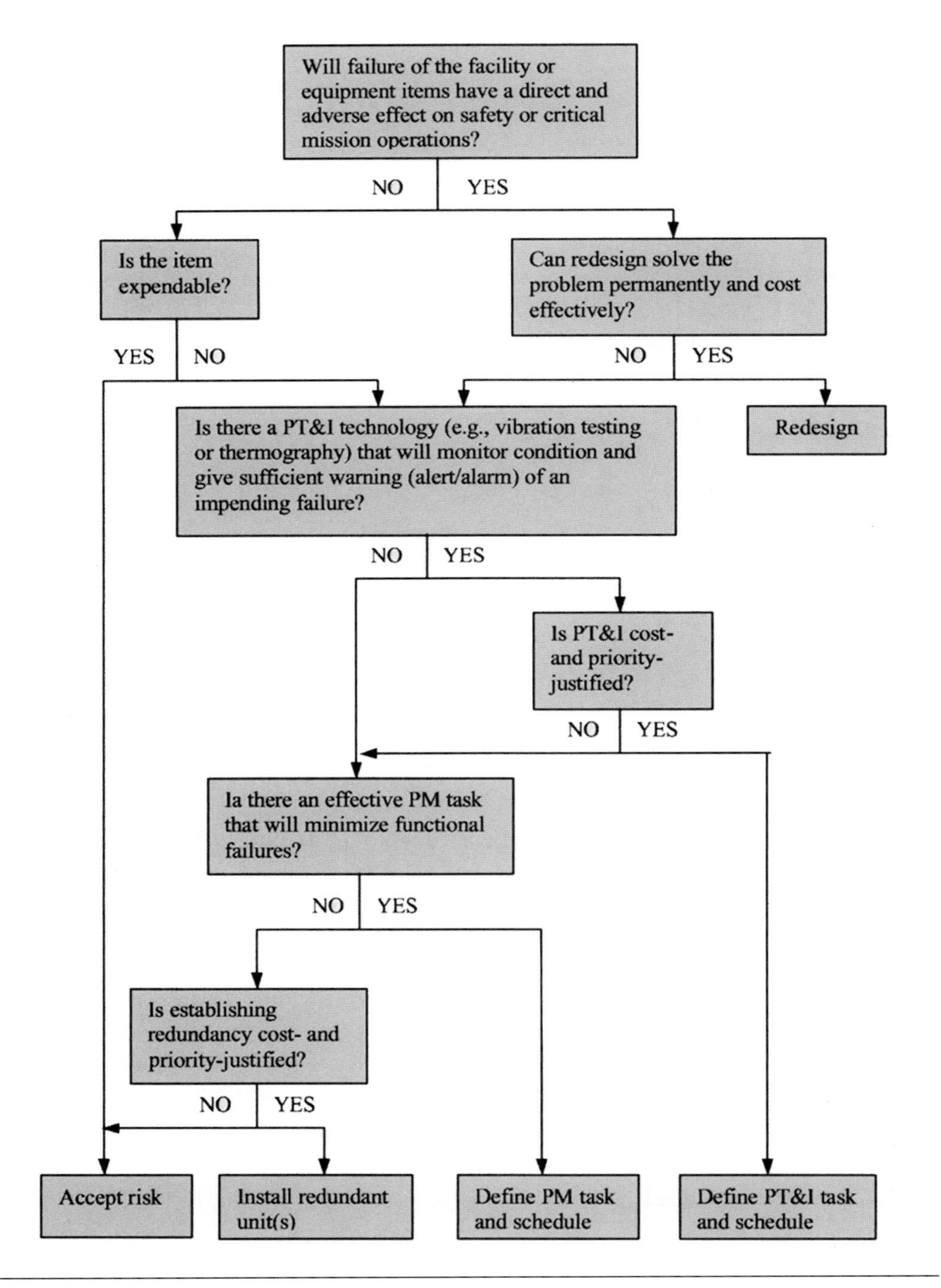


FIGURE 12.3 Reliability-centered maintenance decision tree.

TABLE 12.7 Maintenance costs (Piotrowski, 1996).

| Strategy | Cost ($/hp/yr*) |
|---|---|
| Run-to-fail | 18 |
| Preventive | 13 |
| Condition-based | 9 |
| Reliability-centered | 6 |

*$/hp/yr × 1.33 = $/kW/a.

As additional support to the power of condition-based maintenance, Table 12.8 shows the findings from a survey of 500 plants that have implemented predictive-maintenance methods. Plants included in the survey are from a cross-section of industries.

## MEASURING PERFORMANCE—BENCHMARKING

Benchmarking is the practice of measuring performance against a standard. When applied in the proper context, it will objectively demonstrate whether a facility is being operated efficiently as well as quantifying progress and identifying what areas need improvement. Both industry benchmarking, where performance is measured against others in the same industry, as well as best-practices benchmarking, where performance is measured against industry leaders regardless of business, are useful tools for effective maintenance management. The practices provided here are general guidelines, considered to be industry norms, and have been proven to lead to superior performance. Table 12.9 provides wastewater industry specific benchmarks, summarizing information gathered from 90 plants in the United States, with capacities ranging from 15 140 to 3 251 300 $m^3/d$ (4 to 859 mgd) and a median capacity of 208 175 $m^3/d$ (55 mgd) during 1997.

Table 12.10 provides examples of best-practices benchmarks. Note that there are many variables associated with these values and absolute adherence is cautioned. Data

TABLE 12.8 Predictive maintenance cost reductions (adapted from Mobley, 2001).

| Benchmark | Value |
|---|---|
| Maintenance cost reduction | 50% |
| Unexpected machine failure reduction | 55% |
| Repair time reduction | 60% |
| Inventory reduction | 30% |
| Mean time between failure increase | 30% |
| Equipment uptime increase | 30% |

TABLE 12.9 Wastewater plant industry benchmarks[a] (Benjes and Culp, 1998).

| Parameter | Median value[b] | Best-cost producer[c] |
|---|---|---|
| Maintenance staff/mgd[d] capacity | 0.4 | 0.2 |
| Secondary treatment | 0.4 | |
| Secondary treatment–nitrification | 0.4 | |
| Secondary treatment–nitrification–filtration | 0.45 | |
| Annual maintenance costs/mgd capacity | \$31 816 | \$9688 |
| Number of major assets/maintenance staff | 73 | 177 |
| Annual work orders/maintenance staff | 124 | 293 |
| Work order backlog (days) | 18 | 3.7 |
| Percentage overtime work | 3.30% | 0.80% |
| Percentage maintenance work planned | 50% | 90% |
| Maintenance planners/100-mgd capacity | 2.2 | 0.8 |
| Cost of spare parts/mgd capacity | \$10 294 | \$1991 |

[a]Results are from a 1997 survey of 90 wastewater plants.
[b]Median value—50% of plants higher, 50% of plants lower.
[c]Best-cost producer—10th best in survey.
[d]mgd × 3785 = $m^3$/d.

TABLE 12.10 Best-practices benchmarks.

| Parameter | Benchmark |
|---|---|
| Material dollars installed/labor dollars expended | 0.5–1.5 |
| Maintenance costs/equipment replacement value | 1.5–2.0% |
| Number mechanics/total site workers | 15–25% |
| Repairs requiring rework | <0.5% |
| Equipment downtime/scheduled run-time | 0.5–2.0% |
| Failures addressed by root-cause analysis | >75% |
| Predictive/preventive schedule compliance* | >95% |
| Ratio corrective work orders/P/PM inspections | 1:6 |
| Ratio proactive to total work | 75% |
| Ratio reactive to total work | 25% |
| Ratio predictive PM to total PM | 60% |
| Maintenance overtime percentage | 5–10% |
| Paid maintenance hours captured by work order charges | 100% |
| Percent work orders planned | 90–97% |
| Inventory dollars/equipment replacement value | <1% |
| Inventory turns per year | 2–3 |

*P = predictive and PM = preventive maintenance.

comparison is a useful tool, and much good can come from identifying where improvements are needed.

## RECORDKEEPING AND MAINTENANCE MEASUREMENT

Existing competitive business environments are driving all competitive organizations towards optimized asset-management practices. Asset optimization provides maximum equipment availability and overall equipment effectiveness at the least cost. Computerized maintenance management systems (CMMS) are fundamental in the asset-optimization process, allowing the user to track equipment and the work associated with it. This type of information system is essential for planning, scheduling, establishing baselines, measuring improvements, and providing quantifiable decision support. Assigning failure codes can be quite useful in root-cause failure analysis and can help monitor the extent and cost of downtime. A good CMMS will provide standard reports, such as equipment repair history, and can help identify equipment that is costing too much. The many benefits of accurate work order tracking include better reliability and more accurate repair–replace decisions.

Selection of a CMMS should be based on user-developed objectives, goals, and measurements. For the most part, all CMMS packages available in the market provide the same basic maintenance management functions. A well-thought-out vision for the maintenance process is an essential starting point for a CMMS purchase. Dahlberg (2004) provides a listing of 52 companies that provide CMMS packages.

## REFERENCES

American National Standards Institute (1999) American National Standard Mechanical Vibration–Balance Quality Requirements of Rigid Rotors, Part 1: Determination of Possible Unbalance, Including Marine Applications; American National Standards Institute: Washington, D.C.

Benjes, H.; Culp, G. (1998), Benchmarking Maintenance. *Oper. Forum*, **15** (9), 29.

Dahlberg, S. (2004) Maintenance Information Systems. *Maint. Technol.*, **17** (7), 1.

International Organization for Standardization (2003) Mechanical Vibration–Balance Quality Requirements for Rotors in a Constant (Rigid) State–Part 1: Specification and Verification of Balance Tolerances, 2nd ed.; ISO 1940–1; International Organization for Standardization: Geneva, Switzerland.

Lockheed Martin Michoud Space Systems (1997a) Laser Alignment Specification for New and Rebuilt Machinery and Equipment; LMMSS Specification A 1.0-1977; July.

MIL-STD-167-1 (SHIPS) (1974) Department of Defense Test Method, Mechanical Vibrations of Shipboard Equipment. Department of the Navy; Naval Ship Systems Command, 1 May.

Mobley, R. K. (2001) *Plant Engineers Handbook;* Butterworth-Heinemann: Woburn, Massachusetts.

National Aeronautics and Space Administration (1972) Information regarding experiments using ultrasonics for early bearing fault detection; NASA Technical Brief, Report B72-10494; Technology Utilization Office, NASA, Code KT, Washington, D.C.

National Aeronautics and Space Administration (2000) *Reliability Centered Maintenance Guide for Facilities and Collateral Equipment*, February; www.hq.nasa.gov/office/codej/codejx/rcm-iig.pdf (accessed Mar 2006).

Nowlan, F. S.; Heap, H. F. (1978) *Reliability Centered Maintenance;* United Airlines, San Francisco, California; Report Number AD-A066-579, Dolby Access Press, Controlling Office, Office of Assistant Secretary of Defense, Washington, D.C.

Oberg, E.; Jones, F. D.; Horton, H. L.; Ryffel, H. H. (1996) *Machinery's Handbook;* 25th Ed.; Industrial Press, Inc.: New York.

Piotrowski, J. (1995) *Shaft Alignment Handbook;* Marcel Dekker: New York.

Piotrowski, J., Turvac, Inc., Oregonia, Ohio (1996) Personal communication regarding maintenance costs.

## SUGGESTED READINGS

Alberto, J. (2003) Critical Aspects of Centrifugal Pump and Impeller Balancing. *Pumps & Syst.*, **11** (5), 18.

Alcalde, M. (1999) Keep Your Rotating Machines Healthy. *Chem Eng.*, **106** (11), 106.

American National Standard Institute (1999) American National Standard for Centrifugal Pumps for Design and Application. Hydraulic Institute; ANSI/HI 1.3-2000, Parsippany, New Jersey.

Berger, D. (2003) Know the Score: Our Maintenance Performance Metrics Study Shows Many Plants are Poor at Keeping Score. *Plant Serv.*, **24** (10), 29.

Bernhard, D (1998) *Machinery Balancing;* 2nd ed.; Rich II Resources, Ltd. (distributor): Centerburg, Ohio.

Brandlein, J.; Eschmann, P.; Hasbargen, L.; Weigand, K. (1985) *Ball and Roller Bearings: Theory, Design, & Application*; Wiley & Sons: New York.

Infraspection Institute (1993) *Guidelines for Infrared Inspection of Electrical and Mechanical Systems;* Shelburne, Vermont.

IRD Mechanalysis (1987) The Use of Ultrasonic Diagnostic Techniques to Detect Rolling Element Bearing Defects, Technical Paper No. 121; Columbus, Ohio.

IRDBalancing (year unknown) Balance Quality Requirements of Rigid Rotors; IRD Balancing Technical Paper 1; Columbus, Ohio.

Jacobs, K. S. (2000) Applying RCM Principles in the Selection of CBM-Enabling Technologies. Lubrication & Fluid Power, **1** (2), 5.

Lockheed Martin Michoud Space Systems (1997b) Vibration Standard for New and Rebuilt Machinery and Equipment; LMMSS Specification No. V 1.0-1977; July.

Lundberg, G.; Palmgren, A. (1947) Dynamic Capacity of Roller Bearings. *Acra Polytech.; Mech.Eng. Ser.1, R.S.A.E.E.,* **3,** 7.

Nicholas J.R. and Young, K., *Understanding Reliability Centered Maintenance;* 2nd Ed.; A Practical Guide to Maintenance. A Text to Accompany the RCM Course Maintenance Quality Systems, LLC. Millersville, Maryland.

# Chapter 13

# Odor Control

# INTRODUCTION

**PURPOSE AND BACKGROUND.** This manual is intended to be a reference of practice for operators involved in managing air emissions from wastewater conveyance, treatment, and residuals processing facilities. *Control of Odors and Emissions from Wastewater Treatment Plants* (WEF, 2004) presents a comprehensive review of available information on a variety of subjects concerning odor control in wastewater conveyance and treatment systems and should be referenced by operators in addition to this Manual of Practice.

The collection, treatment, and disposal of municipal and industrial wastewater often emit odors and other air contaminants. In the past, odors commonly associated with wastewater treatment plant operation were accepted by the public. This is no longer the case. There is increasing public concern and intolerance of odors and other air contaminants from wastewater treatment facilities, and management of air emissions has become a significant activity at most treatment plants. Furthermore, federal, state, and local regulations require that a number of air pollutants be contained and controlled. Today, odors, whether a health and safety issue or a nuisance condition, are not tolerated by the community. In fact, odors have been rated as the first concern of the public regarding wastewater treatment facilities. Once a facility has been marred for any reason, including odor, it is very difficult to change the image of the treatment facility in the mind of the public.

Factors contributing to the prevalence of odor concerns include construction of large, regional facilities with long septic travel times in the collection system; urbanization that moves people and industry closer to plant sites; increasing amounts of more complex industrial wastewater discharges; environmental regulations requiring a higher level of treatment; wastewater treatment plant design; facility operation and maintenance; and a more informed citizen who is concerned with the health and safety aspects of public services. Consequently, present wastewater treatment facility design, construction, and operation must address odor concerns as a high priority. In addition, corrosion problems are usually associated with odor detection. Solving an odor problem can minimize corrosion of concrete, including sewer pipe walls, exposed metal, and paints. Odors, particularly hydrogen sulfide, also contribute to the corrosion of galvanized structures and electronics.

Today, in addition to operating wastewater treatment facilities to meet water-quality objectives that protect and preserve limited natural water resources, facilities must provide adequate odor control that the public finds acceptable. This is a difficult task because the realm of odor control is subjective, and the final judge, according to the law or public acceptance, is the human nose.

Controlling air emissions encompasses a broad range of activities used to treat or modify the compounds in the liquid phase, minimize release into the ambient air, provide treatment through an odor control system, and increase the dispersion and dilution of the released gases. Proper odor control is closely coupled to the design and operation of collection systems, liquid treatment facilities, and solids processing systems.

Successful control of air emissions requires proper planning, proficient management, appropriate design, and attentive operation and maintenance. Managers of wastewater treatment facilities must perform odor measurement and characterization, understand modes of odor generation throughout the treatment process, apply the appropriate odor control methods and technologies, and operate and maintain odor control equipment to successfully control air emissions from the facility. Proactive efforts with regulatory agencies and the public should also be considered to avoid unwanted legal and political activities.

## WHAT IS ODOR?

Of the five senses, the sense of smell is the most complex and unique in structure and organization. The human olfactory process has both a physiological and psychological repose to odors.

**THE OLFACTORY PROCESS.** ***Physiological Response.*** The brain and the nose work together to create what an individual identifies as an odor. Odor perception begins well up in the nasal cavity where humans are outfitted with a collection of highly specialized receptor cells. As individual odorous molecules are drawn into the nasal cavity, a portion is dissolved in the mucous film that covers these specialized detectors (Minor, 1995). Once an odorous molecule is captured in the system, it will become attached to more or more of the individual receptor cells based on a shape match. Depending on the molecule, it may be captured by one or several of the specifically shaped receptors. Once a receptor has been stimulated, an electrical signal is transmitted to the brain and the amazing process of identifying odors begins (Campbell, 1996).

Once a signal is generated, the brain takes over and a person responds. When smelling an odor, the reaction may be to flee because of an association with danger or it may be to linger because of the perceived desirable situation. It has been asserted that human beings can detect over 10 thousand different odors even though humans can identify only a small percentage of these (Mackie et al., 1998; Minor, 1995). This sense of smell is much more precise than is our ability to describe the odor we have perceived.

***Psychological Response.*** The psychological response to odor is more complex and far less understood than the physiological process discussed above. Evidence suggests that each individual learns to like or dislike certain odors. Children like almost all smells (Campbell, 1996). It is only as we mature and begin to talk about the odors that we develop a sense of likes and dislikes. Obviously, individuals react differently to the smell of any one odor source.

Perceptions about an odor are based on personal experiences that people have had throughout their life. For some individuals, this experience includes exposure to wastewater treatment, and therefore some level of odor may be acceptable to them. For others who may not have been exposed to such situations, any wastewater treatment odor may be perceived as very offensive and unacceptable.

The human nose provides the accepted standard for detecting and determining odor intensity. No machine has yet been developed that can simulate the human response. The human nose can distinguish differences in more than 5000 odors and detect some compounds with concentrations as low as 0.14 $\mu g/m^3$ (0.1 ppb) (WEF, 2004). In addition, most individuals can discern odors better with the left nostril, women are typically more sensitive to odors than men, and odors can alter or create moods. Simply put, "the nose knows".

**ODOR GENERATION.** Odors can be generated and released from virtually all phases of wastewater collection, treatment, and disposal. Most odor-producing compounds found in domestic wastewater and in the removed solids result from anaerobic biological activity that consumes organic material, sulfur, and nitrogen found in the wastewater. These odor-producing compounds are relatively volatile molecules with a molecular weight of 30 to 150. Domestic wastewater normally contains enough organic and inorganic compounds to cause an odor problem.

Odorous compounds include organic or inorganic molecules. The two major inorganic odors are hydrogen sulfide and ammonia. Organic odors are usually the result of biological activity that decomposes organic matter and forms a variety of extremely malodorous gases including indoles, skatoles, mercaptans, and amines. A list of common, malodorous, sulfur-bearing compounds is presented in Table 13.1. The odors produced, whether strong and persistent or merely a nuisance, are likely to be objectionable at the plant or in the community.

**HYDROGEN SULFIDE ODORS.** Hydrogen sulfide ($H_2S$) is the most common odorous gas found in wastewater collection and treatment systems. Its characteristic rotten-egg odor is well known. The gas is corrosive, toxic, and soluble in wastewater. Hydrogen sulfide results from the reduction of sulfate to hydrogen sulfide gas by bacteria under anaerobic conditions. Desulfovibrio bacteria, which are strict anaerobes, are responsible for the majority of the reduction of sulfate to sulfide, according to the following equation:

$$SO_4^{2-} + 2C + 2H_2O \rightarrow 2HCO_3^- + H_2S \tag{13.1}$$

An illustration of the sulfur cycle is presented in Figure 13.1. At a pH of approximately 9, more than 99% of the sulfide dissolved in water occurs in the form of the nonodorous hydrosulfide ion ($HS^-$). Consequently, odorous amounts of hydrogen sulfide gas will not be released if a pH above 8 is maintained. Below this pH value, hydrogen sulfide gas is released from the wastewater. In comparison, odorous ammonia gas is released primarily at a pH greater than 9.

**Table 13.1** Odorous compounds in wastewater (AIHA, 1989; Moore et al., 1983; Jiang, 2001; U.S. EPA 1985).

| Substance | Formula | Molecular weight | Odor threshold, ppb | Odor descriptions |
|---|---|---|---|---|
| **Odorous nitrogen compounds in wastewater** | | | | |
| Ammonia | $NH_3$ | 17.03 | 17000 | Sharp, pungent |
| Methylamine | $CH_3NH_2$ | 31.05 | 4700 | Putrid, fishy |
| Ethylamine | $C_2H_5NH_2$ | 45.08 | 270 | Ammonia-like |
| Dimethylamine | $(CH_3)_2NH$ | 45.08 | 340 | Putrid, fishy |
| Pyridine | $C_6H_5N$ | 79.10 | 660 | Pungent, irritating |
| Skatole | $C_9H_9N$ | 131.2 | 1.0 | Fecal, repulsive |
| Indole | $C_2H_6NH$ | 117.15 | 0.1 | Fecal, repulsive |
| **Odorous sulfur compounds in wastewater** | | | | |
| Allyl mercaptan | $CH_2=C\text{-}CH_2\text{-}SH$ | 74.15 | 0.05 | Strong garlic |
| Amyl mercaptan | $CH_3\text{-}(CH_2)_3\text{-}CH_2\text{-}SH$ | 104.22 | 0.3 | Unpleasant, putrid |
| Benzyl mercaptan | $C_6H_5CH_2\text{-}SH$ | 124.21 | 0.2 | Unpleasant, strong |
| Crotyl mercaptan | $CH_3\text{-}CH=CH\text{-}CH_2\text{-}SH$ | 90.19 | 0.03 | Skunklike |
| Dimethyl sulfide | $CH_3\text{-}S\text{-}CH_3$ | 62.13 | 1.0 | Decayed vegetables |
| Dimethyl disulfide | $CH_3\text{-}S\text{-}S\text{-}CH_3$ | 94.20 | 1.0 | Decayed vegetables |
| Ethyl mercaptan | $CH_3CH_2\text{-}SH$ | 62.10 | 0.3 | Decayed cabbage |
| Hydrogen sulfide | $H_2S$ | 34.10 | 0.47 | Rotten eggs |
| Methyl mercaptan | $CH_3SH$ | 48.10 | 0.50 | Decayed cabbage |
| Propyl mercaptan | $CH_3\text{-}CH_2\text{-}CH_2\text{-}SH$ | 76.16 | 0.50 | Unpleasant |
| *tert*-Butyl mercaptan | $(CH_3)_3C\text{-}SH$ | 90.10 | 0.1 | Skunk, unpleasant |
| Thiocresol | $CH_3\text{-}C_6H_4\text{-}SH$ | 124.24 | 0.1 | Skunk, rancid |
| Thiophenol | $C_6H_5SH$ | 110.18 | 0.1 | Putrid, garliclike |
| **Other odorous compounds in wastewater** | | | | |
| **Acids** | | | | |
| Acetic acid | $CH_3COOH$ | 60 | 0.16 | Vinegar |
| Butyric acid | $CH_3(CH_2)_2COOH$ | 74 | 0.10 | Rancid |
| Valeric acid | $CH_3(CH_2)_3COOH$ | 102 | 1.8 | Sweaty |
| **Aldehydes and ketones** | | | | |
| Formaldehyde | $HCOH$ | 30 | 370 | Acrid, suffocating |
| Acetaldehyde | $CH_3CHO$ | 44 | 1.0 | Fruity, apple |
| Butyraldehyde | $CH_3(CH_2)_2CHO$ | 72 | 4.6 | Rancid, sweaty |
| Isobutyraldehyde | $(CH_3)_2CHCHO$ | 72 | 4.7 | Fruity |
| Valeraldehyde | $CH_3(CH_2)_3CHO$ | 86 | 0.10 | Fruity, apple |
| Acetone | $CH_3COCH_3$ | 58 | 4580 | Fruity, sweet |
| Butanone | $CH_3(CH_2)_2COCH_3$ | 86 | 270 | Green apple |

Key parameters that affect sulfide generation include

- Concentration of organic material and nutrients,
- Sulfate concentration,
- Temperature,

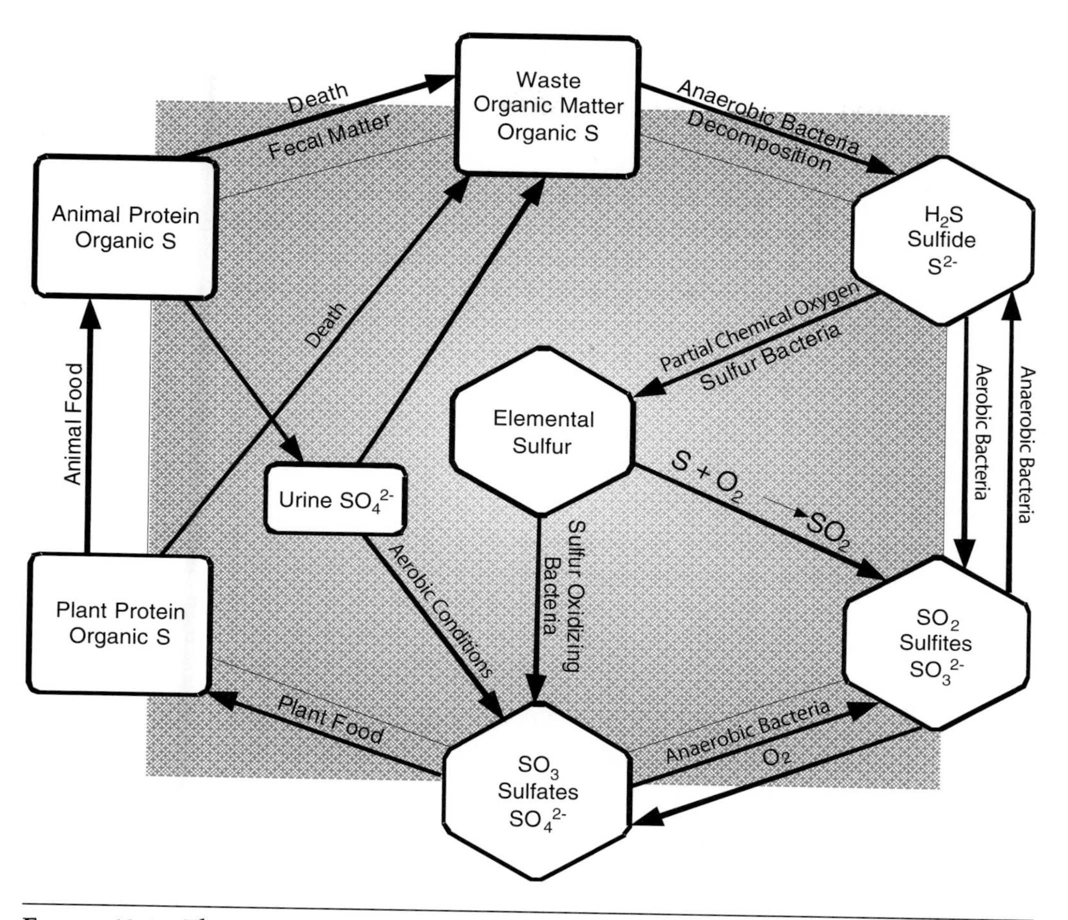


**FIGURE 13.1** The sulfur cycle (U.S. EPA, 1985).

- Dissolved oxygen, and
- Detention time.

The characteristics of hydrogen sulfide gas are summarized in Table 13.2.

Sulfate is reduced to hydrogen sulfide by sulfate-reducing bacteria. Some sulfate is assimilated by organisms to form cell components. Upon mineralization, organic sulfur is converted to hydrogen sulfide. Hydrogen sulfide is transformed to elemental sulfur and then to $SO_4^{2-}$ by sulfide-oxidizing bacteria. Anoxygenic phototrophic bacteria also convert hydrogen sulfide to $SO_4^{2-}$ via elemental sulfur. Elemental sulfur may be transformed back to $H_2S$ by sulfur-reducing bacteria.

**TABLE 13.2** Characteristics of hydrogen sulfide (WEF, 2004).

| | |
|---|---|
| Molecular weight | 34.08 |
| Vapor pressure, −0.4 °C | 10 atm[a] |
| 25.5 °C | 20 atm |
| Specific gravity (vs air) | 1.19 |
| Odor detection threshold | 0.5 ppb |
| Odor character | Rotten eggs |
| Typical 8-hour time weighted average (TWA) exposure limit | 10 ppm |
| Imminent life threat | 300 ppm |

[a]atm × 101.3 = kPa.

**SAFETY AND HEALTH CONCERNS.** The first priority in any aspect of wastewater treatment plant operation is safety. There are two major safety and health concerns in the area of odor control. First, confined-space entry procedures must be used. Second, several odorous gases are toxic. These gases must never be treated with an odor control method that would mask their presence and render them less noticeable to an employee's senses. The Occupational Safety and Health Administration (OSHA) has provided maximum allowable exposure levels of several air pollutants for workers. These values are listed in Table 13.3. *Safety and Health in Wastewater Systems* (WEF, 1994) should be consulted for both confined-space entry and odor control methods.

Hydrogen sulfide is a colorless, toxic gas with a characteristic rotten egg odor. It is considered a broad-spectrum poison, meaning it can poison several different systems in the body. Breathing very high levels of hydrogen sulfide can cause death within just a few breaths. Loss of consciousness can result after fewer than three breaths.

Exposure to lower concentrations can result in eye irritation, a sore throat and cough, shortness of breath, and fluid in the lungs. These symptoms usually go away within a few weeks. Long-term, low-level exposure may result in fatigue, loss of appetite, headaches, irritability, poor memory, and dizziness. The health effects typically seen at various hydrogen sulfide levels are listed in Table 13.4. The OSHA permissible exposure limits for hydrogen sulfide are 10 ppm (time-weighted average) and 15 ppm (short-term exposure limit).

**PUBLIC RELATIONS.** Odors typically pose a long-term problem that affects the community. This situation requires a good public relations program with community residents. The community constitutes a part of the wastewater treatment organization and deserves honest communications. Key elements of a communications system include one specific person to act as the source for information; an odor complaint telephone hotline; an odor complaint investigation form that is completed when each com-

**TABLE 13.3** Maximum allowable exposure of air pollutants for workers.

| Substance | OSHA standard, ppm | |
|---|---|---|
| | TWA[a] | Ceiling[b] |
| Ammonia | 50 | – |
| Bromide | 0.1 | – |
| Carbon dioxide | 5000 | – |
| Carbon monoxide | 50 | – |
| Chlorine | – | 1 |
| Chlorine dioxide | 0.1 | – |
| Ethyl mercaptan | 0.5 | – |
| Hydrogen peroxide (90%) | 1 | – |
| Hydrogen sulfide | 10 | 15 |
| Iodine | – | 0.1 |
| Liquefied petroleum gas | 1000 | – |
| Methyl mercaptan | 0.5 | – |
| Ozone | 0.1 | – |
| Propane | 1000 | – |
| Pyridine | 5 | – |
| Sulfur Dioxide | 5 | – |

[a] Time-weighted average.
[b] A ceiling value is the employee's exposure that shall not be exceeded during any part of the day. For all other values given, the employee's exposure (in any 8-hour work shift of a 40-hour work week) should not exceed the 8-hour TWA shown.

**TABLE 13.4** Health effects of hydrogen sulfide exposure.

| Concentration, ppm | Health effect |
|---|---|
| 0.03 | Can smell. Safe for 8-hour exposure. |
| 4 | May cause eye irritation. Mask must be used as exposure damages metabolism. |
| 10 | Maximum exposure for 8-hour period. Kills sense of smell in 3 to 15 minutes. |
| 20 | Exposure for more than 1 minute may cause eye injury. |
| 30 | Loss of smell, injury to blood brain barrier through olfactory nerves. |
| 100 | Respiratory paralysis in 30 to 45 minutes. Needs prompt artificial resuscitation. Will become unconscious quickly (15 minutes maximum). |
| 200 | Serious eye injury and permanent damage to eye nerves. Stings eyes and throat. |
| 300 | Loses sense of reasoning and balance. Respiratory paralysis in 30 to 45 minutes. |
| 500 | Asphyxia. Needs prompt artificial resuscitation. Will become unconscious in 3 to 5 minutes. |
| 700 | Breathing will stop and death will result if not rescued promptly, immediate unconsciousness. Permanent brain damage may result unless rescued promptly. |

plaint is received and investigated; daily odor tours through the community by plant personnel; and regular public meetings or newspaper articles to inform the public of progress toward solutions to odor problems. Furthermore, treatment plants frequently provide tours of the facility, and many of those individuals taking the tour have the ability to influence the future of the sewer district. It is vital that they leave the tour with the best impression possible. Their impression of the treatment facility will persist into the future regardless of what explanation may be offered about the odors at the treatment plant.

## ODOR MEASUREMENT, CHARACTERIZATION, AND DISPERSION

**SAMPLING METHODS.** Solving any odor problem begins with sampling and analyzing gases to identify and characterize the odors. The principal tools for diagnosing an odor problem are the techniques used for odor quantification and characterization. In general, odors can be quantified by direct sensory measurement of their concentration and intensity, using the human olfactory sense as the odor detector. Alternatively, chemical analysis of odor constituents could be performed. This is an indirect method, because the results of a chemical analysis still need to be related to odor concentration and intensity in some way. Odor regulations vary from location to location and may involve sensory, analytical, or both types of sampling and testing. In both cases, the samples must be carefully prepared to prevent sample contamination.

**QUALITATIVE METHODS—SENSORY ANALYSIS.** Sensory testing involves evaluating odors with the human nose. Odor concentration, intensity, character, and hedonic tone need to be determined for full characterization of odors (Figure 13.2). Although the nose provides only a subjective response to the presence or absence of an odor, several recently developed techniques quantify the human response.

***Odor Panel Evaluations.*** Odors are usually defined as physiological or psychological responses resulting from stimulation of the human nose. An odor panel, the most common method used to evaluate odor nuisances, includes a group of people, typically eight or more, divided equally between men and women. Samples of odorous gas are collected, diluted several times, and delivered to the odor panel for sniffing. The average person can report the presence or absence of an odor with more certainty than its characteristics or objectionability. Therefore, an odor panel member responds either yes or no to the presence of an odor for each of the dilutions sniffed. A variety of sniffing devices can provide static or dynamic (continuous flow) portions of the sample for

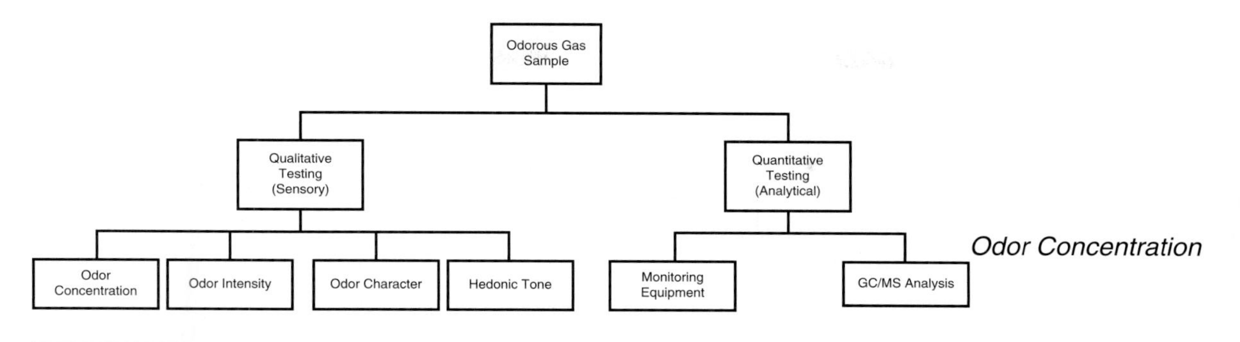


**FIGURE 13.2** Odor sensory evaluation methods.

evaluation. The equipment to perform this testing, called an olfactometer, is commercially available. Testing can also be performed by an odor sciences laboratory.

***Odor Concentration.*** The strength of an odor is determined by the number of dilutions with odor-free air necessary to reduce the odor to a previously defined, detectable level. One method of characterizing odors is to assign an odor threshold number representing the lowest concentration at which the human nose can detect some sensation of odor: The stronger the odor, the higher the odor threshold number. The establishment of odor thresholds requires several evaluations by an odor panel to determine when the odor is no longer detectable. Table 13.1 presents specific odorous compounds found in wastewater, their odor threshold numbers, and characteristic smells.

A second approach to odor strength includes determining the threshold odor concentration or the effective dose number ($ED_{50}$) with an odor panel test. The $ED_{50}$ represents the number of dilutions necessary for the odor to be detectable by 50% of the panel members. In other words, the stronger the odor, the higher the value of the $ED_{50}$. The $ED_{50}$ is expected to approximate the minimum concentration detectable by the average person. A more detailed discussion of odor concentration can be found in *Control of Odors and Emissions from Wastewater Treatment Plants* (WEF, 2004).

***Odor Intensity.*** Odor intensity is measured using an odor intensity reference scale. The intensity scale in use in this country is based on *n*-butanol as the reference substance. The scale is composed of a series of solutions of *n*-butanol in air. Odor intensity is determined by alternately sniffing the odor with unknown intensity and the *n*-butanol odor at successive steps on the scale, in an effort to find the best intensity match. In using the scale, an effort is made to ignore the differences in odor character and the attention is focused on perception of intensity alone (WEF, 2004). A more detailed discussion of odor intensity can be found in *Control of Odors and Emissions from Wastewater Treatment Plants* (WEF, 2004).

***Odor Character.*** Odor character is used to distinguish among different odors. There are three types of descriptors: general, such as "sweet," "pungent," "acrid," etc., a reference to an odor source such as "skunk," "sewage," "paper mill;" or a reference to a specific chemical, such as "ammonia," "hydrogen sulfide," or "methyl mercaptan." The ability to distinguish odor characteristics is helpful for determining the source of an odor in ambient odor monitoring or in odor complaint investigations (WEF, 2004). A more detailed discussion of odor character can be found in *Control of Odors and Emissions from Wastewater Treatment Plants* (WEF, 2004).

***Hedonic Tone.*** Hedonic tone of an odor relates to the degree of pleasantness or unpleasantness of the sensation. It is measured by means of category estimate scales or magnitude estimate techniques. Such measurements, however, are usually confined to laboratory research or investigations and involve primarily odors that are intended to be pleasant such as perfumery. While this property of odor, by definition, seems to determine objectionability, in reality the frequency, location, time, intensity, character, and our previous experience with the specific type of odor determines our response (WEF, 2004). A more detailed discussion of hedonic tone can be found in *Control of Odors and Emissions from Wastewater Treatment Plants* (WEF, 2004).

**QUANTITATIVE TESTING—ANALYTICAL METHODS.** Only a few instruments have been developed for continuous measurement of specific gases. These gases include hydrogen sulfide, chlorine, oxygen, and sulfur dioxide. Gas chromatography (GC) can be used on many odorous organic compounds, but the analysis is complex and expensive. A variety of portable gas-monitoring devices are discussed later in this chapter.

Gas chromatography requires an instrument for the separation of minute quantities (ppb levels) of organic and inorganic substances in a gas or liquid. The GC analysis separates a gas sample into its relative components but does not quantify the concentration of each component. Influent and effluent gas measurements indicate the relative change of specific compounds in a gas stream. These changes can be used to assess the effectiveness of a particular odor control technology. Levels as low as parts per billion of separated components can be specifically identified through mass spectrometry (MS). A complete identification is termed a GC–MS analysis. However, test results from wastewater treatment facilities can be difficult to interpret because the threshold concentrations of many odorous compounds are less than the detection levels of the GC–MS equipment used for analysis.

Portions of the GC-separated gas components can be directed to a sniffing port, allowing a technician to develop an odorgram for the sample. The technician assigns odor numbers to the results from the GC analysis and prepares a graph called an odorgram from these numbers. The odorgram identifies the specific compounds contributing most to the odor and allows comparison of the levels of the same compound in different samples.

## ODOR GENERATION AND RELEASE IN WASTEWATER TREATMENT SYSTEMS

Odors can be generated and released from virtually all phases of wastewater collection, treatment, and disposal. The potential for the initial release or later development of odors begins at the point of wastewater discharge from homes and industries. It continues with collection and movement of wastewater in gravity sewers, lift stations, and force mains, ending with the actual wastewater treatment and solids handling and disposal at the plant or disposal site.

Hydrogen sulfide gas, a major odor source in wastewater treatment systems, results from septic conditions in the wastewater or solids. Metallic sulfide compounds in the wastewater produce a black color, indicating the presence of dissolved sulfide. Ammonia and organic odors are also common. Odors from wastewater and its residuals become significantly more intense and develop much higher concentrations of odorous compounds when the oxygen in the waste is consumed and anaerobic conditions develop. For this reason, much of the discussion of odor generation focuses on the anaerobic conditions that can develop in sewer systems upstream of the wastewater treatment plant as well as unit processes such as primary clarifiers, gravity thickeners, and sludge storage tanks in which anaerobic conditions are likely to develop.

Figure 13.3 illustrates locations in the collection or treatment system where odors initially develop and later become worse due to poor design, such as insufficient ventilation or excessive turbulence. *Design of Municipal Wastewater Treatment Plants* (WEF, 1998) addresses odor problems that can be controlled through proper design considerations. Housekeeping procedures, operational practices including process control and chemical treatment, facility maintenance limitations, and enforceable regulatory policies can also affect odors. Regulatory policies include the development and enforcement of local sewer use ordinances and industrial pretreatment regulations.

**COLLECTION SYSTEMS.** Wastewater in the collection system may contain a variety of odor-causing substances from many sources. While small gravity collection systems generate and release little odor, larger more complex systems often emit nuisance

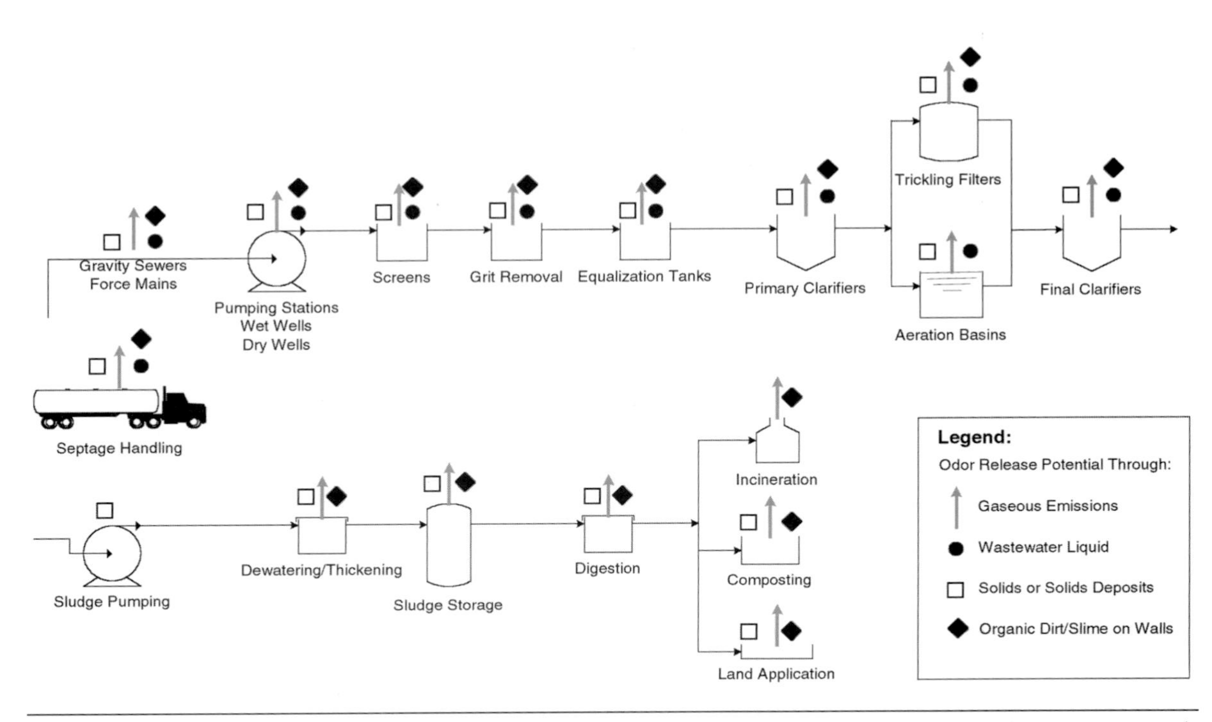


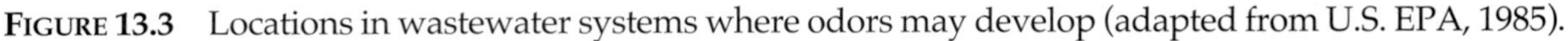

**FIGURE 13.3** Locations in wastewater systems where odors may develop (adapted from U.S. EPA, 1985).

odors. This results from a combination of factors, including the presence of force mains, large gravity interceptors, long detention times, inverted siphons, industrial discharges, and so forth. Solutions to such problems can be costly and include chemical addition to the wastewater, injection of air or oxygen, and collection and treatment of the odorous air to prevent release into surrounding areas.

***Industrial Discharges.*** Industrial wastewater discharges can contain odor-causing substances or contribute to the development of conditions resulting in odor release in the sewer or at the plant. Both conditions can be controlled at the source by enforcing industrial source regulations, summarized in Table 13.5. Smaller plants can control industrial contributions by developing and enforcing local sewer-use ordinances; larger plants can also implement an industrial pretreatment program as part of the discharge permit process. Existing regulations can be strengthened or additional regulatory requirements adopted to obtain end-of-pipe control.

***Gravity Sewers.*** Odors in collection systems result principally from hydrogen sulfide gas and other reduced-sulfur compounds produced by anaerobic sulfate-reducing bacteria in sludge deposits and slime layers. Preventing or limiting the source conditions

TABLE 13.5 Regulatory parameters for industrial discharges.

| Regulated parameter | Unregulated effect |
|---|---|
| pH limitations | At a pH below 8.0, sulfide changes to molecular hydrogen sulfide gas and is available for release. |
| Temperature | Higher temperatures increase microbial action of anaerobic bacteria and increase the release of volatile organic compounds from the liquid to the gaseous phase. |
| Slug loads | Large organic slugs reduce the oxygen concentration of the wastewater by exerting high oxygen demand. |
| Toxic discharges | Inhibits or kills microorganisms involved in biological treatment systems. |
| Flash point or lower explosive limit | Safety and odor problems resulting from organic solvent and gasoline discharges. |
| Fats, oils, and greases | Can collect and anaerobically degrade in wet wells or on tank water surfaces. |
| Chemical discharges | Odorous gases result. |

that allow the bacteria to grow will reduce the odors. General design and operational control measures include

- Designing sewers with sufficient slope to maintain flow velocities adequate to prevent solids deposition,
- Minimizing hydraulic detention times in pipes and wet wells,
- Minimizing the use of drop manholes and other structures that result in significant turbulence,
- Maintaining clean pipes and walls through proper operations and maintenance (O&M) practices such as sewer inspections,
- Maintaining proper hydraulic flows, and
- Ensuring proper collection and treatment of odorous air.

Flow maintenance can be improved through the sewer connection permit process, control of infiltration and inflow, and industrial discharge permits.

***Force Mains.*** Force mains with long hydraulic detention times can allow wastewater to become anaerobic, releasing odors at air release valves and at the force main outlet. In developing areas, temporary flows below the design flow can lead to long residence times. If anaerobic conditions develop in force mains, hydrogen sulfide gas and other odorous compounds will form, creating odors and causing corrosion. Procedures to control sulfide formation include increasing the oxygen level (proper O&M procedures, and air or oxygen injection) and chemical addition (chemical oxidation, inhibition of sulfate reduction, and pH control). Proper hydraulic designs will help maintain sufficient flow velocities and reduce stripping of odorous compounds through turbulence. Often, vapor-phase treatment of odorous compounds is required at pressure-relief valves.

***Pumping Stations.*** The causes of and cures for odor problems in pumping stations are similar to those in the collection system. However, the pumping station wet well provides an ideal location for the release and collection of gases previously trapped in the sewer and for solids deposition due to low velocity. Consequently, additional odor concerns for pumping stations include

- Adequate ventilation of the wet-well area to ensure the health and safety of employees;
- Recirculation of fresh, nonodorous outside air throughout the ventilation system;
- Pump sequencing to maintain proper hydraulic detention times;
- Proper wet-well design to avoid dead areas, including weekly flushing to prevent solids deposition and accumulation;
- Aeration in the wet well to minimize anaerobic conditions;
- Frequent cleaning to remove slime growths on walls or trapped floating fats, oils, and greases; and
- Collection and treatment of the vented odorous air.

**LIQUID TREATMENT PROCESSES.** Influent wastewater entering the treatment facility from the collection system may contain high concentrations of sulfide and volatile organic compounds (VOCs). In general, the preliminary and primary treatment processes are often major sources of emissions. Secondary treatment processes are not normally significant sources of offensive odor as long as sufficient oxygen is supplied to keep the various processes aerobic. Facilities downstream of the secondary treatment processes (filtration and disinfection) are not normally strong sources of odor.

***Septage Disposal.*** Septage is reasonably dilute (approximately 97% water), malodorous, high in nitrogen (up to 500 mg/L total N), highly putrescible (2000 to 5000 mg/L biochemical oxygen demand [$BOD_5$]), and likely to contain large numbers of harmful viruses, bacteria, and other microorganisms (WEF, 2004). Septage collected from portable toilets, septic tanks, or dockside pump-out stations may also contain chemicals that can upset biological treatment processes.

The uncontrolled discharge of relatively large volumes of septage into a wastewater treatment plant would probably result in a rapid dissolved oxygen depletion and subsequent odor generation. Also, septic wastewaters containing sulfides can lead to activated sludge bulking conditions involving filamentous organisms such as *Thiothrix* sp., *Beggiatoa,* and Type 021N (Jenkins et al., 1986).

The handling and receiving of septage at wastewater treatment plants can emit noxious odors and contribute to conditions that result in odor release in downstream treatment processes. Anaerobic septic tank wastes are typically received on a periodic, unscheduled basis.

To minimize upsets and odors in biological treatment systems, the septage discharge should contain less than 10% of the volatile solids in the plant influent, measured concurrently with the septage discharge (WEF, 2004). Excess septage, based on that limit, should be stored or pretreated to provide a lesser, more uniform discharge into the plant influent. An odor control system on the storage tank may be necessary. Chemicals to raise the pH, such as lime or caustic, can be metered into the septage during discharge.

Other possible septage control measures include

- Enforcing existing sewer-use ordinances and changing them if necessary,
- Adopting and implementing chain-of-custody practices, and
- Laboratory testing such as pH or oxygen uptake rate monitoring to prevent unregulated septage discharges of industrial wastewater that could inhibit downstream biological processes and cause odors.

***Preliminary Treatment.*** Trapped or dissolved odorous gases in raw wastewater can escape to the atmosphere at the plant headworks. Wastewater turbulence, designed for headworks areas to ensure mixing and prevent solids deposition, accelerates the release of the gases. If the odors cannot be controlled upstream in the collection system, then the total headworks structure or the immediate area of turbulence may need to be enclosed for collection and treatment of the odorous air. Prevention of odor generation within the headworks itself requires clean influent structure walls and channels, including flow equalization basins, achieved through scheduled washing of slime and sludge accumulations.

High organic septic sidestreams from solids-processing operations, such as filtrates and digester supernatants, which are returned to the head of the plant, may also release odorous gases.

Odors in preliminary treatment emanate primarily from the screening and grit collection processes. Organic matter will adhere to the screens and produce odors if the screening solids or drainage become septic. Accumulated screening solids need daily removal, and units must be kept clean and odor-free. Closed containers store the collected material until it can be transported to a landfill for burial or dewatered for incineration. Drainage systems for screening should quickly collect drainage and return it directly to the raw wastewater instead of allowing it to flow along the floors.

Grit chambers can be cleaned manually or mechanically. Some are aerated, while others depend on proportional weirs or other devices to regulate velocity and grit separation. Aerated grit chambers may release odorous gases during aeration. Because the organic material collected with the grit is the major cause of grit system odors, removal of this material may require grit washing to prevent septic odors generated in the plant and at the grit disposal site. Prevention of the problem requires proper regulation of air or controlled water velocities to remove heavier inorganics while allowing the lighter organics to remain in the wastewater stream.

***Primary Clarification.*** Primary clarifiers can be a major source of odor emissions, particularly if the influent wastewater contains significant levels of dissolved sulfide. Primary clarifiers represent a large odor-emitting surface area, and the character of the odor is often considered objectionable. The turbulent effluent launders often represent a significant portion of the total clarifier odorous emissions. The greater the free fall over the effluent weirs, the greater the release of odorants. Covered effluent troughs and ventilation to a vapor-phase odor control system can minimize odor release resulting from excessive detention times or turbulence as the effluent flows to the next treatment process. Covering the entire basin and treating in an odor control system can reduce off-site odor impacts of primary clarifiers.

Accumulations of scum on the tank water surface and of organic matter on the effluent weirs can generate odors. Scum collection and conveyance equipment needs frequent cleaning, using hot water and abrasion to remove grease and slime buildups. Good odor prevention practice includes skimming floating solids and grease from primary tanks at least twice daily, and immediately removing floating sludge. Tank walls, weir troughs, sludge pits, and open channels need frequent cleaning. The operator should cover the sludge pits whenever possible.

Primary sludge odors will also result if settled solids remain for long periods on the tank floor or in the sludge collection area. Therefore, frequent sludge removal from operating tanks is necessary to prevent the production of gases that could cause the sludge to rise to the surface, releasing odors and inhibiting solids settling. Close control of the sludge removal frequency is required, particularly when concentrated digester supernatant, filtrate, or secondary sludge are returned to primary tanks. Septic conditions can be prevented by providing minimum hydraulic detention times, increasing the frequency of settled solids scraping, and providing the sludge-pumping rates necessary for effective thickening while minimizing detention time. Some plants continually withdraw a thin sludge and pump it to a separate thickener to minimize primary clarifier odors. Sludge collectors need adjustment to allow delivery of most of the sludge to the hoppers on each pass. Generally, sludge is removed several times a day.

Many facilities are designed to provide excess primary clarifier capacity and the tendency is to operate all available clarifiers. Based on the design flows and loads for a given facility, it may be possible to remove primary clarifier tanks from service that are not required for the current flowrates. This prevents excessive detention time across the primary clarifiers, decreasing $H_2S$ generation.

Primary settling tanks removed from service can become a major source of odor unless odor prevention procedures are followed. When any tank is removed from service for two days or more, all contents should be drained and the tank cleaned. If the tank is to remain out of service temporarily, it should be filled with chemically treated (typically chlorinated) water to prevent algal and bacterial growth. If the tank is to remain out of service permanently, it should be cleaned and left empty.

***Secondary Treatment Processes.*** *Fixed-film processes.* Fixed-film processes, such as trickling filters or rotating biological contactors, generate odors if the air supply to the biological film is insufficient to maintain aerobic conditions. Both processes require a continuous, uniform distribution of the raw wastewater over the film area, together with an air supply sufficient to maintain proper slime thickness. Hydraulic overloading of either process or plugging of the trickling filter media or underdrains can reduce air circulation, allowing septic conditions to develop. Odors may also be generated when these processes are highly loaded in terms of BOD. Covering the processes and ventilating the odorous air to a control device provides effective odor control.

*Activated sludge process.* Biological aeration tanks and secondary settling tanks must be aerobic to function efficiently. Keeping these units aerobic is the most important odor control consideration. The two major sources of objectionable odors in activated sludge tanks are development of anoxic or anaerobic conditions in the aeration tank and entry of dissolved odorous compounds into the aeration tank.

As the first odor source, poor mixing or inadequate dissolved oxygen in the mixed liquor impairs the aerobic environment, causing settled deposits with resultant anaerobic conditions at the bottom of the tank. Because most diffusers can become clogged, routine cleaning is necessary. If the plant is equipped with swing-type, air-diffusion piping, operators can change clogged diffusers and flush air headers without emptying the aeration tanks. Mechanical aerators with the highest turbulence levels typically produce the highest odor and VOC emission rate, followed by coarse-bubble diffusers. Because of the low level of turbulence, fine-bubble diffusers typically have the lowest emission rate assuming all other factors (wastewater characteristics, mixed liquor suspended solids, and surface area) are equal (WEF, 2004).

Solids may settle in aeration basins as a result of low turbulence near the points where recirculated activated sludge enters the basins. Such deposits, combined with a limited oxygen supply, will usually generate objectionable odors. Aeration tank walls require regular cleaning to remove deposits.

As the second odor source, even with aerobic conditions, dissolved oxygen compounds may enter the tank with the primary clarifier effluent or with anaerobic return sludge flows; aeration then releases the compounds to the atmosphere. Aerosols formed during the aeration process can be carried by air currents far beyond the local tank environment. Both of these problems require control at the source.

Even a well-operated aeration basin emits a slight organic odor. The characteristic earthy, musty, organic odor of activated sludge results from the volatilization of complex organics and the production of intermediate compounds.

*Secondary clarification.* The causes of odors in secondary clarifiers are similar to those for primary tanks. Accumulations of scum on tank water surfaces, sludge accumulations on tank walls, and organic matter on effluent weir troughs can produce odors without proper daily housekeeping. Frequent, mechanical high-pressure water spraying will minimize algae growth, minimizing odor release.

Secondary sludge odors will develop if the settled solids remain on the tank floor or in the sludge collection area for long periods. Anaerobic conditions resulting from slow removal of settled sludge may produce rising sludge that hinders settling. Floating sludge can also cause violations of effluent suspended solids limitations.

Although an adequate return sludge concentration must be maintained, odors will be minimized if the settled mixed liquor is removed from the tank floor as quickly as possible and returned to the aeration tanks. Consequently, some secondary clarifiers have been designed with hydraulic, rapid sludge-removal systems to minimize the solids retention time on the tank floor. Similar to the control of primary clarifiers, sludge blanket control in secondary clarifiers attempts to maximize thickening and minimize solids retention.

If secondary sludge withdrawal rates are too slow, odors can develop and the septic sludge will exert an additional oxygen demand when it is returned to the aeration tanks. Septic sludge is also difficult to thicken and dewater. The operator controls the solids retention time in the secondary clarifiers to prevent a negative oxidation–reduction potential allowing sulfate-reducing bacteria to produce hydrogen sulfide gas. In addition to chlorinating the return sludge for control of filamentous organisms (Jenkins et al., 1986), some plants chlorinate it intermittently for odor control during odor problem periods or continuously as background for odor control during the warmer months.

***Disinfection.*** The addition of excessive amounts of a disinfecting agent such as chlorine or ozone can cause residual odors, along with concerns for employee health and safety. Using automatic controls to pace the chlorine dosage to flow avoids adding excess chlorine and minimizes odor emissions.

**SOLIDS TREATMENT PROCESSES.** Solids-handling systems are typically a significant source of odors in wastewater treatment plants. Odors from solids-handling processes are typically complex mixtures of reduced-sulfur compounds including hydrogen sulfide, mercaptans, dimethyl sulfide, dimethyl disulfide, and others. Ammonia may also be present. Organic acid odors are also a concern when handling solids.

The incidence and concentrations of reduced-sulfur compounds is dependent on not only the type of treatment process, but also on how the unit processes are operated. For example, excessive detention times in sludge holding or blend tanks can cause high levels of non-$H_2S$ sulfur compounds to be formed. It is important to process activated sludge without holding for more than 12 hours to prevent septicity.

***Sidestreams Treatment.*** Sidestreams from solids processing, generally returned to the headworks of the plant, can contain high concentrations of BOD, chemical oxygen demand (COD), total suspended solids, and ammonium nitrogen. Returned sidestreams can cause odors directly through the release of odorous gases at the headworks structure, or indirectly from organic overloading and the consequent rapid dissolved oxygen depletion of the receiving wastewater.

Material balances within the plant must account for sidestream contributions. Solids in thickener underflows that are returned to the primary clarifier can alter the primary-to-secondary ratio of the feed sludge into downstream heat treatment or dewatering processes. Some sidestreams contain active anaerobic bacteria that enhance septicity if the sidestreams combine with settled, primary solids to produce various gases that hinder sludge settling as they rise.

Recycled sidestreams are generated periodically if thickening or dewatering processes operate intermittently. Consequently, flow equalization and storage of sidestreams may be necessary to supplement odor control equipment. Pretreatment of sidestreams to minimize odors includes aeration, biological treatment, and chemical addition.

***Transfer Systems.*** Equipment used to transfer sludge or solids includes pumps, screw or belt conveyors, and vacuum systems. Most of these transfer systems emit an odor, regardless of the condition of the sludge. Consequently, the objective in controlling such odors is to minimize them. When possible, transfer equipment odors should be contained, collected, and treated. Enclosed screw conveyors or the pumping of sludge through pipes will minimize odors. This requires proper daily housekeeping practices to ensure that equipment, walls, and floors remain clean from sludge, solids, and supernatant.

***Thickening.*** Thickeners, including gravity units, dissolved-air flotation units, gravity belts, or centrifuges, need odor-controlled rooms or covers with odorous air collection and treatment. The sludge blanket on flotation units needs prompt removal during operation and complete removal during shutdown. The feeding of fresh primary solids to thickening processes will minimize the release of dissolved or trapped odorous gases. Secondary sludge, specifically waste activated sludge, has an earthy odor that is lower in intensity than the odor of primary sludge. It is important to process waste activated sludge without holding to prevent septicity.

Gravity thickeners treating primary sludge can be a significant source of objectionable odors. Adding elutriation water or chlorine solution to keep the supernatant from becoming septic, minimizing sludge blanket levels to prevent sulfide generation in the settled sludge, and frequent flushing of the surface and weir troughs can help reduce odor emissions (WEF, 2004). Co-thickening biological and primary solids results in the combination of a hungry population of organisms and a surplus of food in an oxygen-deficient environment and has the potential to create severe odors. The solution to this problem is to thicken primary and biological solids in separate tanks, if possible. If only one tank is available, it may be necessary to add chemicals along with the influent solids to treat sulfide as it is formed.

Dissolved air floatation (DAF), typically used for thickening waste-activated sludge, uses a high aeration rate to float solids, so any sulfide or odors present in the secondary solids will be stripped out in the DAF thickener. As with gravity thickeners, chemical pretreatment may be needed to control odorous emissions.

Chemical addition to the sludge prior to thickening reduces odor generation and release during the thickening process. Covering and ventilating odorous air to a vapor-phase odor control system improves indoor air quality and reduces the risk of the odors traveling off-site.

***Blending and Holding.*** Holding tanks are used to collect and store sludge before treatment, or to blend primary and secondary sludge before downstream dewatering. Maintaining the proper solids retention time (typically less than 12 hours) will minimize odor production. Holding times in excess of 24 hours can cause a significant increase in the production of reduced-sulfur compounds. This not only increases odor emissions from holding or blending tanks, but also increases emissions from subsequent dewatering processes. If possible, the tanks should be covered and the odorous gases collected and treated. Chemical additions will also control hydrogen sulfide release. Tank mixing during the off-shifts will minimize the release of trapped gas during the day.

***Stabilization.*** Stabilization processes include chlorine oxidation, anaerobic digestion, aerobic digestion, composting, lime treatment, and alkaline chemical-fixation processes. With any of these processes, odor problems can develop if adequate stabilization of organic material is not attained, if residual odors are generated by a change in pH levels or specific chemical additives, or if housekeeping practices are poor.

Aerobic digestion processes generate odors similar in character to those from activated sludge aeration basins, if operated with sufficient dissolved oxygen concentrations. Anaerobic digestion occurs in closed vessels, and there is limited opportunity for odorous gas to escape. Points of odorous gas escape include open or inadequately sealed overflow boxes, annular spaces around floating digester covers, and unlit flares. Care should be taken to avoid these conditions.

Specific odor concerns include

- Hydrogen sulfide gas from upset anaerobic digesters;
- Overloaded aerobic digesters;
- Ammonia generation during lime stabilization;
- Residual chlorine odors from chlorine oxidation systems; and
- Release of ammonia during an alkaline, chemical-fixation processes.

***Dewatering.*** Most dewatering processes used today are mechanical systems such as belt filter presses or centrifuges. Relatively few (typically smaller) facilities use drying beds. The characteristics of the odor released from these processes are a function of the characteristics of the feed solids. For example, a belt filter press dewatering aerobically digested solids from a facility with no primary clarifiers has significantly less odor than a belt press handling a blend of thickened primary and waste-activated sludge that has been stored for several days.

Turbulence created in any dewatering process will release the odorous gases generated by septic solids. Solids must be kept fresh, with hydrogen sulfide controlled by chemicals or by pH levels greater than 8 to prevent gas formation. Chemicals used for conditioning may also produce residual odors. For example, polymers often release a fishy odor during the dewatering process. Thus, polymer selection should consider odor potential as well as cost-effectiveness for coagulation.

Chemical addition effectively reduces odor release during dewatering. This method in combination with ventilating odorous air to a control device is the most common practice at the majority of treatment facilities today.

***Composting.*** When the composting process was first developed in the early 1970s, odors were of little concern. When the process was implemented at many wastewater plants, however, it became apparent that odor emissions from composting operations were causing complaints. Many composting facilities were subsequently shut down because of odors. Even today, composting facilities continue to wrestle with odor issues, and the cost to control the odors significantly increases the overall costs of the process.

In most cases, odor release is the direct results of agitation (mixing, pile stacking, loading, etc.) of the composting piles. Most operations occur in a pole-barn-type structure (roof with no walls), providing little impediment to odor transmission off-site. Currently, many composting systems operate with negative aeration and discharge the process air to an odor control system, often a biofilter.

***Thermal Drying.*** Thermal drying processes are advantageous because of the high volume reduction they achieve. However, control of both odor and particulate emission is necessary for most drying processes. Thermal oxidation processes are often used to treat the process off-gas because of the relatively high temperature and the complex constituents of the air stream.

***Incineration.*** Incineration processes include multiple-hearth and fluidized-bed incineration, flash drying, wet oxidation, and pyrolysis. The primary source of odors is the escape of incompletely oxidized gases. Avoidance or control of this problem depends on designing and maintaining proper operating temperatures, turbulence, contact time, and oxygen levels.

***Land Application.*** Odor emissions from land application of solids are a function of the characteristics of the material and the method of application. The greater the odor of the applied solids, the higher the odor emissions rate and potential for downwind complaints. Spray application is likely to have the highest odor emission rate. The lowest odor emissions are associated with application methods such as subsurface injection in which the material is immediately incorporated to the soil. Regardless of the method of application, the highest odor emission rates occur during application. Odor tends to dissipate to background levels within 24 to 48 hours.

## ODOR CONTROL METHODS AND TECHNOLOGIES

Odor control is a complex and time-consuming challenge, often requiring a combination of methods for treating odorous gases and for removing or reducing the potential causes of the odors. If an odor problem is severe enough to affect the community, an emergency response and solution to the problem(s) must be carried out quickly.

Table 13.6 lists several odor control methods and technologies that can reduce an odor problem. The approach for selecting an odor control method or technology, summarized in Figure 13.4, includes the following steps:

1. Identify the odor source and characteristics through sampling and analysis.
2. List and assign priorities to controlling a specific odor problem, recognizing considerations such as cost, plant location, future upgrading of various wastewater processes, severity of the odor problem, and the nature of the affected area.
3. Select one or more odor control method or technology for implementation to meet the objectives of steps 1 and 2, taking into consideration the advantages and disadvantages of each.
4. Monitor odor emissions from the treated air for process adjustments and for feedback to evaluate the solution's effectiveness.

When odor problems develop, plant personnel sometimes lack the resources necessary to implement many of the control methods and technologies listed in Table 13.6. The containment, collection, and treatment alternatives may require design services, capital expenditures, and a best-available control technology (BACT) review by the local regulatory office.

The capital and O&M costs of the various BACT alternatives differ significantly. However, the costs associated with odorous gas capturing and handling (typically the major expenditures) will be the same regardless of the technology selected.

*Control of Odors and Emissions from Wastewater Treatment Plants* (WEF, 2004) provides an in-depth discussion of the theory, design, and application of the various odor

**TABLE 13.6** Odor control methods and technologies.

| | |
|---|---|
| **Vapor-phase odor control methods** | Atmospheric discharge and dilution<br>Masking agents and counteraction chemicals<br>Chemical wet scrubbing<br>Activated carbon adsorption<br>Biofiltration<br>Biotrickling filters<br>Air ionization<br>Thermal incineration<br>Aeration system treatment |
| **Liquid-phase odor control methods** | Chemical addition |
| **Operational odor control methods** | Source control<br>Housekeeping improvements<br>Process or operational changes |

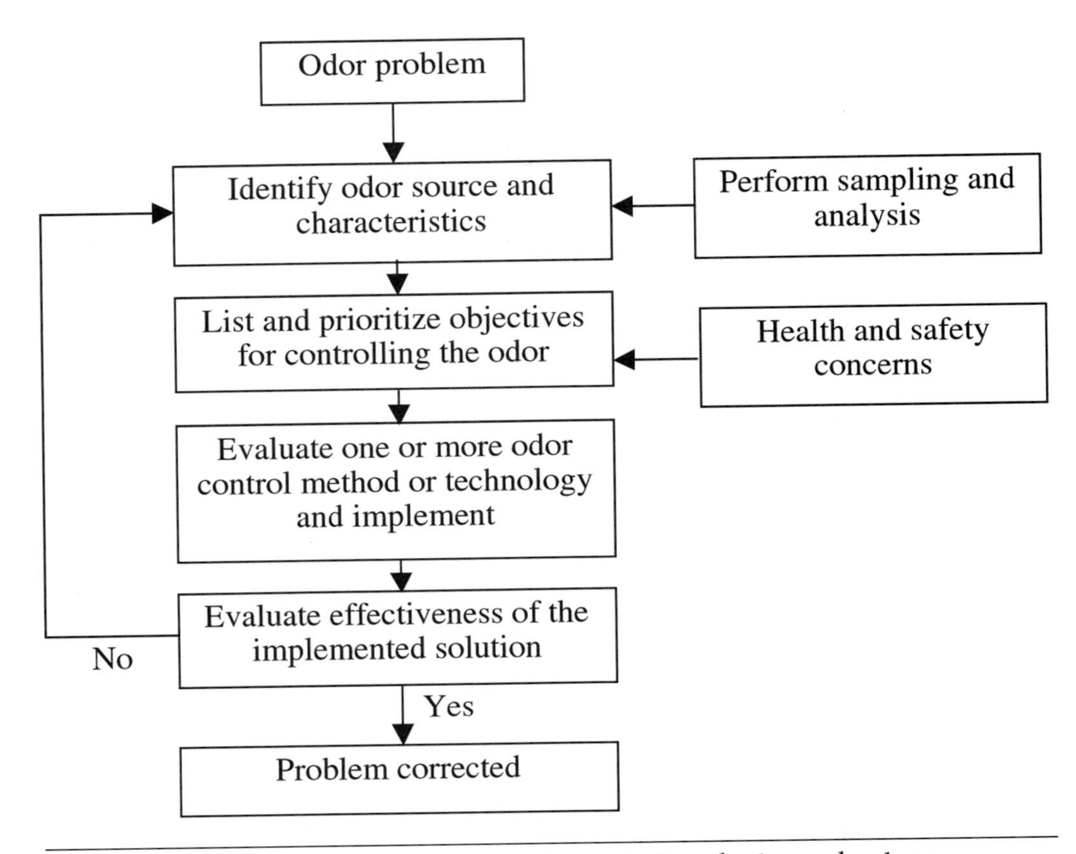


FIGURE 13.4 Flowchart for selecting an odor control method or technology.

control technologies. The following sections discuss odor control methods and technologies that apply to the problems facing a plant operator with normal, limited, day-to-day resources of materials, labor, and money.

**CONTROLLING HYDROGEN SULFIDE GENERATION.** Hydrogen sulfide production can be controlled by maintaining conditions that prevent the buildup of sulfides in the wastewater. The presence of oxygen at concentrations of more than 1.0 mg/L in the wastewater prevents sulfide buildup because sulfide produced by anaerobic bacteria is aerobically oxidized to thiosulfate, sulfate, and elemental sulfur. Maintaining an aerobic environment also inhibits the anaerobic degradation process contributing to the generation of hydrogen sulfide.

The dissolved oxygen concentration in collection systems can be increased by several techniques, including

- Providing air access by such items as manholes with air vents and plumbing vents in structures,
- Minimizing oxygen depletion by O&M procedures,
- Periodic cleaning of manholes and wet wells, and
- Injecting air, pure oxygen, or chemical oxidant into the wastewater stream.

Chemical addition can control sulfides by chemical oxidation, sulfate reduction, inhibition by providing an additional oxygen source, precipitation, or pH control. Minimizing septicity depends on maintaining sufficient velocities to prevent solids deposition and providing, to the extent possible, short detention times in gravity sewers and pumping station wet wells.

Chemical addition at the headworks can control sulfides in the wastewater treatment plant, but sulfides should typically be controlled upstream in the collection system. This approach also protects concrete and metals from corrosion. Preaeration may add the necessary oxygen if $H_2S$ does not already exist in the influent flow. Proper operation and maintenance of all wastewater treatment processes are necessary, with special emphasis on

- Providing sufficient turbulence to prevent solids deposition and to ensure complete mixing (excessive turbulence will release odors already generated but maintained in the liquid phase);
- Maintaining at least 1.0 mg/L of dissolved oxygen in the aeration tanks;
- Keeping settled sludge fresh through adequate return rates;
- Ensuring proper hydraulic and solids detention times in all tanks;
- Adhering to typical process control ranges; and
- Developing an aggressive industrial pretreatment program to control the discharges of industrial wastewaters containing substances with high COD and BOD or odor-contributing compounds.

**OPERATIONAL CONTROL METHODS.** Offensive, localized odors often result from poor operational or housekeeping practices. Control of such odors depends on preventing the conditions leading to odor generation. Preventive operational measures include

- Maintaining adequate dissolved oxygen concentrations in the wastewater;
- Providing sufficient velocity (not turbulence) to ensure complete mixing, thus preventing organic solids deposition;

- Preventing sludge accumulations in tanks or excessive sludge aging through proper mixing, withdrawal rates, and process control parameters;
- Preventing overloading by recirculating, equalizing flows, or putting additional tanks into operation; and
- Ensuring uniform organic loading of biological processes.

Preventive housekeeping measures include

- Regular hosing of channels and walls to prevent solids accumulations, especially during tank dewatering operations.
- Use of hot water, cleaning solutions, and abrasives to remove slimes and greases.
- Rapid and complete drainage and flushing of tanks. Dewatered, clean concrete tanks can still emit offensive odors resulting from sorption of sludge by the concrete. This can be controlled by flushing with a strong chlorine solution.
- Regular skimming of floating scum and solids on water surfaces.

**CHEMICAL ADDITION.** Chemical addition can control odors in gravity sewers, force mains, and the wastewater treatment plant by preventing anaerobic conditions or controlling the release of odorous substances. These chemicals fall into four basic groups based on their mechanisms for odor control, shown in Table 13.7:

1. Chemicals to oxidize odorous compounds into more stable, odor-free forms;
2. Chemicals to raise the oxidation–reduction potential and prevent the chemical reduction of sulfates to hydrogen sulfide;
3. Bactericides to kill or inactivate anaerobic bacteria that produce malodorous compounds; and
4. Alkalis to raise the pH and thereby keep sulfides in their ionized form, rather than hydrogen sulfide.

The effectiveness of chemical addition as an odor control technology depends on such variables as cost, dosage, presence of odor-causing compounds, effects of chemical accumulations in sludge and process waters, equipment maintenance, space limitations, and safety or toxic substance concerns. Typical odor-control applications include collection systems, headworks, primary clarifiers, process sidestreams, aeration tanks, solids-handling applications, and storage lagoons. In general, it is more cost-effective to treat odors in the liquid phase than in the vapor phase.

Common chemical agents used to control odors include iron salts, hydrogen peroxide, sodium hypochlorite (chlorine), potassium permanganate, nitrates, and ozone.

TABLE 13.7 Chemicals used for liquid-phase odor control.

| Chemical | Effective against |
|---|---|
| **Oxidizers** | |
| Ozone | Atmospheric hydrogen sulfide only |
| Hydrogen peroxide | Hydrogen sulfide, also acts as an oxygen source |
| Chlorine | Hydrogen sulfide and other reduced sulfur compounds |
| Sodium and calcium hypochlorite | Hydrogen sulfide and other reduced sulfur compounds |
| Potassium permanganate | Hydrogen sulfide and other reduced sulfur compounds |
| **Raising the oxidation–reduction potential** | |
| Oxygen<br>Nitrate<br>Hydrogen peroxide<br>Chlorine | Higher temperatures increase microbial action of anaerobic bacteria and increase the release of volatile organic compounds from the liquid to the gaseous phase. |
| **Bactericides** | |
| Chlorine<br>Hydrogen peroxide<br>Potassium permanganate<br>Chlorine dioxide<br>Sodium hypochlorite<br>Oxygen | Kill or inactivate anaerobic bacteria |
| **pH modifiers** | |
| Lime<br>Sodium hydroxide | Prevent offgassing of hydrogen sulfide; at a very high pH acts as a bactericide on sewer wall slimes |

***Iron Salts.*** Iron salts react with sulfide to form insoluble precipitates and are widely used for control of sulfide at wastewater facilities. Both ferric and ferrous chemicals provide effective sulfide control, but ferric chloride also enhances primary settling and reduces downstream process loadings. The use of the ferric chemical will also enhance phosphorous removal. The lowest treatment costs are usually obtained by purchasing the product from a local supplier to minimize transportation costs. One advantage of iron salts over oxidation chemicals is that the salts do not react with wastewater organics and they are fairly sulfide-specific. Effective treatment has been obtained with an iron-to-sulfide dosage rate of 3.5 kg/kg (3.5 lb/lb). One disadvantage of iron salts is their absorbance of light, which can impair ultraviolet disinfection.

***Hydrogen Peroxide.*** Hydrogen peroxide is a commonly used oxidant that chemically oxidizes $H_2S$ to elemental sulfur or sulfate depending on the pH of the wastewater. Like other oxidant chemicals, however, peroxide reacts with organic material in the wastewater, so higher dosages are required. For some applications, successful treatment has been reported at a peroxide-to-sulfide ratio of 2:1, but other applications have required as high as 4:1 (Van Durme and Berkenpas, 1989). Peroxide is fast acting, which makes it useful for addition immediately upstream of problem locations. However, it is also quickly consumed, so multiple injection sites are required to treat long reaches of collection systems.

***Chlorine.*** Chlorine is a relatively inexpensive, powerful oxidant, and equipment required for its use is inexpensive and widely available. Chlorine is available as pure gas, hypochlorite solution, or hypochlorite granules or tablets. Commercially available solutions of sodium hypochlorite or calcium hypochlorite are the most common forms. Gaseous chlorine is used at many wastewater treatment plants for disinfection, but storage and handling requirements make it less desirable for collection system odor control applications.

Chlorine is an oxidant, so unlike iron salts it is not sulfide specific and much of the chemical is consumed through reaction with organics in the wastewater. Actual practice has shown that, depending on pH and other wastewater characteristics, between 5 and 15 parts by weight of chlorine are required per pound of sulfide (U.S. EPA, 1985). Chlorine can also act as a bactericide because it is a strong disinfectant. Depending on the point of application and the dose, it can kill or inactivate many bacteria that cause odors. However, because chlorine is nonselective, it will also kill organisms beneficial to the wastewater treatment processes. Care should be exercised when chlorine is added in the collection system near wastewater treatment plants.

***Potassium Permanganate.*** Potassium permanganate is a solid chemical that is easy to handle and has been cost-effectively applied for pretreatment of both solids thickening and dewatering processes. For treatment of solids processes, such as a belt press, permanganate dosage rates have been found to be as high as 16:1. Permanganate is an oxidant that treats a wide range of other odorous compounds. Reducing the odor in the dewatered cake will decrease the amount of odor that escapes when the solids are transported off-site.

***Nitrates.*** Nitrate addition controls dissolved sulfide by prevention and removal. For prevention, nitrate is added to wastewater to be used as a substitute oxygen source. This results in the projection of nitrogen gas and other nitrogenous compounds, rather than sulfide. For removal, nitrate can be added to septic wastewater to remove dissolved sulfide by a biochemical process, converting sulfide to sulfate.

***Ozone.*** Ozone is an extremely powerful oxidant that can oxidize $H_2S$ to elemental sulfur. It is also an effective disinfectant when bacteria levels are low. Although ozone reacts with nearly everything in wastewater, including dissolved sulfide, its principal usage has been to treat odorous gas streams. Ozone is unstable and must be generated on-site. It is also potentially toxic to humans at concentrations of 1 ppm or greater in air. There are no applications that suggest that ozone treatment of dissolved sulfide in wastewater collection systems is economical on a long-term basis.

The selection of chemicals to control odors can be a trial-and-error process. Each of the chemicals has distinct advantages and disadvantages to be considered when seeking the best solution for a specific odor control problem. In many cases, a combination of chemicals, process changes, or other odor control methods may be necessary to solve the problem effectively.

**CONTAINMENT.** There is a full-range of available odor cover and containment systems alternatives. These are commonly used in conjunction with an odorous gas treatment process, such as wet scrubbers, activated carbon, or biofiltration. The selection of the appropriate cover and containment system will depend on several factors, including

- Site-specific unit processes that need to be controlled,
- Area climate,
- Worker safety,
- Ease of construction,
- Staff needs,
- Aesthetics,
- Effectiveness,
- Durability, and
- Cost.

Covers or domes contain gases emitted by odorous units. This method of odor containment works well on equipment such as headworks structure, screens, grit removal equipment, channels, effluent launders, primary clarifiers, aeration tanks, processing units, holding tanks, and screw pumps. When considering odor containment, focusing on the specific areas of odor release can minimize the actual area to be covered. The odorous air is exhausted from the containment and treated with an appropriate odor control technology.

The odorous gases contained under covers are often extremely corrosive or toxic. Flat, low-profile covers should be specified whenever possible to minimize headspace. Minimal headspace also reduces air-exchange volumes and related odor control equipment capacities. All electrical controls should be explosion proof and located outside the covered areas. Cover selection considerations include minimizing condensation problems and supporting snow and ice loads in northern regions.

Covers should have corrosion-resistant materials (although even aluminum will pit severely without adequate air collection); lifting lugs and apparatus where necessary to prevent injuries; and nonslip surfaces for flat cover areas that serve as walkways.

Aluminum and fiberglass covers have been used in numerous wastewater applications for covering small as well as large structures. Aluminum and fiberglass covers are known to be corrosion resistant and durable and can be assumed to have at least a 20-year service life. In many past wastewater applications, dome covers, shown in Figure 13.5, have been used to cover large circular basins. In recent years, flat covers (Figure 13.6) have been considered for many wastewater applications because the volume of air requiring treatment can be reduced. Minimizing that air space is critical to the ultimate sizing of the transfer duct work, blower capacity, and odor control technology. The major disadvantage of flat covers is the limited operator visibility and accessibility.

There are several manufacturers that produce fabric covers, shown in Figure 13.7, which have been used in wastewater applications. The fabric covers are relatively durable, as the manufacturer suggests using a 10-year life for the purposes of life-cycle costing.

**MASKING AGENTS AND COUNTERACTION CHEMICALS.** Odor problems can be addressed at the source or at the symptom. Masking (odor modification) and counteraction (odor neutralization) can be effective as an emergency, short-term solution for the symptom, but generally, long-term control of the odor problem at

**FIGURE 13.5** Full dome covers (Photo by Gayle Van Durme).

**FIGURE 13.6** Full flat covers (Photo by Alicia D. Gilley).

its source will be necessary. Masking and counteraction are usually applied to the air outside of buildings and enclosures or to odorous air discharged through building exhaust vents, including pumping stations. For example, masking or counteraction chemicals can be applied to a building exhaust system for short-term odor control until a permanent odor-control technology is installed or the odor problem is controlled at the source. Neither masking nor counteraction is completely effective if several

**FIGURE 13.7** Flat fabric covers (Photo courtesy of ILC Dover).

odorous compounds exist or if their threshold concentrations vary. Regardless of the application, neither technique should ever be used to cover the presence of potentially dangerous gases.

The effectiveness of masking is difficult to predict because of varying odor characteristics and changing weather conditions. Masking agents usually consist of organic aromatic compounds such as heliotropin, vanillin, eugenols, benzyl acetate, and phenylethyl alcohol. Masking agents, used primarily where the level of odor is relatively low, always increase the total odor level. Without any chemical reaction, the individual constituents of the odor remain unchanged. The main advantages of masking agents are their low cost and nonhazardous nature. A disadvantage is their tendency to separate from the odor downwind. Sometimes, the fragrance is as offensive as the original odor. Applying high concentrations of masking agents for short periods of time provides a simple method to track and identify a specific odor source in the affected area of the community.

**VAPOR-PHASE CONTROL TECHNOLOGIES.** This section provides a general review of vapor-phase treatment technologies that are appropriate for wastewater applications. The main types of vapor-phase technologies are described in the following paragraphs. These include chemical wet scrubbers, activated carbon adsorption, biofiltration, biotrickling filters, and activated sludge treatment.

***Chemical Wet Scrubbers.*** Chemical wet scrubbers, shown in Figure 13.8, operate by passing the odorous air through a chemical solution spray to remove the odorous constituents prior to discharge to the ambient air. The odorous gas is absorbed in the liquid phase, followed by a chemical reaction. The effectiveness of a chemical scrubber is dependent on adequate liquid–gas contact and proper maintenance of the chemical environment.

Wet scrubbers use pH-controlled absorption and chemical oxidation to remove odorous compounds from an air stream. A variety of chemical scrubbing solutions can be used, including sodium hydroxide (caustic), sodium hypochlorite (bleach), chlorine dioxide, hydrogen peroxide, and potassium permanganate. Acidic scrubbing solutions are typically used to remove ammonia and amines, while caustic solutions are used for hydrogen sulfide. Sodium hypochlorite is the most commonly used oxidant to treat $H_2S$ and other odorous compounds prevalent at wastewater applications. If ammonia and/or mercaptans and hydrogen sulfide must be removed, a two-stage scrubbing system is usually implemented using caustic and sodium hypochlorite or another oxidant.

***Activated Carbon Adsorption.*** Another type of air emission treatment commonly used for wastewater applications is activated carbon adsorption, shown in Figure 13.9.

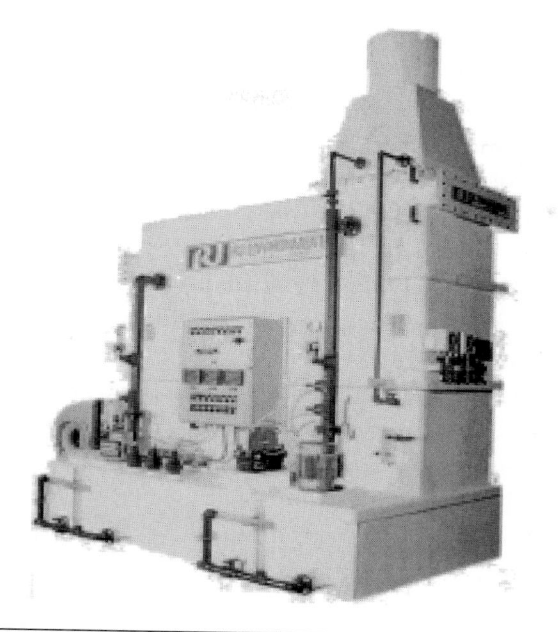

**FIGURE 13.8** Chemical wet scrubber (Courtesy of USFilter–RJ Environmental products).

**FIGURE 13.9** Activated carbon adsorption (Photo by Alicia D. Gilley).

Activated carbon is used to remove pollutants by surface adhesion. The highly porous structure of the carbon supplies a large surface area per unit volume. Systems typically consist of a stainless steel or fiberglass vessel containing a single bed or dual beds of granular activated carbon, through which the odorous air is discharged. Conventional carbon units are sized for face velocities of 203 to 380 mm/s (40 to 75 ft/min) and typically use a media depth of 0.9 m (3 ft).

There are four main types of carbon media: virgin, caustic impregnated, water washable, and high-capacity carbon. Virgin carbon has a greater capacity for removing VOCs than impregnated carbon. It also removes $H_2S$, but has about one-third the capacity of impregnated carbon. Caustic impregnated carbon has a greater $H_2S$ capacity, but a diminished ability to remove other odorous compounds. Caustic carbon can be chemically regenerated in situ, but this is not recommended because of chemical handling concerns. Water washable media has a lower initial $H_2S$ capacity than caustic carbon, but it can be regenerated multiple times with water. Water washable media costs more than caustic carbon initially, but it is more cost-effective over the long term. High capacity carbon has a very high $H_2S$ removal capacity, about double that of caustic carbon. It is designed to be used once and replaced.

***Biofiltration.*** Biofiltration, shown in Figure 13.10, is a biological process using soil, compost, or other media as a substrate for microbes that remove odorous contaminants from an air stream as it travels through the media. Sufficient residence time must be provided for microbes to accomplish effective contaminant treatment, so biofilters must use a low air velocity. For $H_2S$ removal, typical residence times range from 40 to 60 seconds. Longer residence times may be necessary for airstreams containing high concentrations. Because the biodegradability of compounds differs, some compounds

**FIGURE 13.10** Biofiltration (Photo by Gayle Van Durme).

require several minutes of residence time for effective removal. Biofilters can also be designed to remove odorous VOCs and specific compounds.

The key elements in biofilter design are media, air distribution, and moisture. Biofilter media can be composed of organic materials, such as wood chips, compost, soil, peat moss, or some combination of these. Improper media can compact or become too wet, whereas a properly formulated media will last longer. Media porosity is important to minimize the head loss in the bed. Bed depths are typically 0.9 to 1.5 m (3 to 5 ft) with face velocities of 10 to 40 mm/s (2 to 8 ft/min). The low velocity results in very large footprint. Organic biofilter media degrade during use and must be replaced. The lifecycle is dependent on contaminant concentration, but typically ranges from 2 to 3 years. Inorganic media are also available through several vendors and provide a longer media life of 10 years. The media are approximately three times the cost of organic media, but provide a much longer lifecycle.

Because of their larger size and labor-intensive construction, biofilters can cost significantly more to build than wet scrubbers or activated carbon units. The higher initial costs may be offset by lower operational costs, depending on the $H_2S$ concentrations being treated. Pre-engineered biofilters, which reduce labor requirements for construction, are becoming more prevalent.

***Biotrickling Filters.*** In general, biotrickling filters, shown in Figure 13.11, consist of a containment vessel with some type of inorganic media to support microbial growth. Foul air is introduced at the bottom of the media while water is sprayed over the top of

**FIGURE 13.11** Biotrickling filter (Photo by Gayle Van Durme).

the media. The water flows downward through the media and provides a moist environment that encourages microbial growth. The water also serves to flush the metabolic byproduct of sulfuric acid from the system. At wastewater treatment plants, non-potable water can be used to supply sufficient nutrient for the microbes. For collection system applications, potable water is used and supplemental nutrients are required to support the bacteria.

***Activated Sludge Treatment.*** The activated sludge treatment (AST) odor control technique routes foul air through the air diffusers in an aeration basin system for treatment by the activated sludge microorganisms. One issue with AST is that the foul-air flows requiring treatment must match the aeration basin process air requirements. Activated sludge treatment offers several advantages: It can perform well even at high concentrations of $H_2S$; the system is very simple to operate and maintain; it does not require storage, handling, and containment of hazardous chemicals; and it does not have the visual impact that a tall scrubber and exhaust stack present. One of the disadvantages of AST is the corrosion-resistant material required for the aeration basin piping and blowers.

# ODOR CONTROL STRATEGIES FOR THE OPERATOR

Previous sections presented basic information on odor problems, odor development in wastewater collection and treatment systems, and available odor control methods and technologies. This section presents an odor control strategy for the plant operator, using previously presented information coupled with recent operational experiences in solving odor problems at wastewater treatment facilities.

**OPERATOR'S APPROACH TO SOLVING AN ODOR PROBLEM.** When an operator first becomes aware of an odor problem, it is important to evaluate all simple solutions quickly. Otherwise, the release of odors will have adverse effects on-site and possibly in the community. Generally, the operator's available resources restrict corrective approaches to improving housekeeping practices, implementing process or operational changes, or adding chemicals to the wastewater and solids streams. The operator should immediately determine the source of the odors and the specific odor-causing compounds. These sources must be determined before selecting a solution for the problem. Liquid and solid process control measures are listed in Tables 13.8 and 13.9, respectively. The sequence of steps to be considered when approaching an odor control problem is illustrated in Figure 13.12.

Table 13.10 presents a summary of odor control methods used by operators of several wastewater treatment facilities and lists specific effects and problems. The list pre-

TABLE 13.8 Liquid process operational emissions control (Rafson, 1998).

| Process | Problems | Control measures |
|---|---|---|
| **Preliminary treatment** | | |
| Coarse bar screens<br>Fine bar screens | Influent sulfide and VOCs are stripped by turbulence inherent to these processes. | Upstream chemical addition. Recycle return activated sludge (RAS) to headworks.<br>Containment and ventilation to a vapor-phase control system. |
| Parshall flumes | Turbulence causes stripping. | Use ultrasonic or magnetic flow measurement.<br>Containment and ventilation to a vapor-phase control system. |
| Grit basins | Aerated grit basins induce stripping. | Vortex and horizontal-flow grit basins are less turbulent.<br>Containment and ventilation to a vapor-phase control system. |
| **Primary treatment** | | |
| Primary clarifiers | Sulfide formed in basins during holding. Sulfide and VOCs stripped at weirs. Sulfide forms in settled solids. | Remove unneeded basins from service. Raise water level in flume to decrease weir drop. Pump sludge more often. Avoid cosettling of sludge. Add iron salts directly or upstream.<br>Containment and ventilation to a vapor-phase control system. |
| Flow equalization basins | Odor from residual solids in flow equalization basins. | Install collection and removal equipment and flush solids with high-pressure hoses after each basin dewatering. |
| **Secondary treatment** | | |
| Trickling filters<br>Rotating biological contactors (RBCs) | Influent sulfides stripped at distributors. Sulfide formed when overloaded or oxygen deficient. | Add iron salts upstream. Limit loading. Provide power ventilation. Slow distributors or increase wetting rate to maintain a thin aerobic film. |
| Aeration basins | Influent sulfide and VOCs stripped at head of basin. Sulfides form when oxygen deficient. | Decrease aeration at head of basin. Fine-bubble diffusers cause less stripping than coarse bubble. Pure oxygen causes lowest odor emissions. |
| **Disinfection** | | |
| Chlorination | Volatile chlorinated by-products formed during disinfection. | Use automatic controls to pace chlorine dosage to flow. Convert to ultraviolet disinfection. |
| **Advanced treatment** | | |
| Effluent filters | VOCs formed by chlorine used during backwashing to clean and control algae. | Minimize filter backwashing. Cover filters to block sunlight and control algae. |

**TABLE 13.9** Solids process operational emissions control (Rafson, 1998).

| Process | Problems | Control measures |
|---|---|---|
| **Thickening** | | |
| Gravity thickeners | Co-thickening biological and primary sludge causes sulfide generation. Long detention under anaerobic conditions is problematic. | Avoid co-thickening, if multiple basins are available. Use direct chemical treatment to reduce sulfide formed during thickening. Provide containment and ventilation to a vapor-phase control system. |
| Dissolved air flotation | Aeration strips sulfide and odors from sludge. | Use chemical pretreatment to remove sulfide from sludge before processing. Provide containment and ventilation to a vapor-phase control system. |
| **Dewatering** | | |
| Belt presses | Pressing strips sulfide and VOCs from feed sludge into belt press room. | Potassium permanganate or hydrogen peroxide will treat sulfide and other odorous VOCs. Provide containment and ventilation to a vapor-phase control system. |
| **Stabilization** | | |
| Anaerobic digesters | The $H_2S$ formed in the process corrodes combustion equipment. Air quality is a concern, because $H_2S$ is converted to sulfur dioxide during combustion. | Maintain proper temperature and pH in the process. Add iron salts directly to the digester, at headworks, or at primary clarifiers. |
| Aerobic digesters | Odorous compounds form when process is overloaded or oxygen-deficient | Provide adequate aeration and mixing to maintain aerobic conditions. Feed at uniform organic loadings. |
| Lime stabilization | Ammonia is released due to high pH. | Vent ammonia to outside unless concentrations are very high or the site is in a sensitive area. |
| Composting | Decomposition of organics at high temperature produces a wide variety of odorous compounds. | Aerate piles. Draw air down through pile and treat. Avoid overly wet sludge. Use wood ash as an amendment. Apply pile surface treatment chemical. Provide containment and ventilation to a vapor-phase control system. |
| **Storage** | | |
| Short-term | Storage of combined biological and primary sludge generates sulfide. | Avoid storing combined sludge. Provide mixing to maintain aerobic conditions. Add iron salts directly to storage tanks to treat sulfide formed. Provide containment and ventilation to a vapor-phase control system. |
| Long-term | Lime-stabilized biosolids may become odorous in long-term storage. Breakdown of storage piles releases odor. | Add supplemental lime to maintain pH during long-term storage. Limit open face exposure, and cover when possible. Break down piles under favorable weather conditions. |
| Land application | Odor is released during application of biosolids to a large land area. | Separate high odorous biosolids for further processing or other disposal. Use subsurface injection to minimize odor. |

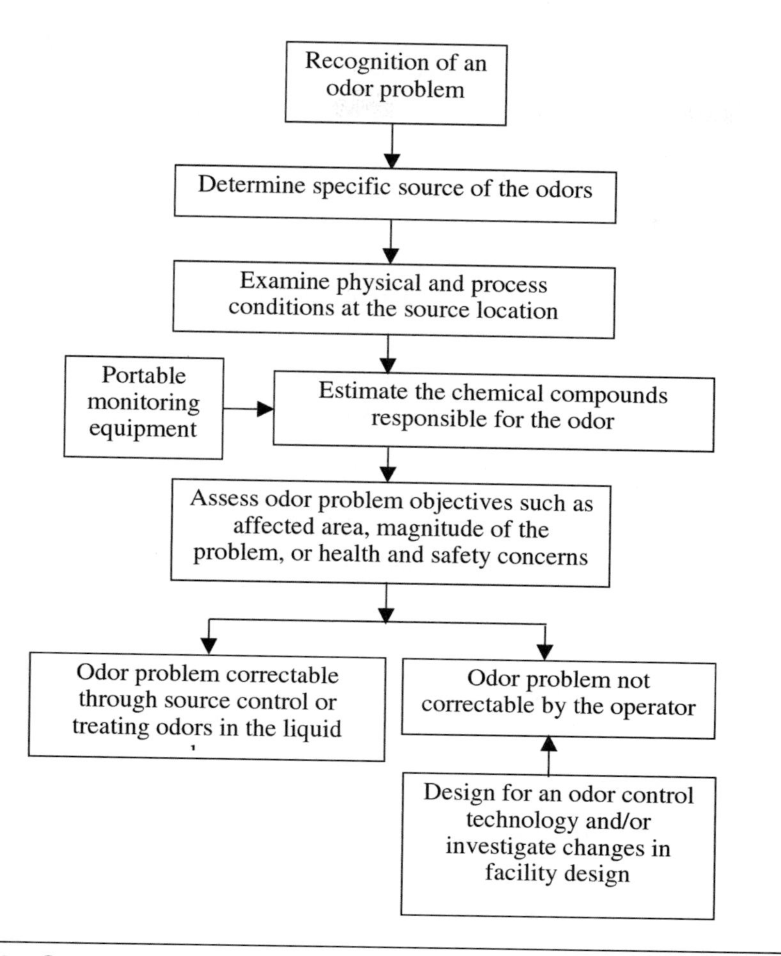


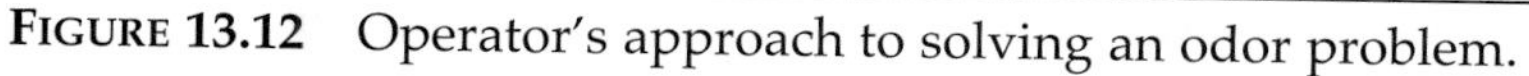

**FIGURE 13.12** Operator's approach to solving an odor problem.

sents the results-oriented methods controlled by the operator. The methods requiring design and significant capital investments include biofilters, chemical wet scrubbing, activated carbon adsorption, and combustion.

**MONITORING.** Many odors at wastewater treatment plants can be identified using portable monitoring equipment, described in Table 13.11, combined with the operator's sense of smell. Regular monitoring of treatment processes can prevent many odor releases as well as provide valuable information on operating procedures. It is important to continually monitor wind direction and speed and maintain a record for later reference, particularly in relation to odor complaints. When the wind is not blow-

**TABLE 13.10** Summary of odor control technology applications at wastewater treatment facilities.

| Methods | Effects | Problems |
|---|---|---|
| **Operating practices** | | |
| • Industrial process changes | | |
| ○ Lower waste temperature | Hydrogen sulfide evolution much retarded | |
| ○ Pretreat to remove odorous organics | | |
| • Collection system | | |
| ○ Mechanical cleaning | Hydrogen sulfide reduction | |
| ○ Aeration | | |
| ○ Ventilation | | |
| • Grit chamber | | |
| ○ Daily grit washing | General odor reduction | |
| • Primary clarifiers | | |
| ○ Increase frequency of solids and scum removal | General odor reduction | |
| • Aeration tanks | | |
| ○ Remove solids deposits | | |
| ○ Increase aeration to maintain dissolved oxygen at 2 mg/L | General odor reduction | |
| • Trickling filters | | |
| ○ Increase recirculation rate | | |
| ○ Keep vents clear | General odor reduction | |
| ○ Check underdrains for clogging | | |
| • Anaerobic digesters | | |
| ○ Check waste gas burner | General odor reduction | |
| ○ Relief valves should close tightly | | |
| • Aerobic digesters | | |
| ○ Maintain constant loading | General odor reduction | |
| ○ Maintain adequate aeration | | |
| **Liquid-phase control alternatives—Chemical addition** | | |
| • Ozone | Oxidizes water-insoluble odorants into water-soluble projects | Requires onsite regeneration |
| • Iron | Controls slime growth; precipitates sulfide; enhances settling | Increases solids |
| • Nitrates | Inhibits production of sulfides | Costly |
| • pH adjustment | | |
| ○ Alkali: NaOH | pH 8 hinders bacterial growth in sewers and retards evolution of hydrogen sulfide. | |
| ○ Acid: HCl or $H_2SO_4$ | Acid combines with basic ammonia and amines | |
| • Chlorine (gas and hypochlorite) | Inhibits growth of sulfate-reducing bacteria in sewers; oxidizes hydrogen sulfide and ammonia | |
| • Potassium permanganate | Reacts with sulfide and other organics to reduce odors | |

TABLE 13.11 Portable monitoring equipment for odor control.

| Equipment | Monitoring use | Comments |
|---|---|---|
| **Gas-phase monitoring** | | |
| Oxygen meter | Oxygen (percent volume in air) | Applicable to confined space entry monitoring; replace sensor yearly. |
| Hydrogen sulfide meter | Hydrogen sulfide gas, ppm or ppb | Meter should have calibration capability with a known gas. |
| Colorimetric tubes | Many organic and inorganic compounds, including hydrogen sulfide, ammonia, and mercaptans | Some analyses subject to interference, applicable to many different concentration levels. |
| Combustible gas meters | Total combustibles as a lower explosive limit | Applicable to confined space entry monitoring. |
| **Liquid-phase monitoring** | | |
| Oxygen meter | Dissolved oxygen (mg/L) | Monitor aerobic conditions in wastewater. |
| pH meter | Hydrogen ion (pH units) | pH levels control the form of hydrogen sulfide, ammonia and other odorous molecules. |
| Oxidation–reduction potential meter | Oxidation–reduction potential (mV) | Indicates if a wastewater is in an oxidizing or reducing state. |
| Portable wastewater testing kits | Dissolved sulfide (mg/L) | Colorimetric comparison. |

ing in the direction of the community, or at certain times of the day or days of the week, the detected odor may come from a source other than the wastewater treatment plant.

**TIME OF DAY, WEEK, OR YEAR.** Time is important when planning potential odor-causing O&M procedures at the wastewater treatment plant or conducting short-term operations when a higher degree of odor control must be provided. The scheduled work time will influence the number of community members who detect an odor and their location with respect to the wastewater treatment facility. Most people work during the day and follow established traffic patterns. Thus, potentially odorous work should not be planned for the early morning or late afternoon and, if possible, should be scheduled during the off-shifts. The day of the week will determine whether people are at work or are outside during the weekend or on a holiday. Yearly time periods will show traditional vacation periods, standard industrial shutdowns, or planned community events for the general public.

**WEATHER CONCERNS.** Regardless of how efficiently a wastewater treatment plant is operated, local weather conditions, particularly seasonal characteristics, will greatly affect how and where odors are perceived by the community. Most plants have an odor season that typically coincides with warmer weather. When higher temperatures volatilize larger amounts of odorous compounds from the wastewater, local residents will readily detect odors because they have open windows in their homes and worksites and spend more time outside than they do in colder seasons. High humidity will also increase an odor's persistence and perception.

Inversion conditions with low wind speeds can temporarily reduce the dissipation of odors to the atmosphere and contain them closer to the ground. These conditions often occur before a rainstorm. The earthy, musty smell of a well-operated, activated sludge plant can become a nuisance to local residents if the odors become concentrated above the aeration basins and then are slowly moved by gentle breezes.

**ODOR COMPLAINTS.** The simplest and most common odor evaluation technique determines an odor's characteristics—what does it smell like? Plants with odors that affect the community can establish an odor telephone hotline to receive citizen complaints. Each complaint should be investigated, and an odor-complaint form should be completed. An example of an appropriate complaint form is shown in Figure 13.13. The form will improve public relations efforts and develop the information necessary to determine the odor source. The form includes information on characteristic smells, intensity, time of day, and weather conditions.

The plant also needs a wind direction meter and recorder. While investigating a complaint, the operator can examine the recorder to determine whether the wind direction matches that of the complaint. If not, the odor is probably coming from a source other than the wastewater treatment facility. During potential odor-problem periods, an odor tour through the community during each shift can help to identify problems early and minimize their effect.

## SUMMARY

Odor perception is subjective, and solving odor problems may seem to be more of an art than a science. However, the pathway from problem to solution should follow a logical evaluation process if it is to succeed. Initially, the most important concern is to identify the odor source and characterize it. Ultimately, the odor control method selected will depend on the class of chemical compounds responsible for the odor.

When an odor problem arises, the operator should consider two simultaneous problems: the symptom and the cause. First, the symptom of the problem (the smell)

**Date** ____________ **Time** ____________

**Location of Odor Complaint** ____________

____________

**Smells Like**

| | **Intensity** 1 Very Weak | 2 Moderate | 3 Strong | 4 Very Strong | 5 Unbearable |
|---|---|---|---|---|---|
| ____ 1.Burnt, smoky | | | | | |
| ____ 2.Ammonia like | | | | | |
| ____ 3.Petroleum | | | | | |
| ____ 4.Sweet, solvent-like | | | | | |
| ____ 5.Rotten eggs | | | | | |
| ____ 6.Garlic, onion | | | | | |
| ____ 7.Metallic | | | | | |
| ____ 8.Garbage Truck | | | | | |
| ____ 9.Sharp, biting, acid | | | | | |
| ____ 10.Musty, mouldy grainlike | | | | | |
| ____ 11.Outhouse | | | | | |
| ____ 12.Chemical, disinfectant | | | | | |
| ____ 13.Other (specify) | | | | | |

**Investigation**

**Date** ____________ **Time** ____________

**Comments** ____________

____________

**Name** ____________ **Temperature**

**Wind Direction** ____________ **Wind Speed** ____________

FIGURE 13.13 Sample odor complaint investigation form.

must be addressed immediately with a short-term solution. Typically, this involves adding chemicals, improved housekeeping procedures, or changes in process control or operational procedures. The operator should never mask toxic gas odors that could become a safety problem. In most situations, these short-term actions will correct the problem. If they do not, then more involved situations must be investigated. Generally, long-term solutions can be costly and may require changes in plant design or construction of an odor control technology.

The problems associated with an odor event depend primarily on the affected area. On-site odors may affect employee safety and health or create unpleasant working conditions. Odors that leave the wastewater treatment site can become a community problem and impact public relations. Characteristics such as topography, weather conditions, and the proximity of neighborhoods play a dynamic role in how off-site odors are perceived by the community.

## REFERENCES

American Industrial Hygiene Association (1989) *Odor Thresholds for Chemicals with Established Occupational Health Standards;* American Industrial Hygiene Association: Akron, Ohio.

American Society of Civil Engineers (1989) *Sulfide in Wastewater Collection and Treatment Systems,* Manuals and Reports on Engineering Practice No. 69; American Society of Civil Engineers: New York.

Bishop, W. (1990) VOC Vapor Phase Control Technology Assessment. Water Environment Research Foundation Report 90-2; Water Pollution Control Federation: Alexandria, Virginia.

Campbell, N. A. (1996) *Biology,* 2nd ed.; The Benjamin/Cummings Publishing Company, Inc.: Fort Collins, Colorado; pp 1011–1044.

Divinny, J.; Deshusses, M. A.; Webster, T. (1999) *Biofiltration for Air Pollution Control;* Lewis Publishers: New York.

Gilley, A. D.; Van Durme, G. P. (2002) Biofilter: Low Maintenance Does Not Mean No Maintenance. Paper presented at Water Environment Federation Odors and Toxic Air Emissions Specialty Conference, Albuquerque, New Mexico.

Henry, J. G.; Gehr, R. (1980) Odor Control: An Operator's Guide. *J Water Pollut. Control Fed.,* **52**, 2523.

Jiang, J. K. (2001) Odor and Volatile Organic Compounds: Measurement, Regulation and Control Techniques. *Water Sci. Technol.,* **44** (9), 237–244.

Jenkins, D.; Richard, M. G.; Diagger, G. T. (1986) *Manual on the Causes and Control of Activated Sludge Bulking and Foaming.* Water Research Commission South Africa: Pretoria, South Africa.

Mackie, R. I.; Stroot, P. G.; Varel, V. H. (1998) Biochemical Identification and Biological Origin of Key Odor Components in Livestock Waste. *J. Animal Sci.*, **76**, 1331–1342.

Minor, J. R. (1995) A Review of Literature on the Nature and Control of Odors from Pork Production Facilities. Executive Summary for the Odor Subcommittee of Environmental Committee of the National Pork Producers Council; Des Moines, Iowa.

Moore, J. E., *et al.* (1983) Odor as an Aid to Chemical Safety: Odor Thresholds Compared with Threshold Limit Values and Volatilities for 214 Industrial Chemicals in Air and Water Dilution. *J. Appl. Toxicol.*, **3**, 6.

Rafson, H. J. (1998) *Odor and VOC Control Handbook*; McGraw Hill: New York.

U.S. Environmental Protection Agency (1985) Odor and Corrosion Control in Sanitary Sewage Systems and Treatment Plants; EPA-625/1-85-018; Center for Environmental Research Information: Cincinnati, Ohio.

Van Durme, G. P.; Gilley, A. D.; Groff, C. D. (2002) Biotrickling Filter Treats High $H_2S$ in a Collection System in Jacksonville, Florida. Paper presented at Water Environment Federation Odors and Toxic Air Emissions Specialty Conference, Albuquerque, New Mexico.

Van Durme, G. P.; Berkenpas, K. (1989) Comparing Sulfide Control Products. *Ops. Forum*, **6** (2), 12.

Water Environment Federation (1994) *Safety and Health in Wastewater Systems*, Manual of Practice No. 1; Water Environment Federation: Alexandria, Virginia.

Water Environment Federation (1998) *Design of Municipal Wastewater Treatment Plants*, Manual of Practice No. 8; Water Environment Federation: Alexandria, Virginia.

Water Environment Federation (2000) Odor and Corrosion Prediction and Control in Collection Systems and Wastewater Treatment Plants. *Proceedings of the 73rd Annual Water Environment Federation Technical Exposition and Conference Workshop*; Anaheim, California, Oct 14–18; Water Environment Federation: Alexandria, Virginia.

Water Environment Federation (2004) *Control of Odors and Emissions from Wastewater Treatment Plants*, Manual of Practice No. 25; Water Environment Federation: Alexandria, Virginia.

# Chapter 14

# Integrated Process Management

# INTRODUCTION

**OVERVIEW.** Individual processes cannot be operated in a vacuum. The operator must be cognizant of the interrelationships among the equipment, biological kinetics, and chemistry that make up a process and the potential effects of upstream processes and effects on downstream processes. Note that specific equipment control strategies are not discussed in this chapter. An expansive discussion of this subject is presented in the Water Environment Federation (WEF) Special Publication *Automated Process Control Strategies* (1997) and is not repeated here.

A wastewater treatment plant (WWTP) consists of many separate processes. Process performance is related primarily to flow characteristics (flowrate, biochemical oxygen demand [BOD], total suspended solids [TSS], nitrogen, and phosphorus) and equipment operation. To optimize operation of a WWTP, the operator must be knowledgeable about the individual treatment units and familiar with the interrelationships that exist among the processes. For example, optimization of a primary clarifier would result in more solids and potentially higher quantities of BOD removed. This would reduce the oxygen demand in the aeration basins, lowering costs by lowering oxygen requirements (and power costs) and increasing the ratio of primary sludge to secondary sludge, which may also improve dewatering characteristics. On the other hand, though the primary treatment system would be optimized, subsequent systems may suffer adverse effects. For example, thickening and digestion facilities may be overloaded or inadequate quantities of food may be available for the secondary system, particularly facilities that practice denitrification or biological phosphorus removal. This will require adjustments in sludge pump settings, mixed liquor concentrations, blower output settings, etc. The plant is interconnected and interdependent such that a change in one area affects other processes, thus necessitating an integrated approach. Many plants are large and an operator may work in only one part. It is easy to overlook the operator's contribution to plant performance as a whole. Nevertheless, operators' actions can and do affect other processes and overall facility performance. Operators should be aware of how other processes affect their work area and the way the process they operate affects other areas.

**INTEGRATED PROCESS MANAGEMENT.** It is not enough to understand how a pump operates, a compressor runs, or even how a specific process works. The operator must know how a process works and how each process in turn affects every other. Consequently, the operator must be intimately familiar with every process and auxiliary facilities such as chemicals, scrubbers, and energy recovery systems. An example approach for assessing process interrelationships follows. For demonstration purposes, a schematic of a secondary treatment facility is presented as Figure 14.1. Recycle flows and chemical-addition points are included.

Based on the schematic, a table can be developed to consider the effects on and from each process. The performance of one process and the characteristics of its sidestreams affect subsequent processes. Within the plant, each process has an interrelationship with other processes that may not be readily apparent. For example, anaerobic digestion typically bears the brunt of poor dissolved air flotation (DAF) performance resulting from low solids concentrations. The quality of the DAF return overflow is often overlooked but, nonetheless, can have a sizable effect on overall plant performance. Recognizing the importance of these relationships, Table 14.1 highlights the effect that one process has on another. This table is not meant to be definitive, but presents some of the more relevant effects to highlight the importance of the process interactions.

To create a table like Table 14.1, the operator should take the following steps. Start at the beginning of the plant and choose one process to evaluate. Write the process

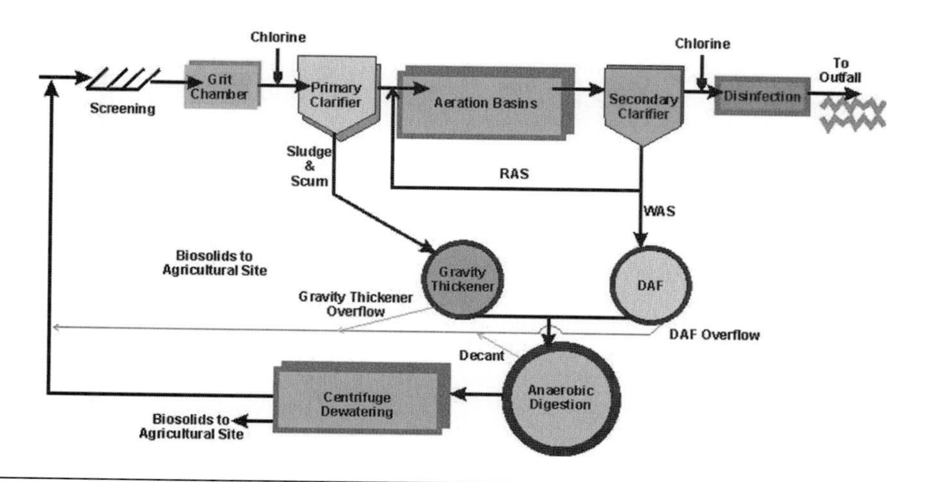


**FIGURE 14.1** Plant schematic (DAF = dissolved air flotation, RAS = return activated sludge).

**TABLE 14.1** Integrated process management considerations.

| Process | Effects from upstream processes | Downstream process effects |
|---|---|---|
| Screenings | Collection system constituents (flow, total suspended solids, sulfides, pH) may create an overload or caustic condition.<br>High flows may tax units in service, add additional unit, or open manual bar screen. | Large objects may clog pipes and pumps. Tank volumes (digesters in particular) may be reduced and detention times decreased. |
| Grit removal | High effluent flows may bring increased amounts of grit and sand.<br>Excessive screening may settle on diffuser equipment and diminish effectiveness. | Channel and tank volumes may be reduced by the accumulation of grit.<br>May plug up channel air diffusers.<br>Excessive grit content may scour, abrade, or obstruct downstream process equipment or piping.<br>May cause excessive prechlorination rates if insufficient grit is removed.<br>Low solids content in dewatered grit may create handling and disposal problems. |
| Primary clarification | High solids or hydraulic loadings may be caused by in-plant return flows from mechanical dewatering, DAF overflow, gravity thickener overflow, or digester decant, etc.<br>Tank volumes may be reduced by the accumulation of grit.<br>Excessive grit content may scour, abrade, or obstruct process equipment and pipes. | Primary clarifier constituents (BOD, ammonia nitrogen, and phosphorus) determine level of mixed liquor required for the secondary system.<br>Primary sludge concentration greater than 2% will settle better in the gravity thickener. Concentrations greater than 5% will tax centrifugal pumps. May cause septicity and generate organic acids.<br>Septic sludge containing hydrogen sulfide may increase digester gas |
| Activated sludge | High influent flows may wash out biomass.<br>Inadequate screening or grit removal may block air diffusers.<br>Primary clarifier effluent constituents (BOD, ammonia nitrogen, and phosphorus) determine level of mixed liquor required for the secondary system.<br>High solids loadings present in the collection system or in plant recycle flows (DAF overflow, gravity thickener overflow, Centrate, or digester decant). | Low dissolved oxygen will cause filamentous organism growth and poor solids separation in the secondary clarifier, diminishing effluent quality.<br>Solids retention times greater than 10 days may be needed for full nitrification. Partial or full nitrification may cause solids separation problems unless secondary clarifier was designed for this mode. |

**Table 14.1** Integrated process management considerations (*continued*).

| Process | Effects from upstream processes | Downstream process effects |
|---|---|---|
| Activated sludge *(continued)* | Excessive detention time in the primary clarifier may create septic conditions and formation of organic acids, which could cause excessive filamentous growth.<br>Poor scum removal in primary clarifier may contribute to filamentous growth. | |
| Secondary clarifier | High effluent flows may wash solids into the effluent.<br>Poor solids separation caused by insufficient or excessive dissolved oxygen, filamentous organisms, hydraulic or solids overload.<br>*Nocardia* may cause foaming removal problems. | Poor solids separation may increase BOD, TSS, and turbidity, creating disinfection problems.<br>Excessive waste activated sludge rates may overload the DAF. |
| Dissolved air flotation | Excessive waste activated sludge rates may overload the DAF and increase solids in the return overflow. | High recycled return overflow constituents may increase primary and secondary treatment loadings.<br>Low thickened sludge concentration will diminish digester performance by hydraulically overloading the unit or making it difficult to maintain temperatures. |
| Gravity thickener | Septic primary sludge will create settling problems, which may increase solids and sulfides in the return overflow. | High recycled return overflow constituents may increase primary and secondary treatment loadings.<br>Low thickened sludge concentration will diminish digester performance by hydraulically overloading the unit or making it difficult to maintain temperatures.<br>Septic sludge containing hydrogen sulfide may increase digester gas hydrogen sulfide concentrations. |
| Anaerobic digestion | Tank volume may be reduced by the accumulation of grit or rags.<br>Low thickened sludge concentration will diminish digester performance by hydraulically overloading the unit or making it difficult to maintain temperatures. | Low solids concentration increases dewatering equipment operation and chemical usage.<br>Decant constituents (BOD, COD*, TSS, hydrogen sulfide) are recycled to the head of the plant and may increase plant loadings. |

**TABLE 14.1** Integrated process management considerations (*continued*).

| Process | Effects from upstream processes | Downstream process effects |
|---|---|---|
| Anaerobic digestion (*continued*) | Septic sludge containing hydrogen sulfide may increase digester gas hydrogen sulfide concentrations.<br>Excessive thickened sludge feed rates may hydraulically overload the unit, making it difficult to maintain temperature and diminish detention time to less than required for stabilization. | |
| Mechanical dewatering | Poor screenings and grit removal may damage equipment. Large objects may clog pipes and pumps.<br>Low solids concentration increases dewatering equipment operation and chemical usage. | High centrate constituents are recycled to the head of the plant and increase plant loadings.<br>Low cake solids may hamper transportation and subsequent application to agricultural or disposal sites. |
| Disinfection | Poor solids separation in the secondary clarifier may increase BOD, TSS, and turbidity, creating disinfection problems. | Receiving water quality may be degraded if low dissolved oxygen is present, coliform counts are high, or BOD/TSS concentrations are excessive. |

*COD = chemical oxygen demand.

name in the first column. Look at all possible flows and conditions entering the process. Enter in the second column the consequences that each of these flows might have on that specific process. Finally, determine what the ramifications of process failure would be on downstream processes. Enter those in the third column. Do the same for each process.

The elements of creating an integrated process management strategy include

- Developing and implementing unified process control plans,
- Reviewing and evaluating process performance and control data, and
- Assessing changes in process control strategies.

Plant complexity frequently determines the tools necessary to implement the most appropriate operational strategy for given circumstances. Consequently, many techniques are outlined in this chapter with detailed explanations.

## DATA COLLECTION

The basis of good operational decisions is built on a foundation of sound information. At a minimum, this includes flow measurement, analytical data, process control data, and maintenance reports.

**FLOW MEASUREMENT AND QUANTIFICATION.** Flow measurement facilities enable accurate information to be obtained concerning flowrates and volumes. When combined with analytical data, flow measurements are used to quantify mass constituents that are being treated and removed and are an essential tool for good plant control and optimization. Adding flow meters is not an expensive investment and is well worthwhile. Pump speed is a poor substitute for flow measurement.

Loading to the plant varies over the time. Consequently, flows should be measured continuously so that the total volume at a certain time period (day, month, or year) can be determined. Further, wastewater samples should be taken frequently as flow-weighted composite samples. If that is not possible, the monitoring program has to be optimized to determine the most representative time for sampling to occur. Typically, those samples may be taken at high flow, as that is when the processes are under the heaviest loads. The optimum sampling periods may differ between days of the weeks or seasonally.

An application of the use of flow measurement and quantification would be in process management, a change in the feed sludge volumes (thickened primary and waste activated sludge), or the ratio of the two sludges that would affect digestion, dewatering, and biosolid reuse operations.

**ANALYTICAL ANALYSIS.** Best results are obtained with sound sampling practices and techniques. To obtain a representative sample, collect them at points where there is a high turbulent flow to ensure good mixing. If obtaining samples from a pipeline, the line should be flushed before taking the sample. If the flow is not mixed properly, the suspended solids and other constituents may be unequally distributed in the water column, which may cause considerable error. Sampling location, samplers, and sampling apparatus must be cleaned regularly to avoid excess contamination by sludge, bacterial film, etc. Chapter 17, Characterization and Sampling of Wastewater, should be reviewed for analytical and sampling practices used in the treatment process.

Samples may be taken as grab or composite and then transported to an on-site laboratory or sent to a commercial laboratory for analyses. Some analyses such as dissolved oxygen and chlorine residual must be taken in situ. Many constituents have specially designed analyzers that provide continuous analyses that can be transmitted to a central location and displayed, stored in a central database, used in part of the system control logic, or simply monitored. The WEF publication, *Instrumentation in Wastewater Treatment Facilities* (1993), provides instruction in the fundamentals of instrumentation and addresses specific sensors, theories of operation, and distinctions between online and laboratory instrumentation.

**MAINTENANCE REPORTS.** Maintenance and plant operations are inseparable. Regular review of preventive, corrective, and predictive maintenance reports can be instrumental in maintaining effective plant operation, reducing costly repairs, and in reducing unscheduled downtime. The well-operated plant should schedule maintenance on structures, gates, valves, pumps, electrical devices, and instrumentation. A review of maintenance reports also serves as a tool for equipment replacement.

Information on equipment may be drawn from plant control systems that have the capacity to track run times. Review of process graphs may also serve as an indicator of equipment performance. A sudden drop-off in pump capacity can be an indication of a partially blocked (closed) line or valve or a pump obstruction.

**PROCESS CONTROL DATA.** Process control data consist of direct readings (such as flows and analytical data) or generated data such as organic and hydraulic loadings, feed rates, and detention times. These values are tracked over time for changes or compared design values for excessive or insufficient values. It is important to look at every piece of information and determine its value. For example, pump revolutions per minute may not be useful by themselves; however, if that same piece of information is expressed as a percent of maximum, the operator may be alerted to an overload condition. Many plant control and monitoring systems (i.e., supervisory control and data acquisition [SCADA]) are programmed to collect and generate some of this information. The SCADA system could calculate a flowrate if the revolutions per minute and pump capacity curve were input values. Comparing the calculated flowrate with the flow meter would also act as an operational check that would send an alarm if the difference between the two readings becomes excessive. The SCADA system can also take care of the tedious, but important, jobs such as graphing information. Once graphs are defined, the SCADA system can be programmed to automatically pop up operational screens or activate an alarm when unusual trends are detected.

Computer modeling, once restricted to the design, research, and development arena, is creeping into many plant control systems. Process modeling and computer-based technologies offer a predictive component to confirm that current operational strategies are appropriate. These same models can act as simulators that allow the operator to run "what-if" scenarios for units that are taken out of service or to predict the consequences or the effect events would have on current operation. Once added to the plant control system, the model is calibrated against actual plant operation. This provides improved diagnosis, forecasting, and control of wastewater treatment processes. The result is improved plant efficiency, leading to large savings in capital and operating costs.

## STANDARD OPERATING PROCEDURES

Standard operating procedures (SOPs) are detailed definitions of each segment of the process control plan. These procedures, developed by plant personnel or outside consultants, require an objective, independent thought process. The SOP format should encourage new ideas and creative solutions. Innovation is necessary and the "we have always done it that way" syndrome should be avoided.

An example of a SOP written for a certified operator is presented below in a simple checklist form.

### EXAMPLE STANDARD OPERATING PROCEDURE—MECHANICAL BAR SCREEN PUMPING STATION NORMAL OPERATION.

✔ The number of screens in operation is dependent on the flow to the pumping station. Screens SC-1 and SC-2 can each handle flows up to approximately 6.97 $m^3/s$ (159 mgd), whereas screens SC-3 and SC- 4 can each handle flows up to approximately 6.22 $m^3/s$ (142 mgd).

✔ During wet weather, the screens in service should normally be operated continuously in the manual mode by placing their control (Hand-Off-Auto) selector switches in the **Hand** position and pressing their **Forward** pushbuttons.

✔ During dry weather, the screens in service should normally be operated in the automatic mode by placing their control (Hand-Off-Auto) selector switches in the **Auto** position, thus allowing them to be under timer-controlled operation. The cycle timer should be set to operate the screen rake mechanism for a duration that allows the rake to travel 2.5 revolutions.

✔ If station flows permit, the screen channels should be alternated to equalize wear on equipment and maintain standby equipment in proper operating condition. The sluice gate at the inlet of the standby screen channels should be closed.

 Screening equipment and screen channel walls should be washed down with a hose whenever a channel is taken out of operation (when channels are alternated).

Some organizations prefer a more intensive SOP that is more detailed. An example of this is shown below.

**EXAMPLE DETAILED STANDARD OPERATING PROCEDURE.** The following is an example detailed SOP.

**Big Red Valley Sewer District – Ira A. Sefer WWTP**
**STANDARD OPERATING PROCEDURE #1 – 600 - 0002**

## TITLE

Aeration/Clarification Basins—Sludge Wasting

## GENERAL DESCRIPTION

An activated sludge secondary treatment process has been designed to treat the wastewater. The biological process is designed to remove carbonaceous BOD and the physical process is designed to remove total suspended solids. The discharge limits are as follows:

| Effluent limits | | |
|---|---|---|
| Characteristic | Weekly | Monthly |
| BOD, mg/L | 40 | 30 |
| TSS, mg/L | 40 | 30 |

The treatment process is described as an aeration/clarification basin. The basin is configured as an oxidation ditch and the clarifier portion is incorporated to the oxidation ditch. The clarifier is known as an intrachannel clarifier.

The aeration/clarification basin creates the environment necessary to allow biological reactions to occur and then allow settling to take place ending with a treated wastewater.

The components of the treatment process include an A/C splitter box, four aeration/clarification basins, waste activated sludge pumping station, scum wasting system, and aeration blowers.

Each basin is designed to treat an average daily flow of 10 mgd * and can handle a peak hydraulic loading of 20 mgd. The total average daily flow is 40 mgd, with a total peak hydraulic capacity of 80 mgd.

*mgd × 3785 = $m^3/d$.

## EQUIPMENT AND SYSTEM OPERATING CHARACTERISTICS

| | |
|---|---|
| Splitter box | |
| Number | 1 |
| Designation | A/C splitter box |
| Sluice gates | |
| Number | 4 |
| Designation | SGE – 24, SGE – 25, SGE – 26, SGE – 27 |
| Aeration/clarification basins | |
| Number | 4 |
| Designation | basin no. 1, basin no. 2, basin no. 3, basin no. 4 |
| Blower | |
| Smaller capacity blowers | |
| Number | 3 |
| Designation | TBU – 1, TBU – 2, TBU – 3 |
| Type | multistage centrifugal |
| Capacity, scfm | 7560 |
| System operating pressure, psi | 8.5 |
| Motor, hp | 400 |
| Motor, V | 4160 |
| Larger capacity blowers | |
| Number | 2 |
| Designation | TBU – 4, TBU – 5 |
| Type | single-stage centrifugal |
| Capacity, scfm | 18 000 |
| System operating pressure, psi | 8.5 |
| Motor, hp | 800 |
| Motor, V | 4160 |
| Diffuser assemblies | |
| Number per chamber | 3 |
| Capacity per assembly, scfm | 105 |

| | |
|---|---|
| Air flow rate per chamber, scfm | |
| Maximum | 315 |
| Minimum | 100 |
| Waste activated sludge (WAS) pumps | |
| Number, two per station | 4 |
| Type | Submersible |
| Capacity, gpm | 2200 |
| Horsepower | 20 |

Note that cfm $\times$ 4.719 $\times$ $10^{-4}$ = $m^3$/s; psi $\times$ 6.895 = kPa; hp $\times$ 745.7 = W; and gpm $\times$ 6.308 $\times$ $10^{-5}$ = $m^3$/s.

## PROCEDURES

### General

Wasting sludge is needed to maintain the target mixed liquor suspended solids (MLSS) concentration in the aeration/clarification basins. This is a daily procedure and ideally needs to occur over the entire day. The waste sludge volume and the pumping capacity of the waste sludge pumps do not always allow this to occur. In those cases, evaluate the timeframe and determine if it would be better to waste sludge several times a day. Sludge wasting can be determined by two methods, solids retention time (SRT) and constant MLSS.

*Solids Retention Time Waste Sludge Operating Method*

One method used to determine sludge wasting is the SRT method. The SRT method is based on holding the biomass in the activated sludge system for a set period of time (days). Target sludge wasting rates are determined using a modified version of the SRT equation. The equation is as follows:

$$\text{WAS, lb/d} = \frac{\text{MLSS, mg/L} \times 8.34\,\dfrac{\text{lb/mil. gal}}{\text{mg/L}} \times \text{Aeration basin volume in service, mil. gal}}{\text{Target SRT}_N\text{, days}}$$

Note that lb/d $\times$ 0.453 6 = kg/d; lb/mil. gal $\times$ 0.119 8 = mg/L; and mil. gal $\times$ 3.785 = $m^3$.

This operating method typically works best when the process organic loading is greater than 75% of the plant design loading. The gallons per day to waste is shown later in the text. Always remember that the waste rate should not be varied by more than 10% from one day to the next.

*Constant MLSS*

One method that can be used to determine the target wasted sludge volume is the constant MLSS method. The equation that can be used to determine the waste sludge volume is as follows.

$$\text{WAS, lb/d} = (\text{Measured MLSS, mg/L} - \text{Target MLSS, mg/L}) \times 8.34\,\frac{\text{lb/mil. gal}}{\text{mg/L}} \times \text{Aeration basin volume in service, mil. gal}$$

Note that lb/d × 0.453 6 = kg/d; lb/mil. gal × 0.119 8 = mg/L; and mil. gal × 3.785 = $m^3$.

*Determining the WAS Volume, gpd*

After the pounds of sludge to waste has been determined by either method, the gallons of sludge to be wasted needs to be determined. A modified version of the equation used to determine pounds of solids and the return sludge concentration is used. The equation to use is as follows:

$$\text{WAS, gpd} = \frac{\text{WAS, lb/d}}{\text{Return activated sludge concentration, mg/L} \times 8.34\,\dfrac{\text{lb/mil. gal}}{\text{mg/L}}} \times 1\,000\,000$$

Note that gpd × 3785 = $m^3$/d; lb/d × 0.453 6 = kg/d; and lb/mil. gal × 0.119 8 = mg/L.

*Determine the WAS Frequency, min/d*

The wasting rate is determined by adjusting the speed of the pump. The maximum capacity of each pump is 120 gpm. Three pumps can be used to waste sludge. In some cases only one pump may have to be used, at other times multiple pumps will have to be used. Continuous wasting is the most desirable operating condition. The following equation can be used to obtain a value of fewer than 1440 minutes per day.

$$\text{WAS pumping, min/d} = \left(\frac{\text{WAS, gpd}}{\text{Pump capacity, gpm}}\right)\left(1400\,\frac{\text{min}}{\text{d}}\right)$$

Note that gpd × 3785 = $m^3$/d and gpm × 5.451 = $m^3$/d.

*Determine the WAS Frequency, min/h*

If the WAS frequency, minutes per day, is fewer than 1440 minutes per day, it may be desirable to waste sludge each hour of the day. The minutes per hour at the selected pumping rate can be determined as follows.

$$\text{WAS pumping, min/h} = \frac{\text{WAS pumping, min/d}}{24\ \text{h/d}}$$

*Description of Operating Procedures*

Once the volume of sludge to waste has been determined, the operating procedures can be implemented. One set of waste sludge pumps is used by basins 1 and 2 and basins 3 and 4. The waste sludge flow meter is on the main line between the waste sludge pumping stations and the gravity sludge thickeners. This arrangement requires wasting sludge from the basins one at a time to accurately measure the WAS flow. The systems are operated independently so there is no method to waste sludge from a common process.

The sludge wasting has to be initiated manually but is stopped automatically by the plant control system.

The waste sludge flows by gravity into the WAS pumping station wet well. The pumps pump the sludge to the waste sludge meter vault and into the gravity thickeners.

### Step-by-Step Procedures

*Step 1*

Select the basin that will be used for service. The following electrically operated valves will be used to waste sludge.

| Aeration/clarification basin | Electrically operated valve |
|---|---|
| No. 1 | 2WS/PV – 2 |
| No. 2 | 2WS/PV – 1 |
| No. 3 | 2WS/PV – 4 |
| No. 4 | 2WS/PV – 3 |

The pump isolation valve needs to be open.

*Step 2*

Set the time delay control on the plant operating screen to allow the waste sludge valve to remain open for the specified time.

*Step 3*

Open the valve from the selected aeration/clarification basin. The pump On-Off-Auto selector switch should be in the **Auto** position.

*Step 4*

The pump will begin operation when the wet well sludge level actuates the pump controls.

*Step 5*

The plant control system will annunciate an audible alarm when the wasting time nears the end of the cycle. The operator should acknowledge the alarm.

*Step 6*

The plant control system will annunciate an audible alarm when the wasting cycle is completed. The operator should acknowledge the alarm. The amount of sludge wasted should be check against the target volume. If acceptable, the next wasting period for the reaming basins should be initiated.

*Step 7*

Record the volume of sludge wasted each day on the daily log and charts.

**End of Procedure**

---

**OPERATOR LOGS AND DATA ENTRY FORMS.** Records are needed for ongoing process management, compliance reports, and establishment of historical trends. Necessary records include visual observations; sampling and analytical results; process calculations; and logs, diaries, and bench sheets (see Chapter 6, Management Information Systems [Records and Reports]). Data should be assembled and recorded in permanent log books. Diaries and bench sheets should be designed to provide easy transfer to regulatory report forms.

An example of a form used to gather information is shown below in Figure 14.2. Note that this form shows, at a glance, the daily activities that have taken place at the dewatering centrifuge.

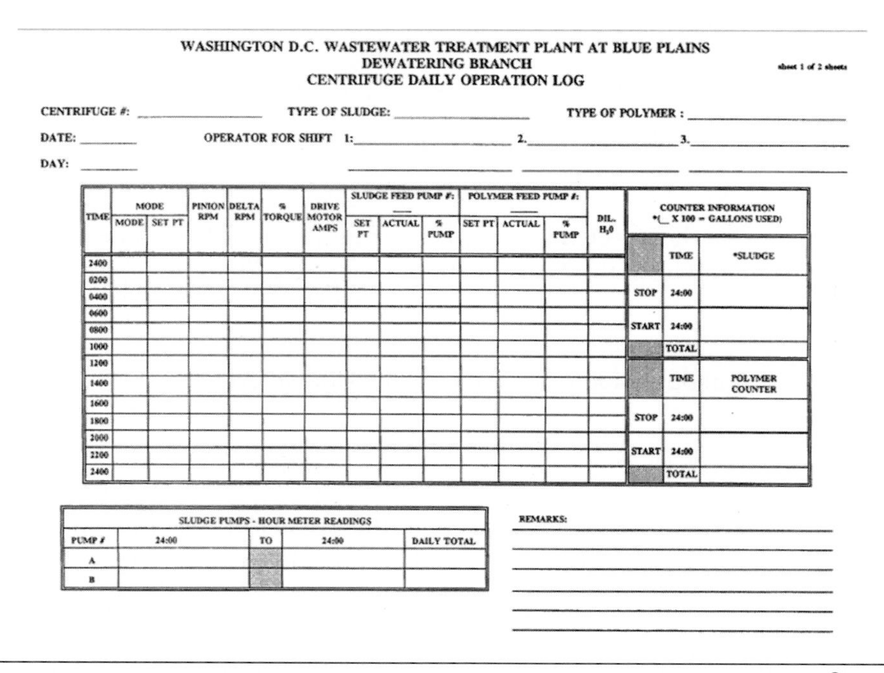

WASHINGTON D.C. WASTEWATER TREATMENT PLANT AT BLUE PLAINS
DEWATERING BRANCH
CENTRIFUGE DAILY OPERATION LOG

sheet 1 of 2 sheets

CENTRIFUGE #: ______ TYPE OF SLUDGE: ______ TYPE OF POLYMER : ______

DATE: ______ OPERATOR FOR SHIFT 1: ______ 2. ______ 3. ______

DAY: ______

| TIME | MODE | | PINION RPM | DELTA RPM | % TORQUE | DRIVE MOTOR AMPS | SLUDGE FEED PUMP #: ___ | | | POLYMER FEED PUMP #: ___ | | | DIL. H,0 |
|---|---|---|---|---|---|---|---|---|---|---|---|---|---|
| | MODE | SET PT | | | | | SET PT | ACTUAL | % PUMP | SET PT | ACTUAL | % PUMP | |
| 2400 | | | | | | | | | | | | | |
| 0200 | | | | | | | | | | | | | |
| 0400 | | | | | | | | | | | | | |
| 0600 | | | | | | | | | | | | | |
| 0800 | | | | | | | | | | | | | |
| 1000 | | | | | | | | | | | | | |
| 1200 | | | | | | | | | | | | | |
| 1400 | | | | | | | | | | | | | |
| 1600 | | | | | | | | | | | | | |
| 1800 | | | | | | | | | | | | | |
| 2000 | | | | | | | | | | | | | |
| 2200 | | | | | | | | | | | | | |
| 2400 | | | | | | | | | | | | | |

| COUNTER INFORMATION *(_ X 100 = GALLONS USED) | | |
|---|---|---|
| | TIME | *SLUDGE |
| STOP | 24:00 | |
| START | 24:00 | |
| | TOTAL | |
| | TIME | POLYMER COUNTER |
| STOP | 24:00 | |
| START | 24:00 | |
| | TOTAL | |

| SLUDGE PUMPS - HOUR METER READINGS | | | | |
|---|---|---|---|---|
| PUMP # | 24:00 | TO | 24:00 | DAILY TOTAL |
| A | | | | |
| B | | | | |

REMARKS:

FIGURE 14.2 Operator work log: centrifuge dewatering (gpd × 0.003 7 = $m^3/d$).

# PROCESS CONTROL CONCEPTS

Poor plant performance often results from failure to apply known process control concepts. Therefore, the process control plan is one of the plant's most important management systems. Specifically, a detailed written description of how the process should be controlled must be developed that

- Assigns responsibility for individual processes,
- Presents methods for checking performance, and
- Explains procedures for changes (fine tuning) necessary to produce the required results.

One person, typically the chief operator, should be responsible for setting the process operating parameters for the facility. Operational parameters to be considered include operating directions, process guidelines, units in service (based on design criteria), sampling schedule, analysis, and calculations for optimizing performance. That person must then approve, in writing, any changes in the parameters or guidelines. Once the process control plan is developed or changed, all employees need to be informed of the plan's integrated process management elements through a workable communications system.

Testing and monitoring the process control parameters requires planning and organization so that performance trends departing from targeted goals are clearly noticeable. Control elements can be brought back to the operator's established limits before permit violations or process upsets occur.

## PROCESS GOALS AND PLANNING

A process control plan (Table 14.2) establishes specific wastewater treatment goals, defined in part by the National Pollutant Discharge Elimination System (NPDES) permit, local ordinances, plant policies, and contracts regarding delivery of byproducts. Land application of biosolids, digester gas use, and effluent reuse are also typically included. The plan identifies routine process control parameters (BOD, TSS, alkalinity, etc.) and potential problems associated with each wastewater treatment process. A range of performance indicators can be set by analyzing historical data for a specific process. Finally and most importantly, the plan includes the actions necessary to keep the process parameters within the established limits.

The manager should review data daily with respect to the process control plan and determine the cause of deviations from the defined target and values that fall beyond the expected range. The manager may then adjust controls as necessary to bring the unit process closer to the set targets. Interactions among processes, including those from recycle streams and their byproducts, need to be considered. Sidestream beneficial use must also be accounted for. The plan and process control strategy must be practical, responsive, and viable for implementation.

Process control planning must provide valid process information by ensuring that flow metering, online sensors, and laboratory analyses provide accurate and timely data necessary for operating decisions. The sampling and analysis plan discussed in Chapter 17, Characterization and Sampling of Wastewater, should be reviewed to ensure that sampling points, frequencies, and procedures will provide representative samples necessary for laboratory testing and analyses. The tests also require quality assurance and quality control (QA/QC) to ensure they will provide all data necessary for process control.

The plant's operations and maintenance (O&M) manual, design criteria, facility-as-built drawings, manufacturer's literature, WEF manuals of practice, and U.S. Environmental Protection Agency technical documents provide useful information for forming the plan. Valuable assistance in preparing the plan may be obtained from the design engineering firm and its O&M group, regulatory agencies, and managers of similar plants.

**TABLE 14.2** Primary clarifier process control plan.

Goals
1. Provide stable food source for secondary treatment process
2. Remove 60% of the influent total suspended solids
3. Remove 25% of the influent biochemical oxygen demand

Process guidelines
1. Influent biochemical oxygen demand = 180 mg/L
2. Influent total suspended solids = 160 mg/L
3. Surface loading rate = 1200 gpd/sq ft[a]
4. Detention time = 2 hours

Calculations

$$\text{Removal efficiency} = \frac{\text{In} - \text{out}}{\text{In}} \times 100$$

Surface loading rate,

$$\text{gpd/sq ft} = \frac{\text{Influent flow, gpd}^{b}}{\text{Primary clarifier area, sq ft}^{c}}$$

Detention time,

$$\text{hours} = \frac{(\text{Volume, gal}^{d})(24\ \text{h/d})}{\text{Flow, gpd}}$$

Analytical data and schedule

| Test | Schedule | Type |
|---|---|---|
| Suspended solids | D | C |
| Settleable solids | D | C |
| Total and volatile solids | D | C |
| Biochemical oxygen demand | D | C |
| pH | Continuous | |
| Grease and oil | M | G |

Contact the chief plant operator for clarification or in an emergency.

Operating instructions
1. Four of six units in service
2. Ensure that drives are running (D)
3. Inspect helical scum skimmer (D)
4. Check pump schedule (D)
5. Check process guidelines (D)
6. Pump down scum well (W)
7. Clean baffles and weirs (W)

Troubleshooting

| Observation | Check |
|---|---|
| Poor solids removal | Hydraulic overloading; activate another clarifier. |
| Sludge hard to remove from hopper | Check pump operation. Check grit removal facility for poor operation. |
| Sludge has low solids content | Check sludge pump timers for excessive pump operation. Hydraulic overloading; activate another clarifier. Check for short-circuiting. |
| Short-circuiting of tank flow | Ensure that weir settings are even. Damage to inlet baffle. |

| | | | | | |
|---|---|---|---|---|---|
| D | = | Daily | C | = | Composite |
| M | = | Monthly | G | = | Grab |

[a] gpd/sq ft × 40.74 = L/m²·d.
[b] gpd × 0.003 7 = m³/d.
[c] sq ft × 0.092 9 = m².
[d] gal × 3.785 = L.

In summary, the process control plan should detail in a step-by-step fashion how the particular process operation is set and who is responsible for the implementation. The following areas should be addressed:

- Assignment of responsibilities for controlling individual processes within established guidelines;
- Identification of technical support sources, including experts from neighboring plants, design engineering firms, technical schools, regulatory agencies, and training centers;
- Establishment of a reference library; and
- Provision for continuous process management where possible, including off-duty call lists for plants with 24-hour staffing.

## PROCESS CONTROL MANAGEMENT

Once developed, the process control plan must be managed and carried out. This entails clarification, correction, improvement, and updating, all of which require effective communications.

**COMMUNICATIONS.** Process management requires a timely response to process changes. This indicates the need for communications links that promote passage of information. Such linkage might be as simple as a verbal message at shift change or a posted process parameter adjustment. Those who are responsible for process control must have immediate access to these messages and process status reports. Some facilities use a more structured approach where a "process control change order" is issued with signatures of both of the responsible persons at the shift change. To accomplish this, an overlapping time between the operator starting and leaving the shift is needed.

**ACCURATE MEASUREMENTS.** Accurate information is needed for process control decisions. Nonrepresentative samples or inappropriate sample points lead to errors in control decisions, with possible severe consequences. The information must be timely. For example, five-day BOD test results are much too late for most daily process decisions. In this case, surrogate testing, such as chemical oxygen demand (COD), may be used as a timely indicator if a consistent relationship exists between the two analyses.

**LABORATORY AND PROCESS RELATIONSHIP.** Timely and accurate laboratory results supplemented by visual and sensory observations are necessary if process control is to succeed. The laboratory capacity and schedules must conform to the NPDES monitoring requirements and generate process control data without conflicts. Process control tests do not have to meet quality requirements of NPDES analyses with respect to QA/QC. Thus, some shortcuts, such as microwave ovens for drying solids samples, can be used to facilitate timely results. Surrogate methods or testing should be checked periodically to ensure that reliable data are obtained.

Timely feedback is very valuable. Ideally, the relevant laboratory result should be posted immediately or, in larger facilities, the information could be shared electronically (e-mailed or with a shared spreadsheet) to the pertinent process areas upon completion. More commonly, either the turn-around time is poor or the laboratory data go to managers and rarely to the operators who can also use them. As a result, the operators typically ignore the laboratory and use a microwave or similar device to follow the process and drive their decision making. These valuable data are, at best, entered into a log book and not shared with anyone outside of the area. Logged into a common spreadsheet and shared across a network, both the process area and laboratory data could be viewed and used to evaluate process performance.

**OVERVIEW ANALYSIS.** Individual process units must be integrated to an efficient wastewater treatment system. The chief operator should learn to recognize key indicators of the performance of the overall plant and monitor them daily. Cause-and-effect relationships in the plant can be detected only by a complete analysis of all data.

Among the chief operator's responsibilities is cost control. The overview analysis allows identification of high-cost processes and will perhaps disclose cost-saving measures.

**DISCRETE PROCESS TRACKING.** Plant loadings and the resultant treatment efficiency must be analyzed to identify any weak links. Computerized data systems can provide this information quickly. With manual tracking, at least a monthly analysis is needed.

Should a single process unit show signs of distress, the operator should determine whether operating procedures or excessive loading is the cause. In either event, the deficiency requires correction.

As a guidance mechanism for process control, upper and lower operating limits should be established for each unit process. These limits provide operations personnel with the guidelines for holding the process on target. As best practice, data from the process control testing are plotted on trend charts with the upper and lower limits are shown. Facilities that include SCADA controls are able to designate these values as part of the programming logic and, based on the specific values, the control logic may be altered or suggested or alarms can be generated.

**COMMON PROBLEM AREAS.** Certain problems such as the following occur in most plants:

- Solids may accumulate within the plant as a result of inadequate storage, inadequate dewatering facilities, or weather-induced restriction of land application. The process control plan should identify the means of preventing or reducing this problem.
- Process instruments frequently fail. This results in erroneous data or no data and, as a consequence, incorrect process decisions. If a pH meter is important, then redundancy is necessary. Provide redundant meters or install two meters in parallel, then set alarms if the reading or the response is unreasonable. The operator, once notified, can set a course of action while the instrument calibration and operation is checked. As plants become more highly automated, the shortage of instrument repair technicians grows, and plants will need to provide training to upgrade the skill sets of their staff.
- Dependence on traditional data may result in late problem diagnosis. This can be avoided by using surrogate or signal performance such as the movement of organisms in mixed liquor. The operator should refer to the individual process chapters for more details.
- Intermittent heavy loadings from plant recycle may upset the plant process. Digester supernatant and filter backwash can overload a plant if flows are not distributed over a long period. Draining plant processes may also inadvertently add a load to the plant.
- Odors quickly arouse indignation and quickly convince the public that the plant is poorly operated, regardless of whether this is the case. A thorough discussion of the causes of plant odors and methods for their prevention and control is presented in Chapter 13, Odor Control, and in *Control of Odors and Emissions from Wastewater Treatment Plants* (WEF, 2004). The World Wide Web has made a lot of information available; much of it is credible, and some is not so good. Professional organizations such as the Water Environment Federation (www.wef.org) also provide forums for discussion groups and may be a good source of information.

## PROCESS CONTROL TOOLS

Four steps are essential to manage any process: gather information, evaluate the data, develop and implement a proper response, then reevaluate. First, information is collected in the form of analytical data, hour meters, visual observations, and other miscellaneous facts. Information is only as good as the device used for the analysis. Reliable instrumentation that has been carefully selected for the application and properly maintained will provide the best information. Much information is in the form of numerical data. Making sense of all this information can be daunting. Several tools such as graphs, mass balance analysis, and basic statistics can make the job of evaluation much easier. The WEF–ASCE publication, *Design of Municipal Wastewater Treatment Plants* (1998), provides a good explanation of mass balance.

Once the information is analyzed, it is then a matter of applying operational principles of a given process to formulate a proper response. Finally, it is necessary to reevaluate how plant performance and conditions have changed since the operating strategy was implemented.

**GATHER INFORMATION.** Gathering information is a basic troubleshooting task that requires knowledge of the specific plant, waste constituents received, and specific processes. The following guidelines will assist in gathering information:

- Determine whether a problem exists or the test results are in error.
- Find out when the problem was first observed and what was done about it.
- Examine the biomass with a microscope to evaluate the state of the microorganisms, their diversity, motility, and numbers.
- Determine whether the operation of the plant contributes to the problem. Check the upstream unit for contributory problems, including the sewer system and recycle streams.
- Complete the flow and mass-balance analysis. Check each unit process separately because one weak process can overload the whole plant.
- Use a quality assurance program to improve the accuracy of test results.
- Check for pipeline leaks that can divert flows, including chemicals.
- Check for changes in treatment chemicals.
- Ensure that air-delivery systems work.
- Determine whether mechanical or electrical failures are causing problems. Check for such failures when problems occur. A brief power outage can stop equipment, requiring manual restart.
- Separate the symptoms from the problem. Listen to everyone's observations but mistrust everyone's interpretation. Identify the root cause, not the symptom. Things are not always as they first appear; do not be too quick to jump to conclusions.

The operator should prepare a checklist for the plant and add to it whenever a unique situation is encountered. This information should be part of the plant SOP and can be incorporated to the specific process troubleshooting guides.

***Process and Instrumentation Diagrams.*** Process and instrumentation drawings (P&IDs) are typically developed during the construction phase of a project and illustrate the relationship and interconnections between the mechanical equipment and instrumentation. These documents also define and locate equipment tag numbers, which facilitate developing wiring diagrams, bills of materials, purchasing, receiving, and installation. The P&IDs are a complement to the other forms of process control logic documentation.

The Instrumentation Society of America has defined standards for producing P&IDs. Within a set of construction plans, the first instrumentation drawing contains information that defines the meaning of the identification letters, symbols, and bubbles.

A P&ID diagram for the flotation thickened sludge skimmer drive is presented as Figure 14.3. The P&ID contains the following information about the DAF drive:

Each instrument is part of the equipment control loop 628.
A local instrument: weight switch (WS).

Located on variable-frequency drive (VFD) control panel:

- Start/stop handswitch (HS-$1^{A/M}$);
- Automatic/manual handswitch (HS-$2^{S/S}$);
- Timer, indicator, and control (SIK);

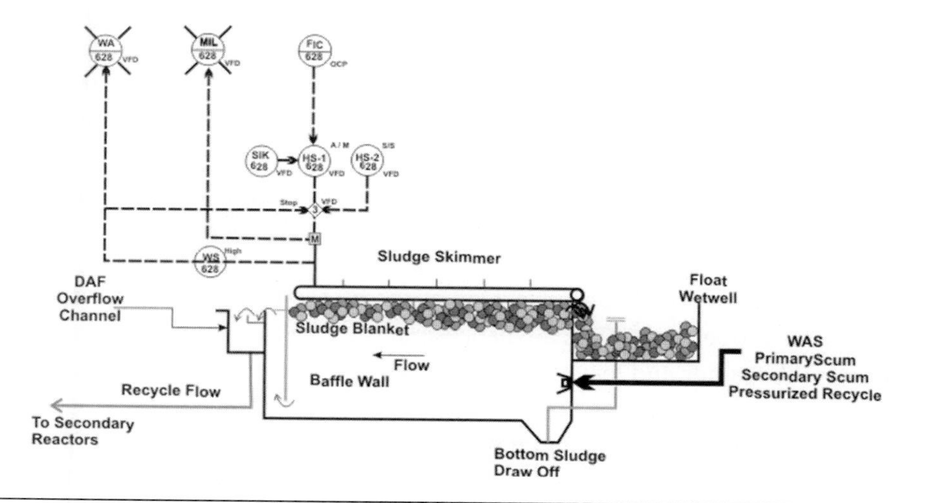


**FIGURE 14.3** Example P&ID diagram.

- Weight alarm (WA); and
- Moisture indicator light (MIL).

Located on the operator control panel (OCP): flow indicator and controller.

Field-mounted instruments are shown as plain circles (no line within the circle). Primary location instruments, typically located on a control panel, are shown as a circle bisected by a line. The location of the instrument is noted in the bottom right-hand side of the circle. Example: FIC-628 shows the "OCP" in that location. Consequently, the instrument is located on the OCP. If the instrument is located locally, no notation is needed. Dotted lines are drawn between the instrument "bubbles" and indicate that this is instrumentation logic and the direction it flows, input or output. If the line is continuous, the instrument is located on the front of a panel. Once the basics of reading a P&ID are understood, the reader can put together the "story" that is being conveyed. The DAF sludge skimmer control logic is as follows:

1. There are two hand switches located on the VFD control panel that control skimmer operation.
2. There is an interlock (the diamond with a "3") that required three conditions to be met—the weight alarm cannot be activated, the VFD must be started, and the there must be a signal from the mode selector switch (HS-$1^{A/M}$).
3. A moisture sensor is located in the skimmer drive motor; when moisture is detected, a light illuminates on the VFD control panel (MIL-628).
4. When the mode selector switch (HS-$1^{A/M}$) is in **Auto**, it receives a signal from FIC-628 that flow is being sent to the DAF. The skimmer operates based on the timer (SIK-628) unless the weight switch high setting is reached. If the high setting is reached, an alarm goes off on the VFD panel (WA-628) and the skimmer stops.
5. When the mode selector switch (HS-$1^{A/M}$) is in **Manual**, skimmer operation is controlled manually with speed adjusted at the VFD control panel.

***Hydraulic and Mass Solids Balance.*** One of the best tools for understanding how a plant or process is operating is mass-balance analysis. The objective is to account for all solids or flow that enter and leave a given process. This concept is readily adapted for use as an evaluation and process control tool.

Flow diagrams are developed to represent a process, using conditions appropriate to the problem that needs to be resolved. For example, the accuracy of a metering system needs periodic verification. The operator can simply check the sum of the subme-

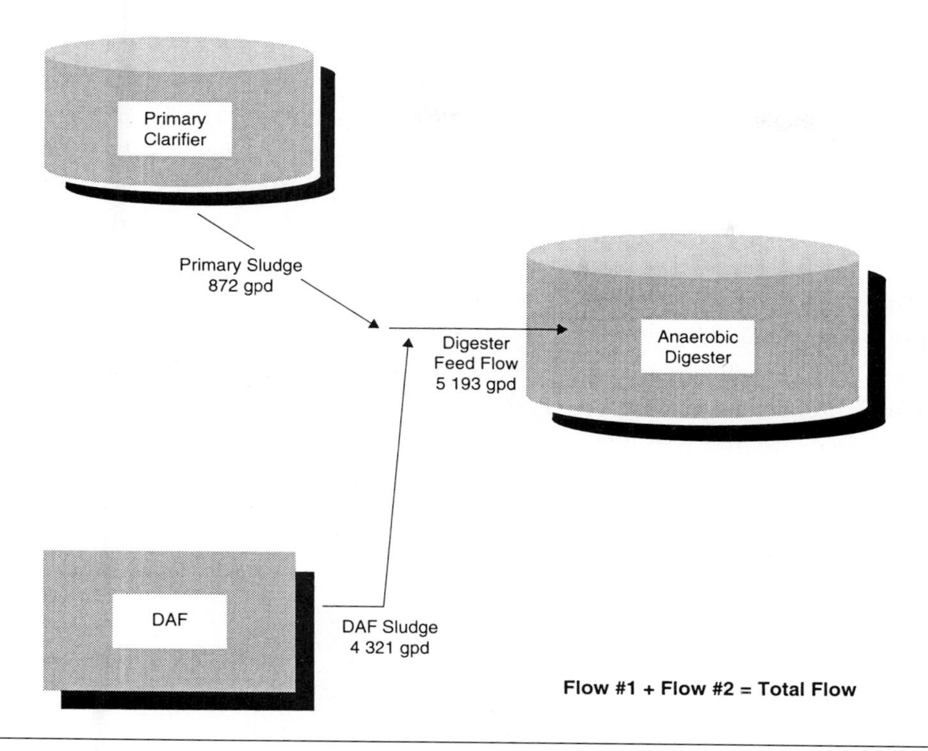


**FIGURE 14.4** Hydraulic flow balance (gpd × 0.003 7 = $m^3/d$).

ter readings against the master meter to perform a hydraulic balance. An anaerobic digester is fed primary and thickened sludge using a gravity belt thickener. Given the configuration and flows from Figure 14.4, the sum of the two flows should be fairly close to the meter that reads the combined digester feed flow. If these two values differ significantly, all three meters need to be calibrated. Because some processes lag (such as the effluent meter compared to the influent meter), choosing the appropriate meter and time interval may be important in the overall analysis.

Using the same logic, any process flow stream can be evaluated. The balancing method is simple to use. From a hydraulic basis,

$$\text{Flow no. 1} + \text{Flow no. 2} = \text{Total flow}$$

Similarly, from a solids or constituent perspective,

$$\text{Solids in} = \text{Solids out}$$

Balancing solids within process units and throughout the plant can help the operator predict process operation, troubleshoot problems, and plan a course of effective treatment. Mass solids balance may be used on one process, such as thickening, or may be used on a plant-wide basis. Regardless, the methodology remains the same. With the availability and power of the personal computer, setting up a spreadsheet with cells for information about the plant can make quick and accurate work of both hydraulic and mass-balance analysis.

To put mass solids balance to use, the operator should first sketch a diagram of the process or plant to be evaluated and write down the flow characteristics. The next step is to calculate the desired constituent on a kilograms-per-day (pounds-per-day) basis and prepare to evaluate. In Figure 14.5, primary clarifier solids entering the clarifier are compared with those leaving the clarifier (primary sludge and primary

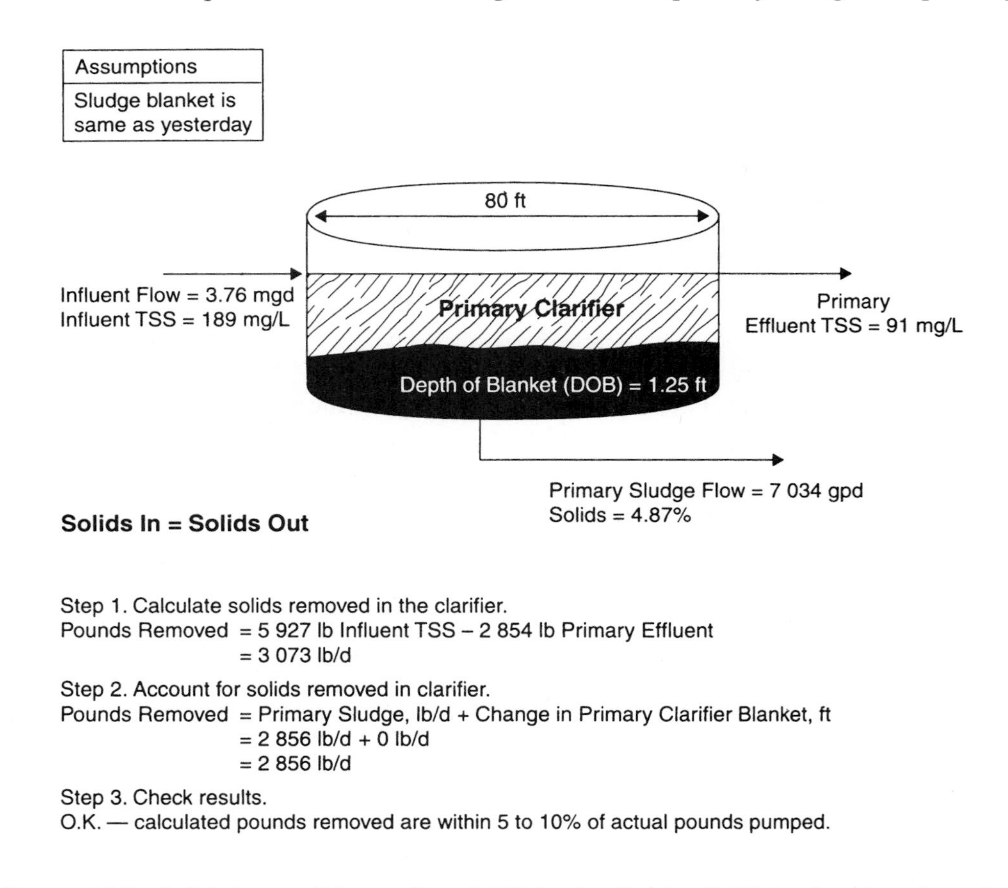


**FIGURE 14.5** Solids in = solids out (lb × 0.453 6 = kg; lb/d × 0.453 6 = kg/d; and mgd × 43.83 = L/s).

effluent). Primary scum need not be considered unless a substantial amount of solids is incorporated to the flow. From this exercise, the following conclusions can be drawn:

- Solids entering the clarifier are about the same as the combined solids leaving the clarifier because the blanket level has not changed.
- Primary sludge pump rates are sufficient for these conditions.
- Primary sludge pumps are working.
- The primary sludge flow meter does not need calibration because metered values are within 5 to 10% of each other.
- Laboratory analysis is being performed consistently because calculated values are within 5 to 10% of each other.
- Expect that the solids removal efficiency falls within design parameters.
- Expect the clarifier surface to be clear with no signs of rising sludge.

Rising sludge may indicate excessive sludge in the clarifier or equipment failure; check flow meter and review analytical data.

Mass balance is a powerful tool in setting and evaluating process operations. Problems typically manifest themselves in the most unlikely places. A solids separation problem in the gravity thickener may not actually be rooted in the thickener. In essence, it is only a symptom of the problem. Recycle flows, conversion of soluble BOD to biomass, solids in the plant effluent, and conversion of solids to liquid and gas must all be taken into account. Determination of the correct cell yields, alpha and beta factors, conversion rates, etc., should be part of the process design memorandum for a specific facility. If unavailable, the values are presented in *Design of Municipal Wastewater Treatment Plants* (WEF and ASCE, 1998).

**EVALUATE DATA.** Quantitative data collection for the plant is so voluminous (analytical data, meter readings, and expenses) that the data defy analysis and interpretation unless they are organized. Small- to medium-sized treatment plants may collect most of these data manually. Larger or more sophisticated treatment schemes may have automated control systems that monitor, collect, and store the information in a database. Once collected, the data must be organized, displayed, and evaluated to be useful. One way to make sense of the data is to organize them in a tabular format. The information can also be organized visually, creating a picture by constructing a graph. There are many types of graphs, including line, bar, scatter, and pie. Graph choice depends on the data and how the information will be used. In general, a graph is a chart used to show relationships between two or more factors.

***Line Graph.*** The simplest picture is the line graph, which is used to represent factors on two different scales. It is particularly suitable for recording historical data of a parameter that constantly changes.

The graph in Figure 14.6 shows changes in plant influent flows during the month. A line has been drawn across the graph to show the design average dry weather flow. Values above or near this line may indicate that downstream processes are taxed. Used in this way, the graph is sometimes referred to as a trend chart. The *x*-axis contains the days of the month. The range for the *y*-axis was selected based on the peak plant capacity of 0.8766 $m^3/s$ (20 mgd).

Line graphs can be developed to summarize performance characteristics such as determining the pump output of a variable-speed pump. The range of the axes in Figure 14.7 is determined by the purpose of the graph; in this case, pump capacity (gpm) and the percent speed of the variable-frequency drive (VFD). To develop this graph, the analyst first determines the actual volume pumped at several points (a minimum of three or four). The pump was run at 20, 40, 60, 80, and 100% speed and liquid volume measures. Then the analyst plots the points on the two axes (VFD speed on the *x*-axis; volume of liquid pumped on the *y*-axis). Finally, the points are connected to establish a curve. The user can now easily determine the flowrate at any given VFD setting.

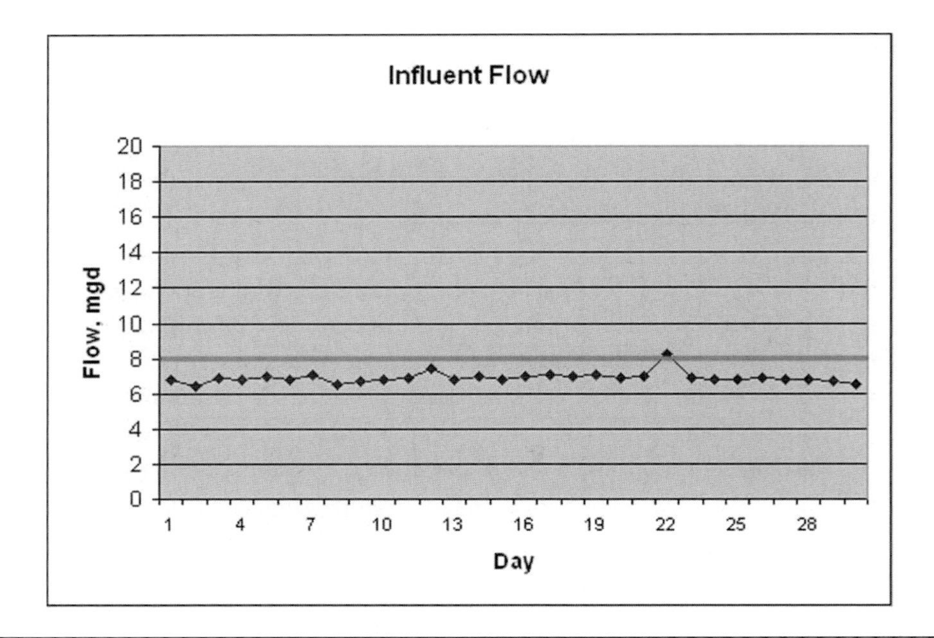


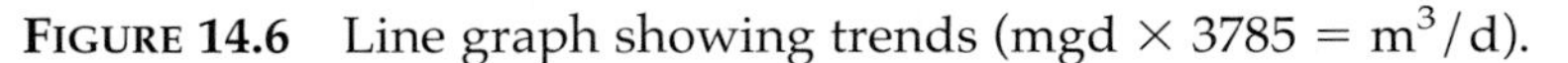
**FIGURE 14.6** Line graph showing trends (mgd × 3785 = $m^3/d$).

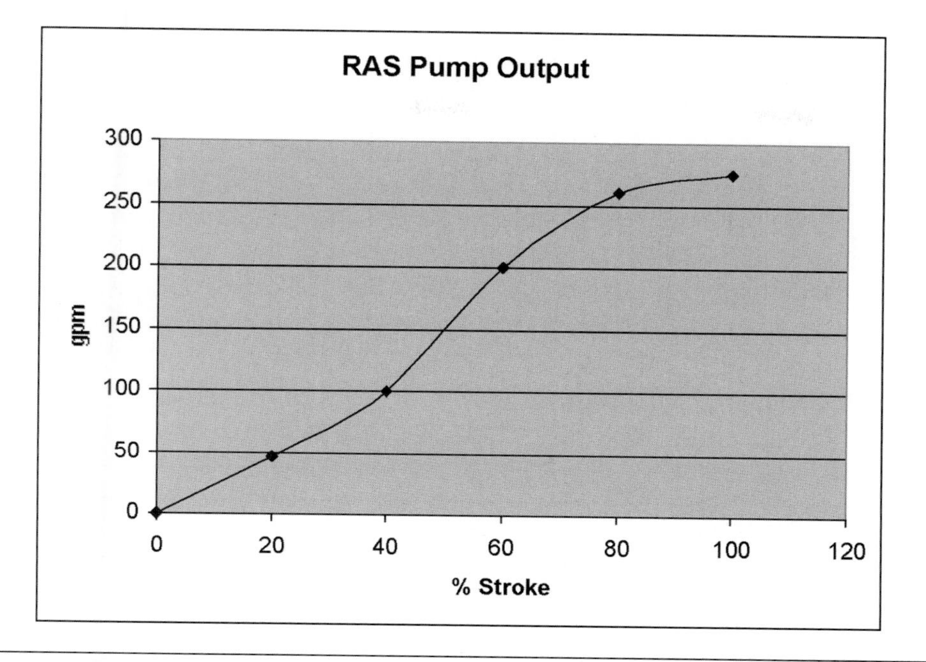


FIGURE 14.7 Line graph summarizing performance (RAS = return activated sludge; gpm × 0.063 08 = L/s).

***Bar Graph.*** Bar graphs use parallel bars to compare amounts of the same type of measurement and are best used to summarize data. The bar graph in Figure 14.8 depicts electrical usage ($x$-axis) at the WWTP during a two-year period on a monthly basis ($y$-axis).

The height of the bar indicates the electrical kilowatt usage. The horizontal axis indicates the specific month the usage occurred. Bars with different patterns or colors were used to distinguish between the two sets of data. A solid bar indicates the year 1 data and a gray bar indicates the year 2 data.

A second type of bar graph, called a stacked bar graph, shows different types of values by "stacking" a bar on top of the previous bar to form a single bar. Different colors or patterns can be selected to differentiate among parts of the bar.

In Figure 14.9, a facility experiencing high influent solids elected to sample each of the flow streams that entered the plant. In addition to the collection system influent, plant sidestreams were analyzed, which makes this a "confluent sample" versus an "influent sample". Sidestreams and recycle flows from the plant may make a significant contribution when the constituents are examined independently as seen in this figure. Each sample stream has a different bar pattern. The $x$-axis displays data by year;

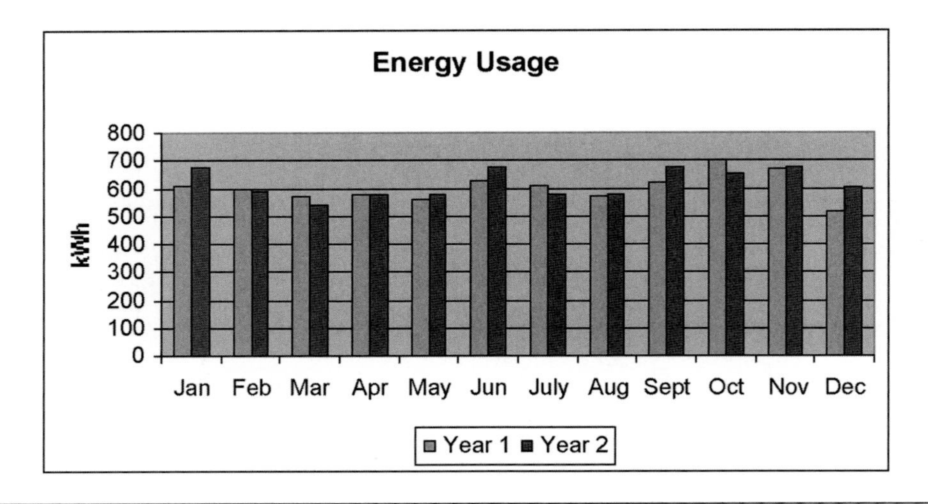


FIGURE 14.8 Bar graph.

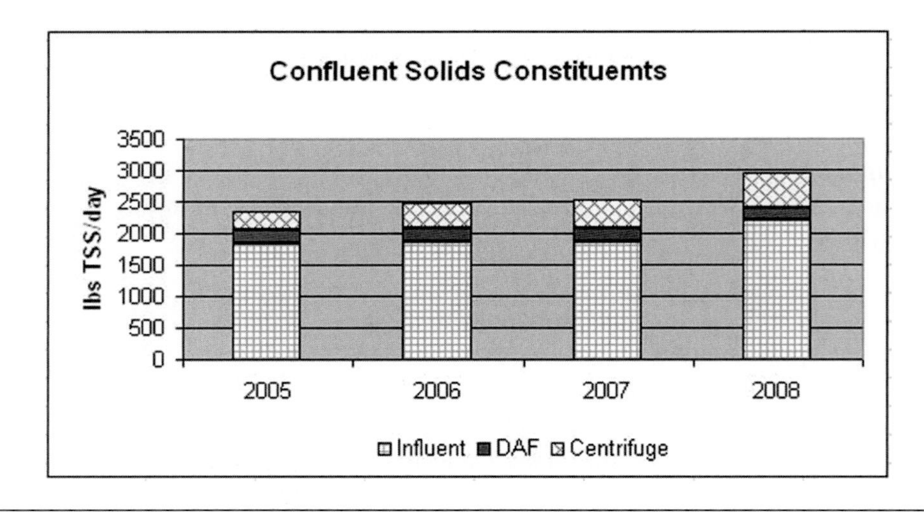


FIGURE 14.9 Stacked bar graph: (1) the bottom hatched pattern represents the solids contribution from the collection system, the "plant influent"; (2) the middle solid pattern represents the contribution of solids from the DAF; (3) the top cross-hatch pattern on the top portion of the bar reflects the solids contribution from centrifuge dewatering (when data are viewed in this manner, it is evident that, over the last four years, the contribution from the centrifuge dewatering operation is increasing) (lb/d × 0.453 6 = kg/d).

the $y$-axis shows the total quantity of solids entering the plant. A stacked bar graph may assist in evaluating and troubleshooting process problems.

***Pie Chart.*** A pie chart resembles a pie, divided into "slices" according to quantities. The pie graph is generally developed on a percentage basis. The whole pie represents 100% of the value. Each slice represents the percentage that its value contributes to the total.

A pie chart is ideal for representing a facility budget (Figure 14.10). Each slice of the pie represents the proportion of the budget allocated for each of the categories identified in the legend. The total budget allocation of $4,371,308 is noted on the figure. Each budget item is divided by the total budget amount (times 100) to obtain its percentage share. After all items are calculated, they should add up to 100%. The pie is then divided proportionally.

***Scatter Graph.*** A scatter graph, also called an $x$–$y$ graph, is used to show how values change relative to other values. It is generally invoked to determine if there is a relationship between the two sets of data. The "$x$" data are the control parameter, a value that may be manipulated such as MLSS, detention time, or loading rates. The "$y$" data are associated with performance or the result, such as BOD, TSS, or ammonia level. A line of best fit, the regression line, may be drawn or calculated for the data and correlation analysis performed to determine the strength of the relationship between the two variables. This subject is discussed in greater detail later in this section.

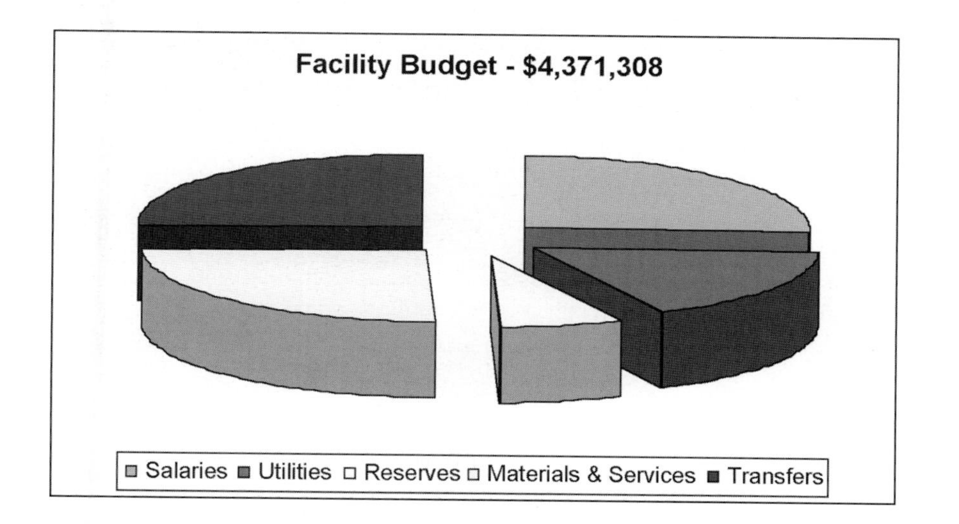


**FIGURE 14.10** Pie chart.

As seen in Figure 14.11 using the scatter graph, primary effluent BOD quality is plotted on the $y$-axis and primary effluent COD is plotted on the $x$-axis. The data pairs are plotted and a regression line is drawn through the points to determine a line of best fit.

From these data, one might expect that at a COD value of 200 mg/L, the BOD would be 115 mg/L. If the relationship between the COD and BOD remains the same, then good estimates of primary effluent BOD can be made.

***Elementary Statistics.*** Monitoring performance using tables, graphs, and mass balance analysis is only the beginning in the art of process control. Reviewing and evaluating plant data are an overwhelming task with the amount of analytical data and process performance information available. It is important to track plant performance parameters, such as effluent quality characteristics, to demonstrate compliance with regulatory permits. However, in the pursuit of high-quality effluent, selecting the variables that most directly influence process performance can be difficult.

Typically, operators review historical data to help develop future process control strategies. Given the multitude of associations to check (control parameters versus

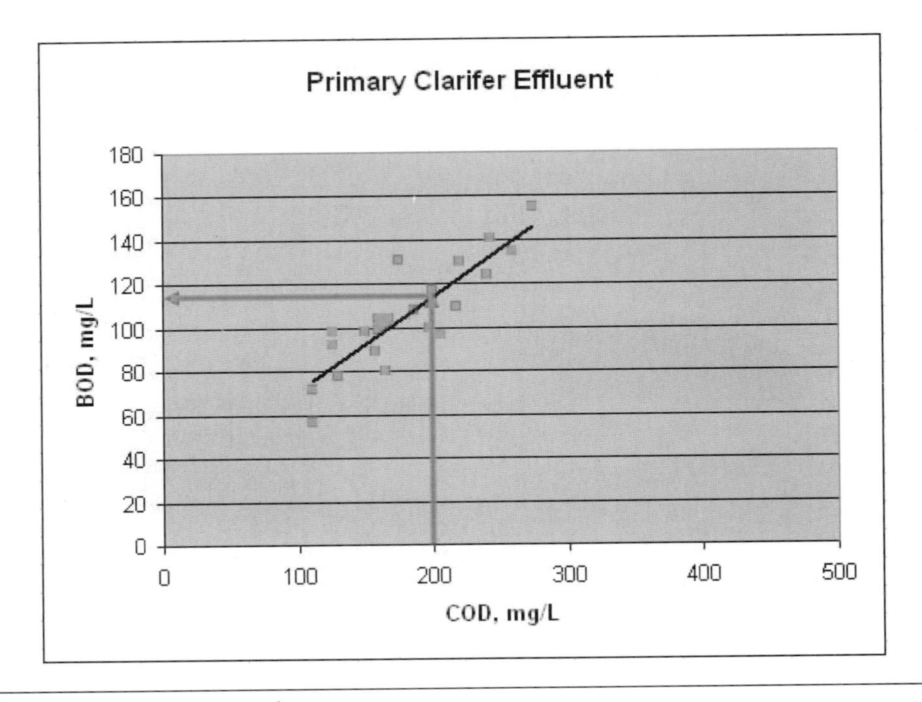


**FIGURE 14.11** Scatter graph.

process performance), a few statistical tools can help limit the parameters to monitor. In choosing criteria to monitor, the operator should periodically review the validity of tracking a particular parameter. This is nothing more than verifying that a connection exists between a variable and how the process performs. Statisticians call this a correlation. Several basic statistical methods can be used to determine how strong the relationship is between factors; these include linear regression, correlation analysis, and the attainment frequency method. The first two relationships are discussed below. The attainment frequency method was developed to address dynamic systems, which tend to produce data with a high degree of scatter (points are not clustered tightly about a straight line) and low correlation coefficients. The attainment frequency method was developed to meet a defined performance within a defined frequency, thereby accounting for and quantifying data scatter. Using this method produces quantitative results that can be applied directly to process control in treatment facilities. See Cochrane and Hellweger (1994) for more information on this method.

***Line of Best Fit—Linear Regression.*** Scatter graphs plot variables of two different measurements. Should the data appear to be randomly scattered about a line, some insight may be drawn regarding the relationship between the two variables. If the points are clustered tightly about a straight line, the data may be used as a prediction tool for process control. A portion of a WWTP'S MLSS and the resulting effluent BOD values for the same day are shown in Table 14.3. Mixed liquor suspended solids compose the control parameter ($x$) and BOD is the process performance indicator ($y$). One value is plot-

**TABLE 14.3** Line of best fit data.

| | Mixed liquor suspended solids, mg/L $x$ | BOD, mg/L $y$ |
|---|---|---|
| | 2 708 | 25 |
| | 2 489 | 18 |
| | 2 465 | 14 |
| | 2 418 | 10 |
| | 2 629 | 19 |
| | 2 633 | 17 |
| | 2 603 | 14 |
| | 2 220 | 12 |
| | 2 412 | 10 |
| | 2 598 | 17 |
| | 2 653 | 27 |
| | 2 599 | 21 |
| Sum | 30 427 | 204 |

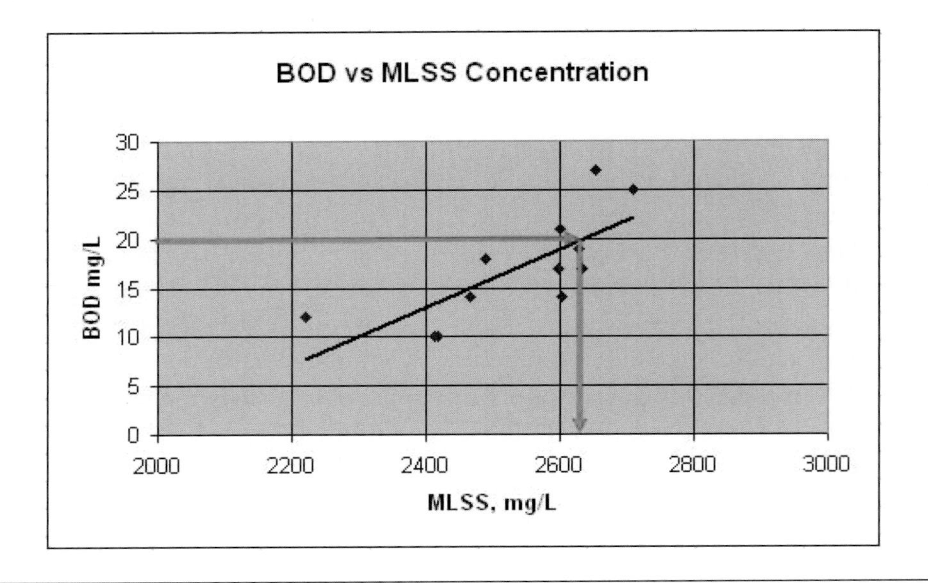


FIGURE 14.12 Line of best fit.

ted for each pair of values. These values and the calculated regression line are plotted in a scatter graph as Figure 14.12.

Although these values do not fall in a straight line, they are clustered around the drawn line. A straight line was drawn through the points in such a manner as to pass as close to the data points as possible. Using this line, rough estimates of the effluent BOD based on the aeration tank mixed liquor solids level may be extrapolated. To produce an effluent BOD less than 20 mg/L, the mixed liquor concentration should be approximately 2600 mg/L. Placing the line "by eye" is quite subjective and, while adequate for quick estimates, leaves a lot of room for argument. A more accurate method would be to calculate the line from the data given by expanding Table 14.3 as seen in Table 14.4 using the following set of formulas:

$$\hat{Y} = mx + b$$

Where

$\hat{Y}$ = the predicted process indicator value,
$m$ = the slope of the line,
$x$ = the control parameter, and
$b$ = a constant (the $y$-intercept).

Where

$$m = \frac{\sum xy - (\sum x)(\sum y)/n}{\sum x_2 - (\sum x)^2/n}$$

and

$$b = \frac{1}{n}(\sum y - m\sum x)$$

The symbol "$\sum$" is mathematical shorthand for "the sum of". For example: $\sum xy$ means that you multiply the value of $x$ by the value of $y$ for each data pair, then add up the products to arrive at the sum.

The symbol $n$ stands for the number of observations. Using the data in Table 14.3, the number of observations ($n$) equals 12.

To set the line, select two values of the control parameter and calculate the value of $\hat{Y}$ and plot the results on the graph, drawing a straight line between the two points. This line becomes the line of best fit, or the regression line.

Using the sums in Table 14.4 to fill in the formulas:

The values of $m$ and $b$ are calculated to be 0.029 6 and −58.14, respectively. Substituting those values into the prediction formula, the equation is now:

$$\hat{Y} = (0.0296)(x) - 58.14$$

**TABLE 14.4** Line of best fit—regression data.

| | Mixed liquor suspended solids, mg/L | Biochemical oxygen demand, mg/L | | |
|---|---|---|---|---|
| | $x$ | $y$ | $xy$ | $x^2$ |
| | 2 708 | 25 | 67 700 | 7 333 264 |
| | 2 489 | 18 | 44 802 | 6 195 121 |
| | 2 465 | 14 | 34 510 | 6 076 225 |
| | 2 418 | 10 | 24 180 | 5 846 724 |
| | 2 629 | 19 | 49 951 | 6 911 641 |
| | 2 633 | 17 | 44 761 | 6 932 689 |
| | 2 603 | 14 | 36 442 | 6 775 609 |
| | 2 220 | 12 | 26 640 | 4 928 400 |
| | 2 412 | 10 | 24 120 | 5 817 744 |
| | 2 598 | 17 | 44 166 | 6 749 604 |
| | 2 653 | 27 | 71 631 | 7 038 409 |
| | 2 599 | 21 | 54 579 | 6 754 801 |
| Sum | 30 427 | 204 | 523 482 | 77 360 231 |

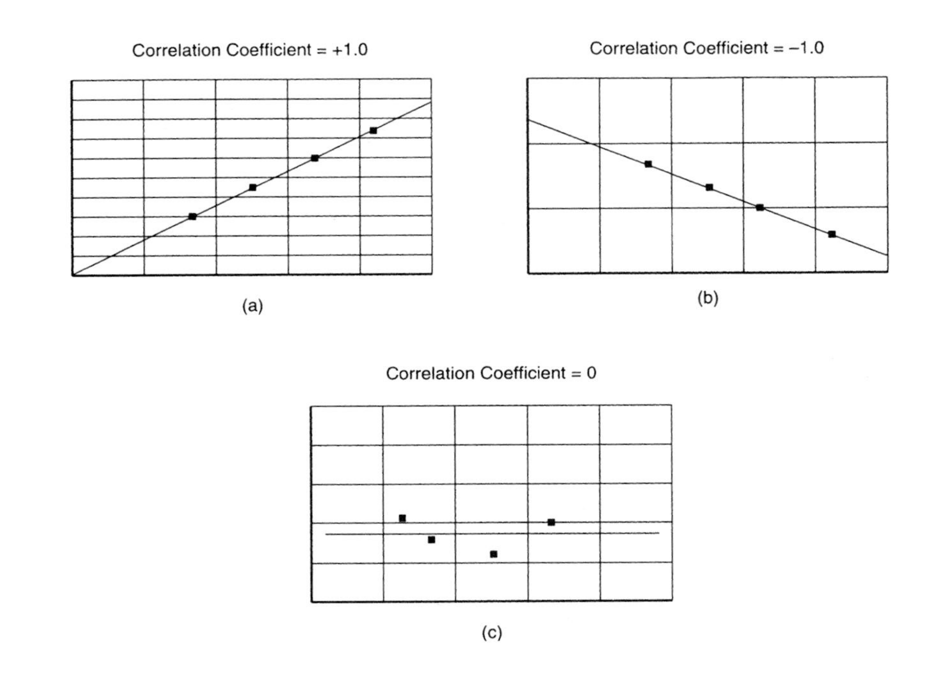


**FIGURE 14.13** Correlation analysis.

Then select two or more MLSS values to determine the line. Using MLSS values of 2400 and 2600 for $x$, the predicted BOD values are 13 and 19 mg/L, respectively.

***Correlation Analysis.*** Correlation analysis takes the equation of a straight line, previously developed, and calculates a correlation coefficient between the data pairs. The correlation coefficient $r$ gives a numerical rating of how closely the data plots to a straight line. A perfect correlation results in a coefficient of $+1.0$ or $-1.0$ and results in a straight line, as seen in Figures 14.13 (a) and (b), with a positive (line rises from left to right) or negative (line falls from left to right) slope, respectively.

The closer the value is to 1, the more meaningful the relationship. Because wastewater treatment plants are dynamic in nature, there is a high variability or "scatter" in the plant data. A correlation coefficient of 0 indicates that the data are distributed around a horizontal line and therefore have neither a positive nor a negative relationship, as in Figure 14.13 (c). For dynamic biological systems such as wastewater treatment facility systems, a good correlation coefficient will range between 0.6 and 0.7. The correlation coefficient is calculated as follows:

$$r = \frac{n\sum xy - (\sum x)(\sum y)}{\sqrt{n\sum x^2 - (\sum x)^2}\sqrt{n\sum y^2 - (\sum y)^2}}$$

Where

$n$ = the number of observations,

$x$ = the control parameter (MLSS, SRT, flowrate), and

$y$ = the process performance indicator (effluent TSS, BOD, percent volatile reduction).

The easiest way to calculate $r$, the correlation coefficient, is to expand the previous table, find the needed sums, and substitute them into the formula. The data set in Table 14.5 is used in the illustration.

Using the sums in Table 14.5 to fill in the formula, $n = 12$, the number of observations.

$$r = \frac{n\sum xy - (\sum x)(\sum y)}{\sqrt{n\sum x^2 - (\sum x)^2}\sqrt{n\sum y^2 - (\sum y)^2}}$$

$$r = \frac{12(523\ 482) - (30\ 427)(204)}{\sqrt{12(77\ 360\ 231) - (30\ 427)^2}\sqrt{12(3794) - (204)^2}}$$

$$r = \frac{74\ 676}{\sqrt{2\ 520\ 443}\sqrt{3912}}$$

$$r = 0.752$$

Considering the dynamic systems involved, a 0.752 correlation would be considered good. A strong relationship exists between the control parameter and the process performance indicator. From this, one may choose to use MLSS as a predictor of effluent BOD as a reliable process control tool.

***Data Handling Summary.*** There are many conventional tools available for handling data generated at WWTPs. Time plots containing several parameters on the $y$-axis versus time on the $x$-axis are useful in evaluating data trends. Sometimes, especially with dynamic biological systems, change occurs more gradually. It may be more useful to use a multiday running average in evaluating process performance. Examining relationships between two variables is best assessed using a scatter graph to show the direction of the potential relationship. From this information, regression lines may be developed for use as a predictive tool and their validity checked using correlation analysis. Selecting the proper tool for the right job to obtain the desired results requires knowledge of the strengths and weaknesses of each analytical method.

TABLE 14.5 Correlation analysis.

| Mixed liquor suspended solids, mg/L | Biochemical oxygen demand, mg/L | | | |
|---|---|---|---|---|
| $x$ | $y$ | $xy$ | $x^2$ | $y^2$ |
| 2 708 | 25 | 67 700 | 7 333 264 | 625 |
| 2 489 | 18 | 44 802 | 6 195 121 | 324 |
| 2 465 | 14 | 34 510 | 6 076 225 | 196 |
| 2 418 | 10 | 24 180 | 5 846 724 | 100 |
| 2 629 | 19 | 49 951 | 6 911 641 | 361 |
| 2 633 | 17 | 44 761 | 6 932 689 | 289 |
| 2 603 | 14 | 36 442 | 6 775 609 | 196 |
| 2 220 | 12 | 26 640 | 4 928 400 | 144 |
| 2 412 | 10 | 24 120 | 5 817 744 | 100 |
| 2 598 | 17 | 44 166 | 6 749 604 | 289 |
| 2 653 | 27 | 71 631 | 7 038 409 | 729 |
| 2 599 | 21 | 54 579 | 6 754 801 | 441 |
| Sum 30 427 | 204 | 523 482 | 77 360 231 | 3794 |

**FORMULATE A RESPONSE.** Before making any process change, the operator should complete the fact-finding phase, evaluate pertinent analytical data, and review SOPs. In addition, individual process troubleshooting guides contained in the process chapters may be of assistance.

In formulating process changes, select one parameter to change. Multiple changes may have unanticipated effects when combined and may cloud future evaluation when trying to determine the effectiveness of a given strategy. Most wastewater treatment plants are biological in nature. As such, environmental changes are slow to manifest. A good method is to wait two to three mean cell retention time (MCRT) periods before making any further changes. That allows enough time for the biosystem to acclimate, giving the operator a firm foundation to make subsequent adjustments.

After determining a course of action, the changes should be documented. Process control parameters should be entered into a permanent log for review by all personnel. Verbal communications ensure an opportunity for questions regarding operational philosophy so that a consistent strategy may be followed through all shifts.

**REEVALUATE.** After adopting a change, the unit process control parameters should be checked daily and evaluated accordingly. In addition to effects on the specific process, effects on downstream processes should be evaluated to determine the overall consequences on plant operations. If changes appear ineffective after several MCRTs, the operator should go back to the beginning and start the process again. This time, all staff members should be asked to help determine possible effects that may have been misinterpreted or overlooked in the previous evaluation.

## REFERENCES

Cochrane, J. J.; Hellweger, F. L. (1994) Interpreting Treatment Plant Performance Using an Attainment Frequency Methodology. *Proceedings of the 67th Annual Water Environment Federation Technical Exposition and Conference*, Chicago, Illinois, Oct 15–19; Water Environment Federation: Alexandria, Virginia.

Water Environment Federation (1993) *Instrumentation in Wastewater Treatment Facilities*, Manual of Practice No. 21; Water Environment Federation: Alexandria, Virginia.

Water Environment Federation (1997) *Automated Process Control Strategies*, Special Publication; Water Environment Federation: Alexandria, Virginia.

Water Environment Federation (2004) *Control of Odors and Emissions from Wastewater Treatment Plants*, Manual of Practice No. 25; Water Environment Federation: Alexandria, Virginia.

Water Environment Federation; American Society of Civil Engineers (1998) *Design of Municipal Wastewater Treatment Plants*, Manual of Practice No. 8; Water Environment Federation: Alexandria, Virginia.

# Chapter 15

# Outsourced Operations Services and Public–Private Partnerships



## GENERAL OVERVIEW—OPTIONS AVAILABLE

The option to outsource operations of a public (or private) wastewater or water treatment facility to an entity other than the owner has been in existence for over 30 years. In its original form, it was simply identified as *contract operations,* and generally referred to any situation where an outside company was contracted to perform, in whole or part, the operations of a facility. These contracts were typically three to five years in duration, with the contractor accepting only limited risk for facility operations in terms of labor, consumables, and routine repair or replacement of the operating equipment. With the revision of the federal tax law in 1997, often referred to as 97-13 (Internal Revenue Service [IRS] Revenue Procedure 97-13), which removed the time duration limit on a private for-profit entity operating a publicly funded facility for more than five years, the private outsourcing companies were now able to assume a much longer term role and have the ability to assume more risks in the full cost of the facility operation.

These contract arrangements are considered public–private partnerships (P3) and can include many options in terms of contract services, full permit compliance, capital repair or replacement, capital upgrades, facility expansion, asset management, project financing, the design–build–operate (DBO) approach, and full turnkey services. In general, there is a much higher transfer of risk across many issues from the public sector to the private contractor in a P3. Each project presents its own requirements and challenges with the accompanying benefit opportunity available for the owner.

There are several key areas for consideration in forming any P3, the first of which should be to determine what the problem is to be resolved by the action. The primary motivator often is the potential economic benefits to be gained through forming a P3. The efficiencies available from a core business provider through leveraged purchasing, personnel training, application of technology, and broad industry knowledge can far surpass those found at the typical municipal level. Other factors may include regulatory compliance, financing capability, introduction of new technology, or be political in nature. A second consideration is the procurement process for retaining the services of a service provider. The process must include evaluation of multiple key factors, assigning appropriate weight to each, to arrive at a proper selection, which will deliver a sound, successful, long-term partnership. The third element is the agreement itself. The agreement must be carefully crafted, absent any significant ambiguity, providing a solid platform on which to base the P3. Apportionment of risk must be carefully defined and adequate controls created to protect both parties. The contractor must be afforded the opportunity to operate and manage the facility, without undue hindrance or encumbrance by the public owner, if they are to successfully bring to bear the expertise and capability for which they have been selected.

The term *contract operations* is generally used to describe a project where there are no significant capital projects or improvements included, or private project financing or ownership. This service can be narrowly defined to limit the involvement of the private-sector partner, similar to the original style contract relationship. However, there can be, and often is, significant risk transferred from the owner to the private sector in these projects. The extended time duration of, in some cases, over 20 years, affords the private contractor the ability to level out operating cost considerations, such as major equipment replacement, that otherwise are not possible. This can provide the opportunity for the municipal sector to transfer significant risk to the contractor, realizing such benefits as long-term rate stabilization.

A turnkey approach can be coupled with outsourcing. This involves the contractor providing design and construction services for a new or rehabilitation project, with the owner providing financing. The contractor would then also provide operations services, at a predetermined cost for the project. This is often referred to as a DBO project and is becoming more popular in the marketplace.

The application of a DBO solution in the execution of a municipal project is a method that requires unique understanding, procedures, and approaches to ensure success. The use of this approach is commonly driven by the need to complete the project in a more timely fashion at lower costs than are typically associated with the traditional design–bid–build approach. There are numerous advantages delivered in a DBO solution, including the amalgamation of risk and responsibility to one party, the design–build contractor, and a design driven from an operations focus with a vertically integrated project design effort through all disciplines. The resulting facility's components have been selected with life cycle cost analysis, as opposed to "lowest bid", and often provide a project delivering unique solutions to the owner's problem. This approach may be used for an entirely new facility or an upgrade or expansion of an existing one.

The DBO contractor's team will, in its most effective configuration, have the operations sector involved in the design development from the very outset, with continued input throughout the project execution. In some situations, the team is actually led by the operations sector, which perhaps delivers the most cost-effective solution to the owner. This is assured if the operator is required to operate the facility for a significant period of time (20 years), at some level of guaranteed operating cost, and must deliver the project capital component at a competitive cost.

Developer financing further involves the private sector and can be used as a standalone, service-delivery option or in conjunction with either turnkey services or outsourcings. Developer financing is often undertaken when a developer has a vested interest in seeing a facility built or a system expanded, to ensure that development plans move along reasonably quickly. The other factor can be debt-issuance limitations of the municipal entity, which requires an outside funding source. In the normal condition, it is almost always true that municipal debt can be issued at more favorable rates than that of a private company.

A fourth form of public–private partnership is full privatization. This term is often used to include those arrangements previously discussed and involves the private sector with all aspects of the project or system, including, most significantly, ownership through the sale of the asset to the private contractor. This approach can provide the public body with broad-scope solutions to regulatory pressure to construct, expand, or upgrade a facility. Full privatization may also be a response to economic pressures in the community to moderate rate or tax increases.

In the privatization of a new wastewater treatment facility, the private sector can be involved in the design, construction, financing, ownership, and operation of the facility. The public sector may be involved in various components of this service delivery option, including participation in the selection of the technology or other decisions associated with facility design and construction.

Privatization, as applied to an existing facility, may serve to upgrade or expand it or simply to reduce and contain operating costs. Where the public entity has financial demands in other community areas, privatization of an existing facility, which is the sale of the facility (assets) to the private sector, may serve as a cash source to address other demands. This can be accomplished with a true ownership transfer or a long-term leaseback arrangement.

The economic scope that must be considered in a privatization effort is much broader than basic outsourcing. Specific issues include the existing debt on the facility and the type of debt involved. The present worth or book value of the facility will establish the selling price and debt level that the existing rate structure will support. These factors must be analyzed carefully to determine the benefits of privatization. Opportunities exist for a municipality to reap the rewards of privatization of wastewater facilities, but it must be accomplished with great care.

The final P3 is a merchant facility. Perhaps the best example of a merchant facility is the siting, construction, and operation of a water reuse facility by the private sector. In this example, the private sector takes the initiative in constructing facilities and maintains ownership and operation responsibilities of the facilities providing service to all customers, processing wastewater to provide water for industrial, irrigation, and other demands. Unlike a privatization option, a merchant facility serves more than one customer and typically does not restrict service to those who have fixed contracts with the merchant facility owner or operator.

The P3 service delivery options may also be referred to as build–own–operate–transfer (BOOT) and build–operate–transfer (BOT), or DBO. These arrangements contemplate that the facility ownership will transfer from the contractor to the municipal client at some point in the contract, typically at the end. The terms of transfer can be defined in the contract to address requirements of procurement laws, bond and tax issues, and project financing.

## SERVICE DELIVERY INDUSTRY

The service delivery industry, in its simplest form and in its core business offering, is considered contract operations. This approach is not a new concept in providing non-core-business services or in providing water and wastewater treatment services for municipalities. Contract operations has, for many years, been widely used outside the utilities industry, in such areas as transportation, transit services, prisons, schools, hospitals, airports, fleet maintenance, janitorial services, street sweeping, and solid waste collection and disposal. In the water and wastewater treatment industry, outsourcings arrangements have existed for many years, and currently there are well in excess of 1400 agreements for the operation and maintenance of wastewater treatment plants (WWTPs) by the private sector, which represents only approximately 3 to 5% of the market. This number includes approximately 40 agreements for large facilities greater than 440 L/s (10 mgd) and 350 contracts for mid-size facilities, ranging from 44 to 220 L/s (1 to 5 mgd). Within these projects, there have been approximately 40 agreements for a term of 10 years or longer, with many of them set at 20 years. It is estimated that there are thousands of outsourcing agreements for small or package WWTPs located throughout the country.

These numbers do not take into account the multitude of moonlighting or technical supervision jobs typically performed by licensed municipal operators, which are also valid operations contracts. These figures also exclude the hundreds of outsourcings agreements in place for the operation of components of WWTPs and for private-sector service arrangements for functional activities, such as meter reading, customer billing, meter repair, and testing.

The current trend toward outsourcings and other P3 service delivery options might well be a result of the enhanced regulatory environment in which utilities operate. The requirements of the Clean Water Act, the Safe Drinking Water Act, and other environmental regulations are forcing utility providers to make significant capital investments in more complex and technologically challenging facilities, to attract and retain qualified staff to operate these facilities, and to do so with the ongoing pressures of cost control and concerns for customer rate increases. In the past, when such pressures existed, the local government sector had a federal partner in the form of a grant-funding program available for wastewater facilities. With the demise of this program and the increasing competition for state revolving fund monies, the economic effect of the new facilities designed to meet federal environmental mandates falls squarely on the shoulders of the local government. All these factors point to the increased need and involvement, in many situations, of professional service providers to assist communities in meeting the increasingly complex challenges associated with water and wastewater service.

The services delivery industry has undergone a significant transition since the enactment of 97-13, which enabled a broader scope service delivery and greater risk transfer to the service provider. There has been a consolidation of the companies, with a number of the smaller providers being merged with larger companies. This was in response to the need for stronger financial resources to accept a larger risk transfer and support financing of capital projects or asset transfer and an emerging recognition that the demand for privatization capability would continue to develop. In addition, the company size brings added benefit to the municipal entity, with their leveraged buying position in the marketplace for chemicals, supplies, and support services. Core competencies, such as training, standard operating procedures (SOPs), operating system support, and performance optimization all benefit from a larger operating base. The pairing of either engineering or construction and/or process equipment capability has also emerged to respond to the broader scope encountered in the full-service P3 contracts.

The outsourcings industry provides core services along a variety of lines, including the following:

- Full system operation;
- Management oversight;
- Laboratory analysis;
- Plant startup and testing;
- Operator training;
- Emergency maintenance;
- Contract management;
- Diagnostic operational reviews;
- Billing, collection, and customer services;
- Groundwater remediation;
- Predictive and preventive maintenance;
- Pipeline inspections and repair; and
- Facility design, construction, and commissioning.

Under full-system operation, the private sector may be retained to design, build, operate, maintain, and repair an entire water or wastewater utility system or portions thereof. What is often more common is to segment system components or functional activities for outsourcings, typically referred to as *inside-the-fence* or *outside-the-fence* operations. Inside-the-fence operations include all or part of the functional responsibilities of the treatment plant processes performed within the confines of the treatment facility. Outside-the-fence operations include operation of the water distribution or wastewater collection system, pumping stations, well fields, or biosolids disposal sites.

Depending on the services for which a service provider is retained, companies can range from small local firms to large operators working on national and international projects. The industry might be divided into three categories: international/national, regional/national, and regional/local. This division is somewhat subjective, but can be useful in assessing the capabilities of the various providers to undertake projects of varying complexity and size. Issues that are relevant when considering service providers include past performance, operations experience in similarly sized projects, financial stability, bonding capability, environmental compliance history, support capability, and successful completion of similar capital projects.

National firms may be affiliated with large domestic or foreign engineering design firms operating on a national or international scale. Regional/national providers are often well-established firms providing services within a broad region of the United States. New entrants in this area include foreign-outsourcings firms that have just begun to provide services or acquired the capability in the United States. The third set of providers includes regional or local firms that may be well-established in a specific geographic region and perhaps even in a specific functional area, such as pumping station maintenance and repair.

## ADVANTAGES AVAILABLE IN OUTSOURCING

The issues that lead municipal entities to consider outsourcing their wastewater system operations are diverse. However, there are a number of key elements that include continually increasing environmental regulations and standards, increased pressure from served constituents to maintain rates, competition for a continually decreasing tax/rate base, aging infrastructure, and general operational challenges. Outsourcing these services can be viewed as an economical, effective alternative to the traditional service-delivery options of local government ownership and operation. Typical reasons that a municipality considers outsourcing either operations or project delivery include the following:

- Solve compliance problems or address increasing regulatory requirements,
- Resolve labor relations problems,
- Reduce operating costs and/or capital project costs,
- Expedite capital upgrade delivery,
- Access capital project funding,
- Assure availability of qualified staff,
- Respond to capital and/or capacity upgrades, and
- Transfer risk to private service provider.

While these are often cited as the reasons for evaluating outsourcing as a service-delivery option, they may also serve as hurdles or barriers to the outsourcing process. For example, labor relations might be a positive or negative factor. The presence of a strong union or organization may be a significant obstacle to implementation. While this issue has precluded outsourcing from serious consideration in some communities, it has also been dealt with effectively in many situations by local government leaders and service providers. A key component in successfully addressing labor-related issues is the development and distribution of a well-structured employee transition plan. The uncertainty of existing employees when outsourcing operations is being considered can often be dealt with by timely and informative communication. For example, employees want assurance that, if they decide to accept employment with the service provider, they will do so with some security and certainty regarding wages and benefits. Employees typically look for security in provisions that prohibit their termination, without cause, within a specified period of time or put limits on the ability of the service provider to transfer employees to other facilities during a given time frame. However, it is not a good idea to impose employment terms in the contract beyond five years. These issues can be addressed effectively in the operations agreement discussed later in this chapter. At the same time, the local government must plan for those employees wishing to stay with the local government. This plan can include the development of retraining programs for former utility employees for a smooth transition to other local government duties.

In a practical sense, cost savings must be achieved if a local government is to accept and implement a privatization action. A contractor is able to avoid considerable expense in the procurement of goods and services because of its freedom from the bidding requirements of most public agencies. The avoided expense is primarily in the cost of the purchasing procedures and the time lost while complying with them in addition to their leveraged size in the marketplace. Contractors may also be able to avoid union restrictions (closed shop, work jurisdictional claims, and feather bedding) and can respond more readily to market forces affecting their ability to maintain a workforce to meet their needs. Other areas that contractors leverage is focused operations expertise, standardization of operational systems, multiskill employee development, and introducing automated systems and controls.

Perhaps the only exception is a situation where the U.S. Environmental Protection Agency (U.S. EPA) or a state agency has issued a compliance order. In this case, the local government may have no choice but to retain private-sector services to ensure compliance. Even in this situation, however, cost savings are likely to be achieved over time.

The element of risk and the apportionment of it in the contractual relationship are not as tangible as other items cited previously, but are just as valuable to the municipal entity and should be carefully evaluated when considering a privatization action. The main risk areas to be considered include environmental compliance, cost of operation, cost of maintenance, cost of major capital projects, and general project performance. These risks can all be quantified and equitably transferred or shared in a carefully crafted contract that balances them equitably and assigns them to the correct party in the best position to manage them.

## ECONOMIC FEASIBILITY OF OUTSOURCING

Certainly, one of the foremost reasons for considering outsourced operations or a P3 is the potential for cost savings. The opportunities for cost savings through outsourced operations might depend on the following:

- Complexity of the facility process and/or operation;
- Size of the facility or system;
- Complexity of capital project or upgrade;
- Method of contracting for capital projects (BOOT/DBO);
- Location of the facility (proximity to other operated facilities);
- Regulatory compliance history and potential;
- Opportunities for rationalization; and
- Labor history and condition.

Individually, these factors influence the ability of a service provider to achieve both cost savings for the local government and acceptable internal profit margins. In situations where services are provided for a small facility employing only a few people, cost-saving opportunities may be quite different from those in larger facilities with a significant operating staff and budget.

The facility's location (remote versus urban) can influence the ability of the service provider to achieve the necessary economies of scale in management and operational areas to produce material cost savings. A remote facility with only one or two employees presents a much different set of circumstances and opportunities for cost savings compared with a larger facility or system in an urban area. With the presence of other communities in close proximity, all of which might also have WWTPs or similar facilities, the contractor may be able to achieve economies of scale for the benefit of all involved parties through a regionalization approach.

Projects that include a capital upgrade or new facility BOOT/DBO approach will generally benefit from a more performance-based contract as opposed to one that is highly prescriptive in defining the ultimate technical project solution. Flexibility in the contract will create the opportunity for the provider to apply maximum creativity in delivering the project capital solution.

The design standards incorporated to a BOOT/DBO project are not necessarily consistent with those found in a typical municipally based traditional project. The municipal standard is one that has its roots in many elements, not the least of which are those applied by U.S. EPA in the early grant programs, which funded much of the wastewater infrastructure and, in some cases, actually rewarded participating parties for escalating project costs. That is not to say these standards were not good ones. However, the product delivered by a typical DBO project will be configured more closely to general industrial standards found in practice today in private industry. These standards are driven by life-cycle cost considerations, reasonable levels of fit and finish for the facility, and acceptable equipment redundancies governed by practical standards. When configuring the contract, the owner should consider what level of standards in the project he or she would like. If there are certain "must haves" in construction materials, types, or configuration, or equipment selection and redundancy, then they should be identified in the request for proposals (RFP) package. However, the more closely constrained the DBO contractor is in this regard, the less benefit the owner may derive in project delivery and cost effectiveness. It is also recommended that the owner be flexible in accepting alternatives that may be offered to the standards prescribed and resist the temptation to universally apply standards used in the past. This approach will deliver the greatest economic benefit in a DBO solution.

Aside from the cost savings often associated with outsourcing, a local government may not be able to comply with the regulations affecting the WWTP under the current operational mode. This situation may provide the incentive and motivation for the local government to consider outsourcings. It is also likely that, in complying with requirements, cost savings may be achieved.

In attempting to quantify cost savings, a local government should take care when comparing the projected private-sector costs and costs of service provided by the local government utility. All too often, significant costs are not included in determining the costs of local government operations. Many indirect costs, such as those associated with the local government's finance and legal departments or the full capture of benefits and other related costs, are often excluded from the analysis. To make an appropriate comparison between an outsourcing proposal and the current cost of service, it is vital to capture all service-related costs. Furthermore, whether under outsourcing or under continued local government ownership, there will be costs associated with administering and overseeing the respective services. These costs may be significantly different under either option and an attempt should be made to estimate them before comparing current operations with the private sector.

In comparing local government service and the outsourcing provider, there is often a failure to compare precise service levels. This problem can be remedied by carefully examining the services currently provided under local government operation and, as appropriate, adjusting those services before issuing a request for qualifications or request for proposals. A local government must thoroughly understand the current services it provides, assess the need for those services, and develop a comparable scope of services for prospective service providers. In estimating the cost of current service delivery by a local government, costs should reflect the scope of services or service levels to be included in the request for proposal. This means that the local government must carefully analyze its current method of providing service and make the necessary adjustments to ensure that service is provided as effectively and efficiently as possible. This internal examination or management/performance review will enhance the ability of the local government to remain competitive with the private sector. Even if the local government decides not to pursue outsourcing, the exercise will benefit the municipality by providing the required services more efficiently and effectively.

Comparing current local government costs is one step in the evaluation process. Another step is to have the local government division or section submit a bid to continue the service being considered for outside bid. This form of managed competition has been used by several local governments when considering outsourcing and affords existing employees the opportunity to compete objectively against the private sector. To assist employees in the preparation of their bid, the local government can provide internal support services or retain outside consulting services. However, this form of managed competition will not be viewed positively by outsourcing companies and will likely discourage enthusiastic participation in the procurement.

Managed competition has produced mixed results in the ability of existing employees to win the competitive bid process. Services awarded to the local government group through this process should carry the same contract provisions and be held to the same standards that would be imposed on the private sector, such as performance measures and reporting, fixed fee arrangements, and cost savings and sharing incentives. Issues that must be considered for an accurate comparison include the availability of performance bonds, liability insurance, permit compliance, technical support, and the accompanying costs associated with these. An evaluation should also consider if the employees are unable to operate the utility within the bid price; the local government will incur the added costs. With an outsourced provider, the inability to operate within the bid price is absorbed by the shareholders of the company. If economic pressures are too great, there is a risk the contractor may petition for fee increases, file contract claims, or pursue litigation to obtain relief.

## PROCUREMENT PROCESS

The procurement process for selecting an outsourcing contractor, regardless of the scope of service, must be carefully crafted to assure delivery of a complete service responding to the project requirements. In addition, the process must be consistent with procurement laws, which vary greatly from state to state.

In many locations, communities have the option of going directly to an RFP or taking the intermediate step of issuing a request for qualifications (RFQ) statement from service providers. The two approaches may require different amounts of time and may influence the number and quality of submitted proposals. Vendors typically prefer a two-step RFQ/RFP process, because the cost of preparing a proposal can be significant. It may range from tens to hundreds of thousands, if not millions, of dollars, depending on the size of the operating contract. Accordingly, most vendors prefer to initially submit qualifications statements, from which the municipality can develop a short list of qualified bidders and ask only those bidders (typically three to five firms) to submit formal, detailed proposals. This approach will enhance the procurement opportunity, narrowing the competitive field. However, a limiting factor may be state or local laws governing the procurement of services by a municipality or public agency. These laws can relate to the bid process associated with procuring either construction or professional services. In some instances, the procurement requirements associated with professional services provide the latitude needed to undertake a two-step RFQ/RFP process.

Having received statements of qualification and technical and cost proposals from vendors, the next step is evaluation and selection. Listed below are several broad evaluation categories that can be used in reviewing submissions:

- Corporate profile,
- Financial strength,
- Corporate experience,
- Key project staffing,
- Project performance references,
- Proposed project pricing, and
- Project technical approach.

Corporate profile is a general criterion requesting vendors to submit information on the background and primary nature of the company, such as who the proposer is, what services it provides, and where it is located. Knowledge of affiliated companies and organizations can be important in evaluating the experience of the proposer. It is also important to understand the financial resources available to the company and its ability to undertake the project. Requiring such information as annual financial statements and reports can provide this kind of information. The goal is to evaluate the ability of the vendor to undertake the project, provide certain performance guarantees, and ascertain historical data on any past bankruptcy filings or past and pending litigation.

The corporate experience of the proposer is probably the most important criterion in selecting a vendor. Corporate experience typically includes a description of similar projects that the proposer has undertaken in terms of size, climate, regulatory environment, and technology. The proposer's ability to provide evidence of its capability to supply the full range of requested services is important. When soliciting proposals for a WWTP requiring a particular treatment technology or process, it is important to ascertain whether the proposer has successfully operated that technology or process. The project or plant manager to be assigned is also important. As with corporate experience, proposers should provide information on the experience of the key operators and plant manager in working with processes in similar climates and conditions.

The key to ascertaining the qualifications and experience of the proposer is primarily through the reference check and, if the community can afford it, visiting one or more facilities currently being operated by the proposer. Reference checks should involve more than a cursory telephone call soliciting general information. Questions should be asked about the vendor's ability to meet project schedules, the success of its safety program (i.e., www.osha.gov, U.S. Occupational Safety and Health Administration citations and days lost ratio), and its ability to work with the municipality under extraordinary or extenuating circumstances, such as natural disasters. The amount of cooperation exhibited by the service provider in dealing with customer problems and the ability to work with local, state, and federal regulatory agencies are also important considerations. Through this process, it is important to identify failed or problem projects. Why a project failed or did not meet the client's expectations is important. Lastly, and of significant importance to elected officials, is the price of the service provider's proposed services.

The proposal structure required for a BOOT/DBO project will include submittal of information beyond the normal scope for either a traditional engineering design package or a typical municipal public construction or operations proposal. As previously discussed, the cost structure can be configured to respond to the project requirements. If project financing were included, then details of how the financing will be accomplished would be included. This would identify the debt issuing structure and entity, redemption schedule of the debt, risk apportionment for interest rates, and sale of bonds, if appropriate.

In addressing the technical requirements for a BOOT/DBO, the proposer will typically include design development drawings representing approximately a 40% design as part of the proposal package. These would include basic drawing information depicting site layout, facility plan, process and instrumentation diagrams, process flow diagrams, primary process equipment layout, building elevations, standard finish and hardware schedules, and others, depending on the project. The supervisory control and data acquisition (SCADA) system will be depicted to a planning level key element schedule. It is not unusual to have a submittal that includes up to 100 drawings for a major project. In addition, there will be process descriptions identifying process function, sizing of primary equipment, manufacturer, and basic standards of fabrication.

The proposer will typically include a project schedule that will cover from the "Notice to Proceed" through commissioning. The design/permitting process is an important one and should be depicted in enough detail to demonstrate a reasonable understanding of the process. As part of a BOOT/DBO project, the design effort is compressed, with time frames of eight months considered normal. Project permitting efforts occur concurrently with the design effort.

In addition to technical requirements, it is typical for the RFQ/RFP to require submittal of DBO team qualifications, including project execution history, financial condition, bonding capability, and experience specifically relevant to the project. These criteria should be quantitatively evaluated to rank the proposers in the selection process. A process that involves "scoring" and "weighing" to reflect the importance of each criterion, given equal weight with the project cost, is advisable to formalize the process and protect the owner's interest. Equal weighting assigned between price and the full technical considerations in the ranking of a team is suggested for the owner to select the best-qualified, most cost-effective proposer providing the best value to the owner.

While no single evaluation method is universally accepted, a numerical matrix that assigns points to the various evaluation parameters is typical. Using the previously discussed evaluation categories, a matrix that expands or reduces the categories into smaller, definable items and assigns maximum points to each one allows the proposal reviewers to assign a numerical score. Totaling the scores of each vendor establishes the ranking. The development of a completely objective evaluation process is a noteworthy, but often impractical, goal. Some element of subjectivity is acceptable and important to reflect specific local issues and goals and complete the process in a timely manner. Table 15.1 is a sample evaluation matrix.

## KEY CONTRACT TERMS

The outsourcing agreement is the foundation upon which the relationship between a local government and a service provider is built. One of the most frequent criticisms of outsourcing, and P3 service-delivery operations in general, is that the community loses control of the contracted operation. The operations agreement can provide greater levels of control and accountability than may currently exist under the traditional service option. Specifically, the contract can definitively outline the scope of services to be provided and compensation to be paid and provide for periodic (monthly, quarterly, or annual) performance reviews. Performance reviews require strict guidelines and standards to measure results. These reviews may also include the state regulatory body responsible for enforcing permit requirements. In many cases, these are standards of per-

**Table 15.1** Example of a Sample Evaluation Matrix. *(continued on next page)*

| Criteria | Ranking[a] and description | | Rank | Assigned value | Assessed value | Comments |
|---|---|---|---|---|---|---|
| (1) Number of years providing relevant services | HA | 5 or more | | 10 | | |
| | A | 3 | | | | |
| | NA | 2 | | | | |
| | U | None | | | | |
| (2) Client references | HA | 100% positive reference checks | | 20 | | |
| | A | 80 to 100% positive | | | | |
| | NA | 50 to 80% positive | | | | |
| | U | >50% positive | | | | |
| (3) Experience with services of similar capacity and/or process (design, construction, startup, and operations services) | HA | 3 or more | | 5 | | |
| | A | 2 | | | | |
| | NA | 1 | | | | |
| | U | None | | | | |
| (4) Number of projects with similar treatment technology in operation | HA | 3 or more | | 10 | | |
| | A | 2 | | | | |
| | NA | 1 | | | | |
| | U | None | | | | |
| (5) Number of projects or resources in proximity capable of providing technical and staff support | HA | 3 or more | | 10 | | |
| | A | 2 | | | | |
| | NA | 1 | | | | |
| | U | None | | | | |
| (6) Litigation, permit enforcement actions, and/or fines from similar projects | HA | None in last 3 years | | 5 | | |
| | A | None in last 2 years | | | | |
| | NA | None in last 1 year | | | | |
| | U | | | | | |
| (7) Detailed project plan | HA | Addresses all project requirements | | 10 | | |
| | A | | | | | |
| | NA | | | | | |
| | U | Incomplete | | | | |
| (8) Detailed corporate safety program addressing operations and construction | HA | Comprehensive program | | 5 | | |
| | A | | | | | |
| | NA | | | | | |
| | U | Incomplete | | | | |

*(continued)*

**TABLE 15.1** Example of a Sample Evaluation Matrix. (*continued from previous page*)

| Criteria | Ranking[a] and description | | Rank | Assigned value | Assessed value | Comments |
|---|---|---|---|---|---|---|
| (9) Identifies strategy for providing operations support: SOPs, operations and maintenance manuals, facility evaluation, maintenance management approach | HA | Comprehensive program | | 5 | | |
| | A | | | | | |
| | NA | | | | | |
| | U | Incomplete | | | | |
| (10) Provides in-house specialized support: process control, instrumentation and control, SCADA, and power management | HA | Comprehensive program | | 5 | | |
| | A | | | | | |
| | NA | | | | | |
| | U | Incomplete | | | | |
| (11) Corporate training programs | HA | Comprehensive program | | 5 | | |
| | A | | | | | |
| | NA | | | | | |
| | U | Incomplete | | | | |
| (12) Financial resources and stability | HA | Meets all RFP requirements | | 5 | | |
| | A | | | | | |
| | NA | | | | | |
| | U | Incomplete | | | | |
| (13) Completeness of proposal | HA | Meets all RFP requirements | | 5 | | |
| | A | | | | | |
| | NA | | | | | |
| | U | Incomplete | | | | |
| Total | | | | 100 | | |

[a]Ranking: HA = highly advantageous = 100% of point value; A = advantageous = 75% of point value; NA = not advantageous = 50% of point value; and U = unacceptable = 0% of point value.

formance or criteria that a municipality does not currently use to evaluate its own employees and operations.

A carefully crafted outsourcing agreement provides the framework under which services are to be provided. It clearly delineates the scope of services, the extent to which the operator has responsibility and authority to act, and the compensation provided to the service provider. Contracts need a certain degree of specificity; however, they also need flexibility because they are often long-term agreements. Figure 15.1 illustrates the components of an outsourcing agreement.

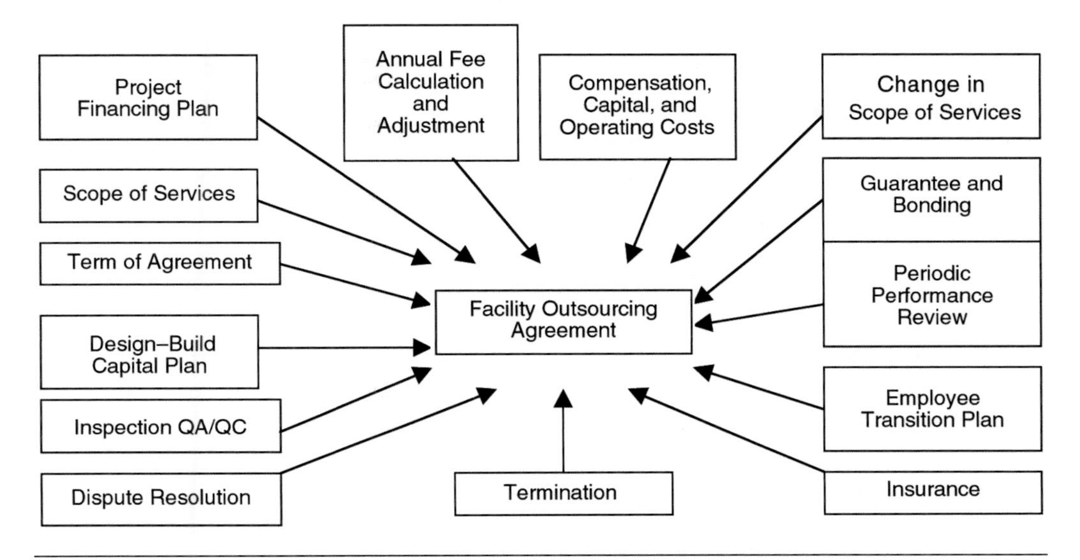


**FIGURE 15.1** Sections of an outsourcing agreement (QA/QC = quality assurance/ quality control).

Employee issues are critical to the success of outsourcing operations. Thus, the agreement must clearly address transition issues for existing employees. A formal transition plan typically includes provisions for addressing termination, transfer, compensation, and benefits. This is particularly true in union situations, but is important in any situation where existing employees are involved. The transition plan should define how employees can potentially be re-employed with the service provider's company, provide for employee counseling, and provide for economic analysis associated with pension plans and the like.

As was previously identified, the IRS changed the tax code, and it is now possible for private contractors to operate facilities for extended contract periods. This approach, incorporated to an outsourcing agreement, enhances the ability of the contractor to participate in the funding of capital facilities and the amortizing of such costs over an extended period of time. This situation can work to the benefit of the local government by mitigating rate increases and providing a financing source.

Most outsourcing agreements provide for a fixed payment by the local government to the service provider. The amount is generally paid on a monthly or quarterly basis and relates to a baseline service level and associated operation and maintenance costs based on a predetermined production level. The fixed payment also contains certain baseline costs for utilities, such as electricity or natural gas, with the utility rates adjusted on an annual basis. Annual adjustments can be allowed, based on a predetermined consumer price or general inflation index. Care should be taken when establishing a cost-adjustment index to ensure that it properly reflects local conditions and the effect inflation would have on the costs involved. A national consumer index may have little relation to the cost of chemicals, materials, and supplies in a particular region. The past performance of the index and how it correlates with changes in actual costs over a period of time are important to assessing the relevance of a particular index.

In addition to general inflation adjustments, the compensation amount can be adjusted to the extent that flows to the facility are more than or less than pre-established ranges. Such adjustments can also involve price increases or decreases for utilities or other production-related expenditures.

In crafting the compensation portion of the agreement, care should be taken to ensure that the service provider has adequate incentive to reduce costs, while still effectively operating the facilities. The incentive can be in the form of allowing the service provider to retain a portion of the savings achieved, with the balance accruing to the benefit of the local government. It should also be recognized that the proponent might have incorporated these anticipated savings to the base proposal.

Another key component of the compensation agreement is an annual maintenance fund, funded by the local government and agreed to and planned by both parties. This component ensures that a maintenance program is established and funded. The plan should have monitoring or reporting requirements associated with it, and, at the end of the year, any remaining funds should revert to the local government. To the extent there were no out-of-scope activities performed and maintenance costs exceeded the fund allowance, those excess costs would typically be borne by the service provider. This approach should also be applied to capital replacement, which includes the major equipment replacement as it reaches the end of its useful life. In a 20-year term, the new replacement assets will remain the property of the owner. Requiring the contractor to finance these major replacements throughout the contract will force accelerated depreciation of these costs against the contract in time frames significantly reduced below normal useful life. The result will be higher-than-necessary operating costs.

In projects involving capital improvements, certain considerations must be included. In an owner-financed project, payment schedules are typically tied to the project critical path schedule and may be milestone-based, which provides a relatively easy system to verify payment progress on a project. In the case of a contractor-financed project, the contractor and trustee will control the release of funds.

To have expedient project negotiations and award, it is strongly recommended the owner include contract terms and conditions in the RFQ/RFP and request comments from the proposer. An approach that establishes a fair and reasonable contractual relationship, without creating any unnecessary risk assignment, will provide the greatest benefit to the owner. There are many issues to consider when crafting a contract that would warrant a much broader review, but the following sections are points that are either unique to the DBO process or provide special challenges.

**LIABILITY.** The nature of a DBO procurement combines three categories of liability that require some form of resolution and assignment. The first is in the design category, which would typically be addressed through professional liability insurance for errors and omissions. This coverage is typically required of a DBO contractor, and, depending on the teaming arrangement, may be provided by an engineering subcontract into the project. Construction liability is generally addressed through the normal performance and payment bonds and typical builder's risk and general umbrella liability insurances associated with any public works project. Operations liability is readily addressed through general liability insurance and a performance bond or letter of credit typically reflecting the value of one or two years of the facility's operating cost. In addition, it may be appropriate to require that the operator carry hazard insurance on the facility assets, depending on which party is the facility owner during the course of the contract. Care should be taken to assign values to each individual coverage, reflecting the relative value of the associated risk, and not simply the total value of the project, which encompasses all three elements.

The long-term nature of these contracts merits consideration of additional liability coverage reflecting the scope of the services provided. This requirement should not be unlimited in nature, but should provide a value that reflects the real-world cost effects for project failure. An unlimited guarantee can really only be pledged once by a corporation to have a real value and is obviously subject to the net worth of the corporation. Requiring acceptance of unlimited liability is not acceptable to large corporations, as it requires "betting the bank", and is not a responsible action to owners and shareholders of the corporation. In addition, it places an unreasonable burden on the company, with an accompanying extraordinary cost. However, a bonded guarantee that pledges, as an example, $20,000,000 from a $3 billion corporation has a real and intrinsic value and is one to which underwriters can subscribe. This approach provides absolute quantification of the protection afforded the owner and removes any exposure associated with corporate success, bankruptcy, or ownership change of the company.

**CONTRACT BUYOUT.** It is sometimes deemed desirable to the owner to provide for contract buyout or call provisions. If this is included, the terms for the buyout must recognize the value of expenditures associated with project startup, ongoing operations, and lost opportunity for the contractor. In addition to these issues, the inclusion of call provisions has an effect on the ability to finance the project. If a bond issue is anticipated to finance the project, the marketability of the bonds and their accompanying discount rate will be negatively affected. It is suggested that, if the contractor is performing properly under the contract, there is really no need for the contract buyout provision. If performance does not meet requirements, then the owner may act on contract breach terms and exercise termination options.

**PROJECT FINANCING.** If the contractor is to provide project financing, the issue of interest rates is an important one. Given the time it often takes between the bid of a project and the award, it is not possible to lock in a bond issue interest rate for this duration. It is recommended that the interest rate used in the bid basis be adjusted at the contract award to provide the best value to the owner in terms of project financing. This also reduces the risk the proposer must assume with regard to project financing and therefore the cost passed on to the owner.

**CONTRACT TERMINATION.** The conditions for contract termination should be based on nonperformance or breach. Termination for convenience is not reflective of the long-term partnering relationship, which is desired in a DBO contract, and can be a disincentive for proposers to consider committing resources to the development of the project.

**BONDING REQUIREMENTS.** Bonds are an instrument used to mitigate project risk for the owner in the project. As previously discussed, the bonding requirements for a DBO are not overly different from typical contracting requirements; they are just combined in one contract body, as opposed to separate contract actions. Care should be exercised in determining the individual bond value requirements to avoid unnecessarily escalating bonding costs and duplication of coverage.

## SERVICE PROVIDER/OWNER INTERFACE

Once a facility is under contract operation, the degree and frequency of owner participation can have a direct effect on the success of the venture. A review of privatized or contract-operated facilities will reveal both success stories and failures. The success or failure can typically be related to the effectiveness of the contract and the type and degree of owner participation.

The type and degree of owner participation varies widely from contract to contract. It is incumbent on the owner not to under- or over-manage. The owner should not take an out-of-sight, out-of-mind attitude just because a contractor operates the facility, unless it has, in fact, transferred the permitting requirements and ownership of the facility. The owner is still ultimately responsible to the customers and state and federal regulators. Sufficient detail must be provided to verify that the facility is being operated within permit requirements and in accordance with the contract provisions and should be based on objective criteria established in the contract. It is also of utmost importance that the owner verifies that contract conditions are met with respect to maintenance and replacement of equipment and facilities to ensure protection of the owner's original investment. The other extreme is overmanagement, where overzealous owner involvement borders on harassment. Overmanagement is more likely to occur at facilities where employees strenuously opposed the move to privatization. Often, the facility owner will attempt to placate dissenting factions by establishing onerous reporting mechanisms, placing owner employees in the facilities as inspectors, or establishing oversight committees to ensure performance of the contract. In most cases, these actions serve only to increase the cost to the owner and limit effectiveness of the contract operator.

The unique requirements of a BOOT/DBO project also extend to the owner and the owner's project execution team. An approach enlisting the same procedures found in traditional municipal capital projects will not suit a DBO project and will likely cause significant problems in its execution. The owner's team must be project-specific, including all necessary disciplines, and be empowered to make project decisions. As the project moves from proposal through negotiations and award to actual execution, the requirements for interface and exchange will continue. It is typical for the BOOT/DBO team to meet with the owner on at least a biweekly basis beginning in negotiations; transition to review progress in the design effort; and share information on products, equipment, hardware, and other project details. In this process, it must be recognized that the contractor has already selected much of the project material at the time of proposal, and this "design" effort is a refining and coalescing process. As such, the level of "submittals" typically seen for approval is, in fact, submitted for the owner's information and record. Should a product prove objectionable to the owner, depending on its significance, the owner should be prepared to absorb any significant cost effect through a project-change order. An important coordination requirement exists for the owner's team with regard to other existing infrastructure or ongoing projects. This will continue throughout the project's duration.

As the project moves from the design and permitting phase to actual construction, the owner may wish to assign a full-time, on-site representative. This will provide direct interface with the contractor and enable quick response to project requirements and issues. In addition, this entity may serve as the quality assurance/quality control (QA/QC) oversight for the owner in assuring a project is constructed to those standards originally presented. Maintaining project records, as-built drawings, and recording progress are typically accomplished by the DBO contractor, and the owner's agent can serve as an active affirmation of this. The owner's representative should have a detailed list of responsibilities and a clear definition of role and authority in the project. He or she would participate in project meetings and keep the owner's management informed of project progress and involve them in ongoing key project decisions. Typically, the contractor provides services of an independent testing agency to provide QA/QC on construction materials, such as concrete or pressure vessel welding and painting, with the results made available to the owner.

The key to a successful public–private partnership is letting the contractor do his or her job and having a contract that clearly identifies the scope of the project and the role of the owner. Privatization, to whatever level accomplished, is a partnership between the owner and the service provider and will work only through establishing clear, reasonable expectations and cooperation by both parties.

# Chapter 16

# Training

# INTRODUCTION

Training is part of a well-run organization, and it can be a great asset or a great liability. The difference between success and failure typically depends on a utility's attitude toward training as a part of its culture. The wrong approach can nearly guarantee failure, despite a large budget and a skilled, determined training staff. The right approach can lead to success, even when means and personnel are limited.

# THE ORGANIZATION AND THE TRAINER

Trainers occupy a special place in any organization, and any trainer or training group should be organized to preserve and capitalize on this separate status. Training decisions must be made without undue constraint or direction from laypeople, so the trainer should operate as freely within the organization as its culture can allow. There are several reasons for this separation, the most important of which is that if training personnel lack the freedom to identify and address training needs and to refuse to address what other staff mistakenly believe to be training needs, training inevitably becomes a subset of the organization's management, to the detriment of both training and management.

Ideally, a trainer or training group would report to the head of the organization so they have reasonable access to the resources of all departments and freedom from any competing pressures in the organization.

## THE STARTING POINT: SUPPORT

It is easy to identify an organization that does not support its training team. Even if the training budget is large and training policies exist, the organization has managerial stresses and inefficiencies, poor or spotty job performance in important departments, too-frequent equipment breakdowns, and stress fractures in the organization itself. If an organization supports a vibrantly effective training group, such problems will shrink (though they never entirely disappear from any organization).

Support means more than a program budget and policy statements. The people responsible for training should be free to design, develop, and deliver training that meets the group's actual needs, and they should have access to the human and physical resources necessary for the job. More important, the training group must have a voice in the organization, be seen as a vital part of it, and have a defining role in its future direction.

## THE REAL GOALS

Any modern training system should create a learning culture in the organization. In a learning culture, all participants share the responsibility for their own development, and they have a responsibility to help their fellow employees develop.

To achieve this goal, the training team must implement a technical training system that is project- rather than curriculum-based. Such a system allows the trainers to focus on "just-in-time" training, in which instruction is available when needed and delivered only to those who truly need it. This process maximizes results and provides the most training flexibility. Each training need is a separate project rather than a step in a regimented system.

Project-based training also eliminates the need to create, maintain, and update a large body of seldom-used curricular material, relying instead on the trainer's ability to develop material for immediate, clearly defined needs. This is both a benefit and a disadvantage. Without an overall curriculum, trainers need a comprehensive reference system of standard operating procedures and other job aids to ensure that appropriate information is available to guide employees whenever needed.

Ideally, an effective, project-based training program would only provide information that is fully relevant to both the job and the participant. It also would consistently use the training media best suited to the material and the participants' skills and knowledge.

A project-based training program may require a higher standard for trainers than one that uses a set curriculum, but because training is need-specific, personnel will spend fewer total hours being trained. So, trainers should be proficient in all aspects of training-program development and delivery, and they must either have expertise in the jobs they are teaching or have access to employees who do.

This approach to technical training is driven by the organization's performance needs and tied to bottom-line results rather than predetermined course material. If analysis shows that training is not the solution to a performance deficit, trainers can help find other cost-effective solutions.

## WHAT IS TRAINING?

A trainer's real job is to identify deficiencies in skill and knowledge sets that could cause poor performance or nonperformance and correct them as quickly and inexpensively as possible. Training is *not* presenting a set curriculum via prespecified methods to groups of people identified by administrative policies and procedures. Rather, training provides the information people need to do the organization's work, when and where they need it, in ways that can be easily assimilated and put to use.

Of course, each participant should be expected to demonstrate that they have mastered the material, and each training class on a topic should match others on the same topic. Supervisors also should monitor employees' work to ensure that they are implementing what they learned.

Outside authorities may issue requirements that conflict with the local approach to training. This is particularly true of safety training, where state and federal safety agencies often dictate periodic repetition of many topics and local legal counsel may require retraining for various reasons. These variances are typical at governmental or quasigovernmental organizations, and training personnel routinely accommodate them.

Sometimes repetition of training can be useful. For example, periodic "refreshers" on key duties can help keep people sharp. If an important part of the job is only done during one season (e.g., cold-weather operations in northern states or rainy-season-flow emergencies in southern states) retraining may be the only workable solution. This kind of repetition is valid training, because skills will lapse through disuse during the seasons when they are not needed.

## THE TRAINER'S ROLE

Trainers are vital to utility operations. They may simultaneously implement new administrative policies, help test and start up new facilities, create new work methods for existing jobs, maintain a central compendium of job-related information, introduce new people to the workplace, determine soft-skill training needs, and prospect for outside trainers to meet them. The job requires range and versatility, as well as freedom within the organization and administrative support.

Too often, training is assigned to the person who is most proficient at a given job, who may or may not have the skills needed to train others. This approach works well for straightforward tasks that can be taught via demonstration. However, the most skilled person may not communicate well with trainees, and ongoing personal relationships also may affect the results.

Trainers should be proficient at the specific job or able to learn it from experts. They should know how to organize a presentation and adapt it to their trainees' abilities while delivering it.

## THE TRAINER AS SURGEON

Surgeons do not outline the hospital's food-service menus, sterilize laboratory equipment, or administer daily doses of medication. They perform a specific set of functions—ones only they can do and have experience with. Surgeons are not called on to do things outside their area of expertise, but are the only ones considered when a patient has internal problems to correct.

In an ideal organization, trainers look for deficiencies in employee performance and devise ways to improve them. The goal is to make the entire organization work better by finding and fixing specific problems. This organization does not use trainers as a substitute for managerial solutions, but over time, finds it has fewer management problems because people are more confident and motivated on the job. The problems that may have prevented them from performing well (or even rewarded poor performance) are disappearing and the entire workplace functions more effectively.

Just as surgery is not the solution to all of a patient's problems, trainers may be vitally important in solving some performance deficiencies and irrelevant in correcting others.

## WHEN TRAINING WILL SUCCEED—WHEN TRAINING WILL FAIL

Training will succeed when given as close to the actual need as possible. If delivered so far in advance that the participants will not retain the information or cannot see an immediate advantage, it will fail. If given so late that the participants already believe they know the material, it will fail.

Training will succeed when the material is fresh and new. It will fail when it duplicates previous training.

Training will succeed when it addresses one topic. It will fail when it attempts to cover multiple topics at once.

Training will succeed when the participants believe the new material will help them do a better job or make the job easier by solving a problem, teaching a useful new technique, or otherwise improving life in the workplace. If participants think the training has no immediate payback, it will fail.

Training will succeed when responding to participants' need for skills and knowledge that they currently do not have. It will fail when it is a substitute for more appropriate forms of performance feedback or punishment for a job failure.

In other words, training will succeed when it gives the participants an immediate advantage on the job (e.g., training on new facilities, process equipment changes, policy changes, and safety). If the topic has no immediate benefit and does not address a known deficiency, it will fail.

Trainers can help ensure training success by ensuring that participants know from the start that their needs will be met. They should begin by explaining why the training is being given and what advantages it will provide in the workplace.

## PROGRAM DEVELOPMENT

When developing a training program, trainers should consult both rank-and-file employees and subject-matter experts.

Logically, the first step in developing a training curriculum is to break down every job into its overall responsibilities, then into individual duties, and then into the specific steps to follow when performing those duties. Once the job analysis, task analysis, and standard operating procedures are completed for every aspect of every job, trainers would develop a training package for each job and deliver it to everyone who may be required to perform those duties. However, at wastewater utilities, the analysis and procedural development alone could take years, the delivery more years, and maintaining the file library required for it would be a bureaucratic nightmare. One large wastewater utility began such an effort in 1993, only to discover that it would take more than 8000 work hours just to develop the procedures for plant operations alone. If that program had been continued, the utility would have discovered that, once the training was finished, much of the material would seldom be needed. The program would have been valid, but too cumbersome to use in the real world.

Training needs to be more practical. A project-based training system is simpler and more valid because it delivers the training that is actually needed, when it is needed, and in the form that is most effective for the subject matter.

This is not to say that other methods should be excluded. Each has its place, and the emphasis should be on using the most practical methods available for any part of a training program. Although the emphasis should be on project-based training, sometimes general training may be necessary. For example, when operators are preparing for state certification tests, necessarily the training is based on a curriculum because the required topics are known in advance and everyone must learn all the material whether or not it is relevant to their facility.

## PROGRAM ELEMENTS

No two training programs are alike, but there are elements common to all. Among these are the basic methods and approaches proven successful in training adults, the policy and procedural documents developed before training is given, skill verification after every training, and the use of onsite expertise when developing both training and procedures.

**BASELINE.** The initial introduction to utility rules and practices that all new employees receive no matter what their job duties, may or may not be part of the training program. For the purposes of this chapter, operations and maintenance training begins by introducing employees to specific equipment and how it operates. This is an area in which peer training can be particularly valuable, though most larger utilities develop a system in which trainees are initially supervised by a trainer. This training enables new employees to have direct, hands-on experience with the equipment in a setting that does not require adherence to specific routines. Either individually or as a group, they will learn startup, shutdown, control methods, and other duties (e.g., sampling and reporting).

An easy way to outline this new-hire program, while giving ownership to experienced employees, is to do the following:

- Call together a group of employees for a 30-minute meeting, preferably without a supervisor present. Bring the following tools: a marker and either paper and tape or a self-sticking note pad.
- Ask someone to name any job duty, write it down, and then stick the paper to the wall.
- As other duties are named, write them down and stick them up too. At the same time, ask for the main parts of each duty. For example, after the plant headworks is named, notes for bar screens and grit chambers would be placed below it. More specific duties (e.g., clearing grit piping) would be placed with the grit chamber note.

An outline will emerge and be fleshed out so rapidly that a second note-taker may be needed.

After the session, put the outline on paper and ask a lead worker or supervisor to review it and note any details that may have been missed. Invariably, this outline will be a list of every job duty that a new person needs to learn, produced by the people who know the job best, and every item on the list can be taught peer-to-peer in the plant.

This technique can be used to outline any job in the organization, including those of its board of directors.

**SAFETY.** Safety cannot be handled simply or by a person who is not thoroughly versed in both safety methods and the applicable state and federal laws. The legal consequences of an on-the-job injury or death are too great to allow anything but the utility's best effort. At the very least, the training should be developed and delivered by skilled vendors who specialize in safety training.

Following are some of the topics that a comprehensive safety program might include. A safety professional will add others that are plant-specific.

- Safety and health policies;
- Basic safety and health for wastewater professionals;
- First aid and cardiopulmonary resuscitation;
- Wastewater hazards;
- Industrial hygiene;
- Respiratory protective equipment;
- Hand and power tool safety;

- Fire prevention and control;
- Accident and illness reporting;
- Accident investigation;
- Hazardous energy control (lockout-and-tagout) procedures;
- Confined space entry;
- Safe work permitting procedures;
- Emergency response;
- Safety for hydrogen sulfide and other gases;
- Chemical handling;
- Hazard communication and labeling (material safety data sheets);
- Hearing, eye, and face protection;
- Right-to-know laws for employees and the community;
- Housekeeping and safe equipment storage; and
- Electrical hazards (e.g., hot work and basic awareness).

For more safety information, see Chapter 5.

**OTHER.** While baseline training and safety are the two largest concerns, an overall training program also may include the following:

- Cross-training and rotational job assignments.
- Individual training goals for employees seeking to advance into other positions in the utility.
- Professional organizations (e.g., the Water Environment Federation and its state and local affiliates). Many utilities pay employee dues for participation in such organizations.
- An operator certification training program, either locally developed or using the accredited certification training courses available in most states.
- Formal education, which can be limited to courses brought to the plant or extended to local colleges and regional occupational programs.

## THE TRAINING SYSTEM

Training begins with management. A system must be in place to manage training development, maintain complete records of every employee's training, and allocate training resources (money and time) throughout the organization.

People matter; the policies and procedures for developing and delivering training are as important as those for operating and maintaining the plant. These standards should include organizing and maintaining training materials and courses, and standard formats for training documents (e.g., work procedures, lesson plans, certifications, qualification guides, and attendance rosters).

Many organizations now find that training has become such a complex discipline that they need training information management software. Such software can allow small and large utilities to manage their training programs efficiently and maintain all the necessary employee-development records.

Just as a critical assessment of a training session helps to improve later classes, an audit of the training program can help identify its weaknesses and strengths. Such audits should be done periodically (typically at 5-year intervals). An audit will include an overview of the personnel and money allocated for training and how it is divided among the utility's departments. Auditors will scrutinize the methods and standards currently in use and attempt to gauge the utilitity's satisfaction (or dissatisfaction) with the program. The audit may also include benchmarking visits to other utilities to see their training facilities and compare them to those of the utility being audited.

## CHOOSING TRAINERS

The best trainer is not necessarily the one who knows the subject best (although subject-matter experts should be part of the training development process). Just as architects probably would not write travel brochures for the hotels they designed, subject-matter experts are more likely to be resources for the trainer rather than trainers themselves. In today's workplace, when a training need is identified, the trainer will seek appropriate experts inside or outside the organization and mold their knowledge into an effective package.

Formally educated trainers tend to have certain characteristics in common. For example, versatility—the ability to gain competence quickly in unfamiliar surroundings—is very important. A modern trainer must adapt to new developments in the workplace, because technology rapidly changes the nature of the job. Versatility in delivery is also necessary. A class can be turned from failure to success by the trainer's ability to alter the presentation while giving it so it better matches the needs and expectations of a particular group of participants. Technical competence is highly desirable in a technically complex workplace (e.g., operations and maintenance work). Experience in the organization helps give potential trainers familiarity with existing systems, personnel, and methods.

## TRAINING DOCUMENTS

Hard-skill (and most soft-skill) training is pointless if the presentation is not backed by descriptive documents. In a complex work environment, participants simply cannot retain all the information that is pertinent to the job, and these documents provide both reinforcement and references.

Consider, for example, a newly installed activated sludge influent pumping station with variable-speed drives, a constant-speed backup pump, and controls that can govern pump operation via wet-well level, flow, or manual adjustment. It has local and some limited remote control, plus automated alarm and emergency shutdown sequences. Power is available through the plant grid or from a backup generator that can autostart or be run manually. Operators must be able to "black start" this station, run it in any mode, switch between modes or shut down as needed, and recognize the conditions that trigger each activity. Years from now, incoming personnel will need to attain the same proficiency.

Because no one, including the trainer, knows how to operate this station, the utility must create an operating manual and specific operating procedures for every activity associated with its operation. It would be impossible to develop meaningful training for the station without them. After startup, these documents will serve as a reference that will guide operators for as long as the station exists. Training for the maintenance staff will require the same level of diligence and its own backup documentation.

Also, all contents of training presentations and documents must be official requirements of the organization, backed by management approval. If the organization does not have a stated policy or procedure for the training subject (especially those with potential for equipment damage, personal injury, or resulting public hazards), the trainer may be held legally liable for any negative results. No trainer should have to accept that level of responsibility, and no organization should require it.

**POLICY.** ***Definition.*** A policy is an administrative document governing a single phase of a facility's, department's, or work group's operations. It outlines management and workplace practices, sets performance expectations, assigns responsibilities, and describes the processes to be followed in fulfilling them. A policy should deal with one particular topic—not a range of topics. If more than one topic will be addressed, a policy should be written for each. Policies must be easily recognizable, clearly written, and well-organized. A written policy may have legal implications, so it should not be open to misinterpretation and released only after administrative approval.

***Content.*** Almost any written policy must include the following items:

- Title,
- Issue and revision dates,
- Summary,

- Purpose,
- Objectives,
- Definitions,
- Criteria and constraints,
- Enforcement and penalties,
- Statement, and
- Authority.

A given policy may not require all these items, but the following guidelines should be considered when composing policy and all required items should be included. Minimum standards would require the title, applicable dates, purpose and objectives, policy statement, and issuing authority. (The headings shown in this section are used for demonstration purposes only.)

*Title.* The title is short, preferably five words or less. It describes the policy and should include any policy numbering system in effect. A revision number should be added when applicable.

*Issue and Revision Dates.* The policy must be dated, because some are issued before their effective dates. If the issue and effective dates are not identical, however, the effective date should be included in the text.

*Summary.* The summary is a concise description of the policy, intended for executive use. It is equivalent to the executive summary in a staff report. This item should only be included in extended policy documents.

*Purpose.* The purpose is a statement listing the reason(s) why a formal policy is necessary. For example, "These written policies should increase our mutual understanding of [utility] expectations, minimize the need for personal decisions on matters of [utility-wide] policy, and help to assure uniformity in program administration."

*Objectives.* Each objective should be a simple, declarative sentence (in active voice) stating what the policy is intended to accomplish. If a policy has several objectives, they should be written as a list, and each objective should begin with an active verb. For example, "This policy is intended to do the following:

- Establish parking enforcement authority within plant boundaries,
- Designate parking areas for each employee group,
- Provide guidelines for the use of handicapped parking spaces, and
- Delineate zones for rideshare parking."

*Definitions.* Any terms that are not in common use should be identified and defined. For example, "*Sunset*—a provision that limits the effective life of a policy by providing a specific end date or other terminating criteria, such as the end of a construction project."

*Criteria and Constraints.* Criteria and constraints explain when the policy is in effect and when it is not. They also should list affected departments or employee groups, as well as any limitations, special cases, or exclusions. Following are examples of criteria and constraints:

- Criteria: "This policy will apply to selected employees who have completed more than 15 years of service, without respect to age, health, or job position. It will not apply to employees who have already retired or who have filed retirement documents."
- Constraints: "This policy will apply to laboratory personnel and operations personnel taking samples from plant process streams."
- Limitations: "This policy will not apply to contract personnel doing approved work on agency property."
- Special cases or exclusions: "Employees currently under disciplinary suspension are not eligible for the early retirement option."

*Enforcements and Penalties.* Monitoring requirements and noncompliance penalties should be clearly stated. For example, "Employees violating this policy will be subject to disciplinary action by the agency, including, but not limited to, termination of employment."

*Statement.* A policy statement describes the process to be followed to fulfill the policy, including any details required to clarify the utility's intentions. Delegations of responsibility and authority should be made clear (to departments and job positions, not to individuals by name). For example, "The senior operations supervisor will have the power to set work schedules for plant operations personnel, under direction of the chief operator."

Also, the required compliance standards should be given. For example, "Results of the calculation will be expressed in standard engineering units."

Reporting responsibility and frequency, along with a description of the reporting process, should be included. For example, "The operations engineer will develop plans to train supervisors in all areas of the plant and will submit an outline and recommendations to the director of operations in December of each year."

The specific policy requirements should be stated. For example, "Employees may not copy or duplicate, or permit anyone else to copy or duplicate, any physical or magnetic version of the computer programs, documentation, or information ."

All critical dates should be included. If there are sunset provisions [review dates, expiration dates, or other criteria (e.g., completion of a project or issuance of another document)], these should be included in the text. Also, the effective date should be provided. For example, "This policy's limitations on aeration deck access will be terminated upon completion of the P1-36-2 construction project."

Any new policy statement may affect or even eliminate existing policies. Any necessary eclipse provision for other affected policies should be included; if other policies are changed or eliminated, this should be made clear. For example, "Any policy issued previous to this statement, or any portion thereof, which conflicts with this document, is hereby repealed." or "This policy supersedes Policy SP-003, 'Use of Respiratory Protection Equipment by Agency Employees'."

The administrative authority under which the policy is issued should be stated. A policy is not valid without the organization's sanction. The person under whose authority the policy is issued must sign the final draft of the policy statement.

**PROCEDURE.** The procedure is an organized set of instructions used to accomplish a particular task. The task may be as simple as the series of calls to be made and information taken if an odor complaint occurs, or as complex as black-starting an activated sludge plant. The hallmark of a procedure is its limitation to the completion of a single item. Related items are handled in separate documents.

Procedures are written in active voice, with direct instructions on performing each step. Conditional references (e.g., "could", "should", "may", and "ought") must be avoided; these steps are mandated, not suggested.

A procedure should include the following items.

***Ordination.*** The ordination should include

- The procedure name and reference number (and revision number, if appropriate);
- The effective date and latest revision date; and
- The signature of the responsible manager (keep this signed document as a legal requirement of the organization).

***Purpose and Scope.*** This is a succinct statement of why the procedure was written and the exact circumstance it is meant to cover. This should be no more than one or two sentences.

***Definitions.*** Any term, title, or organization that a competent but inexperienced person may not know should be defined.

***Responsibilities.*** This is a listing (by position or department, not individual name) of the exact responsibilities that everyone affected by the procedure will have. If an entire department is affected, it and all affected positions should be listed. This includes a statement of the utility's expectations for action by outside organizations.

***Procedure.*** This is a chronological, extended outline detailing each step, from the initiating event (phone call, alarm, supervisory decision, etc.) through the final disposition of all people, materials, and necessary documentation. However, if any stage of the procedure includes safety considerations, they must be stated before that portion of the procedure so personnel will be prepared.

***References.*** The references should include the titles of all regulations, policies, or laws that reasonably have a bearing on the performance of the procedure or that may have created the need for it.

***Attachments.*** Items that may help staff perform the procedure (e.g., sample documents, maps, blueprints, and contact listings) are included in the attachments.

**OPERATING MANUAL.** The operating manual is a compilation of information on primary equipment or unit processes, with all the policies and procedures needed for both normal and irregular operations. The following list of items is commonly, but not necessarily, included. An operating manual should address the topics needed to provide a complete operational guide for the equipment or process.

(1) Unit description
  (a) Basic process principles
  (b) Operational features
  (c) Capacities
  (d) Location
  (e) Utilities
  (f) Drawings or schematics
(2) Relationship to adjacent units
  (a) Influent sources
  (b) Effluent receivers
  (c) Byproduct removal
  (d) Effects on overall process
(3) Classification and control
  (a) Mechanical operations
  (b) Basis of control
  (c) How control is exerted
  (d) Automatic changes

(4) How equipment responds to conditions
    (a) Standard conditions
    (b) Operator changes
(5) Items that are operator-controllable (how to make control changes)
(6) Major components
    (a) Tanks, piping, channels
    (b) Robotic functions (conveyors, flights, sweeps, etc.)
    (c) Control panels
    (d) Motors and drives
    (e) Chemical systems
(7) Chemicals used
    (a) Schematics and isometrics
    (b) Auxiliary systems
(8) Common operating problems (troubleshooting)
(9) Laboratory controls
    (a) Samples
    (b) Expected quality control results
    (c) Quality control during startup and shutdown
(10) Startup and shutdown
    (a) Black startup
    (b) In-service startup
    (c) Emergency shutdown
    (d) In-service shutdown
    (e) Isolation and lockout
(11) Normal operations
    (a) Recommended rounds
    (b) Expected readings
    (c) Standard settings (gates, valves, switches, etc.)
    (d) Making process adjustments
    (e) Safety
(12) Alarms
    (a) Alarm conditions
    (b) Alarm points
    (c) Where alarms are received
(13) Alternate operations
    (a) Emergency operations
    (b) Secondary operating modes
    (c) Methods for changing modes
    (d) Fail-safes

**OTHER DOCUMENTS.** A simple job aid that can make life easier in the workplace is a "picture procedure"—a printed document posted at the work site. It provides both written instructions and pictures of an individual performing the major steps of the task. A well-done picture procedure can take the place of some training and eliminate the need for refreshers. It places a visual reminder at the employee's disposal, exactly when needed. Some agencies laminate, frame, and hang these documents permanently at the equipment site.

## TRAINING TYPES, ISSUES, AND STYLES

There are several styles of training, and no one method works well in every situation. Also, no matter what the method, the agency must provide sufficient resources and time for the trainee to complete the training.

**INSTRUCTOR-LED VERSUS SELF-PACED.** The most basic choice is whether to have an instructor deliver the training or create a package—either online or a printed document—that will allow participants to train at their own pace. This sounds like a simple choice, but it is not. Using an instructor to deliver training better controls the training process, but takes personnel away from other duties for set periods of time and creates other scheduling problems (e.g., makeup presentations and training for new employees).

A self-directed training package overcomes these problems, but may create others. If an instructor is not present, the training package must contain all the information and reference materials that the trainee will need. This means many hours of preparation for even a simple package. However, once created, the package can be used by large numbers of people and kept as a resource.

Self-directed training can be an excellent tool for specific technical material presented to a large audience (e.g., a new air-monitoring device that will be routinely used in several ways by plant operators and maintenance workers, confined-space entry teams, laboratory personnel, outside construction crews, and air-quality workers). Proper training is a vital safety concern, and staff changes would require many classes to be given later. So, a study package that employees can use at their own pace, and a qualification guide that can help supervisors verify that employees know how to use the equipment properly enables training to be handled without formal classes.

**SITE-SPECIFIC VERSUS VENDOR.** A training vendor is an individual or company that develops and sells training packages on various topics. Because this training is sold to multiple buyers, it must be general and has limited use in operations or maintenance training. Some training (for the most common types of equipment) can be purchased this way, but typically vendor training is best for soft skills, which are more applicable to the utility's office workers.

In the plant, there is no substitute for site-specific training developed onsite, and delivered when needed. A vendor may be able to develop such a package with supervision, but typically this training will be created and delivered by utility staff.

For new facilities and equipment, the equipment manufacturer can be another valuable source for training. This training must be adapted to the facility and the utility's own operating requirements or supported by locally developed training containing this information.

**PEER- AND SUPERVISOR-GUIDED.** The differences between peer- and supervisor-guided training are real but not large. Each can be a real asset, but neither is a cure-all. If a supervisor is doing the instruction, there is the advantage of authority and familiarity with the participant, but there may also be frictions from other sources that carry over into the training session and mar the result. With peer training, personal relationships and the trust of an equal are big advantages; however, because the trainer is an equal, the lack of authority can sometimes lead to trouble.

This type of training is useful but cannot be allowed to proceed piecemeal. The utility is responsible for the outcome of this training, as it is for any other. The peer trainer or supervisor should be provided with all needed reference documents and performance guides that will be completed and retained as part of training-program documentation.

**INTERVAL-BASED.** Training repeated verbatim, year after year, is rarely effective. So, interval-based training is not recommended, because it invariably leads to training people who already know the material, damaging morale and leaving participants with a nonchalant attitude toward the overall training process. Remember, training should improve work performance in areas where the job is not done well because of something the workers do not know. It should not be done merely to put a check mark in an employee record.

Unfortunately, most state and federal requirements—particularly safety requirements—do require retraining at stated intervals, and utilities must comply with them.

**AD HOC.** This is a brief (30 minutes or less) class developed on short notice in response to a new, important piece of equipment or change in utility work methods. Sometimes called a "tailgate training session", its goal is training key personnel on a relatively simple topic quickly.

**CONSTRUCTION.** Construction training is often the most difficult to arrange because it involves new facilities. Design specifications should call for training on installed equipment, for every affected department. The contractor should provide the initial operations and maintenance manuals in cooperation with the facility's design engineer. However, this training is not equipment-specific; the overall facility is a training challenge of its own. Staff also must verify that every operating procedure provided by the manufacturer works under local conditions using the actual equipment installed. This verification is part of normal startup procedures, and includes the trainer as an active participant.

**BASELINE.** Baseline training is given to new hires to bring them up to basic competence. It should be maintained as a separate package and updated as changes in equipment and procedures are implemented. This training is curriculum-based because there are basic duties that every new employee must learn. This is the area in which peer- and supervisor-guided training can be most effective.

## THE JOB AS A TRAINING FACTOR

Operations and maintenance staff face opposing sets of job stresses, which the trainer must take into account when developing and delivering training. Although largely hired and largely promoted based on their ability to recognize and respond to breakdowns and emergencies, operations and maintenance staff have job duties that are mostly routine. An operator may make the same checks on a piece of equipment and observe the same readouts from its instrumentation for years before a problem occurs. These people are justifiably proud of their ability to maintain this kind of vigilance and to use virtually all of their senses as tools of the trade, but such routine takes its toll and eventually can lead to fatigue and burnout. At milder levels, it may reduce the resources an individual has to invest in a training session.

**SHIFT WORK.** The job itself places constraints on training. Ten- and 12-hour shifts are now common, and round-the-clock operation means people must work through the night. Longer shifts and night shifts mean more fatigue, particularly at the end of a workweek, and less tolerance for any training that is not directly job-related. Scheduling training during the appropriate shift helps but is often impractical because of the hours and expense involved. Also, staffing may be too thin to break people free for extended training, even when it can be delivered during their shift.

Shift work also means staggered days off, which further limits the days when training can be scheduled. Staffing limitations can be a psychological barrier to training because employees do not want to leave their colleagues shorthanded. While this sentiment may be commendable, the result is not.

The plant also can be an impediment to certain kinds of training. Process equipment must run and be monitored 24 hours per day, and operators must respond to any irregularities, despite other scheduled activities. Training is further constrained by work rules that discourage the use of overtime to cover training.

These roadblocks cannot be eliminated, but their effects can be reduced by keeping training sessions short, holding them onsite whenever possible, and giving makeup classes.

**SPECIALIZATION AND SKILL STRATIFICATION.** As wastewater collection and treatment facilities incorporate modern operating techniques, department sizes will change. This restructuring can lead to skill stratification, particularly among operators. Such strong variations in skill and knowledge levels within a work group can reduce the organization's ability to adapt to further changes but increases efficiency in the short term. Stratification stems from three causes.

First, reducing the overall group size can lead to specialization and a tendency to make the day shift the central talent pool and carrier of most responsibilities. Then, the off-shifts are considered caretakers whose duty is simply to maintain the operational status quo. This changes the organization's expectations of each group's performance and estimation of their training needs and job-mastery requirements.

Second, larger facilities tend to create area work teams that each focus on a specific process area and the related job skills rather than maintaining proficiency in all areas of plant operation. While area work teams improve overall plant performance in the short term, they promote specialization and change each individual's perceived training needs. Team members gain some skills and lose others.

Third, reinvention often results in "department hopping", typically from maintenance operations. Maintenance specialists enter operations with highly developed skill sets that often can be directly applied to operations work and that experienced operators often lack. So, a trainee may be precociously effective at certain aspects of the work, outperforming his more experienced colleagues. Such individuals will bypass their experienced coworkers because of skills and knowledge that the "old hands" have not had the opportunity to acquire.

It is a trainer's duty to see such trends and find ways to minimize their negative effects and optimize their positive ones.

**POSSIBLE SOLUTIONS.** Individually, any personnel and job-related constraint can be overcome without a debilitating effect on the training program. Collectively, however, they can be problematic. To keep training effective, trainers should do the following:

- Train only in response to a clear and immediate need;
- Help managers assess and solve organizational problems;
- Limit training to topics that will give the trainee an immediate advantage on the job;
- Avoid duplication;
- Keep sessions brief and informal;
- Make maximum use of hands-on training;
- Keep groups small and provide as many makeup sessions as the training budget will allow;
- Focus on quick, familiar training methods and easy-to-use job aids;
- Carefully screen trainers and their training plans before approval;
- Obtain as much rank-and-file participation in session content and planning as time and security will allow;
- Present policy-related training as help for the employee, not as a "decree from above";
- Provide training during a shift when it offers a clear advantage;
- Offer maintenance training to operators;
- Include as many off-shift personnel as possible in all training opportunities;
- Give training in the plant whenever practical; and
- Create and maintain a practical, positive image of the job and the training organization.

## TRAINING DESIGN AND DEVELOPMENT

The following rules should be followed in training design and development:

**RULE 1.** Train only in response to a clear and immediate need. Avoid the trap of delivering training that is not supportable by routine training analysis. Such training may be a substitute for supervisory remedies or a previously established, repeating schedule that does not reflect current training needs. Training time is too valuable to be wasted on unnecessary classes.

**RULE 2.** Decide whether training is the proper response to the perceived need. While not completely independent, trainers must have enough control to be able to insist on training when needed and refuse to provide inappropriate training. Because trainers analyze perceived needs as if they were outsiders, they can help the organization's managers find appropriate solutions, even if they do not involve training.

**RULE 3.** Use practical methods. Training technology can be an integral part of a program's success or a needless expense. Trainers should stay informed of training-technology developments and use the most cost-effective methods that are appropriate to the subject matter.

**RULE 4.** Training builds skills and knowledge; it does not solve organizational problems. Trainers can be a helpful observer and resource in situations that do not lead directly to training, but training cannot substitute for necessary changes in supervisory or management practices.

# ASSESSMENT TOOLS

Training assessment often reveals the need for management changes. In these instances, trainers may help supervisors deal with the problem but typically will not have the authority to make or enforce policy changes. When training is needed, the exact need should be defined so training can be developed simply and logically.

Following are some common methods used to determine whether a performance problem can be repaired by training or if another solution may be more appropriate.

First, trainers should ask the following question: Could the employees in question do this if they had to? If the answer is "yes", then training is not the answer. Something in the workplace is rewarding poor performance or nonperformance. This situation could be caused by one of the following reasons:

- Employees may not have the opportunity to do the task often enough to stay proficient.
- Poor intrastaff relationships or attitudes could cause job duties to be left undone.
- There may be no visible consequences for doing the task poorly or not at all. "It does not matter" is a perfect excuse for inaction.
- Supervisors may not be responding adequately to substandard performance, or working with employees who need coaching.

These situations cannot be corrected by training. If attitudes are the problem, training may actually make things worse by temporarily masking the real source of the trouble.

If the answer to the first question is "no, they would not be able to do this if they had to", answer the second question: "Are they willing to do the task?"

If the employees cannot do the task and are willing to learn how, arrange for either training or coaching by the supervisor or lead worker, depending on the severity of the problem and the number of people affected. Adults in a workplace seldom refuse or resist job duties without a reason. If members of the group cannot do the task and are unwilling to learn how, find out why. Training will fail if the resistance is not eliminated first. Following are some possible reasons for such resistance:

- Attitudes. Some individuals may have poor attitudes toward the task or the job in general. If the attitudes are culturally based, training will not solve them; the only solution may be to excuse those people from doing the task or reassign them to other jobs. Alternatively, the group may feel strongly that the task should be done by someone else, not the group to which it has been assigned. Do not automatically reject this thought; they may be right.
- Poor supervision. Supervisors may disagree on how, or whether the job should be done, causing confusion and discord among the staff. Follow-up on poor performance may be weak, inconsistent, or overly harsh. This may indicate that supervisory training is needed, rather than skill training for the rank-and-file employees.
- The task is not valued. If doing the task poorly or leaving it undone does not have negative consequences, perhaps it does not need to be done at all, and the whole staff knows it. Inspect the advantages and disadvantages of the task, and work with supervisors to eliminate it if unnecessary. If a task needs to be performed, the trainer's first job may be to teach the employees why and then teach them how, if necessary.
- Built-in resistance. Something in the overall job duties may cause resistance to this particular task. For example, it may be so time-consuming that more important duties cannot be done properly, or it may negatively affect another part of the operation.

## TRAINING ADULTS: IT IS NOT A SCHOOL

Adults learn for different reasons than children, and they learn in different ways. While a child may learn for the joy of it, most adults do not. The typical adult approaches a training situation with one goal: to ease some aspect of the job that causes distress or makes the participant feel inadequate. Something affects that individual in an immediate and perceptible way, and he or she is there to fix it.

Because training is geared toward adults, they will have little interest in training to maintain skills unrelated to their current problem(s). This is not lack of motivation or poor attitude; it is simply a facet of the adult mind. So, each person in every session must be reminded how the material will make work-life better for the person who learns it and puts it to use.

## THE IMPORTANCE OF VERIFICATION

Until a new skill can be demonstrated, it has not been learned. This fact has several ramifications for both participants and the organization in which they work.

Demonstrating new skills or knowledge gives an individual immediate confidence and a strong impetus to use them. This obviously helps the organization by increasing workplace efficiency and safety, and it improves the individual's attitude toward the job. Employees who know they can do their jobs well are happier and more motivated than those who have reservations about their performance.

For the utility, skill verification lets supervisors know that their crews can do what is needed. Also, if an operating error results in damage or injury, there may be serious financial and even criminal liabilities if the utility cannot show that those involved were fully competent.

A simple step-by-step qualification guide can eliminate such negative outcomes, help ensure positive ones, and serve as a checklist of employee skills. Rather than repeating classes periodically, employees can work through the related qualification guide with a supervisor or instructor and demonstrate competence without wasting job time and training expenses. In a classroom, the qualification guide could be a written quiz that covers the central points of the class. In the field, it could be a simple checklist of skills that the employee would demonstrate in a "dry run". Each guide should be dated and signed by both the participant and the trainer when the test is complete.

While it may be simple, the qualification guide is a vital record. It should be kept on file along with other important employee records.

## THE IMPORTANCE OF TIMELINESS

The importance of timeliness cannot be overstated when dealing with an adult workforce. Training that is forgotten before it can be put to use is wasted and could damage both equipment and workplace morale.

Even worse is training given after workers have groped their way through hands-on learning of unfamiliar systems. Aside from the obvious problems of potential equipment damage and personal injury, they must unlearn bad practices and replace them with correct operating procedures. This will never completely succeed, and the negative effects may last for years.

It is difficult to develop and deliver "just-in-time" training, but the rewards are clear. It is particularly hard to provide this kind of training in a new facility or process, where nobody has full knowledge of or experience in using it. Trainers must work with available resources and be full participants in the activities of any startup team assigned to work out proper methods.

## ANALYSIS, NOT PARALYSIS

If not carefully governed, the analysis portion of training development can be wasteful and time-consuming. Many utilities have had their training programs grind to a halt while training personnel performed increasingly detailed task analysis of every facet of every job. A practical look at the level of detail required can avoid this problem.

Training development takes as much analysis as that used to create standard operating procedures for trained personnel on the job. The same people do both tasks, and need the same level of support for each. The target audience in each case is an intelligent employee competent in using other plant equipment but unfamiliar with the equipment in question. The level of detail that will let such a person work through the task is appropriate for both procedure and training development.

## WRITING OBJECTIVES

Training objectives are deceptively simple. An objective is a statement specifying the task that the participant will be expected to be able to perform after the training (i.e., "At the end of this training, each participant will demonstrate the ability to . . ."). Objectives are written in the same form every time to facilitate their use in developing training and to help ensure that each is appropriate. Each objective describes only one item, not a list of related items, and notes the level of competence expected. If a percentage of correct performance is required, it should be stated in the objective.

Objectives are the first thing written when creating a training package. When the training package is completed, the performance guide or examination should be written directly from the objectives. This ensures that the objectives are met, and it helps the trainer confirm that the training document is complete.

## BLOOM'S TAXONOMY

During the late 1940s and early 1950s, B. S. Bloom organized and described the levels of cognition that are part of learning (Table 16.1). These levels proceed from the most

**TABLE 16.1** Bloom's taxonomy guidelines.

| Level | Definition | Sample verbs |
|---|---|---|
| Knowledge | Remembers previously learned material; recalls or recognizes information, ideas, and principles in the approximate form in which they were learned. | Write, list, label, name, state, define, describe, identify, and match |
| Comprehension | Grasps the meaning and explains or interprets the material; predicts outcomes and effects; estimates trends. | Explain, summarize, paraphrase, describe, illustrate, convert, defend, distinguish, estimate, generalize, and rewrite |
| Application | Applies learned material to a new situation; uses rules, methods, theories, or principles to complete problems or tasks with little direction. | Use, compute, solve, demonstrate, apply, construct, change, compute, operate, and show |
| Analysis | Understands the organization of material and the relationship of its parts; makes independent conclusions and hypotheses. | Analyze, categorize, compare, contrast, separate, distinguish, diagram, outline, relate, break down, discriminate, and subdivide |
| Synthesis | Combines and integrates learned ideas to create new products, plans, or proposals. | Create, design, hypothesize, invent, develop, combine, comply, compose, and rearrange |
| Evaluation | Appraises, assesses, or critiques on a basis of specific standards and criteria, judges relative values, and supports conclusions with facts. | Judge, recommend, critique, justify, appraise, criticize, compare, support, conclude, discriminate, and contrast |

rudimentary learning (recall of facts) to the most complex (evaluation and assessment based on derived standards and criteria).

When analyzing a potential training package, trainers should first assess the level of mastery required to do the portion of the job covered by the training. Then, they should write appropriate training objectives. By incorporating Bloom's sample verbs into the written objectives, trainers can easily define the competency level that will be achieved. For example, if the application level is desired, an objective might read as follows:

"At the conclusion of this training, each participant will demonstrate the ability to:

- Compute digester efficiency, given laboratory analyses of volatile and non-volatile solids."

If the knowledge level is desired, the objective might state:

- "List the readings taken during rounds of the gas compressor building."

## PITFALLS

Many pitfalls can harm, or even doom, a training program. Following are the most common and the most serious.

**DUPLICATION.** Duplication can creep into a training program in several ways. Most common is simply lax development. If an agency uses a particular type of equipment in several parts of its plant, staff do not need to be trained on that equipment in every instance of its use (including new installations).

Another form of duplication involves multiple topics centered on one piece of equipment. It may be necessary to teach the operation of that equipment once, but not repeatedly for every new operation that requires its use. Only the new material should be taught.

**REPETITION.** Repetition is a drag on a training program. Many utilities require personnel to be trained at stated intervals (typically every year) on the operation or maintenance of certain equipment. This sounds good, but it is not. For example, participants have been pulled in for training on the operation of equipment that they would have been using successfully had they not been interrupted by the training. Such an occurrence may amuse those with an eye for irony, but it damages the training program.

Interval-based training is a grossly inefficient way to run a training organization and should be eliminated wherever possible. One way to achieve the same effect is to use the qualification guide that was part of the original training package. A trainer or lead worker can take each employee through the guide and fix any performance deficiencies on the spot. This eliminates the need to take time off the job for unnecessary training.

**SCATTERSHOT.** Fighting brushfires can render a training organization ineffective. While trainers must respond quickly to some training needs, such responses must not become a pattern that will eventually choke off productive activity. If trainers spend their time responding to one emergency after another, they will not be able to push the program forward. Items start to be handled as training issues, that simple analysis—had there been time for it—would have indicated they were not training issues.

This overload leads to more emergencies, as items that would have been scheduled and developed normally are pushed back until they too become emergencies. In the end, the utility sees its training organization as a response to random emergencies, not a structured group moving toward a goal. When this happens, the training organization ceases to be an asset to the utility and may stop being an organization at all.

**FAILING TO LEAD.** Leadership is necessary and difficult. This is especially true when training has been done for a long time without any overarching plan. Department leaders are accustomed to having their needs met without question and will not respond well when a trainer says "no" to a training request. Nonetheless, to meet the utility's needs the training organization must be able to refuse a request and to insist on training over management disapproval.

Trainers should be free to identify and assess training needs, deliver training when and where needed, and refuse to let training be used as a substitute for more appropriate solutions to the utility's problems.

**SUBSTITUTION FOR MANAGERIAL REMEDIES.** At least 50% of the training at typical utilities is unnecessary, and some of it aggravates the very problems it was intended to correct. This is because training is often used as a substitute for more appropriate, but personally unpleasant, managerial action. If an employee (or group of employees) is failing to perform as needed, training is only appropriate if that person or group actually does not know how to do the job.

Most such failures do not result from a lack of knowledge. More often, they are the result of a situation that fails to reward good performance and may reward poor performance. This reward can be anything that makes life on the job easier or more enjoyable (e.g., a longer break time, relief from the need to climb flights of stairs, or less paperwork). When analyzing a training request, trainers should look for such punishments (anything that produces any kind of discomfort in the workplace) and rewards (anything that adds any sort of comfort). A manager who uses training to avoid the more unpleasant task of changing workplace methods also has fallen victim to this feedback loop.

Trainers can help managers find and permanently correct these problems, or they can fall into the trap of using retraining sessions to flog people into doing things differently, only to watch them fall back into pre-training performance levels.

**TOO EARLY OR TOO LATE.** This pitfall both renders training ineffective and damages the training organization's reputation. It can be difficult to avoid. Scheduling training at the exact time it is needed can be challenging, especially when there are multiple needs. Still, it must be done.

**UNNECESSARY.** Few things damage a training organization's effectiveness more than giving unnecessary training. If most employees already know the material, do not train them. Concentrate on only those who need it. This applies to both locally developed and vendor training. (It is tempting to let vendor training be delivered, especially when offered without charge by a company that does business with the utility.) The effect can be softened by making it optional, so employees do not feel constrained to participate. Overall, though, trainers should evaluate the content of the training and turn it down if it is not needed.

## RECOMMENDED READING

There are thousands of texts on training and adult education, and the degree of repetition within them can be daunting. However, Robert F. Mager's books contain virtually everything truly necessary to build a training program. Sometimes called the "manager six-pack," these books are simple, direct, and often sold as a bound unit. They are a valuable addition to any training organization's library. Their titles are as follows (Mager, 1997a to e; Mager and Pipe, 1997):

- *Analyzing Performance Problems,*
- *Preparing Instructional Objectives,*
- *Measuring Instructional Results,*
- *How To Turn Learners On . . . Without Turning Them Off,*
- *Goal Analysis,* and
- *Making Instruction Work.*

For anyone charged with procedure writing, *Fundamentals of Procedure Writing,* by Carolyn Zimmerman and John J. Campbell (1988), is an excellent resource.

## REFERENCES

Mager, R. F. (1997a) *Goal Analysis.* Center for Effective Performance: Atlanta, Georgia.

Mager, R. F. (1997b) *How To Turn Learners On . . . Without Turning Them Off.* Center for Effective Performance: Atlanta, Georgia.

Mager, R. F. (1997c) *Making Instruction Work*. Center for Effective Performance: Atlanta, Georgia.

Mager, R. F. (1997d) *Measuring Instructional Results*. Center for Effective Performance: Atlanta, Georgia.

Mager, R. F. (1997e) *Preparing Instructional Objectives*. Center for Effective Performance: Atlanta, Georgia.

Mager, R. F.; Pipe, P. (1997) *Analyzing Performance Problems*. Center for Effective Performance: Atlanta, Georgia.

Zimmerman, C.; Campbell, J. J. (1988) *Fundamentals of Procedure Writing*. GP Books: Columbia, Maryland.